GEOGRAPHIC INFORMATION SYSTEMS AND SCIENCE

GEOGRAPHIC INFORMATION SYSTEMS AND SCIENCE

PAUL A LONGLEY
University College London, UK

MICHAEL F GOODCHILD
University of California, Santa Barbara, USA

DAVID J MAGUIRE
Environmental Systems Research Institute, Inc., Redlands, USA

and

DAVID W RHIND
City University, London, UK

JOHN WILEY & SONS, LTD
Chichester • New York • Weinheim • Brisbane • Singapore • Toronto

Other Wiley Editorial Offices

John Wiley & Sons, Inc., 605 Third Avenue,
New York, NY 10158-0012, USA

WILEY-VCH Verlag GmbH, Pappelallee 3,
D-69469 Weinheim, Germany

John Wiley & Sons Australia, Ltd, 33 Park Road, Milton,
Queensland 4064, Australia

John Wiley & Sons (Asia) Pte Ltd, 2 Clementi Loop #02-01,
Jin Xing Distripark, Singapore 129809

John Wiley & Sons (Canada) Ltd, 22 Worcester Road,
Rexdale, Ontario M9W 1L1, Canada

Library of Congress Cataloging-in-Publication Data
Geographic information systems and science / Paul Longley ... [et al.].
      p. cm.
   Includes bibliographical references.
   ISBN 0-471-49521-2 (alk. paper)—ISBN 0-471-89275-0 (alk. paper)
   1. Geographic information systems. 2. Problem solving. I. Longley, Paul.

G70.212 .G44553 2001
910′.285—dc21   2001017870

British Library Cataloguing in Publication Data
A catalogue record for this book is available from the British Library

ISBN 0-471-49521-2 (cloth)
ISBN 0-471-89275-0 (paper)

Typeset in 8.5/10 pt Lucid Sans from the authors' disks by C.K.M. Typesetting, Salisbury, Wiltshire.
Printed and bound in Great Britain by Bath Press Ltd.
This book is printed on acid-free paper responsibly manufactured from sustainable forestry,
in which at least two trees are planted for each one used for paper production.

CONTENTS

FOREWORD

These guys must be mad. First they invite me to write a part of their "Big Book 2" on GIS. In it I was the only one out of the five luminaries who broke ranks and told readers anything useful. All of those intellectuals (myself excepted) presented a cosy conspiracy that GIS was wonderful, and academics and government people would lead us to the promised land. That's not the way the world is, folks – as I said.

Despite that, Messrs Longley, Goodchild, Maguire, and Rhind have invited me – a lowly consultant in the backwoods of GIS – to do an introduction to their new book for students and practitioners. I've now met these guys and they may not be as crazy as I first thought. It helps that they drink a little beer (one of them drinks a lot). But the clincher is that their book is better than the opposition, most of which is produced by techno-freaks or academics with little contact with the reality. Jeez, they've even used applications from the "real world" as they call it, which looks to me to be in the same orbit as Planet Lobley. Mind you, it's not perfect. I like the idea that there are some principles, even maybe

science, behind this stuff. "GIS as button-pushing" is not a smart view of how to change the world. I also think that having stuff on business makes sense – that's where things get done, where people take risks, and make (and lose) their money. It's also where any youngster these days should aim to work. But I guess I'm not as convinced as these guys that academia and government are wising up and becoming more business-like – seems to me that the opposite is happening as business is being loaded with more and more of the bureaucracy of government!

So guys, I reckon you got about 70% to 80% right in this new book of yours. That's quite a mite better than the others I've seen – my rating for them is between 20%, and 40%. I wish you well with it but suspect that you will not dare publish this foreword! That's unless you left it too late to ask some academic or bureaucratic sycophant instead.

Joe Lobley
Senegal, September 2000

PREFACE

There are perhaps 100 other books on geographic information systems (GIS) now on the world market. To the best of our knowledge, there is no other one like ours. One reason for that is that most treat GIS as a largely technical issue. This is reflected in the skills of existing GIS staff and the junior and middle level jobs they occupy. But our philosophy differs more profoundly than simply believing that there is too much emphasis on the technology. We see GIS as providing a gateway to science and problem-solving. Our philosophy is summarized below.

OUR APPROACH

The basic operations of GIS provide secure and established foundations for analysis, although the technology is still evolving rapidly (especially in relation to the Internet, its likely successors, and spin-offs). Better technology will remain a necessary condition for achievement of cheaper, faster GIS and better interoperability – but it is far from a sufficient condition for successful application of such systems.

GIS is fundamentally an applications-led technology, yet science underpins successful applications. Effective use of such systems is impossible if they are simply seen as black boxes producing magic. Understanding the imprecision and uncertainty of our representations of the world, and the consequences of our operations on them, is essential for everything except the most trivial use of GIS. Empirical analysis of the real world can be a messy and analytically inconvenient business and so the science of real-world application is the difficult kind – it can rarely refer to apparently universal truths, such as the laws of gravity. Rather it is one founded on a search for understanding and predictive power in a world where human factors interact with those relating to the physical environment. Social science and natural science are part of what we embrace. In addition, ethics and esthetics – the basis of the most effective graphic displays – can also play an important role.

Geographic information is central to the practicality of GIS. If it does not exist, it is expensive to collect, edit, or update. If it does exist, it cuts costs and time – assuming it is fit for purpose, or good enough for the particular task in hand. It underpins the rapid growth of trading in geographic information (g-commerce). It provides possibilities not only for local business but also for entering new markets or for forging new relationships with other organizations. But it is a foolish individual who sees it only as a commodity like baked beans or shaving foam. Its value relies upon its coverage and on the strengths of its representation of diversity, on its truth within a constrained definition of that word, and on its availability.

Few of us are hermits. The way in which geographic information is created and it and GIS are exploited affects us as citizens, as owners of enterprises, and as employees. It has increasingly been argued that GIS is only a part – albeit a part growing in importance and size – of the Information, Communications, and Technology (ICT) industry. That is a limited perception, typical of the ICT supply-side industry which tends to see itself as the sole progenitor of change in the world (wrongly). It is actually much more sensible to take a balanced demand- and supply-side perspective: GIS and geographic information can and do underpin many operations of many organizations, but how GIS works in detail differs between different cultures. The fact that few Japanese streets have names creates a very different navigation problem there compared with North America. Such underpinning is true whether the organizations are in the private or public sectors. Seen from this perspective, management of GIS facilities is crucial to the success of these organizations – businesses as we term them later. The management of the organizations using our tools, information, knowledge, skills, and commitment is therefore what will ensure the ultimate local – and hence global – success of GIS. For this reason we devote an entire section to management issues. But in so doing we go far beyond how to choose, install, and run a GIS. That is only one part of the enterprise. We try to show how to use GIS and geographic information to contribute to the business success of your organization, and have it recognized as doing just that. To achieve that, you need to know what drives organizations and how they operate in reality in their business environment. You need to know something about assets, risks, and constraints on actions – and how to avoid the last two and nurture the first. And you need to be exposed – for that is reality – to the inter-dependencies in any organization and the trade-offs in decision-making.

Success with GIS only comes from understanding and familiarity with science, technology, people, and institutions. Expertise in one area is not enough.

OUR AUDIENCE

As a team, we have already produced one very different book – the second edition of the "Big Book" of GIS. This reference work on GIS (Longley *et al* 1999) contains 72 chapters written by many of the best GIS people in the world. It was designed for those who were already very familiar with GIS, taking them to the frontiers of research and practice across a huge range of topics. It was not designed as a book for those being introduced to the subject.

This one is. We have in mind those studying at an intermediate stage in the huge range of under-graduate courses that are available throughout the world. The coupling of the book to the second edition of the Big Book (and also its predecessor, Maguire *et al* 1991) also makes it of use to postgraduates in GIS. Such users might desire an up-to-date overview of GIS to locate their own particular endeavors, or (particularly if their previous experience lies outside the mainstream geographic sciences) a fast track to get up-to-speed with the range of principles, techniques, and practice issues that govern real-world application. We have also directed it towards the busy professional, who has many demands on his

or her time but who has heard something of this new wonder and needs to gain familiarity with the subject in a no-nonsense way.

This is not just a textbook for mechanics. There are plenty of those already (some of them good). The market for GIS is huge yet fast-growing and most of the new users do not consider themselves principally to be technicians. We are convinced of the need for high-level under-standing and so our book deals with ideas and concepts – but also with actions. In geographic science, for instance, you need to be aware of the complexities of interactions between people and the environment. In management, you cannot become competent without practice, informed by a wide range of knowledge about issues which might impact your actions. Success in management – and in GIS more generally – often comes from dealing with people, not machines. We seek to prepare you for such situations.

Because of the rapid annual growth rates of GIS, there are far more people new to it than there are existing experts. The two groups need different material.

LEARNING OR BEING TAUGHT?

As the title implies, this is a book about geographic information systems, the practice of science in general, and the principles of geographic infor-mation science (GIScience) in particular. But it is even more than that. It is a deliberate attempt – an early one – to recognize and exploit the fact that information and communications technol-ogies are helping to change the world of learning, as well as business, government, and science. ICT is not really about making the old ways better. In the world of learning, it offers a genuinely different way, and one which has many advantages and some disadvantages.

For this reason, we start off by talking about the purpose of this book – to make you more capable of doing certain desirable things. This ability will come from a greater understanding of the factors involved, expertise in use of the tools at your disposal, recognition of the laws of unintended consequences, and adherence to the set of values (e.g. professional integrity) by which you operate. We follow this up with an overview of the kinds of problems with which GIS and GIScience can help. First, however, we must recognize the impacts of technology and other factors on the way we learn and the ways in which we seek to organize and change the world.

The general model of education – at university level and during in-work training – is changing. There are two interrelated reasons for this. The first is that injections of knowledge are no longer the cornerstone of education. The second is that businesses are becoming a serious player in the most advanced levels of education, long the preserve of state or private universities – and some universities are increasingly acting like businesses in how they run their courses.

The transfer of knowledge, suitably codified and fitted within widely accepted conceptual models, was the basis of the old education. That transfer from teachers to students was part of an implicit contract subscribed to by both the teacher and those taught. Two changes of great importance have occurred. The first is that higher education in most countries has gone from being the preserve of the few to the right of the many. When we authors went to university, for instance, we were part of only 5–10% of people who experienced such education; now the average in the countries surveyed by the Organization for Economic Cooperation and Development (OECD) is around 30% and it is over 50% in the USA. Many of these students are also working part-time as well as studying and have different requirements from those full-timers of the past. The second change is that a rapidly growing amount of information is readily available via the Internet, ever more cheaply and easily. In these circumstances, the value added by universities is to ensure that those being educated have skills of information retrieval, sifting, assessment, and analysis, plus the ability to ferment knowledge from reliable information and apply it to good effect. Thus we are moving rapidly from an era where the creation and inculcation of knowledge was often the primary focus (frequently through traditional but sometimes ineffectual methods like the formal lecture) to one where learning and knowing how to learn, all achieved between other pressing concerns, are the prime drivers.

Information – and sometimes evidence and knowledge – are increasingly available on the Internet. Where to find information, how to analyze it, and how to assess its quality are now more important than memorizing it.

This does not mean that there is only one way of learning or that conversing with a computer is the only way to achieve educational enlightenment. We know that different people have different learning abilities and prefer different learning styles. Whilst some students rate lectures very highly, those who are not familiar with extracting meaning from them often feel alienated. Much the same occurs in some books written in an old academic style. Some research has shown computer-based learning to be the least popular mode. But in a world where education and learning is a mass pursuit – rather than one simply for intellectual elites – the bulk of evidence suggests multiple, non-traditional approaches are more effective (Table 1). This requires a shift from didactic teaching to an active experience of learning through integrating knowledge bases and analyzing the results, all in the interests of solving particular problems yet gaining wider understanding of how to do such tasks better next time.

In a rapidly changing world, our learning also needs to be life-long. How best can we ensure this happens and is successful? To do this, we need to recognize that learning is also a social and cultural process, not just a cognitive one. This leads us to believe that distance-learning methods on their own are unlikely to be successful. It is self-evident that different approaches work best for different stages in education, as suggested in Table 1. We may debate the contents of any one cell in the matrix, but the benefits of tailoring the learning methodology to the task are certain.

There is no one right way to learn – different approaches suit different people at different times and for different tasks.

Across the world the proportion of people attending universities is rising. The expenditure on high-level education is thus large and increasing. It is not surprising therefore that the private sector sees opportunities, not least because studies in various countries have revealed that

Table 1. A matrix of learning steps and learning/teaching technology (Hills 1999). Crosses show where a particular technology does not function well for particular educational needs

	Live teacher	Lab work	Peer group	Print	Audio	Video	Computer off-line	Computer on-line
Motivation	✓	×	✓	✓	✓	✓	✓	×
Information transfer	×	✓	✓	✓	✓	✓	✓	×
Knowledge gestation	✓	✓	✓	×	×	×	✓	✓
Assessment	✓	✓	✓	✓	×	×	✓	×
Prescription of action	✓	×	✓	×	×	×	×	✓

there are major obstacles to the existing providers adapting to new ways of operating – notably the inertia and complacency exhibited by many staff in existing universities. Add to this the recognition by manufacturers of sophisticated software products that education and training of their users is crucial to building up a cohort of skilled, successful (and loyal) customers. And finally, in a world where Continuing Professional Development (CPD) is required to keep people up-to-date, the opportunities for repeat business are great. The result is that the private sector sees education and learning as a golden opportunity. Manifestations of this range from the private University of Phoenix – which commissions courseware from other suppliers, engages staff part-time, and makes great use of distance-learning methods – to software vendors who run successful courses on their own software, preceded by courses on the principles. In GIS, a number of different approaches to distance learning and CPD have been taken by private and publicly funded bodies such as the UNIGIS consortium, ESRI in its Virtual Campus, and at Pennsylvania State University.

How does this book fit with the new learning paradigm?

This is not a traditional textbook because:

- GISystems and GIScience do not lend themselves to traditional classroom teaching. Only by a combination of approaches can such crucial matters as principles, technical issues, practice, management, ethics, and accountability be learned. Thus the book is complemented by a Web site and by a range of other online resources, including modules specially written to accompany the book.

- It attempts for the first time to bring the principles and techniques of GIScience to those learning about GIS for the first time – and as such represents a major new stage in the evolution of GIS.

- The very nature of GIS as an underpinning technology in huge numbers of applications, spanning different fields of human endeavor, ensures that learning has to be tailored to individual or small-group needs.

- We have recognized the need to be driven by real-world needs. Hence a variety of applications and case studies are threaded through the text.

- We have linked our book to online learning resources throughout, notably the ESRI Virtual Campus.

SCIENCE, GOVERNMENT, BUSINESS, AND THE PROBLEMS AMENABLE TO GISCIENCE

Since this book is about GIScience, anything which changes the nature of science is relevant to us. The environment of GIScience is provided by GIS which has, since its inception in the 1960s, frequently been seen and used as a method of solving real-world problems. Thus any developments in problem-solving within business or government are also relevant to us and need to be reflected in the structure and contents of the book. What is happening more generally in business – the largest employer in most countries – is particularly crucial to us. As it turns out, there is something of a convergence between contemporary approaches to problem-solving in science, business, government, and society. This convergence and the (diminishing) tensions between these groups are now explored and the consequences for the book set out.

GIS, and Mode 1 and Mode 2 science

In recent years, the combined effect of technological and societal factors has begun to re-shape science. Much science has become multinational. Growing distrust of stereotypical scientists in white coats has led to demands for much greater openness and accountability. Again, of course, there is no unanimity about what is happening in detail but the general outline of the change is clear. It has been argued that there are two distinct attitudes to the creation and use of knowledge, which is the role of science. Mode 1 approaches still dominate the current scene and are manifested in subject-based universities. However, the business/government/academic convergence is manifested in a Mode 2 approach. Table 2 summarizes the characteristics of the two modes.

The public understanding of science and scientific understanding of the public are becoming key issues. Any method of enhancing communication between the two groups – like GIS – is valuable.

Where does this book fit in? Geography is universal; all physical phenomena and many abstract ones demonstrate spatial differentiation. The differentiation is rarely (if ever) random. Patterns are frequently observable if viewed at appropriate levels of detail (though sometimes these can be spurious artefacts of the way we carry out our analyses). More than that, there is some generality about these geographical patterns. Tobler's First Law of Geography is presented in Chapter 5, and

Table 2. The characteristics of Mode 1 and Mode 2 approaches to science (after Hills 1999)

Mode 1	Mode 2
Subject, and publication specialization, and fragmentation of knowledge	Holistic, not reductionist
Curiosity-driven, often blue-skies research usually within subject. Objectivity and disinterestedness	Mission-oriented, not blue-skies research – usually strategic or applied. Context-driven, not subject-driven. Service of practical interests involving subject values
Sets out to produce general laws or statements – but sometimes fails	Context-specific results – results *must be* obtained
Impersonal attitudes, open publication, and open argument. Progress by conjecture and refutation	Reflexive philosophy rather than absolute judgments
Sometimes (though less and less common) the work of a solitary scholar	Team-work based, not an individual scholar
Leads to convergence, consistency, reliability, but also consolidation of establishment values	Divergent, not convergent
Publications may be single- or multi-authored, homogeneous knowledge bases. Typically published in openly accessible, refereed scientific journals – but often with substantial publication delay	Multi-authored publications, heterogeneous knowledge bases. Some work not published if it provides competitive advantage, e.g., for exploitation of Intellectual Property Rights. Much use made of the Internet
Long-established scientific method, widely accepted within science community	Reflects the world outside academia – the world in which graduates work
Life-long vocation on part of researchers	Professional teams, re-assembled on project basis

illustrates this neatly. On the face of it therefore, Geography is a Mode 1 type activity in seeking universal truth. In practice, that is often far from the case. The great bulk of GIS applications are about problem-solving in a particular context rather than in elucidating general theories or laws. There are several reasons why much Geography – and GIS in particular – has Mode 2 characteristics:

● It is commonplace to have to work with both physical science and social and economic science factors (e.g. in choosing the siting of a nuclear waste plant). This rarely permits the formulation of universal laws but draws upon them.

● Because of this, team working is the norm, with different members contributing different skills (and sometimes different value systems).

● There is rarely a single set of objective criteria with which to measure success. Normally any GIS analyses and subsequent recommendations involve accommodating various trade-offs, each solution having advantages and disadvantages. Put another way, the success criteria are not universal – the team will have

their own means of judging success (e.g. achievement of a solution through rigorous application of certain methodologies and the garnering of adequate data matching the requirements of any statistical tests). But real-estate developers, public officials, environmental scientists, non-governmental organizations, and lay citizens may have quite different success criteria.

● GIS specialists are very likely to be working in different subject domains in successive jobs.

● Publication of results is often manifested in forms other than papers in refereed journals (e.g. as part of reports, as maps, or as Internet Web sites).

Given all of this, we should not pretend that we are always engaged in traditional Mode 1 science. GIS deals with everything from the universal laws of physics, through areas of statistical regularity, to the highly individual. And our results often have to be acceptable – or at least defensible – to different people with different experiences, different views, and (sometimes) different vested interests. That does not mean that we simply produce answers to order – we have an ethical commitment to produce the best, most objective, and replicable results possible.

Working in GIS inevitably involves being in conflict with others at different times. Coping with this involves both scientific and personal skills.

Making governments work better with GIS

Governments still matter. Notwithstanding the growth of business into traditional areas of public provision of services like education, governments at national, state, and local levels have a big influence on the lives of citizens. Yet many governments under-achieve, even though they often have excellent staff. The reasons why this occurs include outdated organizational structures, cultures, and remits dating from much earlier times. Backward-looking organizations tend to look after their own interests (e.g. in terms of opening hours, service provision locations, and what information they require from their clients). They tend to be relatively self-contained, minimizing collaboration with other bodies. In addition, the nature of most governments is that politicians are appointed to run their own departments – but the problems facing government as a whole are rarely capable of solution by a single vertically organized department. For example, reducing poverty in society has employment, educational, housing, social security, tax, and (possibly) health, and transport implications.

Inertia may prevail because there is little incentive (other than political pressure generated by elections) to improve. A focus on inputs (e.g. the budget-setting process is a key activity in most governmental years) rather than outcomes is quite common in government. And risk aversion, sometimes because over-seeing bodies will punish all failure even if a good risk assessment was carried out, is very frequent. Rarely in government is anyone praised or rewarded for succeeding, but punishment for failure is inevitable and often public. All this adds up to lower-than-feasible performance, unresponsiveness to elected politicians and citizens, and costly inefficiencies.

Minimizing these problems is non-trivial, but many governments across the world have begun to tackle them. The ways this has been done include:

● Setting targets, publicizing these, and measuring achievement for different branches of government (which can be effective but tends to atomize government, minimizing interactions and fostering the shifting of problems to other people).

● Creating matrix-like management structures across government and also rotating staff from one department to another to propagate best practice.

● Empowering citizens with information about what the departments are setting out to achieve and the background to judge them, rewarding the departmental successes appropriately.

● Trying to build integrated or "joined-up" government by common use of the same electronic forms and procedures and re-use of information wherever it exists, rather than collecting it and retaining it within silos in each department. In short, taking a corporate approach to information management and to management of the enterprise as a whole.

Where does GIS fit into the new approaches to government? It will be obvious that solving problems in this domain is not straightforward, not least because there are usually multiple objectives and infinite expectations of government. Typically in large and complex organizations like democratically elected governments, some policies may actually contradict and undermine others. Certainly the need to establish corporate approaches across government can easily undermine the local autonomy needed when someone is asked to solve a particular problem. But it is quite extraordinary how many of these problems can be eased by appropriate use of GIS. In particular, use of a common base set of information, including the topography and key datasets (environmental, socio-economic, financial statistics, etc.) can be effective in fostering integrated government. Use of a common GIS language and an ability to sew together datasets from many sources and assess them can be crucial – in the late 1980s, the CORINE environmental GIS created for Europe demonstrated that different sampling and data creation algorithms ensured that the greatest rate of environmental change across that continent often occurred at country boundaries! More than that, empowering citizens is ever easier – at least for those with access to computers – through information provided over the Internet. Use of GIS can make a huge variety of information available in understandable form to the non-expert citizen. In addition, some risk assessments can be carried out effectively within a GIS and reported readily.

In summary, much of government is related to things happening or not happening in a given area. As a minimum its success requires consistent, relevant, and up-to-date information. In a world where more is expected for less, costs must be cut through reduction in staff time, although the political process ensures that multiple solutions to problems with many causes have to be evaluated and exposed. Given all this, it is no great surprise that GIS is increasingly widely used at all levels of government. GIScience may, however, seem irrelevant in an arena where improvements in routine decision-making and information exchange will make things better. In practice, the science is still central to good governance if only because it should help us to avoid capricious or even gerrymandering use of GIS.

Government "of the people, by the people, for the people" is made easier by the widespread use of GIS.

Meeting business needs with GIS

It seems self-evident that business is primarily driven by the bottom line. Sometimes this truly is entirely a short-term affair: stock prices rise and fall according to quarterly profitability in some industries. But other businesses act on a longer-term basis. For instance, Amazon.com and many other .com companies made no profits in their first few years of trading. Even beyond that, however, many businesses seek to operate in a way which is sustainable, environmentally friendly, and takes account of local priorities and issues. To do so is often good business: the reputation of a business for integrity, sensitivity, and fair dealing is hard won and influences potential customers, yet is easily lost. To abide by local laws and customs also reduces business risk. For these reasons, it is grossly simplistic to see firms as simple profit-maximizers. They too need information about business opportunities and risks and how these are geographically distributed. Yet they also need to know how to solve a particular problem at minimum cost but maximum efficacy. The latter does not just mean a narrow concern with efficiency and short-run profitability: it normally means being sensitive to the opinions of customers and government and building long-term relationships. One big part of this is achieved by bringing together the best available contemporary information and analyzing it using the best science. It is obvious that GIS is a superb vehicle to do this.

Making profits is a necessary condition for success in the private sector – but not a sufficient one for long-term success. Use of GIS can help to evaluate alternatives, minimize business risk, share cost, and audit processes for regulators and citizens' groups.

SUMMARY

In this preface we have tried to show that there is growing commonality between the concerns of business, government, and science. The examples in the book of problems tackled through GIS have been chosen deliberately to show this commonality, as well as the interplay between organizations and people from different sectors. Clearly the concerns and commonality will never totally coincide. But concerns with effectiveness, efficiency, bringing together information from disparate sources, acting within regulatory and ethical frameworks, and preserving a good reputation are all common at the meta-level. For this reason, this book combines the basics of GIS with the solving of problems which often have no single, ideal solution – the world of business, government, and Mode 2 science.

There are few absolutes in our world. For instance, the distinction between Mode 1 and Mode 2 approaches can be exaggerated: much of "big science" is now a team-based endeavor, often inter-disciplinary and international in scope, rather than the activity of a lonely scholar in a traditional discipline. That said, the convergence noted above is mostly in relation to Mode 2 approaches to knowledge creation or science: the approaches in government and in business certainly resemble Mode 2 science rather more than the Mode 1 variety. The common characteristics are a search for good solutions within a particular context, achieved within available time and budget, and taking account of all available information (however imperfect) – rather than seeking the truth, however long it takes, and rigid adherence to certain procedures. At the same time, there are many aspects of Mode 1 science – the formulation of hypotheses to be tested, the need to be able to replicate and demonstrate results, plus an acceptance of the need to justify and debate results – which remain essential. In this book therefore we have sought to bring in different strands of scientific approach and to be eclectic in dealing with both physical and social sciences. Throughout the book, and in its accompanying Web site and associated material, we have also tried to escape "silo" or stovepipe mentality in tackling real-world problems of relevance to governments, businesses, and the citizen alike, and to avoid being constrained by traditional disciplinary blinkers. We see our efforts as being to underpin the business activities of those working in the public and private sectors – defining "business" in its widest sense.

In short, we have tried to create a book tuned to the way the world works now, to the way in which most of us increasingly operate as knowledge workers, and to our need to face complicated issues without ideal solutions in our daily lives. As we have said above, it is obvious that this book is an unusual enterprise and product. It has been written by a multinational partnership, drawing upon material from around the world. One of the authors is an employee of a leading software vendor and two of the other three have had business dealings with ESRI over many years. Moreover, many of the illustrations and examples come from the customers of that vendor. We wish to point out however that neither ESRI (or Wiley) has ever sought to influence our content or the way in which we made our judgments, and we have included references to other software and vendors throughout the book. Whilst we make frequent reference to ESRI's Virtual Campus (and some modules of it have been specifically written around this book), we also make reference to similar sources of information in both paper and digital form. We believe we have created something novel but valuable by our lateral thinking in all these respects. Whether we actually have or not is of course for others to judge.

CONVENTIONS USED IN THE BOOK

We use the acronym *GIS* in many ways in the book, partly to emphasize one of our goals, the interplay between geographic information *systems* and geographic information *science*; and at times we use two other possible interpretations of the three-letter acronym: geographic information *studies* and geographic information *services*. We distinguish between the various meanings where appropriate or where the context fails to make the meaning clear, especially in Section 1.6 and in the Epilog. We also use the acronym in both singular and plural senses, following what is now standard practice in the field, to refer as appropriate to a single geographic information system or to geographic information systems in general.

We have used a series of devices to aid navigation of the book, notably through the use of color and symbols. Each section is color-coded by a title bar. The header for each box contains an icon to show whether it is technical, relates to people or is about applications.

HOW TO MAKE THE MOST OF THIS BOOK

Although the book stands alone as a self-contained work, we have intended it to be used in conjunction with the two editions of *Geographical Information Systems*. We include specific references to both editions at the end of each chapter and at points in the text where a reference is particularly

appropriate. Wiley has made chapters and the bibliography of the first edition available on the companion Web sites, www.wiley.com/gis and www.wiley.co.uk/gis. The references give access to greater depth, and more detail on many core topics, and also to reviews of more peripheral topics that we have not been able to cover here. Boxes, each with distinctive icons, are used to describe key techniques, to present illustrative real-world applications, and to summarize the activities of key individuals in the field.

At the end of each chapter we also include references to other, easily accessible, books on special topics. We have not attempted to provide a complete bibliography because extensive reference lists appear in both editions of *Geographical Information Systems*. There are references after each chapter to relevant sections of the materials available online at the ESRI Virtual Campus, campus.esri.com; links to other relevant online information; and references to the two Core Curricula developed by the National Center for Geographic Information, and Analysis (NCGIA), both now online. Throughout the book we have tried to limit references to only the most stable WWW sites but unfortunately it is inevitable that some will disappear through time.

We have organized the book in three major but interlocking sections: after two chapters of introduction, the sections appear as Principles (Chapters 3 through 7), Techniques (Chapters 8 through 15), and Practice (Chapters 16 through 19). We conclude with an Epilog (Chapter 20). It was not always easy to decide whether some topic belonged in principles or in techniques but we have tried to separate the persistent principles – ideas that will be around long after today's technology has been relegated to the museum – from knowledge that is necessary to an understanding of today's technology and likely near-term developments. Much of the accumulated knowledge that existed in many cases long before digital computers, but is now more important than ever, appears in the Principles section, such as the map projections invented by cartographers in centuries past. Some topics such as uncertainty or spatial interpolation clearly belong in both and we have discussed them in general terms as principles and in specific terms as techniques, with cross-references where appropriate and without unnecessary repetition.

ACKNOWLEDGMENTS

As the book went to press, we were saddened to hear of the death of Professor John Estes, the well-known expert in remote sensing. A long-time colleague of Michael Goodchild at Santa Barbara and long-standing friend of David Rhind, he made major contributions to the new global map (Chapter 19) and to many aspects of the interface between remote sensing and GIS. We gratefully acknowledge his support and friendship. At about the same time we also learned of Ian McHarg's death. McHarg's method, in which planners inventory every level of detail about a place and take this into account in development, is the basis to what became GIS. Ian McHarg was a long term associate of, and inspiration to, Michael Goodchild and will also be sadly missed.

We take complete responsibility for all the material contained herein. But much of it draws upon contributions made by friends and colleagues from across the world, many of them outside the academic GIS community. We thank them all for those contributions, and the discussions we have had over the years. We cannot mention all of them but would particularly like to mention the following:

For their input to this project, and for many GIS discussions over the years: Mike Batty, Clint Brown, Nick Chrisman, Keith Clarke, Andy Coote, Danny Dorling, Jason Dykes, Max Egenhofer, Pip Forer, Andrew Frank, Gayle Gaynor, Richard Harris, Les Hepple, Sophie Hobbs, Karen Kemp, Chuck Killpack, Vanessa Lawrence, John Leonard, Bob Maher, David Mark, David Martin, Scott Morehouse, Scott Orford, Peter Paisley, Jonathan Raper, Helen Ridgway, Jan Rigby, Christopher Roper, Garry Scanlan, Karen Siderelis, David Simonett, Andy Smith, Roger Tomlinson, Carol Tullo, Dave Unwin, David Willey, Jo Wood, Mike Worboys.

To Peter Haggett, for guidance (not always heeded!) as to how to prepare a good textbook.

At University College London, where much of the artwork was drawn, and the drafts assembled: Nick Mann, Elanor McBay, Cath Pyke, Sarah Sheppard.

Our friends in Tuscany: General Facciorusso, Col Serino, Lt Col Orru, and Captain Bari of the Italian Geography and Mapping Institute for their hospitality and for showing us their astounding maps and books from the 15th century onwards.

At John Wiley: Rob Garber, Jim Harper, Lou Page, and (for guiding the project from conception to fruition) Sally Wilkinson.

We would be remiss if we did not mention the contribution of Jack Dangermond, a true visionary, whose enthusiasm has done much to inspire the entire GIS enterprise, both at ESRI and around the world.

We would especially like to thank the following people for allowing us to use their material: Academic Press Ltd, Aerial Images Inc./ Sovinformsputnik, the Association of American Geographers, Autodesk, Richard Bailey, Mike Batty, Blackwell Publishers, CA Department of

Fish and Game, Martin Callingham, John Colkins of ESRI and the State of Hawaii for the background image on the front cover, Tom Cova, Peter H. Dana, Daratech Inc., Frank W. Davis of University of California Biogeography Lab., DeLorme Publishing Co. Inc., Daniel Dorling, Durham Herald Company Inc., Earth Science Department of University of Siena in Italy, Ecotrust, ERDAS Inc., EROS Data Center, ESRI, Experian, GDS, the Geographical Magazine, GTCO Calcomp, Richard Harris, Human Settlements Research Center of Tsinghua University in China, Institute of Transport Engineers, Intergraph, Landmark Information Group, Mitch Langford of MRRL (University of Leicester), Lawrence Berkeley National Laboratory, Leica, London Transport Museum, MapInfo, MapQuest.com Inc., David Mark, Matrix Directory and Information Services, Microsoft, NASA/JPL/Caltech, Open GIS Consortium Inc., C. S. Papacostas, Pearson Education Ltd, Jesper Persson of Teleadress Information AB, Pinter Press, Professor J. Radke, San Parks in South Africa, Ashton Shortridge, Andy Smith, Taylor & Francis, Henry Tom of Oracle, USGS Department of the Interior, Richard Webber, and Marcel de Wit.

Whilst every effort has been made to trace the owners of copyright material, in a few cases this has proved impossible and we take this opportunity to offer our apologies to any copyright holders whose rights we may have unwittingly infringed.

Finally, thanks go to our families, especially Amanda, Fiona, Heather, and Christine, who have endured so much during the preparation of this manuscript.

Paul Longley
Michael Goodchild
David Maguire
David Rhind

November 2000

Reference Links

Maguire D J, Goodchild M F, and Rhind D W (eds) 1991 *Geographical Information Systems: Principles and applications*. Harlow, UK: Longman (Text available online from 'Links to Big Book 1' at www.wiley.com/gis and www.wiley.co.uk/gis).

Longley P A, Goodchild M F, Maguire D J, and Rhind D W (eds) 1999 *Geographical Information Systems: Principles, techniques, management and applications*. New York, John Wiley.

References

Hills G 1999 The University of the future. In Thorne M (Ed.) *Foresight: Universities in the future*, London: Department of Trade and Industry, 213-232.

LIST OF ACRONYMS

AA Automobile Association (UK)
ACORN A Classification of Residential Neighbourhoods (UK)
ADRG Arc Digitized Raster Graphics
AGI Association for Geographic Information
AM/FM automated mapping/facilities management
AML Arc Macro Language
API application programming interface
ARPANET Advanced Research Projects Agency Network
AVHRR Advanced Very High Resolution Radiometer
BASIC Beginners All-purpose Symbolic Instructions Code
CAD computer-aided design
CAMA Computer Assisted Mass Appraisal
CASE computer-assisted software engineering
CBD central business district
CD compact disk
CES Centre for Environmental Studies
CGIA Center for Geographic Information and Analysis (North Carolina)
CGIS Canada Geographic Information System
CGM computer graphics metafile
CLM collection-level metadata
COM component object model
COGO coordinate geometry
CORBA common object request broker architecture
COTS commercial off-the-shelf
CPD continuing professional development
CSDGM Content Standards for Digital Geospatial Metadata
DBA database administrator
DBMS database management system
DCW Digital Chart of the World
DDL data definition language
DEM digital elevation model
DIME Dual Independent Map Encoding
DLG digital line graph
DML data manipulation language
DRG digital raster graphics
DWG drawing
DXF drawing exchange format
EC European Commission
EDC EROS Data Center
EPA Environmental Protection Agency
EPS encapsulated postscript
ERDAS Earth Resources Data Analysis System
EROS Earth Resources Observation Systems
ESRI Environmental Systems Research Institute Inc

EU European Union
FEMA Federal Emergency Management Agency
FGDC Federal Geographic Data Committee
FSA Forward Sortation Area
GAM geographical analysis machine
GEM geographical explanations machine
GDT Geographic Data Technology
GIF graphics interchange format
GNIS (USGS) Geographic Names Information System
GPS Global Positioning System
GSDI Global Spatial Data Infrastructure
GUI graphical user interface
HTML hypertext markup language
HTTP hypertext transmission protocol
IBM International Business Machines
ICMA International City/Council Management Association
ICT information and communication technology
ID identifier
IDE integrated development environment
IDW inverse distance weighting
IGN Institut Géographique National
IMS Internet Map Server (as in ESRI's ArcIMS)
IMW The International Map of the World
ISCGM International Steering Committee for Global Mapping
ISO International Standards Organization
ISO TC International Standards Organization Technical Committee
IT information technology
KE knowledge economy
LAN local area network
LIS land information system
MAPEX MAP EXplorer
MAT minimum aggregate travel
MAUP modifiable areal unit problem
MBR minimum bounding rectangle
MCDM multi-criteria decision-making
MMU minimum mapping unit
MOADB mother of all databases
MrSID Multiresolution Seamless Image Database
MSC (US National Research Council) Mapping Science Committee
MSS multi-spectral scanner
NAD27 North American Datum of 1927
NAD83 North American Datum of 1983
NASA National Aeronautics and Space Administration

NCDI National Center for Disaster Information (Honduras)
NCGIA National Center for Geographic Information and Analysis
NGDC National Geospatial Data Clearinghouse
NIMA National Imagery and Mapping Agency
NIMBY not in my back yard
NMD National Mapping Division (former part of USGS)
NMO national mapping organization
NMP (US) National Mapping Program
NSDI National Spatial Data Infrastructure
NTF national transfer format
ODBMS object database management systems
OLE object linking and embedding
OLM object-level metadata
OLS ordinary least squares
OMG Object Management Group
ONC Operational Navigation Chart
ORDBMS object-relational database management systems
PC personal computer
PCC percent correctly classified
PDA personal digital assistant
PLSS Public Land Survey System
PNG portable network graphics
R&D research and development
RDBMS relational database management systems
RMSE root mean squared error
SAP spatially aware professional
SDE Spatial Database Engine (as in ESRI's ArcSDE)
SDTF spatial data transfer standard
SPC State Plane Coordinate
SPOT Système Probatoire d'Observation de la Terre

SQL structured (or standard) query language
SWOT strengths, weaknesses, opportunities, threats
TIFF tag image file format
TIGER Topologically Integrated Geographic Encoding and Referencing
TIN triangulated irregular network
TLA three letter acronym
TM (Landsat) Thematic Mapper
TSP traveling salesman problem
UCGIS University Consortium for Geographic Information Science
UML unified modeling language
UNIGIS UNIversity GIS Consortium
UPS Universal Polar Stereographic
URISA Urban and Regional Information Systems Association
USGS United States Geological Survey
UTM Universal Transverse Mercator
VAR value-added reseller
VBA visual BASIC for applications
VGA video graphics array
ViSC Visualization in Scientific Computing
VPF vector product format
WAN wide area network
WIMP windows, icons, menus, and pointers
WIPO World Intellectual Property Organization
WWF World Wide Fund (for Nature)
WGS84 World Geodetic System of 1984
Win Windows
WTO World Trade Organization
WWW World Wide Web
XML Extensible Markup Language

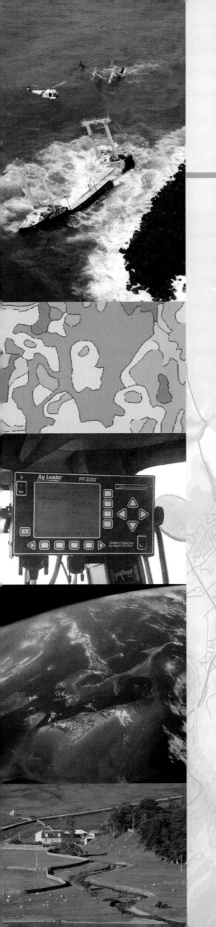

SYSTEMS, SCIENCE, AND STUDY

1

This chapter introduces the conceptual framework for the book, by addressing several major questions:

- What exactly is geographic information, and why is it important? What is special about it?

- What is information generally, and how does it relate to data, knowledge, evidence, wisdom, and understanding?

- What kinds of decisions make use of geographic information?

- What is a geographic information system, and how would I know one if I saw one?

- What is geographic information science, and how does it relate to the use of GIS for scientific purposes?

- How do scientists use GIS, and why do they find it helpful?

Learning objectives

At the end of this chapter you will:

- Know definitions of the terms used throughout the book, including GIS itself;

- Be familiar with a brief history of GIS;

- Recognize the sometimes invisible roles of GIS in everyday life, and the roles of GIS in business;

- Understand the significance of geographic information science, and how it relates to geographic information systems;

- Understand the many impacts GIS is having on society, and the need to study those impacts.

1.1 Introduction: Why Does GIS Matter?

Almost everything that happens, happens somewhere. We humans are largely confined in our activities to the surface and near surface of the Earth. We travel over it and in the lower levels of the atmosphere, and through tunnels dug just below the surface. We dig ditches and bury pipelines and cables, construct mines to extract mineral deposits, and drill wells to access oil and gas. Keeping track of all of this activity is important, and knowing where it occurs can be the most convenient basis for tracking. Knowing where something happens is critically important if we want to go there ourselves or send someone there, to find other information about the same place, or to inform people who live nearby. Geographic information systems are a special class of information systems that keep track not only of events, activities, and things, but also of *where* these events, activities, and things happen or exist.

> Almost everything that happens, happens somewhere. Knowing where something happens is critically important.

Because location is so important, it is an issue in many of the problems society must solve. Some of these are so routine that we almost fail to notice them – the daily question of which route to take to and from work, for example. Others are so monumental in scope that they affect millions –

the recent repartitioning of Bosnia, for example (Figure 1.1). Problems that involve an aspect of location, either in the information used to solve them, or in the solutions themselves, are termed *geographic problems*. Here are some more examples:

● Health care managers solve geographic problems when they decide where to locate new clinics and hospitals.
● Delivery companies solve geographic problems when they decide the routes, and schedules of their vehicles, often on a daily basis.
● Transportation authorities solve geographic problems when they select routes for new highways (Figure 1.2).
● Forestry companies solve geographic problems when they determine how best to manage forests, where to cut, where to locate roads, and where to plant new trees (Figure 1.3).
● Governments solve geographic problems when they decide how to allocate funds for building sea defenses.
● Travelers solve geographic problems when they find their way through airports, give and receive driving directions, and select hotels in unfamiliar cities (Figure 1.4).
● Farmers solve geographic problems when they employ new information technology to make better decisions about the amounts of fertilizer and pesticide to apply to their fields (see Box

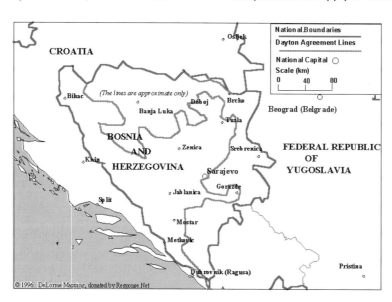

Figure 1.1 The map of Bosnia and Herzegovina drawn up as part of the Dayton Accord of 1995. GIS tools were used to support the discussions and negotiations that led to the agreement

Figure 1.2 The route for a new highway is a geographic problem

Figure 1.4 Navigating and receiving directions is a geographic problem

Figure 1.3 Management of forests is a geographic problem

1.1 on the developing GIS application of precision agriculture).

If so many problems are geographic, what distinguishes them from each other? Here are three bases for classifying geographic problems. First, there is the question of scale, or level of geographic detail. The architectural design of a building is a geographic problem, but only at a very detailed or local scale. The information needed to develop the design is also local – the size and shape of the parcel, the slope of the land, and perhaps the visibility of the building from nearby locations. The repartitioning of Bosnia is a problem at a much broader and coarser scale, involving information about the entire area of Bosnia, including maps of the ethnicity of its residents, historical boundaries, and perhaps topography. Solution of the problem

may also require data from areas outside Bosnia, in the border regions.

Scale or level of geographic detail is an essential property of any GIS project.

Second, geographic problems can be distinguished on the basis of intent, or purpose. Some problems are strictly practical in nature – they must be solved as quickly as possible, at minimum cost, in order to achieve such practical objectives as cost minimization (saving money). Others are better characterized as driven by human curiosity. When geographic data are used to verify the theory of continental drift, or to map distributions of glacial deposits, or to analyze the patterns of early agriculture evident in archaeological discoveries, there is no sense of an immediate problem that needs to be solved – rather, the intent is the advancement of human understanding of the world, which we often recognize as the intent of science.

Although science and practical problem-solving are often seen as distinct human activities, Laudan (1996) argues that there is no longer any effective distinction between their methods. The tools and methods used by a scientist in a government agency to ensure the protection of an endangered species are essentially the same as the tools used by an academic ecologist to advance our scientific knowledge of biological systems. Both use the most accurate measurement devices, use terms whose meanings have been widely shared and agreed, insist that their results be replicable by others, and in general follow all of the principles of science that have evolved over the past centuries.

Box 1.1   *Building better precision agriculture systems for Illinois farmers with GIS*

In the past, farmers have been forced to apply pesticides and fertilizer to crops uniformly, because there was no way to vary applications in response to point-by-point variation in soil conditions, drainage, slope, and other variables. Precision farming combines the Global Positioning System (GPS, see Section 10.2.2.2), GIS, sensors mounted on tractors, and harvesters, and digital images from aircraft, to adjust applications to suit local conditions. This means that applications no longer have to be at a uniform rate determined by the greatest need. By adjusting applications it is possible to reduce the total amounts used, and hence the damage to the local environment (and save money, too).

Figure 1.5 shows parts of the system being used by researchers working for the Illinois Farm Service, Urbana, Illinois, USA.

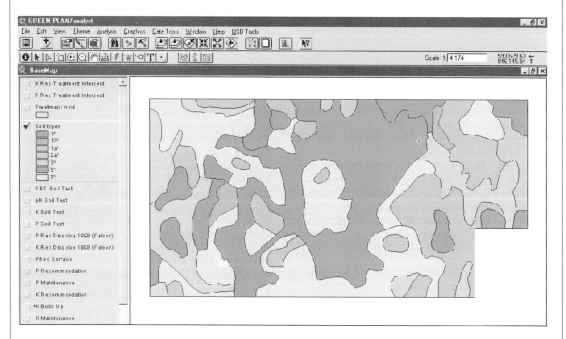

Figure 1.5 Some of the components of a precision agriculture system being used in research by the Illinois Farm Service, Urbana, Illinois, USA (Source: ESRI)

The use of GIS for both forms of activity certainly reinforces this idea that science and practical problem-solving are no longer distinct in their methods, as does the fact that GIS is widely used in all kinds of organizations, from academic institutions to government agencies and corporations. In the Preface, we discussed the changes that are occurring in the role of science and the way it is conducted, in terms of a shift from traditional Mode 1 science to new Mode 2 science. The use of similar tools and methods for pursuing curiosity and solving problems was identified as one of the characteristics of that shift.

At some points in this book it will be useful to distinguish between uses of GIS that focus on design, or so-called *normative* uses, and uses that advance science, or so-called *positive* uses (a rather confusing meaning of that term, unfortunately, but the one commonly used by philosophers of science – its use implies that science confirms theories by finding *positive* evidence in support of them, and rejects theories when negative evidence is found). Finding new locations for retailers is an example of a normative application of GIS, while a social scientist who studies how people make choices between alternative retail outlets is using GIS in a positive application.

> With a single collection of tools, GIS is able to bridge the gap between curiosity-driven science and practical problem-solving.

Third, geographic problems can be distinguished on the basis of their time scale. Some decisions are *operational*, and are required for the smooth functioning of an organization. Others are *tactical*, and concerned with medium-term decisions, such as where to cut trees in next year's forest harvesting plan. Others are *strategic*, and are required to give an organization long-term direction. These terms are explored in the context of business applications of GIS in Section 2.3.3.1. Other problems that interest geophysicists, geologists, or evolutionary biologists may occur on time scales that are much longer than a human lifetime, but are still geographic in nature, such as predictions about the future physical environment of Japan, or about the animal populations of Africa. Geographic databases are often *transactional*, meaning that they are constantly being updated as new information arrives, unlike maps, which stay the same once printed.

Chapter 2 contains a more detailed discussion of some GIS applications, and an idealized (and at times futuristic) view of how GIS plays a role in almost all aspects of our daily lives. Other applications are discussed throughout the book.

1.1.1 Spatial is special

The adjective *geographic* refers to the Earth's surface and near-surface, and defines the subject matter of this book, but other terms have similar meaning. *Spatial* refers to any space, not only the space of the Earth's surface. It is used frequently in the book, almost always with the same meaning as *geographic*. But many of the methods used in GIS are also applicable to other non-geographic spaces, including the surfaces of other planets, the space of the cosmos, and the space of the human body that is captured by medical images. GIS techniques have even been applied to the analysis of genome sequences on DNA. So the chapters on analysis in this book are titled *spatial* analysis (Chapters 13 and 14), not geographic analysis, to emphasize this versatility.

Another term that has been growing in usage in recent years is *geospatial* – implying a subset of spatial applied specifically to the Earth's surface and near-surface. In this book we have tended to avoid geospatial, preferring geographic, and spatial where we need to emphasize generality.

People who encounter GIS for the first time are sometimes driven to ask why geography is so important – why is spatial special? After all, there is plenty of information around about geriatrics, for example, and in principle one could create a geriatric information system. So why has geographic information spawned an entire industry, if geriatric information hasn't to anything like the same extent? Why are there no courses in universities specifically in geriatric information systems? Part of the answer should

Figure 1.6 Maps are a common form of geographic information

Box 1.2   Some technical reasons why geographic information is special

- it is multidimensional, because at least *two* coordinates must be specified to define a location, whether they be x and y or latitude and longitude;
- it is voluminous, since a geographic database can easily reach a terabyte in size (see Table 1.1);
- it must often be projected onto a flat surface, for reasons identified in Section 4.7;
- it requires many special methods for its analysis (see Chapters 13 and 14);
- it can be time-consuming to integrate and analyze the many varied types of geographic information;
- although much geographic information is static, the process of updating is complex, and expensive;
- display of geographic information in the form of a map requires the retrieval of large amounts of data.

Table 1.1. Potential GIS database volumes for some typical applications (volumes estimated to the nearest order of magnitude). Strictly, bytes are counted in powers of 2 – 1 kilobyte is 1024 bytes, not 1000

1 megabyte	1,000,000	Single dataset in a small project database
1 gigabyte	1,000,000,000	Entire street network of a large city or small country
1 terabyte	1,000,000,000,000	Elevation of entire Earth surface recorded at 30 m intervals
1 petabyte	1,000,000,000,000,000	Satellite image of entire Earth surface at 1 m resolution

be clear already – almost all human activities and decisions involve a geographic component, and the geographic component is important. Another reason will become apparent in Chapter 3 – working with geographic information involves complex and difficult choices that are also largely unique. Other, more technical reasons will become clear in later chapters, and are briefly summarized in Box 1.2.

1.2   Data, Information, Evidence, Knowledge, Wisdom

Information systems help us to manage *what we know*, by making it easy to organize and store, access and retrieve, manipulate and synthesize, and apply to the solution of problems. We use a variety of terms to describe what we know, including the five that head this section. There are no universally agreed definitions of these terms, the first two of which are used frequently in the GIS arena. Nevertheless it is worth trying to come to grips with their various meanings, because the differences between them can often be significant.

Data consist of numbers, text, or symbols which are in some sense neutral and almost context-free. Raw geographic facts, such as the temperature at a specific time and location, are examples of data. When data are transmitted, they are treated as a stream of bits; and the internal meaning of the data is irrelevant in the transfer process.

Information is differentiated from data by implying some degree of selection, organization, and preparation for particular purposes – information is data serving some *purpose*, or data that have been given some degree of *interpretation*. Information is often costly to produce, but once digitized it is cheap to *reproduce* and distribute. Geographic datasets, for example, may be very expensive to collect, and assemble, but very cheap to copy and disseminate. One other characteristic of information is that it is easy to add value to it through processing, and through merger with other information. GIS provides an excellent example of the latter, because of the tools it provides for combining information from different sources.

Knowledge can be considered as information to which value has been added by interpretation based on a particular context, experience, and purpose. Put simply, the information available in a book or on the Internet or on a map becomes knowledge only when it has been read and understood. How the information is interpreted and used will be different for different readers depending on their previous experience, expertise, and needs. It is important to distinguish two types of knowledge: *codified*, and *tacit*. Knowledge is codifiable if it can be written down and transferred relatively easily to others. Tacit knowledge is often slow to acquire and much more difficult to transfer. Examples include the knowledge built up during an apprenticeship, understanding of how a particular market works, or familiarity with using a particular technology or language. Because of its

Table 1.2. A ranking of the support infrastructure for decision-making

Decision-making support infrastructure	Ease of sharing with everyone	GIS example
Wisdom	*Impossible*	Policies developed and accepted by stakeholders
↑		
Knowledge	*Difficult, especially tacit knowledge*	Personal knowledge about places and issues
↑		
Evidence	*Often not easy*	Results of GIS analysis of many datasets or scenarios
↑		
Information	*Easy*	Contents of a database assembled from raw facts
↑		
Data	*Easy*	Raw geographic facts

nature, tacit knowledge is often a source of competitive advantage.

Increasingly, *evidence* is considered a halfway house between information and knowledge. It seems best to regard it as a multiplicity of information from different sources, related to specific problems and with a consistency that has been validated. Major attempts have been made in medicine to extract evidence from a welter of sometimes contradictory sets of information, drawn from world-wide sources, in what is known as *meta-analysis*, or the comparative analysis of the results of many previous studies.

Wisdom is even more elusive to define than the other terms. It is normally used in the context of decisions made or advice given which is disinterested, based on all the evidence and knowledge available, but given with some understanding of the likely consequences. Almost invariably, it is highly individualized rather than being easy to create and share within a group. Wisdom is in a sense the top level of a hierarchy of decision-making infrastructure (Table 1.2).

1.3 The Science of Problem-Solving

How are problems solved, and are geographic problems solved any differently from other kinds of problems? We humans have accumulated a vast storehouse of information about the world, including how it *looks*, its *forms*, how it *works*, and its dynamic *processes*. Some of those processes are natural and built into the design of the planet, such as the processes of tectonic movement that lead to earthquakes, and the processes of atmospheric circulation that lead to hurricanes (Figure 1.7). Others are human in origin, reflecting the increasing influence that we have on our natural environment, through the burning of fossil fuels, the felling of forests, and the cultivation of crops (Figure 1.8). Others are

imposed by us, in the form of laws, regulations, and practices. For example, zoning regulations affect the ways in which specific parcels of land can be used.

> Knowledge about how the world works is more valuable than knowledge about how it looks, because such knowledge can be used to predict.

These two types of information differ markedly in their degree of generality. Form varies geographically, and the Earth's surface looks dramatically different in different places – compare the settled landscape of England with the deserts of the US Southwest (Figure 1.9). But processes can be very general. The ways in which the burning of fossil fuels affects the atmosphere are essentially the same in China as in Europe, although the two landscapes look very different. Science has always valued such general knowledge over knowledge of the specific, and hence has valued knowledge of process over knowledge of form. Geographers in particular have witnessed a longstanding debate, lasting centuries, between the competing needs of

Figure 1.7 Natural processes affect the form of the Earth's surface

Figure 1.8 Social processes also modify the Earth's surface

idiographic geography, which focuses on the description of form and emphasizes the unique characteristics of places; and *nomothetic* geography, which seeks to discover general processes. Both are essential, of course, since

Figure 1.9 The form of the Earth's surface shows enormous variability

knowledge of general process is only useful in solving specific problems if it can be combined effectively with knowledge of form. For example, we can only assess the impact of soil erosion on agriculture in New South Wales, Australia if we know *both* how soil erosion is generally impacted by such factors as slope, *and* specifically how much of New South Wales has steep slopes, and where they are located (Figure 1.10).

One of the most important merits of GIS as a tool for problem-solving lies in its ability to combine the general with the specific, as in this example from New South Wales. A GIS designed to solve this problem would contain knowledge of New South Wales's slopes, in the form of computerized maps, and the programs executed by the GIS would reflect general knowledge of how slopes affect soil erosion. The *software* of a GIS captures and implements general knowledge, while the *database* of a GIS represents specific information. In that sense a GIS resolves the old debate

Figure 1.10 Predicting the effects of erosion in an area requires general knowledge of processes and specific knowledge of the area – both are available in a GIS

between nomothetic and idiographic camps, by accommodating both.

GIS solves the ancient problem of combining general scientific knowledge with specific information, and gives practical value to both.

General knowledge comes in many forms. Classification is perhaps the simplest and most rudimentary, and is widely used in geographic problem-solving. In many parts of the USA and other countries efforts have been made to limit development of wetlands, in the interests of preserving them as natural habitats, and avoiding excessive impact on water resources. To support these efforts, resources have been invested in mapping wetlands, largely from aerial photography and satellite imagery. These maps simply classify land, using established rules that define what is, and what is not a wetland.

More sophisticated forms of knowledge include *rule sets* – for example, rules that determine what use can be made of wetlands, or what areas in a forest can be legally logged. Rules are used by the US Forest Service to define wilderness, and to impose associated regulations regarding the use of wilderness, including prohibition on logging, and road construction.

Much of the knowledge gathered by the activities of scientists suggests the term *law*. The work of Sir Isaac Newton established the Laws of Motion, according to which all matter behaves in ways that can be perfectly predicted. From Newton's laws we are able to predict the motions of the planets almost perfectly, although Einstein later showed that certain observed deviations from the predictions of the laws could be explained with his new Theory of Relativity. Laws of this level of predictive quality are few, and far between in the geographic world of the Earth's surface. Market researchers use spatial interaction models, in conjunction with GIS, to predict how many people will shop at each shopping center in a city. There are substantial errors in the predictions, but nevertheless the results are of great value in developing location strategies for retailing. The Universal Soil Loss Equation, used by soil scientists in conjunction with GIS to predict soil erosion, is similar in its relatively low predictive power, but again the results are sufficiently accurate to be very useful in the right circumstances.

Solving problems involves several distinct components, and stages. First, there must be an *objective*, or a goal that the problem-solver wishes to achieve. Often this is a desire to maximize or minimize – find the solution of least cost, or

shortest distance, or least time, or greatest profit; or to make the most accurate prediction possible. These objectives are all expressed in *tangible* form, that is, they can be measured on some well-defined scale. Others are said to be *intangible*, and involve objectives that are much harder, if not impossible to measure. They include maximizing *quality of life* and *satisfaction*, and minimizing *environmental impact*. Sometimes the only way to work with such intangible objectives is to involve human subjects, through surveys or focus groups, by asking them to express a preference among alternatives. A large body of knowledge has been acquired about such human-subjects research, and much of it has been employed in connection with GIS – interested readers are referred to the appropriate texts (e.g. Massam 1980, 1993). For an example of the use of such mixed objectives see Section 14.3.3.

Often a problem will have *multiple objectives*. For example, a company providing a mobile snack service to construction sites will want to maximize the number of sites that can be visited during a daily operating schedule, and will also want to maximize the expected returns by visiting the most lucrative sites. An agency charged with locating a corridor for a new power transmission line may decide to minimize cost, while at the same time minimizing environmental impact. Such problems employ methods known as *multi-criteria decision-making* (MCDM), and again interested readers are referred to the appropriate texts (e.g. Thill 1999).

Many geographic problems involve multiple goals and objectives, which often cannot be expressed in commensurate terms.

1.4 The Technology of Problem-Solving

The previous sections have presented GIS as a technology to support both science and problem-solving, using both specific and general knowledge about geographic reality. But what exactly is this technology called GIS, and how does it achieve its objectives? In what ways is GIS more than a technology, and why has it attracted such attention as a topic for scientific journals and conferences in recent years – far more, for example, than word processing?

Many definitions of GIS have been suggested over the years, and none of them is entirely satisfactory, though many suggest much more than a technology. Today, the label *GIS* is attached to many things: amongst them, a software product that one can buy from a vendor

Figure 1.11 A GIS is a computerized tool for solving geographic problems

to carry out certain well-defined functions (*GIS software*); digital representations of various aspects of the geographic world, in the form of datasets (*GIS data*); a community of people who use and perhaps advocate the use of these tools for various purposes (the *GIS community*); and the activity of using a GIS to solve problems or advance science (*doing GIS*). The basic label works in all of these ways, and its meaning surely depends on the context in which it is used.

Nevertheless, certain definitions are particularly helpful (Table 1.3). A GIS is a *container of maps in digital form*, a particularly helpful definition to give to someone looking for a simple explanation – a guest at a cocktail party, or a seat neighbor on an airline flight (an *elevator pitch* in colloquial US English). We all know and appreciate the value of maps, and the notion that maps could be processed by a computer is clearly analogous to the use of word processing or spreadsheets to handle other types of information. A GIS is also *a computerized tool for solving geographic problems* (Figure 1.11), a definition that speaks to the purposes of GIS, rather than to its functions or physical form – an idea that is expressed in another definition, *a spatial decision*

support system. A GIS is *a mechanized inventory of geographically distributed features, and facilities*, the definition that explains the value of GIS to the utility industry, where it is used to keep track of such entities as underground pipes, transformers, transmission lines, poles, and customer accounts. A GIS is *a tool for revealing what is otherwise invisible in geographic information*, an interesting definition that emphasizes the power of a GIS as an analysis engine, to examine data, and reveal its patterns, relationships, and anomalies – things that might not be apparent to someone looking at a map. A GIS is *a tool for performing operations on geographic data that are too tedious or expensive or inaccurate if performed by hand*, a definition that speaks to the problems associated with manual analysis of maps, particularly the extraction of simple measures, of area for example.

Everyone has their own favorite definition of a GIS, and there are many to choose from.

1.4.1 A brief history of GIS

The last objective of the previous section largely drove the development of the first GIS, the Canada Geographic Information System or CGIS, in the mid-1960s. The Canada Land Inventory was a massive effort by the federal and provincial governments to identify the nation's land resources and their existing, and potential uses. The most useful results of such an inventory are measures of area, yet area is notoriously difficult to measure accurately from a map (Section 13.3). CGIS was planned and developed as a measuring tool, a producer of tabular information, rather than as a mapping tool.

The first GIS was the Canada Geographic Information System, designed in the mid-1960s as a computerized map measuring system.

Table 1.3. Definitions of a GIS, and the groups who find them useful

a container of maps in digital form	the general public
a computerized tool for solving geographic problems	decision-makers, community groups, planners
a spatial decision support system	management scientists, operations researchers
a mechanized inventory of geographically distributed features and facilities	utility managers, transportation officials, resource managers
a tool for revealing what is otherwise invisible in geographic information	scientists, investigators
a tool for performing operations on geographic data that are too tedious or expensive or inaccurate if performed by hand	resource managers, planners, cartographers

A second burst of activity occurred in the late 1960s in the US Bureau of the Census, in planning the tools needed to conduct the 1970 Census of Population. The DIME program (Dual Independent Map Encoding) created digital records of all US streets, to support automatic referencing and aggregation of census records. The similarity of this technology to that of CGIS was recognized immediately, and led to a major program at Harvard University's Laboratory for Computer Graphics and Spatial Analysis to develop a general-purpose GIS that could handle the needs of both applications – a project that led eventually to the ODYSSEY GIS of the late 1970s.

Early GIS developers recognized that the same basic needs were present in many different application areas, from resource management to the census.

In a largely separate development, cartographers and mapping agencies had begun in the 1960s to ask whether computers might be adapted to their needs, and possibly to reducing the costs and shortening the time of map creation. National mapping agencies, such as the UK's Ordnance Survey, France's Institut Géographique National, and the US Geological Survey, and US Defense Mapping Agency began to investigate using computers to support the editing of maps, to avoid the expensive and slow process of hand correction and redrafting. The first automated cartography developments occurred in the 1960s, and by the late 1970s most major cartographic agencies were already partly computerized.

Remote sensing also played a part in the development of GIS, as a source of technology as well as a source of data. The first military satellites of the 1950s were developed and deployed in great secrecy to gather intelligence, but the declassification of much of this material in recent years has provided interesting insights into the role played by the military and intelligence communities in the development of GIS. Although the early spy satellites used conventional film cameras to record images, digital remote sensing began to replace them in the 1960s, and by the early 1970s civilian remote sensing systems such as Landsat were beginning to provide vast new data resources on the appearance of the planet's surface from space, and to exploit the technologies of image classification and pattern recognition that had been developed earlier for military applications. The military was also responsible for the development in the 1950s of the world's first uniform system of measuring location, driven by the need for accurate targeting of intercontinental ballistic missiles, and this development led directly to the methods of positional control in use today (Section 4.6). Military needs were also responsible for the initial development of the Global Positioning System (GPS; Section 10.2.2.2).

Many technical developments in GIS originated in the Cold War.

GIS really began to take off in the early 1980s, when the price of computing hardware had fallen to a level that could sustain a significant software industry and cost-effective applications. Among the first customers were forestry companies and natural resource agencies, driven by the need to keep track of vast timber resources, and to regulate their use effectively. At the time a modest computing system – far less powerful than today's personal computers – could be obtained for about $250,000, and the associated software for about $100,000. Even at these prices the benefits of consistent management using GIS, and the decisions that could be made with these new tools, substantially exceeded the costs. The market for GIS software continued to grow, computers continued to fall in price, and increase in power, and the GIS software industry has been growing ever since.

The modern history of GIS dates from the early 1980s, when the price of sufficiently powerful computers fell below a critical threshold.

The history of GIS is a complex story, much more complex than this brief history. Table 1.4 summarizes the major events of the past three decades. Interested readers who would like to explore this topic further are urged to consult more comprehensive sources, including the text by Foresman (1998) and the Reference Links.

1.4.2 Views of GIS

It should be clear from the previous discussion that GIS is a complex beast, with many distinct appearances. To some it is a way to automate the production of maps, while to others this application seems far too mundane compared with the complexities associated with solving geographic problems and supporting spatial decisions, and with the power of a GIS as an engine for analyzing data and revealing new insights. Others see a GIS as a tool for maintaining complex inventories, one that adds geographic perspectives to existing information systems, and allows the geographically distributed resources of a forestry or utility company

Table 1.4. Major events that shaped GIS

Date	Type	Event	Notes
		The Era of Innovation	
1963	Technology	CGIS development initiated	Canada Geographic Information System is developed by Roger Tomlinson and colleagues for Canadian Land Inventory. This project pioneers much technology and introduces the term GIS.
1963	General	URISA established	The Urban and Regional Information Systems Association founded in the US. Soon becomes point of interchange for GIS innovators.
1964	Academic	Harvard Lab established	The Harvard Laboratory for Computer Graphics and Spatial Analysis is established under the direction of Howard Fisher at Harvard University. In 1966 SYMAP, the first raster GIS, is created by Harvard researchers.
1967	Technology	DIME developed	The US Bureau of Census develops DIME-GBF (Dual Independent Map Encoding-Geographic Database Files), a data structure and street-address database for 1970 census.
1969	Commercial	ESRI Inc. formed	Jack Dangermond, a student from the Harvard Lab, and his wife Laura form ESRI to undertake projects in GIS.
1969	Commercial	Intergraph Corp. formed	Jim Meadlock and four others that worked on guidance systems for Saturn rockets form M&S Computing, later renamed Intergraph.
1969	Academic	"Design With Nature" published	Ian McHarg's book was the first to describe many of the concepts in modern GIS analysis, including the map overlay process (see Chapter 13).
1972	Technology	Landsat 1 launched	Originally named ERTS (Earth Resources Technology Satellite), this was the first of many major Earth remote sensing satellites to be launched.
1974	Academic	AutoCarto 1 Conference	Held in Reston, Virginia, this was the first in an important series of conferences that set the GIS research agenda.
1977	Academic	Topological Data Structures conference	Harvard Lab organizes a major conference and develops the ODYSSEY GIS.
		The Era of Commercialization	
1981	Commercial	ArcInfo launched	ArcInfo was the first major commercial GIS software system. Designed for minicomputers and based on the vector and relational database data model, it set a new standard for the industry.
1984	Academic	"Basic Readings in Geographic Information Systems" published	This collection of papers published in book form by Duane Marble, Hugh Calkins, and Donna Peuquet was the first accessible source of information about GIS.
1985	Technology	GPS operational	The Global Position System, although slow to be taken up, is today a major source of data for navigation, surveying, and mapping.
1986	Academic	"Principles of Geographic Information Systems for Land Resources Assessment" published	Peter Burrough's book was the first specifically on GIS. It quickly became the worldwide reference text for GIS students.
1986	Commercial	MapInfo Corp. formed	MapInfo software develops into first major desktop GIS product. It defined a new standard for GIS products, complementing earlier software systems.
1987	Academic	International Journal of Geographic Information Science introduced	Terry Coppock and others published the first journal on GIS. The first issue contained papers from the USA, Canada, and Germany.
1987	General	Chorley Report	"Handling Geographic Information" was an influential report from the UK government that highlighted the value of GIS.
1988	General	GISWorld begins	GISWorld, now GeoWorld, the first worldwide magazine devoted to GIS was published in the USA.

Continued

Table 1.4. *Continued*

Date	Type	Event	Notes
1988	Technology	TIGER announced	TIGER (Topologically Integrated Geographic Encoding and Referencing), a follow-on from DIME, is described by the US Census Bureau. Low cost TIGER data stimulates rapid growth in US business GIS.
1988	Academic	US and UK Research Centers announced	Two separate initiatives, the US NCGIA (National Center for Geographic Information and Analysis) and the UK RRL (Regional Research Laboratory) show the rapidly growing interest in GIS in academia.
1991	Academic	Big Book 1 published	Substantial two volume compendium "Geographical Information Systems; principles and applications" edited by David Maguire, Mike Goodchild and David Rhind documents progress to date.
1992	Technical	DCW released	The 1.7 GB Digital Chart of the World, sponsored by the US Defense Mapping Agency (now NIMA) is the first integrated 1:1 million scale database offering global coverage.
1994	General	Executive Order signed by President Clinton	Executive Order 12906 leads to creation of US National Spatial Data Infrastructure (NSDI), clearinghouses and Federal Geographic Data Committee (FGDC).
1994	General	OpenGIS® Consortium born	The OpenGIS® Consortium of GIS vendors, government agencies and users is formed to improve interoperability.
1996	Technology	Internet GIS products introduced	Several companies, notably Autodesk, ESRI, Intergraph, and MapInfo release new generation of Internet-based products at about the same time. Leads to rapid expansion of GIS involvement.
1996	Commercial	MapQuest	Internet mapping service launched, producing over 130 million maps in 1999. Later AOL purchases for $1.1 billion.
1999	General	GIS Day	First GIS Day attracts over 1.2 million global participants who share an interest in GIS.
		The Era of Exploitation	
2000	Commercial	GIS passes $7 bn	Industry analyst Daratech reports GIS hardware, software and services industry at $6.9 bn, growing at more than 10% per annum.
2000	General	GIS has 1 million users	GIS now has more than 1 million core users and perhaps 5 million casual users of GI.

to be tracked, and managed. All of these perspectives are clearly too much for any one software package to handle, and GIS has grown from its initial commercial beginnings as a simple off-the-shelf package to a complex of software, hardware, people, institutions, networks, and activities that can be very confusing to the novice. A major software vendor such as ESRI today sells many distinct products, designed to serve very different needs: a major GIS workhorse (ArcInfo), a simpler system designed for viewing, analyzing, and mapping data (ArcView), an engine for supporting GIS-oriented Web sites (ArcIMS), an information system with spatial extensions (ArcSDE), and several others. Other vendors specialize in certain niche markets, such as the utility industry, or military, and intelligence applications. GIS is a dynamic, and evolving field,

and its future is certain to be exciting, but speculations on where it might be headed are reserved for the final chapter.

Today a single GIS vendor offers many different products for distinct applications.

1.4.3   Anatomy of a GIS

1.4.3.1   The network

Despite the complexity noted in the previous section, a GIS does have its well-defined component parts. Today, the most fundamental of these is probably the *network*, without which no rapid communication or sharing of digital information could occur, except between a small group of people crowded around a computer monitor. GIS today relies heavily on the Internet, and on its limited-access cousins, the *intranets* of corporations, agencies, and the military. The Internet was originally designed as a network for connecting computers, but today it is rapidly becoming society's mechanism of information exchange, handling everything from personal messages to massive shipments of data, and increasing numbers of business transactions.

It is no secret that the Internet in its many forms has had a profound effect on technology, science, and society in the last few years. Who could have foreseen in 1990 the impact that the Web, e-commerce, digital government, mobile systems, and information, and communication technologies would have on our everyday lives? These technologies have radically changed forever the way we conduct business, how we communicate with our colleagues, and friends, the nature of education, and the value, and transitory nature of information.

The Internet began life as a US Department of Defense communications project called ARPANET (Advanced Research Projects Agency Network) in 1972. In 1980 Tim Berners-Lee, a researcher at CERN, the European organization for nuclear

research, developed the hypertext capability that underlies today's World Wide Web – a key application that has brought the Internet into the realm of everyday use. Uptake, and use of Web technology have been remarkably quick, diffusion being considerably faster than almost all comparable innovations (for example, the radio, the telephone, and the television). By 2000, one in six people in North America, and Europe used the Internet, there were 5 million Web servers containing 800 million pages, and the worldwide e-commerce market passed $1 trillion (www.internetindicators.com/global.html). Figure 1.12 shows a map of Internet hosts (servers) in 1999. By the end of the last millennium, the Internet was a truly global phenomenon with major activity in North America, Europe, and the Far East. Other Internet usage maps are available at the Atlas of Cybergeography maintained by Martin Dodge (www.geog.ucl.ac.uk/casa/martin/atlas/atlas.html).

Geographers were quick to see the value of the Internet, and in 1993 Steve Putz of the Xerox PARC center published the first Web-based interactive map. Users connected to the Internet could zoom in to parts of the map, or pan to other parts, using simple mouse clicks in their

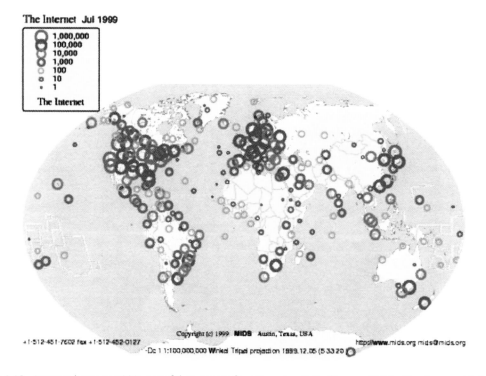

Figure 1.12 Internet hosts in 1999, a useful surrogate for Internet activity (Source: Matrix Directory and Information Services)

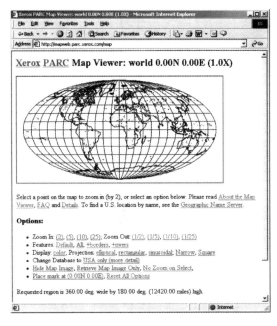

Figure 1.13 The first Internet mapping site, established by Steve Putz at Xerox PARC in 1993 and still accessible

Figure 1.14 A GIS-enabled Swedish electronic yellow pages (Source: Jesper Persson, Teleaddress Information AB)

desktop WWW browsers, without ever needing to install specialized software or download large amounts of data. The site is still running today at mapweb.parc.xerox.com/map (Figure 1.13), and has served over 157 million maps. This research project soon gave way to industrial-strength Internet GIS software products from mainstream software vendors (see Section 8.5).

The use of the WWW to give access to maps dates from 1993.

The recent histories of GIS and the Internet have been heavily intertwined (Harder 1998, Plewe 1997). GIS has turned out to be a compelling application that has prompted many people to take advantage of the Web. At the same time GIS has benefited greatly from adopting the Internet paradigm and the momentum that the Web has generated. Today there are many successful applications of GIS on the Internet, and we have used them as examples and illustrations at many points in this book. They range from using GIS on the Internet to disseminate information – a type of electronic yellow pages – (e.g. www.gulasidorna.se/main/frameset.asp, Figure 1.14), to selling goods, and services (e.g. www.kemperinsurance.com/find_an_agent/us_business_agent.html, Figure 1.15), to direct revenue generation (e.g. www.mapquest.com, Figure 1.16), to helping members of the public

to participate in important local, regional, and national debates (e.g. www.inforain.org, Figure 1.17).

The Internet has proven very popular as a vehicle for delivering GIS applications for several reasons. It is an established, widely used platform and accepted standard for interacting with information of many types. It also offers a relatively cost-effective way of linking together distributed users (for example, telecommuters and office workers, customers and suppliers, students and teachers). The interactive, and

Figure 1.15 Kemper Insurance agent-finder, showing an example agent in Orange County, Florida (Source: Kemper Insurance Company www.kemperinsurance.com)

(A)

Driving Directions Results

FROM:
909 WEST CAMPUS LN GOLETA, CA 93117 US
Save this Address

TO:
1401 DE LA VINA SANTA BARBARA, CA US
Save this Address

Total Distance:
11.5 miles
(18.6 km)
Total Estimated Time
20 minutes

Directions appear below the map
You may click on the map to use MapQuest's online map features.

Use Subject to License/Copyright

(B)

Directions	Distance
1: Start out going North on W CAMPUS POINT LN towards W CAMPUS POINT DR by turning right.	0.2 mile (0.4 km)
2: Stay straight to go onto W CAMPUS POINT DR.	0.2 mile (0.3 km)
3: Turn RIGHT onto SLOUGH RD.	0.1 mile (0.1 km)
4: SLOUGH RD becomes W CAMPUS POINT LN.	0.1 mile (0.1 km)
5: Turn LEFT onto STORKE RD.	1.2 mile (1.9 km)
6: Take the US-101 SOUTH ramp.	0.3 mile (0.5 km)
7: Merge onto US-101 S.	8.4 mile (13.5 kn
8: Take the MISSION ST exit.	0.3 mile (0.4 km)
9: Turn LEFT onto W MISSION ST.	0.3 mile (0.5 km)
10: Turn RIGHT onto DE LA VINA ST.	0.6 mile (0.9 km)
11: Turn RIGHT onto W SOLA ST.	0.0 mile (0.0 km)
Total Distance: 11.5 miles (18.6 km	
Total Estimated Time: 20 minute	

Figure 1.16 (A) Driving directions from the MapQuest GIService. (B) Written driving directions for the route shown in Figure 1.16A, generated by the MapQuest GIService (Source: MapQuest website © 2000 MapQuest.com Inc. Screenshots used with permission)

exploratory nature of navigating linked information has also been a great hit with users. The availability of multi-content site gateways (portals) with powerful search engines has been a further reason for success.

As new and ever more ingenious products and applications have been devised, so Internet GIS has started to change the way we think of delivering geographic information and processing to users. In the past few years the term GIService (yet another use of the three-letter acronym GIS) has been introduced to define this new model in which distributed users access a centralized GIS capability. In some quarters the term Location-

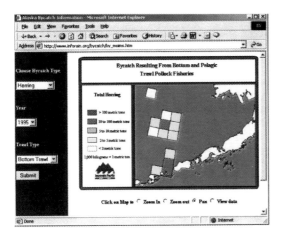

Figure 1.17 GIS in public policy – fishing in Alaskan waters (Source: Ecotrust)

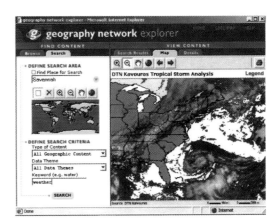

Figure 1.19 The Geography Network (Source: ESRI)

Based Service is taken to mean the same thing, and the term *g-commerce* is also used to describe types of electronic commerce, or e-commerce, that include location as an essential element. This in turn has spawned a new generation of mobile and handheld applications for personal use. Many types of personal devices, from pagers to mobile phones (Figure 1.18) to Personal Digital Assistants, are now filling the briefcases and adorning the clothing of people in many walks of life. These devices are able to provide real-time geographic services such as mapping, routing, and geographic yellow pages. These services can be purchased on a pay-as-you-go or subscription basis, and are beginning to change the business GIS model for many types of applications.

A further interesting twist is the development of themed geographic networks, such as the Geography Network (www.geographynetwork.com, Figure 1.19). The Geography Network is an integrated collection of geographic information providers and users that interact via the medium of the Internet. Online content can be located using the interactive search capability of the portal and then content can be directly used over the Internet. This form of Internet application is explored further in Sections 7.5.3, and 10.4.

The Internet is increasingly integrated into many aspects of GIS use, and the days of standalone GIS are mostly over.

1.4.3.2 The other five components of the GIS anatomy

The second piece of the GIS anatomy (Figure 1.20) is the user's hardware, the device that the user interacts with directly in carrying out GIS operations, by typing, pointing, clicking, or speaking, and which returns information by displaying it on the device's screen or generating

Figure 1.18 Commercial geographic mapping service offered through a cellphone connection to the Internet by KPN in The Netherlands (Source: Royal KPN N.V.)

- Hardware
- Software
- Data
- People
- Procedures
- Network

Figure 1.20 The six component parts of a GIS (Source: ESRI)

meaningful sounds. Traditionally this device sat on an office desktop, but today's user has much more freedom, because GIS functions can be delivered through laptops, personal digital assistants (PDAs), in-vehicle devices, and even cellular telephones. Section 8.5 discusses the currently available technologies in greater detail. In the language of the network, the user's device is the *client*, connected through the network to a *server* that is probably handling many other user clients simultaneously. The client may be *thick*, if it performs a large part of the work locally, or *thin* if it does little more than link the user to the server. A PC or Macintosh is an instance of a thick client, with powerful local capabilities, while devices attached to TVs that offer little more than Web browser capabilities are instances of thin clients.

The third piece of the GIS anatomy is the software that runs locally in the user's machine. This can be as simple as a standard Web browser (e.g. Microsoft Explorer or Netscape) if all work is done remotely using assorted digital services offered on large servers. More likely it is a package bought from one of the GIS vendors, such as Autodesk Inc. (San Rafael, California, USA; www.autodesk.com), Environmental Systems Research Institute (ESRI; Redlands, California, USA; www.esri.com), Intergraph Corp. (Huntsville, Alabama, USA; www.ingr.com), MapInfo Corp. (Troy, New York, USA; www.mapinfo.com), or GE Smallworld Systems Ltd. (Cambridge, England; www.smallworld.co.uk). Each vendor offers a range of products, designed for different levels of sophistication, different volumes of data, and different application niches. Idrisi (Clark University, Worcester, Massachusetts, USA, www.clarklabs.org) is an example of a GIS produced and marketed by an academic institution rather than by a commercial vendor.

Many GIS tasks must be performed repeatedly, and GIS designers have created tools for capturing such repeated sequences into easily executed *scripts* or *macros* (Sections 8.3.3 and 9.2). For example, the agency that needs to predict erosion of New South Wales's soils (Section 1.3) would likely establish a standard script written in the scripting language of its favorite GIS. The instructions in the script would tell the GIS how to model erosion given required data inputs, and parameters, and how to output the results in suitable form. Scripts can be used repeatedly, for different areas or for the same area at different times. Support for scripts is an important aspect of GIS software.

GIS software can range from a simple package designed for a PC, and costing a few hundred dollars, to a major industrial-strength workhorse designed to serve an entire enterprise of net-worked computers, and costing tens of thousands of dollars. New products are constantly emerging, and it is beyond the scope of this book to provide a complete inventory.

The fourth piece of the anatomy is the database, which consists of a digital representation of selected aspects of some specific area of the Earth's surface or near-surface, built to serve some problem-solving or scientific purpose. A database might be built for one major project, such as the location of a new high-voltage power transmission corridor, or it might be continuously maintained, fed by the daily transactions that occur in a major utility company (installation of new underground pipes, creation of new customer accounts, daily service crew activities). It might be as small as a few megabytes (a few million bytes, easily stored on a few diskettes), or as large as a terabyte (a trillion bytes, occupying a storage unit somewhat larger than a shoebox). Table 1.1 gives some sense of potential GIS database volumes.

GIS databases can range in size from a megabyte to several petabytes.

In addition to these four components – network, hardware, software, and database – a GIS also requires management. An organization must establish procedures, lines of reporting, control points, and other mechanisms for ensuring that its GIS activities stay within budgets, maintain high quality, and generally meet the needs of the organization. These issues are explored in the later chapters of this book.

Finally, a GIS is useless without the people who design, program, and maintain it, supply it with data, and interpret its results. The people of GIS will have various skills, depending on the roles they perform. Almost all will have the basic knowledge needed to work with geographic data – knowledge of such topics as data sources, scale, and accuracy, and software products – and will also have a network of acquaintances in the GIS community. We refer to such people in this book as *spatially aware professionals*, or SAPs, and the humor in this term is not intended in any way to diminish their importance, or our respect for what they know – after all, we would like to be recognized as SAPs ourselves! The next section outlines some of the roles played by the people of GIS, and the industries in which they work.

1.5 The Business of GIS

People play many roles in GIS, from software development to software sales, and from teaching about GIS to using its power in everyday activities. This section looks at these roles, and is organized by the major areas of human activity associated with GIS.

1.5.1 The software industry

Perhaps the most conspicuous sector, although by no means the largest in either economic or human terms, is the GIS software industry. While some GIS vendors have their roots in other, larger computer applications (e.g. Intergraph, and Autodesk, which have roots in computer-assisted design software developed for engineering, and architectural applications; and ERDAS (www.erdas.com) and PCI (www.pcigeomatics.com), which have roots in remote sensing and image processing), others began as specialists in GIS. Measured in economic terms, the GIS software industry currently accounts for some $1 billion in annual sales, although estimates vary, in part because of the difficulty of defining GIS precisely. The software industry employs several thousand programmers, software designers, systems analysts, application specialists, and sales staff, with backgrounds that include computer science, geography, and many other disciplines.

> The GIS software industry accounts for about $1 billion in annual sales.

1.5.2 The data industry

The acquisition, creation, maintenance, dissemination, and sale of GIS data also account for a large volume of economic activity. Traditionally, a large proportion of GIS data has been produced centrally, by national mapping agencies such as the Great Britain's Ordnance Survey. A sense of the magnitude of the public sector of the GIS data industry can be gained from a survey conducted in 1993 by the US Office of Management and Budget, which found total annual expenditure of close to $4 billion in federal agencies alone. Another reliable estimate placed total annual revenues from GIS hardware, software, and data sales at $7 billion in 1999.

> In value of annual sales, the GIS data industry is much more significant than the software industry.

In recent years improvements in GIS, and related technologies, and reductions in prices, along with various kinds of government stimuli, have led to the rapid growth of a private GIS data industry, and to increasing interest in data sales to

customers on the part of local governments. Private companies are now licensed to collect high-resolution data using satellites, and to sell it to customers – Space Imaging (www.spaceimaging.com) and its IKONOS satellite are a prominent instance. Other companies collect similar data from aircraft. Still other companies specialize in the production of high-quality data on street networks, a basic requirement of many delivery companies. Geographic Data Technology (Lebanon, New Hampshire, USA; www.gdt1.com) is an example of this industry, employing some 400 staff in producing, maintaining, and marketing its particular line of high-quality street network data.

1.5.3 The GIService industry

The Internet also allows GIS users to access specific functions that are provided by remote sites. For example, the MapQuest site (www.mapquest.com), provides a routing service that is used by millions of people every day to find the best driving route between two points. By typing a pair of street addresses onto a MapQuest form, the user can execute a routing analysis (see Section 14.3.2), and receive the results in the form of a map and a set of written driving directions (see Figure 1.16). This has several advantages over performing the same analysis on one's own PC: there is no need to buy software to perform the analysis, there is no need to buy the necessary data, and the data are routinely updated by MapQuest. The terms introduced in Section 1.4.3.1 to describe this kind of service were GIService, Location-Based Service, and g-commerce. ESRI's Geography Network (www.geographynetwork.com, Figure 1.19) is an example of a site that provides access to both GIS data, and GIServices.

> GIServices are a rapidly growing form of electronic commerce.

At this point the potential for GIServices has barely been tapped. In today's world one of the most important commodities is attention – the fraction of a second of attention given to a billboard, or the audience attention that a TV station sells to its advertisers. The value of attention also depends on the degree of fit between the message and the recipient: an advertiser will pay more for the attention of a small number of people if it knows that they include a large proportion of its target customers. Advertising directed at the individual, based on an individual profile, is even more attractive to the advertiser. Direct mail companies have exploited the power of geographic location to target specific audiences for many years, basing their strategies on neighborhood profiles constructed from census records. But new technologies offer to take this much further. For example, the technology already exists to identify the buying habits of a customer who stops at a gas pump and uses a credit card, and to direct targeted advertising through a TV screen at the pump.

1.5.4 The publishing industry

Much smaller, but nevertheless highly influential in the world of GIS, is the publishing industry, with its magazines, books, and journals. Several magazines are directed at the GIS community, as well as some increasingly significant news-oriented Web sites (see Box 1.3).

Several journals have appeared to serve the GIS community, by publishing new advances in GIS research. The oldest journal specifically targeted at the community is the *International Journal of Geographical Information Science*, established in 1987. Other older journals in areas such as cartography now regularly accept GIS articles, and several have changed their names and shifted focus significantly. Box 1.4 gives a list of the journals that particularly emphasize GIS research.

Box 1.3 Magazines and Web sites offering GIS news and related services

GeoSpatial solutions (published monthly by Advanstar Communications), and see their Web site www.geospatial-online.com

GeoWorld (published monthly by GeoTechMedia), and see their Web site www.geoplace.com. This company also publishes *Business Geographics, GeoWorld, GeoEurope*, and *Geo AsiaPacific*

GI News (a UK publication)

GIS Development (published monthly by GIS Development Pvt. Ltd: www.GISdevelopment.net

ArcNews, and *ArcUser Magazine* (published by ESRI), see www.esri.com

Some Web sites offering online resources for the GIS community:

> www.gis.com
> www.giscafe.com
> gis.about.com
> www.geocomm.com
> www.spatialnews.com
> www.directionsmag.com

Box 1.4 Scholarly journals emphasizing GIS research

International Journal of Geographical Information Science (formerly *International Journal of Geographical Information Systems*)

Cartography and Geographic Information Science (formerly *American Cartographer* and *Cartography and Geographic Information Systems*)

Computers and Geosciences

Computers, Environment and Urban Systems

Photogrammetric Engineering and Remote Sensing

Transactions in Geographic Information Systems

Geographical and Environmental Modelling

Geographical Analysis

GeoInformatica

Annals of the Association of American Geographers

Journal of Geographical Systems (successor to *Geographical Systems*)

1.5.5 GIS education

The first courses in what is now called GIS were offered in universities in the early 1970s, often as an outgrowth of courses in cartography or remote sensing. Today, thousands of courses can be found in universities and colleges all over the world (Figure 1.21). Training courses are offered by the vendors of GIS software, and increasing use is made of the Web in various forms of distance GIS education, and training.

A distinction is often made between education, and training in GIS – training in the use of a particular software product is contrasted with education in the fundamental principles of GIS. In many university courses, lectures are used to emphasize fundamental principles while computer-based laboratory exercises emphasize training. In our view, an education should be for life, and the material learned during an education should be applicable for as far into the future as possible. Fundamental principles tend to persist long after software has been replaced with new versions, and the skills learned in running one software package may be of very little value when a new technology arrives. On the other hand much of the fun and excitement of GIS comes from actually working with it, and fundamental principles can be very dry and dull without hands-on experience.

Figure 1.21 Thousands of courses on GIS are offered all over the world

1.6 GISystems, GIScience, and GIStudies

Geographic information systems are useful tools, helping everyone from scientists to citizens to solve geographic problems. But like many other kinds of tools, such as computers themselves, their use raises questions that are sometimes frustrating, and sometimes profound. For example, how does a GIS user know that the results obtained are accurate? What principles might help a GIS user to design better maps? How can user interfaces be made readily under-standable by novice users? Some of these are questions of GIS design, and others are about GIS data, and methods. Taken together, we can think of them as questions that arise from the use of GIS – that are stimulated by exposure to GIS or to its products. Many of them are addressed in detail at many points in this book, and the book's title emphasizes the importance of both systems, and science.

The term *geographic information science* was coined in a paper published in 1992 (Goodchild 1992). In it, the author argued that these questions and others like them were important, and that their systematic study constituted a science in its own right. Information science studies the fundamental issues arising from the creation, handling, storage, and use of information – similarly, GIScience should study the fundamental issues arising from geographic information, as a well-defined class of informa-tion in general. Other terms have much the same meaning: *geomatics* and *geoinformatics*, *spatial information science*, *geocomputation*, *geo-information engineering*. All suggest a scientific approach to the fundamental issues raised by the use of GIS and related technologies, though they all have different roots and emphasize different ways of thinking about problems (specifically geographic or more generally spatial, emphasizing engineering or science, etc.).

GIScience has evolved significantly over the past eight years. It is now part of the title of several renamed research journals (see Box 1.4), and the focus of the US University Consortium for Geographic Information Science (www.ucgis.org), an organization of roughly 60 research universi-ties that engages in research agenda setting, lobbying for research funding, and related activities. The First International Conference on GIScience was held in the USA in 2000 (see www.giscience.org).

In 1996 the UCGIS institutions held an assembly to identify the most important research topics of GIScience. The 10 topics they chose are shown in Box 1.6. Many are explored in detail in later chapters of this book.

Box 1.5 Sites offering Web-based education and training programs in GIS

ESRI's Virtual Campus at www.esri.com (see Online Resources at the end of each chapter)

Pennsylvania State University Certificate Program in Geographic Information Systems at www.worldcampus.psu.edu

UNIGIS International Postgraduate Courses in GIS at www.unigis.org

Birkbeck Department of Geography MSc in Geographic Information Science by Distance Learning at www.bbk.ac.uk/geog/study/msc_gisc_dl.html

Box 1.6 The 10 research challenges of the US University Consortium for Geographic Information Science (1996), and related chapters in this book

Cognition of geographic information (Chapter 6)

Spatial data acquisition and integration (Chapter 10)

Spatial analysis in a GIS environment (Chapters 13 and 14)

Interoperability of geographic information (Chapter 8)

Distributed computing (Chapter 8)

Future of the spatial information infrastructure (Chapter 19)

GIS and society (Chapter 1)

Uncertainty in geographic data and GIS-based activities (Chapters 6 and 15)

Extensions to geographic representations (Chapter 9)

Scale (Chapter 5)

More detail on all of these topics, and additional topics added at more recent UCGIS assemblies, can be found at www.ucgis.org

Many of the research topics in Box 1.6 are actually much older than GIS. The need for methods of spatial analysis, for example, dates from the first maps, and many methods were developed long before the first GIS appeared on the scene in the mid-1960s. Another way to look at GIScience is to see it as the body of knowledge that GISystems implement and exploit. Map projections, for example, are part of GIScience, and are used and transformed in GISystems. Another area of great importance to GIS is cognitive science, and particularly the scientific understanding of how people think about their geographic surroundings. If GISystems are to be easy to use they must fit with human ideas about such topics as driving directions, or how to construct useful and understandable maps. Box 1.7 introduces David Mark, a GIScientist who has done much to build an interest in cognitive science among the GIScience community, and to build bridges to this discipline.

In the 1970s it was easy to define or delimit a geographic information system: it was a single piece of software residing on a single computer. With time, and particularly with the development of the Internet, and new approaches to software engineering, the old monolithic nature of GIS has been replaced by something much more fluid. The emphasis throughout this book is on this new vision of GIS, as the set of coordinated parts discussed earlier in Section 1.5. Perhaps the *system* part of GIS is no longer necessary – certainly the phrase *GIS data* suggests some redundancy, and many people have suggested that we could drop the S altogether in favor of GI, for geographic information. GI*Systems* are only one part of the GI whole, which also includes the fundamental issues of GI*Science*. Much of this book is really about GI*Studies*, which can be defined as the systematic study of society's use of geographic information, including its institutions, standards, and procedures, and many of these topics are

Box 1.7 David Mark, GIScientist

David Mark works at the University at Buffalo, part of the State University of New York system, where he is Professor of Geography and Director of the Buffalo site of the National Center for Geographic Information and Analysis. His interest in GIScience dates from the 1970s, when he worked as a graduate student on methods for representing terrain, notably the TIN model (see Section 9.2.3.4). Today his teaching and research interests center on the cognitive, and linguistic aspects of geographic information science. For example, he studies how different languages express such geographic concepts as "near" or "across", and how this might affect the design of user interfaces to GIS. Much of his spare time is spent pursuing another of his passions, birding.

"I believe that my main contribution has been to introduce and promote methods, and approaches from cognitive science, and linguistics into geographic information science. Behavioral geography drew mainly on wayfinding and visual perception, whereas my work, especially within NCGIA in collaboration

with Andrew Frank and Max Egenhofer, emphasized linguistic and conceptual aspects, especially spatial relations, and led into my current work on geographic ontology (the study of the basic components used to construct geographic information). My two biggest surprises regarding the development of GIS over the last 20 years have been on the academic side. One is institutional – I am surprised at the success of the UCGIS (Box 1.6), and the rapid credibility it established with US Federal agencies, and politicians. The second great surprise has been the emergence of academic interest in 'Geographic Information and Society', especially the constructive engagement between the social theory and GIS communities in this research area."

Figure 1.22 David Mark, GIScientist

addressed in the later chapters, and in a chapter by Forer and Unwin (see Reference Links). Several of the UCGIS research topics suggest this kind of focus, including *GIS and society*, and *Future of the spatial information infrastructure*. In recent years the role of GIS in society – its impacts and its deeper significance – has become the focus of extensive writing in the academic literature, particularly in the discipline of geography, and much of it has been critical of GIS. We explore these critiques in detail in the next section.

The importance of social context is nicely expressed by Nick Chrisman's definition of GIS (Chrisman 1999), which might also serve as an appropriate final comment on the earlier discussion of definitions: "GIS – Organized activity by which people measure and represent geographic phenomena, then transform these representations into other forms while interacting with social structures". Chrisman's social structures are clearly part of the GIS whole, and as students of GIS we should be aware of the

ethical issues raised by the technology we study. This is the arena of GIStudies.

1.7 GIS and Geography

GIS has always had a special relationship to the academic discipline of geography, as it has to other disciplines that deal with the Earth's surface, including planning and landscape architecture. This section explores that special relationship and its sometimes tense characteristics.

Chapter 2 presents a gallery of GIS applications, and paints a picture of a successful applications-led field built around low-order concepts, that stands in quite stark contrast to the scientific tradition in the academic discipline of geography. Here, the spatial analysis tradition has developed during the past 40 years around a range of more-sophisticated operations and techniques, which

Box 1.8 Brian J L Berry, geographer

Brian J L Berry is a graduate in Geography and Economics of University College, London (UCL) who completed his Ph.D. in Geography at the University of Washington in the early 1960s. There, along with other key figures of the "Quantitative Revolution" in geography such as Waldo Tobler (Box 5.1), Art Getis, John Nystuen, Michael Dacey, William Bunge, and Richard Morrill, he was responsible for seminal work on the development of geographic theory and spatial analysis. His decisive contribution was in helping to establish that there could be a *scientific* human geography based on the theory of location (see, for example, Box 14.1 on the Varignon Frame), and supported by statistical analysis and computation. He subsequently moved to the University of Chicago, where he and Duane Marble edited the 1968 volume *Spatial Analysis: A Reader in Statistical Geography*, which was to define scientific human geography for a generation of researchers, and students. It was also to lay some of the foundations for the establishment of the Computer Graphics Laboratory at Harvard, and this in turn would inspire a further generation of researchers such as Nick Chrisman, Geoff Dutton, Scott Morehouse, and Denis White. Brian subsequently served as Dean at Carnegie Mellon University, and is presently Professor of Political Economy at the University of Texas at Dallas. He is a prominent representative and advocate of geography on the Council of the US National Academy of Sciences.

Reflecting on his career and the present state of academic geography, Brian says:

"When I began my graduate studies in 1955, the mantra of leading geographers was that there could be no science because every region was unique. The remit of science was subsequently accepted by the many, although I regret that the geographic discipline has subsequently retreated from science into what I see as the self-indulgence of postmodernism. I have nevertheless been pleasantly surprised that GIS has flourished within geography, although I suspect that this is partly because so much of the development has occurred and is occurring outside geography. What saddens me about this is that geography as a discipline may be missing a golden opportunity to establish intellectual leadership in an emerging new field, and this in turn is leading GIS to chart an independent course. This is especially true in the modeling of dynamic processes, where GIS can provide the underpinnings for the rapidly evolving field of geocomputation, which applies information technology to model and predict social and environmental changes to the Earth".

Figure 1.23 Brian Berry, geographer

have a much more elaborate conceptual structure (see Chapters 13, and 14). One of the foremost proponents of the spatial analysis approach over this period has been Brian Berry, whose contribution is discussed in Box 1.8. As we will see in Chapters 13, and 14, spatial analysis is the process by which we turn raw spatial data into useful spatial information. For the first half of its history, the principal focus of spatial analysis in most universities was upon development of theory, rather than working applications. Actual data were scarce, as were the means to process and analyze them.

Many of the roots of GIS can be traced to the spatial analysis tradition in the discipline of geography.

In the 1980s GIS technology began to offer a solution to the problems of inadequate computation and limited data handling. However, the quite sensible priorities of vendors at the time might be described as solving the problems of 80% of the customers 80% of the time, and the integration of techniques based upon higher-order concepts was a low priority. This was the right priority for a nascent field, particularly given that the field remained constrained by a lack of spatial data. The data blockage was, however, significantly eroded with the advent of GPS new remote sensing satellites, and other digital data infrastructure initiatives by the late 1990s. New data handling technologies and new rich sources of digital data thus open up prospects for refocusing and reinvigorating academic interest in applied scientific problem-solving.

Although repeat purchases of GIS technology leave the field with a buoyant future in the IT mainstream, there is enduring unease in some academic quarters about GIS applications and their social implications. As we noted in the previous section, much of this unease has been expressed in the form of critiques, notably from geographers. John Pickles has probably contributed more to the debate than almost anyone else, notably through his edited volume *Ground Truth: The social implications of geographic information systems* (Pickles 1993; Curry 1998 is another source on these issues). Several types of arguments have surfaced:

- The ways in which GIS represents the Earth's surface, and particularly human society, favor certain phenomena and perspectives, at the expense of others. For example, GIS databases tend to emphasize homogeneity, partly because of the limited space available and partly because of the costs of more accurate data collection (see Chapters 3, 5, and 9). Minority views, and the views of individuals, can be submerged or *marginalized* in this process, as can information that differs from the official or consensus view.

- Although in principle it is possible to use GIS for any purpose, in practice it is often used for purposes that may be ethically questionable or may invade individual privacy, such as surveillance and the gathering of military and industrial intelligence. As with the debates over the atomic bomb in the 1940s, and 1950s, the scientists who develop and promote the use of GIS surely bear some responsibility for how it is eventually used. The idea that a tool can be inherently neutral, and its developers therefore immune from any ethical debates, is strongly questioned in this literature.

- The very success of GIS is a cause of concern. There are qualms about a field that appears to be led by technology and the marketplace, rather than by human need.

- There are concerns that GIS remains a tool in the hands of the already powerful – notwithstanding the diffusion of technology that has accompanied the plummeting cost of computing and

wide adoption of the Internet. As such, it is seen as maintaining the *status quo* in terms of power structures.

- There appears to be an absence of applications of GIS in *critical* research. This academic perspective is centrally concerned with the connections between human agency, and particular social structures, and contexts. Some of its protagonists are of the view that such connections are not amenable to digital representation in whole or even in part.

- Science, and technology, with which GIS is strongly associated, are viewed by some as fundamentally flawed. More narrowly, there is a view that GIS applications are inextricably bound to the philosophy and assumptions of the approach to science known as logical positivism. As such, the argument goes, GIS can never be more than a positivist tool and a normative instrument, and cannot enrich other more critical perspectives in geography.

Many geographers remain suspicious of the use of GIS in geography.

We wonder where all this discussion will lead. We have chosen a title that includes both systems, and science, and certainly much more of this book is about the broader concept of geographic information than about isolated, monolithic software systems *per se*. We believe strongly that effective users of GIS require some awareness of *all* aspects of geographic information, from the basic principles and techniques to concepts of management and familiarity with applications. We hope this book provides that kind of awareness. On the other hand we have chosen not to include GIStudies in the title. Although the later chapters of the book address many aspects of the social context of GIS, including issues of privacy, the theoretical context of GIStudies is rooted in social theory. GIStudies needs the kind of focused attention that we cannot give, and we recommend that students interested in more depth in this area explore the specialized texts listed in the references.

Questions for Further Study

1. Examine a recent issue of one of the GIS magazines listed in Box 1.3. Which of the definitions of GIS discussed in this chapter fits best to the way GIS is used in the applications described in the issue?

2. What are the distinguishing characteristics of the scientific method? Discuss the relevance of each to GIS.

3. We argued in Section 1.4.3.1 that the Internet had dramatically changed GIS. What are the arguments for and against this view?

4. Compare and contrast the views of David Mark (Box 1.7) and Brian J. L. Berry (Box 1.8) on the interrelationship between geography and GIS.

Online Resources

NCGIA Core Curricula (www.ncgia.ucsb.edu/pubs/core.html):

Core Curriculum in GIScience, Sections 0 (Michael F Goodchild, What is Geographic Information Science?), 4.1 (Stephen J Ventura, Land Information Systems and Cadastral Applications), and 4.2 (material by PrecisionAg.org)

Core Curriculum in GIS, 1990, Units 1, 23, 51–56

ESRI Virtual Campus courses (campus.esri.com):
Turning Data into Information, by Paul Longley, Michael Goodchild, David Maguire, and David Rhind (Module "Introduction")
Geographic Problem Solving, by Francis Harvey

See Section 1.5.4 for a selection of Web sites that feature online news, chatrooms, lists of events, and other resources for the GIS community.

For a summary of the history of GIS see the GIS Timeline at www.casa.ucl.ac.uk/gistimeline/

Reference Links

Maguire D J, Goodchild M F, and Rhind D W (eds) 1991 *Geographical Information Systems: Principles, and applications*. Harlow, UK: Longman (Text available online from 'Links to Big Book 1' at www.wiley.com/gis and www.wiley.co.uk/gis).
Chapter 1, An overview and definition of GIS (Maguire D J).
Chapter 2, The history of GIS (Coppock J T, Rhind D W).
Chapter 3, The technological setting of GIS (Goodchild M F).
Chapter 4, The commercial setting of GIS (Dangermond J).
Chapter 6, The academic setting of GIS (Unwin D J).

Chapter 7, The organizational home for GIS in the scientific professional community (Morrison J L).
Longley P A, Goodchild M F, Maguire D J, and Rhind D W (eds) 1999 *Geographical Information Systems: Principles, techniques, management, and applications*. New York: John Wiley.
Chapter 3, Geography and GIS (Johnston R J).
Chapter 4, Arguments, debates and dialogues: the GIS–social theory debate, and the concern for alternatives (Pickles J).
Chapter 40, The future of GIS and spatial analysis (Goodchild M F, Longley P A).
Chapter 54, Enabling progress in GIS education (Forer P, Unwin D J).

References

Chrisman N R 1999 What does GIS mean? *Transactions in GIS* **3**: 175–186.

Curry M R 1998 *Digital Places: Living with geographic information technologies*. London: Routledge.

Foresman T W (Ed.) 1998 *The History of Geographic Information Systems: Perspectives from the pioneers*. Upper Saddle River, New Jersey: Prentice Hall PTR.

Goodchild M F 1992 Geographical information science. *International Journal of Geographical Information Systems* **6**: 31–45.

Harder C 1998 *Serving Maps on the Internet*. Redlands, California: ESRI Press.

Laudan L 1996 *Beyond Positivism and Relativism: Theory, method, and evidence*. Boulder, Colorado: Westview Press.

Massam B H 1980 *Spatial Search: Application to planning problems in the public sector*. Oxford: Pergamon.

Massam B H 1993 *The Right Place: Shared responsibility and the location of public facilities*. Harlow, UK: Longman Scientific and Technical.

Pickles J 1993 *Ground Truth: The social implications of geographic information systems*. New York: Guilford Press.

Plewe B 1997 *GIS Online: Information retrieval, mapping, and the Internet*. Santa Fe, New Mexico: Onword Press.

Thill J C 1999 *Spatial Multicriteria Decision Making and Analysis: A geographic information sciences approach*. Aldershot, England: Ashgate.

University Consortium for Geographic Information Science 1996 Research priorities for geographic information science. *Cartography and Geographic Information Systems* **23**: 115–127.

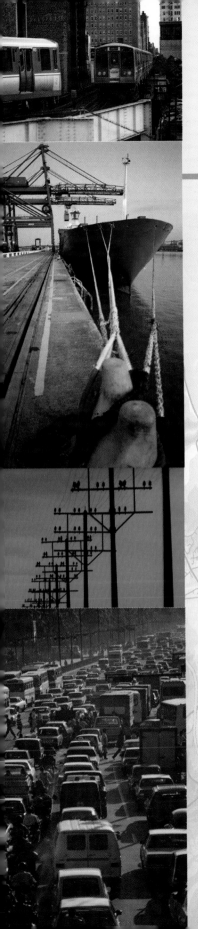

A GALLERY OF APPLICATIONS

2

GIS is fundamentally about workable applications. This chapter gives a flavor of the breadth and depth of real-world GIS implementations. It considers:

- How GIS affects our everyday lives;
- How GIS applications have developed, and how the field compares with scientific practice;
- The goals of applied problem-solving; and
- How GIS can be used to study and solve problems in transportation, the environment, local government, and business.

Learning objectives

After studying this chapter you will:

- Grasp the many ways in which we interact with GIS in everyday life;
- Appreciate the range and diversity of GIS applications in environmental and social science;
- Be able to identify many of the scientific assumptions that underpin real-world applications; and
- Understand how GIS is applied in the representative application areas of transportation, the environment, local government, and business.

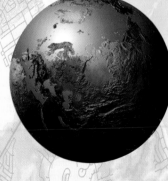

2.1 Introduction

2.1.1 One day of life with GIS

7:00 My alarm goes off ...

The energy to power the alarm comes from the local energy company, which uses a GIS to manage all its assets (e.g. electrical conductors, devices, and structures) so that it can deliver electricity continuously to domestic and commercial customers (Figure 2.1).

7:05 I jump in the shower...

The water for the shower is provided by the local water company, which uses a hydraulic model linked to its GIS to predict water usage and ensure that water is always available to its customers (Figure 2.2).

7:35 I open the mail...

A property tax bill comes from a local government department that uses a GIS to store property data and automatically produce annual tax bills. This has helped the department to peg increases in property taxes to levels below retail price inflation (see Figure 2.8).

There are also a small number of circulars addressed to me, sometimes called "junk mail". We spent our vacation in Southlands and Santatol last year, and the holiday company uses its GIS to market similar destinations to its customer base – there are good deals for the Gower and Northampton this season. A second item is a special offer for property insurance, from a firm that uses its GIS to target neighborhoods with low

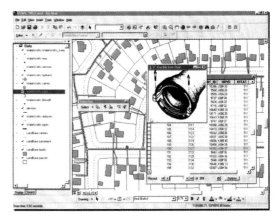

Figure 2.2 Application of a GIS for managing the assets of a water utility (Source: ESRI)

past-claims histories. We receive less junk mail than we used to (and we could opt out of all programs if we wish), because geodemographic and lifestyles GIS is used to target mailings more precisely, thus reducing waste (Figure 2.3).

8:00 My partner leaves for work...

He teaches GIS at one of the city community colleges. As a lecturer on one of the college's most popular classes he has a full workload and likes to get to work early.

8:05 I walk the kids to the bus stop...

Our children attend the local middle school that is three miles away. The school district admin-

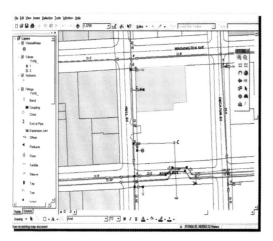

Figure 2.1 Electrical utility application of GIS, showing part of a file of customer records, a detailed map of connections in a neighborhood, etc. (Source: ESRI)

Figure 2.3 The claims of a geodemographic targeting system. Claims such as these should be treated with some skepticism, although the industry is undoubtedly more sophisticated than it was prior to wide use of GIS (Source: Convergent/Schlumberge)

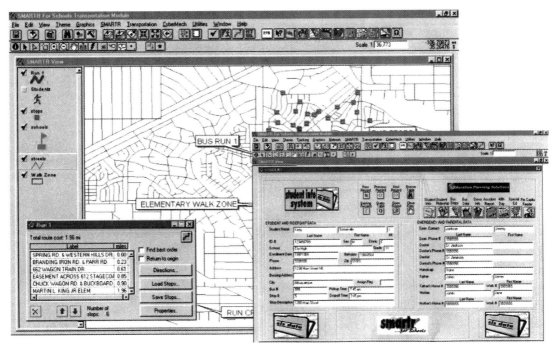

Figure 2.4 A GIS used for school bus routing (Source: ESRI)

istrators use a GIS to optimize the routing of school buses (Figure 2.4). Introduction of this service enabled the district to cut their annual school busing costs by 16% and the time it takes the kids to get to school has also been reduced.

8:15 I catch a train to work...
At the station the current location of trains is displayed on electronic map displays on the platforms using a real-time feed from global positioning (GPS) receivers mounted on the trains. The same information is also broadcast on the Internet so I was able to check the status of trains before I left the house.

8:20 I read the newspaper on the train...
The paper for the newspaper comes from sustainable forests managed by a GIS. The forestry information system used by the forest products company indicates which areas are available for logging, the best access routes, and the likely yield (Figure 2.5).

8:50 I arrive at work...
I am GIS Manager for the local City GIS. Today I have meetings to review annual budgets, plan for the next round of hardware and software acquisition, and deal with a nasty copyright infringement claim.

12:00 I grab a sandwich for lunch...
The price of bread has fallen in real terms in the past decade. In some small part this is because of the increasing use of GIS in precision agriculture. This has allowed real-time mapping of soil nutrients and yield, and means that farmers can apply just the right amount of fertilizer in the right location and at the right time (see Box 1.1).

6:30 Shop till you drop...
After work we go shopping and spend the booklet of discount coupons that arrived in the morning mail. The promotion is to entice customers back to the renovated downtown Tesbury Center. We usually go to SafeMart on the far side of town, but thought we'd participate in the promotion. We actually bump into a few of our neighbors at Tesbury – I suspect the promotion was targeted by linking a marketing GIS to Tesbury's own store loyalty card data.

10:30 The kids are in bed...I'm on the Internet to try and find a new house...
We live in a good neighborhood with many similarly articulate, well-educated folk, but it has become noisier since the new distributor road was routed close by. Our resident association mounted a vociferous campaign of protest, and its members filed numerous complaints to the Web site where

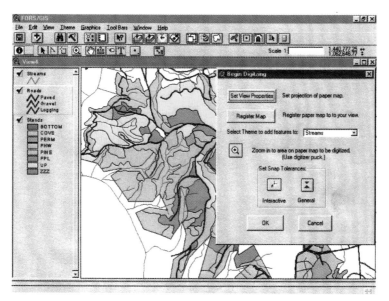

Figure 2.5 Forestry management GIS (Source: ESRI)

the draft proposals were posted. But the benefit–cost analysis carried out using the local authority's GIS demonstrated that it was either a bit more noise for us, or the physical dissection of a vast swathe of lower income housing, and that we would have to grin and bear it. Post GIS, I guess that narrow interest NIMBY (Not In My Back Yard) protests don't get such a free run as they once did. So here I am using one of the free online GIS-powered Web sites to find properties that match our criteria (Figure 2.6). Once we have found a property, other mapping sites provide us with details about the local and regional facilities.

GIS is used to improve many of our day-to-day working and living arrangements.

This diary is fictitious of course, but most of the things described in it are everyday occurrences repeated thousands and thousands of times around the world, and all the examples are real. It highlights a number of key things about how GIS:

● Affects each of us, every day;
● Can be used to foster effective short-, and long-term decision-making;
● Has great practical importance;
● Can be applied to many socio-economic and environmental problems;
● Supports measurement, management, monitoring, and modeling operations;
● Generates measurable economic benefits;
● Requires key management skills for effective implementation;
● Provides a challenging and stimulating educational experience for students;
● Can be used to generate direct income;
● Can be combined with other technologies; and
● Is a dynamic, and stimulating area in which to work.

Figure 2.6 Real estate GIS www.realtor.com (Source: www.realtor.com)

At the same time, the examples suggest some of the elements of the critique that has been leveled at GIS in recent years (Section 1.7). Only a very small fraction of the world's population presently has access to information technologies of any kind, let alone high-speed access to the Internet. Information technology seems to be exacerbating the differences between rich and poor communities, and between developed and less developed nations, across what is often called the *digital divide*. Uses of GIS for marketing often involve practices that border on invasion of privacy, since they allow massive databases to be constructed from what many would regard as personal information. It is important that we reflect on issues like these while exploring GIS.

2.1.2 Why GIS?

Our day of life with GIS illustrates the unprecedented frequency with which, directly or indirectly, we interact with digital machines. Today more and more individuals and organizations find themselves using GIS to answer the fundamental question, *where*? This is because of:

- Wider availability of GIS through the Internet, as well as on organization-wide local area networks.

- Reductions in the price of GIS hardware and software, because economies of scale are realized by a fast-growing market.

- Greater awareness of why decision-making has a geographic dimension.

- Greater ease of user interaction, using standard windowing environments.

- Better technology to support applications, specifically in terms of visualization, data management and analysis, and linkage to other software.

- The proliferation of geographically referenced digital data, such as those generated using Global Positioning System (GPS) technology or supplied by value-added resellers (VARs) of data.

- Availability of packaged applications, which are commercial off-the-shelf (COTS) or "ready to run out of the box".

- The accumulated experience of applications that *work*.

2.2 Science, Geography, and Applications

2.2.1 Scientific questions and GIS operations

As we saw in Section 1.3, one objective of science is to solve problems that are of real-world concern. The range and complexity of scientific principles and techniques that are brought to bear upon problem-solving will clearly vary between applications. Within the spatial domain, the goals of applied problem-solving include, but are not restricted to:

- Rational, effective, and efficient allocation of resources, in accordance with clearly stated criteria – whether, for example, it be physical construction of infrastructure in utilities applications, or scattering fertilizer in precision agriculture;

- Monitoring and understanding observed spatial distributions of attributes – such as variation in soil nutrient concentrations, or the geography of environmental health;

- Understanding the difference that *place* makes – identifying which characteristics are inherently similar between places, and what is distinctive and possibly unique about them. For example, there are regional and local differences in the nature of beer consumption (e.g., the UK geographies of *lager* and *bitter* consumption: Box 6.1), and regional variations in voting patterns are the norm in most democracies;

- Understanding of processes in the natural and human environments, such as processes of coastal erosion or river delta deposition in the natural environment, and understanding of changes in residential preferences or store patronage in the social;

- Prescription of strategies for environmental maintenance and conservation, as in national park management.

Understanding and resolving these diverse problems entails a number of general data handling operations – such as inventory compilation and analysis, mapping and spatial database management – that may be successfully undertaken using GIS.

GIS is fundamentally about solving real-world problems.

GIS has always been fundamentally an applications-led area of activity. The accumulated experience of applications has led to borrowing and creation of particular conventions for representing, visualizing, and to some extent analyzing data for particular classes of applications. Over time, some of these conventions have become useful in application areas quite different from those for which they were originally intended, and software vendors have developed general-purpose routines that may be customized in application-specific ways, as in the way that spatial data are visualized. The way that accumulated experience and borrowed practice becomes formalized into conventions of practice makes GIS essentially an inductive field.

In terms of the definition and remit of GIScience (Section 1.6) these applications conventions are essentially low-order concepts, rather than fundamental generic data-handling issues. Many of the data-handling operations are essentially so routine that some are now even available as adjuncts to popular word-processing packages. They work and they are very widely used, yet they are not really scientific in the sense developed in Section 1.3.

2.2.2 GIScience applications

Today GIScience has moved the world of applications onto a bigger agenda – specifically the deeper conceptual grounding of successful applications. Early GIS was successful in routine and blunt-edged senses, but shied away from most of the bigger questions concerning how the world works.

GIS has always been an applications-led technology, and many applications have had quite modest goals in terms of the science of problem-solving. This is by no means a fatal problem, however, as the test of good science and technology lies in its usefulness for exploring the world around us, and an applications-led field survives or falls against this simple goal. Indeed, no amount of scientific and technological ingenuity can salvage a representation that is too inaccurate, expensive, cumbersome, or opaque to reveal anything new about the world. Yet GIS applications need also to be grounded in sound concepts and theory if they are to resolve any but the most trivial of questions.

GIS applications need to be grounded in sound concepts and theory.

2.3 Representative Application Areas and their Foundations

2.3.1 Introduction and overview

There is, quite simply, a huge range of applications of GIS, and indeed several pages of this book could be filled with a list of application areas. They include topographic base mapping, socio-economic and environmental modeling, global (and interplanetary!) modeling, and education. Applications generally set out to fulfil the *five Ms* of GIS: mapping, measurement, monitoring, modeling, and management.

The four Ms of GIS application are mapping, measurement, monitoring, modeling, and management.

In very general terms, GIS applications may be classified as traditional, developing, and new. Traditional GIS application fields include military, government, education, and utilities. The mid-1990s saw the wide development of business uses, such as banking and financial services, transportation logistics, real estate, and market analysis. The early years of the 21st century are seeing new forward-looking application areas in small office/home office (SOHO) and personal or consumer applications. This is a somewhat rough-and-ready classification, however, because the applications of some agencies (such as utilities) fall into more than one class.

A further way to examine trends in GIS applications is to examine the diffusion of GIS use. Figure 2.7 shows the classic model of GIS diffusion originally developed by Everett Rogers (Rogers 1995). Rogers' model divides the adopters of an innovation into five categories:

- Venturesome Innovators – willing to accept risks and sometimes regarded as oddballs.
- Respectable Early Adopters – regarded as opinion formers or "role models".
- Deliberate Early Majority – willing to consider adoption only after peers have adopted.
- Sceptical Late Majority – overwhelming pressure from peers is needed before adoption occurs.
- Traditional Laggards – people oriented to the past.

GIS seems to be in the transition between the Early Majority and the Late Majority stages. The Innovators who dominated the field in the 1970s were typically based in universities and research organizations. The Early Adopters were the users of the 1980s, many of whom were in government and military establishments. The Early Majority, typically in private businesses, came to the fore in the mid-1990s. The current question for potential users appears to be: do you want to gain competitive advantage by being part of the Majority user base or wait until the technology is completely accepted and join the GIS community as a Laggard?

A wide range of motivations underpins the use of GIS, although it is possible to identify a number of common themes. Applications dealing with day-to-day issues typically focus on very practical concerns such as cost-effectiveness, service provision, system performance, competitive advantage, and database creation, access

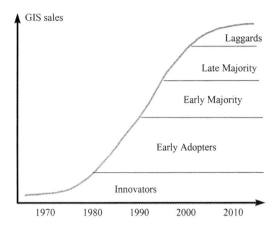

Figure 2.7 The classic Rogers model of innovation diffusion applied to GIS

and use. Other, more strategic applications are more concerned with creating and evaluating scenarios under a range of circumstances.

Many applications involve use of GIS by large numbers of people. It is not uncommon for a large government agency, university, or utility to have more than 100 GIS "seats" and a significant number have more than 1000. Many large organizations have site licences for software products. Once GIS applications become established within an organization, usage often spreads widely. Integration of GIS with corporate information system (IS) policy, planning, and systems is an essential pre-requisite for success in many organizations.

The scope of these applications is best illustrated with respect to representative application areas and in the remainder of this chapter we consider:

1. Local government (Section 2.3.2)
2. Business, and service planning (Section 2.3.3)
3. Logistics (including transportation) (Section 2.3.4)
4. Environment (Section 2.3.5)

We begin by identifying the range of applications within each of the four domains. Next, we go on to focus upon one application within each domain. Each application is chosen, first, for simplicity of exposition but also, second, for the scientific questions that it raises. In this book, we try to relate science and application in two ways. First, we flag the sections elsewhere in the book where the scientific issues raised by the applications are discussed. Second, the applications discussed here, and others like them, provide the illustrative

material for our discussion of principles, techniques, and practices in the other chapters of the book.

A recurrent theme in each of the application classes is the importance of geographic location, and hence what is *special* about the handling of georeferenced data (Section 1.1.1). The gallery of applications that we set out here intends to show how geographic data can provide crucial context to decision-making.

2.3.2 Local government

2.3.2.1 Applications overview

Government users were among the first to discover the value of GIS. Indeed the first recognized GIS – the Canadian Geographic Information System (CGIS) – was developed for natural resource inventory and management by the Canadian government (see Section 1.4.1). Today government users still comprise the biggest single group of GIS professionals, and O'Looney (2000) estimates that 70–80% of local government work involves GIS in some way.

> Of the tasks undertaken by local governments 70–80% are geographically related.

Today, local government organizations are acutely aware of the need to improve the quality of their products, processes and services through ever-increasing efficiency of resource usage (O'Looney 2000: see also Section 15.4). Thus GIS is used to inventory resources and infrastructure, plan transportation routing, improve service response time, manage land development, and generate revenue by increasing economic activity.

Local governments also use GIS in unique ways. Because governments are responsible for the long-term health, safety, and welfare of citizens, wider issues need to be considered, including in-corporating public values in decision-making, delivering services in a fair and equitable manner, and representing the views of citizens by working with elected officials. Typical GIS applications thus include monitoring public health risk, managing public housing stock, allocating welfare assistance funds, and tracking crime (see Greene 2000).

It is convenient to group local government GIS applications on the basis of their contribution to asset inventory, policy analysis, and strategic modeling/planning. Table 2.1 summarizes GIS applications in this way.

These applications can be implemented as centralized GIS or distributed desktop applications. Some will be designed for use by highly trained GIS professionals, while citizens will access others as "front counter" or Internet systems. Chapter 8 discusses the different implementation models for GIS.

2.3.2.2 Case study application: GIS in Tax Assessment

Tax mapping, and assessment is a classic example of the value of GIS in local government. In many

Table 2.1 GIS applications in local government (simplified from O'Looney 2000)

	Inventory Applications (locating property information such as ownership and tax assessments by clicking on a map)	Policy Analysis Applications (e.g. number of features per area, proximity to a feature or land use, correlation of demographic features with geological features)	Management/Policy-Making Applications (e.g. more efficient routing, modeling alternatives, forecasting future needs, work scheduling)
Economic Development	Location of major businesses and their primary resource demands	Analysis of resource demand by potential local supplier	Informing businesses of availability of local suppliers
Transportation and Services Routing	Identification of sanitation truck routes, capacities and staffing by area; identification of landfill and recycling sites	Analysis of potential capacity strain given development in certain areas; analysis of accident patterns by type of site	Identification of ideal high-density development areas based on criteria such as established transportation capacity
Housing	Inventory of housing stock age, condition, status (public, private, rental, etc.), durability, and demographics	Analysis of public support for housing by geographic area, drive time from low-income areas to needed service facilities, etc.	Analysis of funding for housing rehabilitation, location of related public facilities; planning for capital investment in housing based on population growth projections
Infrastructure	Inventory of roads, sidewalks, bridges, utilities (locations, names, conditions, foundations, and most recent maintenance)	Analysis of infrastructure conditions by demographic variables such as income and population change	Analysis to schedule maintenance and expansion
Health	Locations of persons with particular health problems	Spatial, time-series analysis of the spread of disease; effects of environmental conditions on disease	Analysis to pinpoint possible sources of disease
Tax Maps	Identification of ownership data by land plot	Analysis of tax revenues by land use within various distances from the city center	Projecting tax revenue changes attributable to land-use changes
Human Services	Inventory of neighborhoods with multiple social risk indicators; location of existing facilities and services designated to address these risks	Analysis of match between service facilities and human services needs, and capacities of nearby residents	Facility siting, public transportation routing, program planning and place-based social intervention
Law Enforcement	Inventory of location of police stations, crimes, arrests, convicted perpetrators and victims; plotting police beats and patrol car routing; alarm and security system locations	Analysis of police visibility and presence; officers in relation to density of criminal activity; victim profiles in relation to residential populations; police experience and beat duties	Reallocation of police resources and facilities to areas where they are likely to be most efficient and effective; creation of random routing maps to decrease predictability of police beats
Land-use Planning	Parcel inventory of zoning areas, floodplains, industrial parks, land uses, trees, green space, etc.	Analysis of percentage of land used in each category, density levels by neighborhoods, threats to residential amenities, proximity to locally unwanted land uses	Evaluation of land-use plan based on demographic characteristics of nearby population (e.g. will a smokestack industry be built upwind of a respiratory disease hospital?)

Table 2.1 (*Continued*)

	Inventory Applications (locating property information such as ownership and tax assessments by clicking on a map)	Policy Analysis Applications (e.g. number of features per area, proximity to a feature or land use, correlation of demographic features with geological features)	Management/Policy-Making Applications (e.g. more efficient routing, modeling alternatives, forecasting future needs, work scheduling)
Parks and Recreation	Inventory of park holdings/ playscapes, trails by type, etc.	Analysis of neighborhood access to parks and recreation opportunities, age-related proximity to relevant playscapes	Modeling population growth projections and potential future recreational needs/playscape uses
Environmental Monitoring	Inventory of environmental hazards in relation to vital resources such as groundwater; layering of nonpoint pollution sources	Analysis of spread rates and cumulative pollution levels; analysis of potential years of life lost in a particular area due to environmental hazards	Modeling potential environmental harm to specific local areas; analysis of place-specific multilayered pollution abatement plans
Emergency Management	Location of key emergency exit routes, their traffic flow capacity and critical danger points (e.g. bridges likely to be destroyed by an earthquake)	Analysis of potential effects of emergencies of various magnitudes on exit routes, traffic flow, etc.	Modeling effect of placing emergency facilities and response capacities in particular locations
Citizen **Information/ Geodemo- graphics**	Location of persons with specific demographic characteristics such as voting patterns, service usage and preferences, commuting routes, occupations	Analysis of voting characteristics of particular areas	Modeling effect of placing information kiosks at particular locations

countries local government agencies have a mandate to raise revenue from property taxes. The amount of tax payable is partly or wholly determined by the value of the taxable land and property. A key part of this process is evaluating the value of land and property fairly to ensure equitable distribution of a community's tax burden. In the United States the task of determining the taxable value of land and property is performed by the Tax Assessor's Office, which is usually a separate local government department. The Valuation Office Agency fulfils a similar role in the UK. The tax department can quickly be overwhelmed with requests for valuation of new properties, and protests about existing valuations.

> The Tax Assessor's Office is often the first home of GIS in local government.

Essentially, a Tax Assessor's role is to assign a value to properties using three basic methods: cost, income, and market. The cost method is based on the replacement cost of the property and the value of the land. The Tax Assessor must examine data on construction costs and vacant land values. The income method takes into consideration how much income a property would

generate if it were rented. This requires details on current market rents, vacancy rates, operating expenses, taxes, insurance, maintenance, and other costs. The market method is the most popular. It compares the property to other recent sales that have a similar location, size, condition, and quality.

Collecting, storing, managing, analyzing, and displaying all this information is a very time-consuming activity and not surprisingly GIS has had a major impact on the way Tax Assessors go about their business.

2.3.2.3 Method

Tax Assessors, working in a Tax Assessor's Office, are responsible for accurately, uniformly, and fairly judging the value of all taxable properties in their jurisdiction. Details about properties are maintained on a tax assessment roll that includes information such as ownership, address, land and building value, and tax exemptions. The Assessor's Office is also responsible for processing applications for tax abatement, in cases of overvaluation, and exemptions for surviving spouses, veterans, and the elderly. Figures 2.8 shows some aspects of a tax assessment GIS in Ohio, USA.

(A)

(B)

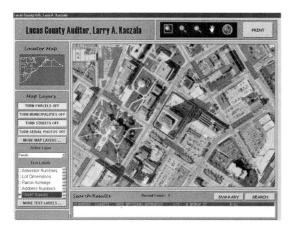

Figure 2.8 Lucas County, Ohio, USA. Tax Assessment GIS: (A) tax map; (B) property attributes and photograph (Source: ESRI)

A GIS is used to collect and manage the geographic boundaries and associated information about properties. Typically, data associated with properties is held in a computer assisted mass appraisal (CAMA) system that is responsible for sale analysis, evaluation, data management, and administration, and for generating notices to owners. CAMA systems are usually implemented on top of a database management system (DBMS) and can be linked to the parcel database using a common key (see Sections 11.2 and 11.3 for further discussion of how this works).

The basic tax assessment task involves a geographic database query to locate all the property sales of a similar type within a specified distance of an owner's property. First the owner's property must be found in the property data using a query. Then a geographic query can locate all comparable properties valued similarly within 1 mile of the owner's property. The selected properties can then be displayed on the screen. The Tax Assessor can then compare the characteristics of these properties (floor area, last sale price, neighborhood status, property improvements, etc.) and determine the valuation of the property that is to be valued.

2.3.2.4 Scientific foundations, geographic principles, and techniques

Scientific foundations
Critical to the success of the tax assessment process is a high-quality, up-to-date geographic database that can be linked to a CAMA system. Considerable effort must be expended to design,

implement, and maintain the geographic database. Even for a small community of 50,000 properties it can take several months to collect the geographic descriptions of property parcels and their associated attributes. Chapters 10 and 11 explain the processes involved in managing geographic databases such as this. Linking GIS and CAMA systems can be quite straightforward providing that both systems are based on DBMS technology and use a common identifier to effect the linkage between a map feature and a property record. Typically, a unique parcel number (in the US) or unique property reference number (in the UK) is used.

> A high-quality geographic database is essential to tax assessment.

Clearly, the system is dependent on a clear and unambiguous definition of parcels, and common standards about how different characteristics (such as size, age, and value of improvements) are represented. The GIS can help enforce coding standards and can be used to derive some characteristics automatically in an objective fashion. For example, using a GIS it is very easy to calculate the area of properties from the boundary information.

Fundamentally, this application, like many others in GIS, is about maintaining an unambiguous and accurate inventory of geographic features. To be effective it must employ methods of description and representation that are clear, understood by every user of the system, and work the same way every time they are used. These are all core objectives of scientific method, and although the application is clearly not driven by scientific curiosity, it nevertheless follows procedures that are much like those used in a scientific laboratory.

Geographic principles

Tax assessment makes the assumption that, other things being equal, properties close together in space will have similar values. This is an application of Tobler's First Law of Geography, as discussed in Sections 3.1, and 5.2. However, it is left to the Assessor to determine the impact of other properties on the valuation of a property. Although this seems relatively straightforward, in practice it can be much more difficult. It assumes that property groups or neighborhoods have similar values and that they can be treated as having similar levels of common attributes (see Section 5.2). This assumption is valid in areas where houses were constructed at the same time according to common standards; however, in older areas where infill has been common, properties vary radically over short distances.

Techniques

Tax assessment actually uses standard GIS techniques such as proximity analysis, and geographic and attribute query, mapping and reporting. Chapters 13, 11 and 12 describe, respectively, proximity analysis, geographic query, and mapping and visualization. All that is required is a good database, a plan for system management and administration, and a workflow design. Although many experienced GIS users take techniques like proximity analysis for granted, their value in tax assessment cannot be overestimated. Manually sorting through paper records, or even tabular data in a CAMA system, is a very laborious and time-consuming task, especially when it must be repeated many times a day. Automating this in a GIS is very cost-effective.

2.3.2.5 Generic scientific questions arising from the application

Tax assessment is a classic operational GIS application. It requires an up-to-date inventory of properties and information from several sources about sales and sale prices, improvements, and building programs.

To help tax assessors understand geographic variations in property characteristics it is also possible to use GIS for more strategic modeling activities. The many tools in GIS for charting, reporting, mapping, and exploring data help assessors to understand the variability of property value within their jurisdiction. Some assessors have also built models of property valuations and have clustered properties based on multivariate criteria (see Section 5.7). These help assessors to gain knowledge of the structure of communities and highlight unusually high or low valuations.

Once a property database has been created, it becomes a very valuable asset, not just for the tax assessor's department, but also for many other departments in a local government agency. Public Works will be interested to know about access points for repairs and meter reading, Housing would like to use it as a basemap for information about the status of the housing stock, and many departments would like to share a common address list for database organization and standardized mailings.

A property database is useful for many purposes besides tax assessment.

2.3.2.6 Practice

Tax assessment is a key local government application because it is a direct revenue generator. It is easy to develop a cost–benefit case for this application and it can pay quickly for a complete department or corporate GIS implementation. Tax assessment is a service offered directly to members of the public. As such, the service must be robust, reliable, and provide a quick turnaround (usually within 1 week). It is quite common for citizens to question the assessed value for their property, since this is a principal determinant of the amount of tax they will have to pay. A tax assessor must, therefore, be able to justify the method and data used to determine property values. A GIS is a great help and often convinces people of the objectivity involved (sometimes over-impressing people that it is totally scientific). GIS can satisfy the local government requirement for efficiency and equitability.

2.3.3 Business and service planning

2.3.3.1 Applications overview

Business and service planning (sometimes called retailing) applications focus upon the use of geographic data to provide operational, tactical, and strategic context to decisions that involve the fundamental question, *where*? *Geodemographics* is a shorthand term for composite indicators of consumer behavior that are available at the small-area level (e.g. census tract, or postal zone), and which have developed out of the work of market researchers such as Richard Webber (Box 2.1).

Box 2.1   Richard Webber, market researcher

After a degree in Economics at Cambridge, UK, Richard Webber completed a Masters in Transportation Design at the University of Liverpool. Working at the London (UK) Centre for Environmental Studies (CES) in the 1970s, Richard developed a software package specifically designed to help the British government identify clusters of neighborhoods for which different types of urban deprivation program were appropriate. With the closure of CES in 1979, Richard Webber moved to the commercial firm CACI, where he linked the census enumeration district (block) classification, which in 1971 comprised 36 neighborhood types, to the UK postcode geography. Marketed as ACORN, this system became widely used in both public and private sector applications as a means of profiling customers based on their postcodes. It provided many marketing personnel with their first practical experience of geodemographics.

The basic classification methodology has been extended to incorporate the use of ancillary data sources (which are often more relevant, detailed, and up-to-date) for discriminating between consumer groups in today's sophisticated, fast-changing markets. GIS routines (such as point-in-polygon, Section 13.4.2) make it possible to compare census (polygon) and postcode (point) geographies, where these do not match one another. In 1985, Richard Webber joined CCN (which subsequently became Experian in 1997), where he developed the system known as MOSAIC. This typology, released in late 1986, drew upon ancillary datasets to which CCN had access as a result of its credit referencing activities, including electoral data, credit applications, and County Court Judgements (CCJs) by postcode.

Today, there are few large consumer-oriented organizations that do not use neighborhood classifications and GIS as key elements in their processes for retail planning, for target marketing, and for customer management. The current version of MOSAIC comprises 12 Groups (High Income Families, Suburban Semis, Blue Collar Owners, Low Rise Council, Council Flats, Victorian Low Status, Town Houses and Flats, Stylish Singles, Independent Elders, Mortgaged Families, Country Dwellers, and Institutional Areas), which in turn are subdivided into 52 Types (see Figure 2.10 for an example).

Figure 2.9   Richard Webber (center), market researcher

Reflecting on his experiences in geodemographic analysis, Webber says:

"What constantly surprises me is how much the focus of GIS has been in assembling data and the qualitative techniques for managing these data, how little on the interpretation of the rich patterns of urban social structure that GIS technology can reveal. The man in the street has an instinctive appreciation of the nuances of urban social structure. Geographers, by contrast, are wary of stereotyping and of introducing personal bias into their interpretation of local cultural differences. Studying the role of these communities as creators and consumers of wealth and the process of the dissemination of new ideas and technological innovation can provide insights relevant to a wide range of public policy issues".

Figure 2.10 illustrates the profile of one geodemographic type from a UK classification called MOSAIC. Geodemographic data are frequently used in business applications to identify geographic variations in the incidences of customer types. The term *market area analysis* describes attempts to assess the distribution of retail outlets relative to the greatest concentrations of potential customers.

Geodemographic data are the basis for much market area analysis.

The tools of business applications typically range from simple desktop mapping to sophisticated

Figure 2.10 A geodemographic profile: Studio Singles (a Type within the Stylish Singles Group of the MOSAIC classification) (Courtesy: Experian)

decision support systems. Tools are used to analyze and inform the range of *operational*, *tactical*, and *strategic* functions of an organization. As noted in Section 1.1, operational functions concern the day-to-day processing of routine transactions and inventory analysis in an organization, such as stock management. Tactical functions require the allocation of resources to address specific (usually short-term) problems, such as store sales promotions. Strategic functions contribute to the organization's longer-term goals and mission, and entail problems such as opening new stores or rationalizing existing store networks. Early business applications were simply concerned with mapping spatially referenced data, as a general descriptive indicator of the retail environment. This remains the first stage in most business applications, and in itself adds an important dimension to analysis of organizational function. More recently, decision support tools used by spatially aware professionals (SAPs, Section 1.4.3.2) have created a mainstream management role for business GIS applications.

Some of the operational roles of GIS in business are discussed under the heading of logistics applications in Section 2.3.5 below. These include stock flow management systems and distribution network management, the specifics of which vary from industry sector to sector. Geodemographic analysis is an important operational tool in market area analysis, where it is used to plan marketing campaigns. Each of these applications can be described as assessing the circumstances of an organization.

The most obvious strategic application concerns the spatial *expansion* of a new entrant into a retail market. Expansion in a market poses fundamental spatial problems – such as whether to expand through *contagious diffusion* across space, or *hierarchical diffusion* down a settlement structure, or to pursue some combination of the two (Figure 2.11). Many organizations periodically experience spatial *consolidation* and branch *rationalization*. Consolidation and rationalization may occur: (a) when two organizations with overlapping networks merge; (b) in response to competitive threat; or (c) in response to changes in the retail environment. Changes in the retail environment may be short term, and cyclic, as in the response to the recession phase of business cycles, or structural, as with the rationalization of clearing bank branches following the development of personal, telephone, and Internet-based banking (see also Section 16.7). Still other organizations undergo spatial *restructuring*, as in the market repositioning of bank branches to supply a wider range of more profitable financial services. Spatial restructuring is often the conse- quence of technological change. For example, a "clicks and mortar" strategy might be developed by a chain of conventional bookstores, whereby their retail outlets might be reconfigured to offer reliable pick-up points for Internet and telephone orders. This enables them better to compete with new, purely Internet-based entrants. A final type of strategic operation involves *distribution* of goods and services, as in the case of so-called "e-tailers", that use the Internet for merchandizing, but must create or buy into viable distribution networks. These various strategic operations require a range of spatial analytic tools and data types, and

entail a move from "what-is" visualization to "what-if" forecasts and predictions.

2.3.3.2 Case study application: promoting petroleum and convenience shopping

Figure 2.12(A) shows the location of a gas (petrol) station near the town of Warwick, UK. Petroleum retail is a high-volume, low-margin business and, in appropriate circumstances, profitability can be increased through the addition of convenience stores to the forecourt. An oil company awarded Spa Marketing Systems, based in Leamington Spa, UK, a small consultancy contract to identify the best way of boosting sales to a new convenience store that was planned to be added to the existing gas station.

> GIS can be used to study the success of a retailer in penetrating a local market area.

The location is in a residential neighborhood, and as such it was anticipated that its customer base would be overwhelmingly local. A budget was allocated for promoting the new establishment, in order to encourage repeat patronage, and it was decided to use a leaflet mail drop as the most effective publicity medium. Figure 2.12(A) was obtained by assigning the in-store responses to postcode sectors (see Section 4.3; coarse postcode sectors in the UK are roughly equivalent to US ZIP codes, while the smaller unit postcodes are roughly equivalent to ZIP + 4 codes) and dividing the number of known customers by the adult population for each zone. The map gives an indication of the *market penetration* of the store in each zone (see Section 13.4.5). The organization principally uses desktop mapping products, which put into practice straightforward (low-order in the terminology of Section 2.2.1) data handling concepts, but not many of the more advanced spatial analytic techniques that are available in more comprehensive software systems or through consultancy organizations. The data resources available to the business analysts included:

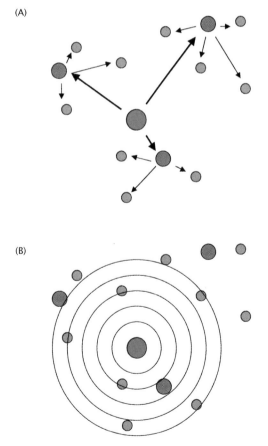

(A)

(B)

Figure 2.11 (A) Contagious and (B) hierarchical spatial diffusion

(a) The results of a gas station forecourt survey, in which customers were asked the frequency of their visits and whether or not they usually made non-full purchases (of newspapers, candy, etc.) from the limited range of goods that had always been available from the station. Each customer was also asked details of occupation, number of cars owned and other general household characteristics, plus full (zip + 4) unit postcodes.

(A)

(B)

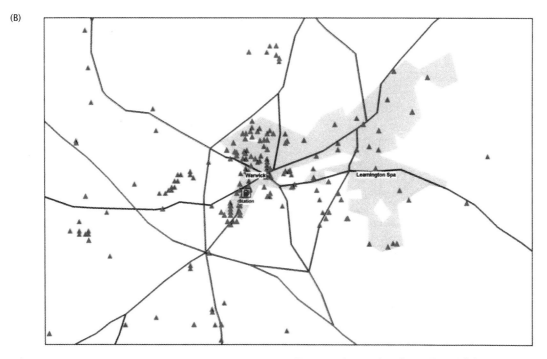

Figure 2.12   Promoting a new gas station convenience store (Framework map data Crown Copyright)

(b) Geographic references for the first mail delivery point in each unit postcode.

(c) Census of Population data, presenting a range of socio-economic variables at the small area census block (enumeration district) scale. The size of a typical UK census block in the study area is 120 households or 260 individuals.

(d) A file containing all of the digital boundary data for the census tracts.

The objective of the analysis is to use information from the existing customer base to identify concentrations of likely customers in the new catchment.

2.3.3.3 Method

The procedures for doing this in a vector GIS are illustrated in Figure 2.13 (similar operations are described by Martin and Longley 1995 and Martin 1996: 119–22):

(i) Existing or potential frequent (four or more times per month) gas purchasers and other potential convenience store customers are deemed the *target group* most likely to use the new convenience store facilities. The post- (zip-) codes of existing frequent gas purchasers (green triangles in Figure 2.12(B), and existing non fuel customers (red triangles in Figure 2.13(B) are extracted from the in-shop survey results (Figure 2.13A).

(ii) The post- (zip-) codes of each existing customer in the study region (as defined in (b) above) are overlain onto a census zone map of the region (Figure 2.13B).

(iii) Socio-economic and car ownership data from the census are attributed to each existing

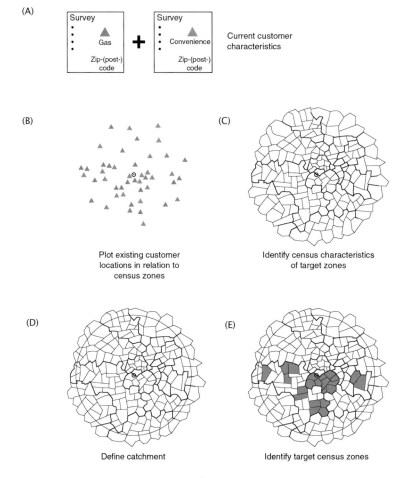

Figure 2.13 Targeting likely gas station customers: (A) accessing existing customer database; (B) identifying neighborhood type of each existing customer; (C) identifying target census area characteristics; (D) and (E) identifying target zones within gas station market area

customer. This enables a file of target census zone characteristics within the market area of the new gas station to be drawn up (Figure 2.13(C); this is likely to include zones with high proportions of residents of working age, and zones with high rates of car ownership, for example).

(iv) A circular distance is drawn on a census map around the location of the newly sited gas station, in order to identify all the smallest-area census blocks that fall within its market area (Figure 2.13D).

(v) Those zones that share the target census zone characteristics are identified for the leaflet drop (Figure 2.13E).

In this instance, the decision to locate a new convenience store outlet had been made, and its location was fixed. A variant on this theme is the calculation of market share for each of the outlets in an existing retail chain, as a means of establishing a benchmark against which the actual performances of individual stores might be compared. It is also possible to envisage strategic variants upon this theme, in which overlay and market area analysis are used to identify possible new store locations. Birkin et al (1999) provide an illustration of a strategic application which uses essentially the same method.

GIS can be used to select new locations in support of a retailer's strategic expansion.

2.3.3.4 Scientific foundations, geographic principles, and techniques

The following assumptions and organizing principles are inherent to this case study.

Scientific foundations

Steps (i), and (ii) above assume that each customer is equally and utterly typical of the census zone in which he or she resides. The individual resident *in an area* is thus assigned the characteristics *of the area*. In practice, of course, individuals within households have different characteristics, as do households within streets, zones, and any other aggregation. The practice of confounding characteristics *of* areas with individuals resident *in* them is known as invoking the *ecological fallacy*. This is inevitable in this instance because census data are only made available at the aggregated scale of the census zone, in order to preserve the confidentiality of respondents (see the point about GIS and the surveillance society at the end of Section 2.2.2 above). But more generally, geography is a science which has very few natural units of analysis. Even if we had disaggregate data we would be uncertain as to whether we should

consider the individual or the household as the basic unit of analysis – sometimes one individual in a household always makes the important decisions, while in others this is a shared responsibility. More generally, what, for example, is the natural unit for measuring a soil profile? We return to this issue in our discussion of *uncertainty* (Chapter 6).

Steps (i), (ii) and (iv) all assume that individual behavior can be related to measurable *predictor variables*, such as employment status and car ownership. Formal statistical ways of doing this are discussed in Chapters 5 and 14. Of course if you do not own a car, you are unlikely to purchase petroleum from a gas station and so, at an individual level, car ownership is likely to be a good predictor variable. However, ownership data alone provides little indication of *frequency* of likely visits to the gas station, or the amount spent, both of which are important in identifying the best prospective customers for a location. Other variables such as age and employment status are likely to be helpful in this context, but alone or in combination are rarely likely to provide perfect predictions of behavior. We are unlikely ever to have access to the full range of possible predictor variables, and even those measures that we can ascribe to individuals are likely to be subject to measurement error. We return to the issue of *unobserved predictor variables* and *measurement error* in our discussion of *uncertainty* (Chapters 6 and 15). More fundamentally still, is it acceptable (in predictive terms) to represent the behavior of consumers using socio-economic variables?

Geographic principles

Step (iii) assumes that a *linear* radial distance measure is intrinsically meaningful in terms of defining market area. In practice, there are several major shortcomings in this. The simplest is that *spatial structure* will distort the circular market area – the market is likely to extend further along the major travel arteries, for example, and will be restricted by physical obstacles such as blocked-off streets and rivers, and traffic management devices such as stop lights (Section 5.5). An alternative solution would be to assume that the maximum extent of the market is determined by the extent of all of the coarse-scale census zones (wards) that register at least a very low market penetration. But even this assumes that an arbitrary administrative aggregation means something in terms of convenience store and petroleum sales. We return to the issue of appropriate distance metrics in our discussion of *geographic query and analysis* (Chapter 13).

The assumed spatial structure requires that everybody in the catchment area be equally likely to patronize the new outlet, and that those resident anywhere outside it have a zero probability of patronage. In reality, distance is likely to have a much more gradual *attenuating* effect upon store patronage, as discussed as an aspect of the *nature of geographic data* (Chapter 5).

The problem assumes that there is no existing *competition* – despite that fact that residents almost certainly already purchase convenience goods and petroleum somewhere! New outlets have also been known to poach customers from the existing sites of the same organization – a process known as *cannibalizing*. The representation thus ignores the *spatial interaction* between competing store outlets. Although it is beyond the scope of this book, Birkin *et al* (1999) discuss how the tradition of *spatial interaction modeling* is ideally suited to the problems of defining realistic catchment areas, and estimating store revenues. In this approach, a study area is divided into a set of residential zones, say postal sectors or ZIP codes, and all centers and outlets are identified. The models then attempt to quantify exactly how many customers in each demand zone will patronize each and every store in the region. Since customers to different stores will come from a variety of demand zones, overlapping catchment areas are anticipated, and are explicitly handled.

> Spatial interaction models can be used to predict the flow of customers from neighborhoods to shopping centers and stores.

Techniques

The assignment of unit postcode coordinates to census tracts is not, in practice, carried out by manual interpretation of overlay analysis. Its digital equivalent is a procedure known as *point in polygon* analysis, which is considered in our discussion of *geographic query and analysis* (Chapter 13).

Finally, the analysis assumes that the principles that underpin consumer behavior in Warwick, UK, are essentially the same as those operating anywhere else on Planet Earth. There is no attempt to accommodate regional and local factors. These might include: adjusting the attenuating effect of distance (see Section 5.5) to accommodate the different distances people are prepared to travel to find a convenience store or gas station (e.g. as between an urban and a rural area); and adjusting the likely attractiveness of the outlet to take account of store ownership, forecourt size and

layout, or other qualitative factors. A range of spatial techniques is now available for making the general properties of spatial analysis more *sensitive to context*.

2.3.3.5 Generic scientific questions arising from the application

Retailing is a fundamental activity in advanced economies, and the investments in location are so huge that they cannot possibly be left to chance. Doing nothing is simply not an option. Intuition tells us that the effects of distance to outlet, and the organization of existing outlets in the retail hierarchy must have an impact upon patterns of store patronage. But, in very competitive consumer-led markets, the important question is not whether, but how much?

> Consumers are complex, but predicting even a small part of their behavior can be important to a retailer.

Consumers are sophisticated beings and their shopping behavior is often complex. Most retail decisions are more difficult to explain or anticipate than the choice of where to fill up with petroleum. The stores that people patronize also tend to have a wider range of attributes, in terms of floor space, range and quality of goods and services, price, and miscellaneous customer services. Different consumer groups find different attributes attractive, and hence it is the mix of *individuals* with particular characteristics that largely determines the likely store turnover of a particular location. Our example illustrates the sometimes crass simplifying assumptions that we may be forced to make using the best available data in order to represent consumer characteristics and store attributes. However, it is important to remember that even blunt-edged tools can increase the effectiveness of operational and strategic target marketing activities manyfold. An untargeted leafleting campaign might typically achieve a 1% hit rate, while one informed by even quite rudimentary market area analysis might conceivably achieve a rate that is five times higher. The pessimist might dwell on the 95% failure rate that a supposedly scientific approach entails, yet the optimist should be more than happy with the fivefold increase in the efficiency of use of the marketing budget!

The general nature of problems of locational analysis would seem to suggest that standard techniques would be applicable to them. However, in practice, our discussion has illustrated that there are serious shortcomings with the straightforward procedure set out above.

Indeed Birkin *et al* (1999) state: "it is our belief that making simple (...) models available for *applied* problems can be as dangerous as not having such modeling power in the first place, mainly because it is unlikely that such simple models will be able to reproduce accurately the complex consumer behavior patterns seen in modern retail markets".

2.3.3.6 Practice

Commenting on the specifics of our gas station example, Paul Cresswell (Director, Spa Marketing Systems Ltd.) says:

"In general, the key issue which drives the sophistication of analysis in a given application is cost. For an exercise such as local convenience store promotion, clients just could not spend the kind of money required to investigate the issues of competition, distance decay, and so forth. It is often very difficult even to get hold of customer data that are reliable! Moreover, leaflet distribution companies are often only able to target fairly coarse units of geography (commonly postcode sectors). This means that what can be achieved is limited by the routing of the final delivery vehicle for the leaflet drop.

In fact, the same client, with Spa's help, has undertaken some sophisticated analysis of the whole network of sites, using turnover prediction and spatial interaction principles. This work was regarded as much more strategic (unlike the Warwick example, which addressed only one local store opening) and therefore it was judged by the powers in charge as justifying far higher expenditure. This is often the kind of problem that we, as consultants, are faced with and what we recommend as worthwhile needs to be set in the context of tactical and strategic organizational objectives. This is not to say that we do not make clients aware of the drawbacks of the more simplistic approaches!"

More generally, the geographic development of most retail and business organizations has taken place in a haphazard way. However, the competitive pressures of today's markets require an understanding of branch location networks, as well as their abilities to compete with both conventional retailers and new entrants (e.g. e-tailers). Space is, without doubt, of ongoing and central importance in retailing, at a variety of scales. Within the automotive industry, for example, Birkin *et al* (1999) present compelling evidence of the relationship between spatial variation in market share and the pattern of dealerships at regional and local scales.

Thus the spatially aware professional is increasingly a mainstream manager alongside accountants, lawyers, and general business managers. SAPs complement understanding of corporate performance in national and international markets with performance at the regional and local levels. They have key roles in such areas of organizational activity as marketing, store revenue predictions, new product launch, improving retail networks and the assimilation of pre-existing components into combined store networks following mergers and acquisitions.

The requirements of market area analysis go well beyond simple mapping of data.

It is clear from the example above that simple mapping packages alone provide insufficient scientific grounding to resolve retail location problems adequately. There is also some divergence of view as to whether standard GIS offer sufficient analytic power to resolve complex and multifaceted business and service planning problems (Birkin *et al* 1999). There is ongoing debate as to whether the best solutions to most problems lie in desktop mapping, in-house use of a GIS by a SAP, or the purchase of external consultancy services. The answer to this problem will depend on the nature of the organization, the range of its goods, and services, and the priority that it assigns to investment.

2.3.4 *Logistics (including transportation)*

2.3.4.1 Applications overview

Knowing where things are can be of enormous importance for the fields of logistics and transportation, which deal with the movement of goods and people (e.g. high school students: Figure 2.4) from one place to another, and the infrastructure (highways, railroads, canals) that moves them. Highway authorities need to decide what new routes are needed and where to build them, and later need to keep track of highway condition. Logistics companies (e.g. parcel delivery companies, and shipping companies) need to organize their operations, deciding where to place their central sorting warehouses and the facilities that transfer goods from one mode to another (e.g. from truck to ship), how to route parcels from origins to destinations, and how to route delivery trucks. Transit authorities need to plan routes and schedules, to keep track of vehicles, and to deal with incidents that delay them, and to provide information on the system to the traveling public. All of these fields employ GIS, in a mixture of operational, tactical, and strategic applications.

The field of logistics addresses the shipping and transportation of goods and services

Each of these applications has two parts: the static part that deals with the fixed infrastructure, and the dynamic part that deals with the vehicles, goods, and people that move on the static part. Of course, not even a highway network is truly static, since highways are often rebuilt, new highways are added, and highways are even sometimes moved. But the minute-to-minute timescale of vehicle movement is sharply different from the year-to-year changes in the infrastructure. Historically, GIS has been easier to apply to the static part, but recent developments in the technology are making it much more powerful as a tool to address the dynamic part as well. Today, it is possible to use GPS (Section 10.2.2.2) to track vehicles as they move around, and transit authorities increasingly use such systems to inform their users of the locations of buses and trains.

GPS is also finding applications in dealing with emergency incidents that occur on the transportation network (Figure 2.14). The OnStar system (www.onstar.com) is one of several products that make use of the ability of GPS to determine location accurately virtually anywhere. When installed in a vehicle, the system is programmed to transmit location automatically to a central office whenever the vehicle is involved in an accident and its airbags deploy. This can be life-saving if the occupants of the vehicle do not know where they are, or are otherwise unable to call for help.

The Internet provides an excellent way of making minute-to-minute maps of traffic conditions, and other aspects of the transportation system available to citizens. Figure 2.15 shows the Web site implemented by the Department of Transportation of the State of Washington, USA (www.wsdot.wa.gov), giving current information on traffic flow on the major highways in the Seattle area. Box 12.4 presents a UK example which has a still broader remit in transportation logistics.

Many applications in transportation and logistics involve optimization, or the design of solutions to meet specified objectives. Section 14.3 discusses this type of analysis in detail, and includes several examples dealing with transportation and logistics. For example, a delivery company may need to deliver parcels to 200 locations in a given shift, dividing the work between 10 trucks. Different ways of dividing the

Figure 2.14 Systems such as OnStar (www.onstar.com) allow information on the location of an accident, determined by a GPS unit in the vehicle, to be sent to a central office and compared to a GIS database of highways and streets, to determine the incident location so that emergency teams can respond (Courtesy: NCGIA)

work, and routing the vehicles, can result in substantial differences in time and cost, so it is important for the company to use the most efficient solution (see Box 14.4 for an example of the daily workload of an elevator repair company). Logistics and related applications of GIS have been known to save substantially over traditional, manual ways of determining routes.

GIS has helped many service and delivery companies to substantially reduce their operating costs in the field.

2.3.4.2 Case study application: planning for emergency evacuation

Modern society is at risk from numerous types of disasters, including terrorist attacks, extreme weather events such as hurricanes, accidental spills of toxic chemicals resulting from truck collisions or train derailments, and earthquakes. In recent years several major events have required massive evacuation of civilian populations, for example, the 800,000 people evacuated in Florida in advance of Hurricane Floyd in 1999 (Figure 2.16).

In response to the threat of such events, most communities attempt to plan. But planning is made particularly difficult because the magnitude and location of the event can rarely be anticipated. Suppose, for example, that we attempt to develop a plan for dealing with a spill of a volatile toxic

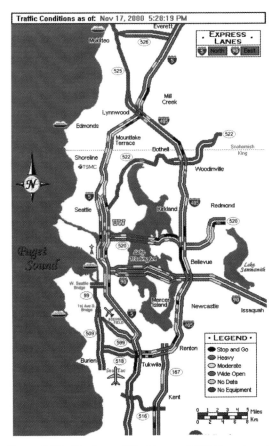

Figure 2.15 The State of Washington (USA) Department of Transportation maintains a Web site with minute-to-minute maps of traffic flow on major highways (www.wsdot.wa.gov) (Source: The Puget Sound Traffic Conditions: North-up System View (sysvert.gif) as published at www.wsdot.wa.gov on Nov. 17 2000 at 5.28.19 pm Pacific Standard Time)

The magnitude and location of a disaster can rarely be anticipated.

To illustrate the value of GIS in evacuation planning, we have chosen the work of Tom Cova, the author of a chapter on GIS in emergency management in the companion volume (see Cova 1999). This work was strongly motivated by the problems that occurred in the Oakland Hills fire of October 1991 in Northern California, USA, which destroyed approximately 1580 acres and over 2700 structures in the East Bay Hills (Figure 2.17). This became the most expensive fire disaster in Californian history, taking 25 lives and causing over $1.68 billion in damages.

Cova has developed a planning tool that allows neighborhoods to rate the potential for problems associated with evacuation, and to develop plans accordingly. The tool uses a GIS database containing information on the distribution of population in the neighborhood, and the street pattern. The result is an evacuation vulnerability map. Because the magnitude of a disaster cannot be known in advance, the method works by identifying the worst-case scenario that could affect a given location.

Suppose a specific household is threatened by an event that requires evacuation, such as a wildfire, and assume for the moment that one vehicle is needed to evacuate each household. If the house is in a cul-de-sac, the number of vehicles needing to exit the cul-de-sac will be equal to the number of households on the street. If the entire neighborhood of streets has only one exit, all vehicles carrying people from the neighborhood will need to use that one exit. Cova's method works by looking further and further from the household location, to find the most important bottleneck – the one that has to handle the largest amount of traffic. In an area with a dense network of streets traffic will disperse among several exits, reducing the bottleneck effect. But a densely packed neighborhood with only a single exit can be the source of massive evacuation problems, if a disaster requires the rapid evacuation of the entire neighborhood. In the Oakland Hills fire there were several critical bottlenecks – one-lane streets that normally carry traffic in both directions, but became hopelessly clogged in the emergency.

Figure 2.18 shows a map of Santa Barbara, California, USA, with streets colored according to Cova's measure of evacuation vulnerability. The color assigned to any location indicates the number of vehicles that would have to pass through the critical bottleneck in the worst-case evacuation, with red indicating that over 500

chemical resulting from a train derailment. It might make sense to plan for the worst case, for example the spillage of the entire contents of several cars loaded with chlorine gas. But the derailment might occur anywhere on the rail network, and the impact will depend on the strength and direction of the wind. Possible scenarios might involve people living within tens of kilometers of any point on the track network (see Section 13.4.1 for details of the buffer operation, which would be used in such cases to determine areas lying within a specified distance). Locations can be anticipated for some disasters, such as those resulting from fire in buildings known to be storing toxic chemicals, but hurricanes and earthquakes can impact almost anywhere within large areas.

CREDIT: NOAA

Figure 2.16 Hurricane Floyd approaching the coast of Florida, USA, September 14, 1999 (Source: NOAA)

vehicles per lane would have to pass through the bottleneck. The red area near the shore in the lower left is a densely packed area of student housing, with very few routes out of the neighborhood. An evacuation of the entire neighbourhood would produce a very heavy flow of vehicles on these exit routes. The red area in the upper left has a much lower population density, but has only one narrow exit.

2.3.4.3 Method

Two types of data are required for the analysis. Census data are used to determine population

Figure 2.17 The Oakland Hills fire of October, 1991, which took 25 lives, in part because of the difficulty of evacuation (Source: Department of City and Regional Planning, University of California)

and household counts, and to estimate the number of vehicles involved in an evacuation. Census data are available as aggregate counts for areas of a few city blocks, but not for individual houses, so there will be some uncertainty regarding the exact numbers of vehicles needing to leave a specific street, though estimates for entire neighborhoods will be much more accurate. The locations of streets are obtained from so-called *street centerline* files, which give the geographic

locations, names, and other details of individual streets (see Section 10.4 for an overview of geographic data sources). The TIGER (Topologically Integrated Geographic Encoding, and Referencing) files, produced by the US Bureau of the Census and the US Geological Survey, and readily available from many sites on the Internet, are one free source of such data for the USA. Many private companies also offer such data, many adding new information such as traffic flow volumes or directions (for US sources, see, for example, GDT Inc, Lebanon, New Hampshire, www.gdt1.com; and Navigation Technologies, Rosemont, Illinois, www.navtech.com).

Street centerline files are essential for many applications in transportation and logistics.

The analysis proceeds by beginning at every street intersection, and working outwards following the street connections to reach new intersections. Every connection is tested to see if it presents a bottleneck, by dividing the total number of vehicles that would have to move out of the neighborhood by the number of exit lanes. After all streets have been searched out to a specified distance from the start, the worst-case value (vehicles per lane) is assigned to the starting intersection. Finally, the entire network is colored by the worst-case value.

people / lane
- 0 - 200
- 201 - 300
- 301 - 400
- 401 - 500
- 500 <

Figure 2.18 Evacuation vulnerability map of the area of Santa Barbara, California, USA. Colors denote the difficulty of evacuating an area based on the area's worst-case scenario (Courtesy: Tom Cova)

2.3.4.4 Scientific foundations, geographic principles, and techniques

Scientific foundations

Cova's example is one of many applications that have been found for GIS in the general areas of logistics and transportation. As a planning tool, it provides a way of rating areas against a highly uncertain form of risk, a major evacuation. Although the worst-case scenario that might affect an area may never occur, the tool nevertheless provides very useful information to planners who design neighborhoods, giving them graphic evidence of the problems that can be caused by lack of foresight in street layout. Ironically, the approach points to a major problem with the modern style of street layout in subdivisions, which limits the number of entrances to subdivisions from major streets in the interests of creating a sense of community, and of limiting high-speed through traffic. Cova's analysis shows that such limited entrances can also be bottlenecks in major evacuations.

The analysis demonstrates the value of readily available sources of geographic data, since both major inputs – demographics and street layout – are available in digital form. At the same time we should note the limitations of using such sources. Census data are aggregated to areas that, while small, nevertheless provide only aggregated counts of population. The street layouts of TIGER, and other sources can be out of date and inaccurate, particularly in new developments, although users willing to pay higher prices can often obtain current data from the private sector. And the essentially geometric approach cannot deal with many social issues: evacuation of the disabled and elderly, and issues of culture and language that may impede evacuation.

Geographic principles

Central to Cova's analysis is the concept of *connectivity*. Very little would change in the analysis if the input maps were stretched or distorted, because what matters is how the network of streets is connected to the rest of the world. Connectivity is an instance of a *topological* property, a property that remains constant when the spatial framework is stretched or distorted. Other examples of topological properties are *adjacency* and *intersection*, both of which cannot be destroyed by stretching a map. We discuss the importance of topological properties and their representation in GIS in Section 9.2.3.2.

The analysis also relies on being able to find the *shortest path* from one point to another through a street network, and it assumes that people will follow such paths when they evacuate. Many forms of GIS analysis rely on being able to find shortest paths, and we discuss some of them in Section 14.3. Many Web sites will find shortest paths between two street addresses (Figure 2.16). In practice, people will often not use the shortest path, preferring routes that may be quicker but longer, or routes that are more scenic.

Techniques

The techniques used in this example are widely available in GIS. They include *spatial interpolation* techniques, which are needed to assign worst-case values to the streets, since the analysis only produces values for the intersections. Spatial interpolation is widely used in GIS to use information obtained at a limited number of sample points to guess values at other points, and is discussed in general in Section 5.5, and in detail in Section 13.4.4.

The shortest path methods used to route traffic are also widely available in GIS, along with other functions needed to create, manage, and visualize information about networks.

2.3.4.5 Generic scientific questions arising from the application

Logistic and transportation applications of GIS rely heavily on representations of networks, and often must ignore off-network movement. Drivers who cut through parking lots, children who cross fields on their way to school, houses in developments that are not aligned along linear streets, and pedestrians in underground shopping malls all confound the network-based analysis that GIS makes possible. Humans are endlessly adaptable, and their behavior will often confound the simplifying assumptions that are inherent to a GIS model. For example, suppose a system is developed to warn drivers of congestion on freeways, and to recommend alternative routes on neighborhood streets. While many drivers might follow such recommendations, others will reason that the result could be severe congestion on neighborhood streets, and reduced congestion on the freeway, and ignore the recommendation. Residents of the neighborhood streets might also be tempted to try to block the use of such systems, arguing that they result in unwanted, and inappropriate traffic, and risk to themselves. Arguments such as these are based on the notion that the transportation system can only be addressed as a whole, and that local modifications based on limited perspectives, such as the addition of a new freeway or bypass, may create more problems than they solve.

2.3.4.6 Practice

GIS is used in all three modes – operational, tactical, and strategic – in logistics and transportation. This section ends with some examples in all three categories.

In *operational* systems, GIS is used:

● To monitor the movement of mass transit vehicles, in order to improve performance and to provide improved information to system users (Figure 2.19);

● To route and schedule delivery and service vehicles on a daily basis to improve efficiency and reduce costs.

In *tactical* systems:

● To design and evaluate routes and schedules for public bus systems, school bus systems, garbage collection and mail collection, and delivery;

● To monitor and inventory the condition of highway pavement, railroad track, and highway signage, and to analyze traffic accidents.

In strategic systems:

● To plan locations for new highways and pipelines, and associated facilities;

● To select locations for warehouses, intermodal transfer points, and airline hubs.

2.3.5 Environment

2.3.5.1 Applications overview

Although it is the last area to be discussed here, the environment drove some of the earliest applications of GIS, and was a strong motivating force in the development of the very first GIS in the mid-1960s (Section 1.4.1). Environmental applications are the subject of several GIS texts, so only a brief overview will be given here for the purposes of

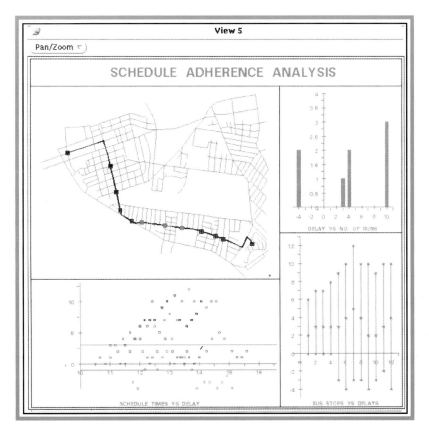

Figure 2.19 Analysis by C S Papacostas, Department of Civil Engineering, University of Hawaii of the adherence of buses to schedules in part of the Honolulu system (Source: © 1995 Institute of Transportation Engineers. Used by permission)

illustration (Bonham-Carter 1994, Goodchild *et al* 1993, Haines-Young *et al* 1993, Johnston 1998).

The development of the Canada Geographic Information System in the 1960s was driven by the need for policies over the use of land. Every country's land base is strictly limited (although the Dutch have managed to expand theirs very substantially by damming and draining), and alternative uses must compete for space. Measures of area are critical to effective strategy – for example, how much land is being lost to agriculture through urban development, and sprawl, and how will this impact upon the ability of future generations to feed themselves? Today, we have very effective ways of monitoring land use change through remote sensing from space, and are able to get frequent updates on the loss of tropical forest in the Amazon basin (see Figure 14.12). GIS is also allowing us to assemble historic information on urban sprawl in the USA (Figure 2.20).

> Competition between alternative uses of land has driven many applications of GIS.

Recently researchers have been using GIS to build dynamic models of urban sprawl, in order to make predictions into the future. Such models are based on historic patterns of growth, together with information on the locations of roads, steeply sloping land unsuitable for development, land that is otherwise protected from urban use, and other factors that influence urban development. Each of these factors is represented in map form, and becomes a layer in the GIS, while the software is designed to simulate the processes that drive growth. These urban growth models are examples of *dynamic simulation models*, or computer programs designed to simulate the operation of some part of the social or environmental system. Many models have been coupled with GIS in the past decade, to simulate such processes as soil erosion, forest growth, groundwater movement, and runoff (Box 2.2).

2.3.5.2 Case study application: gap analysis

Urban sprawl is just one of the factors contributing to a continual loss of natural habitat around the world – some would argue that the rate of loss has been increasing in recent years. Although numerous arrangements have been made to place various areas under protection, in National Parks, Conservation Areas, and other arrangements of varying degrees of land use control, the Earth is losing habitat for wildlife at an alarming rate. Moreover, the locations of protected areas tend to be dictated by factors that have little to do with the need for protection. In North America, for example, there are extensive protected areas in the west but far fewer in the east, and while wildlife is protected on massive tracts of land in Alaska, few comparable areas exist to protect the species of Florida. Figure 2.23 compares the actual

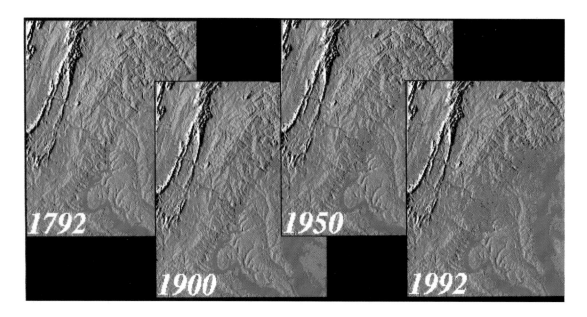

Figure 2.20 Urban sprawl in the Baltimore-Washington area, USA, since the colonial period (Source: USGS)

Box 2.2   Modeling nitrogen runoff in the Rhine basin

PCRaster is a GIS developed at Utrecht University specifically for dynamic simulation modeling (see www.geog.uu.nl/pcraster/). In this application, by Marcel de Wit, PCRaster is used to simulate nitrogen levels in the Rhine under various scenarios (De Wit 1999). Current levels exceed target levels in many parts of the Rhine, so these scenarios investigate various strategies for reducing levels. Data on climate, soils, slopes, hydrology, and other factors were collected for the entire Rhine basin, and used as input to the dynamic simulation model. Also input were sources of nitrogen, including both point sources (typically wastewater treatment plants and manufacturing plants) and diffuse sources (typically agriculture). Figure 2.21 shows the input map of diffuse nitrogen sources for the Rhine basin, and Figure 2.22 shows the result of one run of the model, with colors indicating the contribution of diffuse sources to the total nitrogen load in the river Rhine.

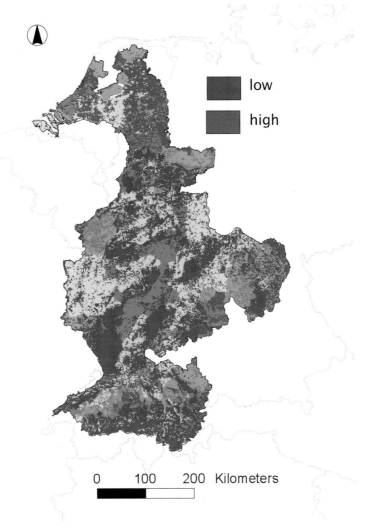

Figure 2.21   Distribution of diffuse nitrogen sources in the Rhine basin. Colors indicate the nitrogen surplus at the soil surface (yellow, orange, and light green denote intermediate values). This surplus can potentially runoff into the river network (Source: Marcel de Wit)

Box 2.2
Continued

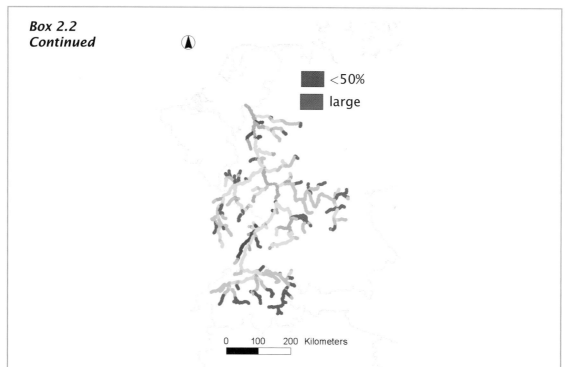

<50%

large

0   100   200   Kilometers

Figure 2.22 The contribution of diffuse sources to the total nitrogen load in the Rhine and its tributaries; colors as for Figure 2.21 (Source: Marcel de Wit)

distribution of the Gila monster lizard in Utah with the locations of protected areas.

> Gap analysis addresses the frequent geographic mismatches, or gaps, between the ranges of endangered species and the areas set aside for their protection.

Gap analysis attempts to draw attention to this problem, and to provide support for additional protection in areas where species are threatened. Since its inception, the program has grown to become a major effort involving academics, government, and the general public, in a scientifically based procedure for identifying suitable areas for protection. Today, the National Gap Analysis Program is administered by the Biological Resources Division at the United States Geological Survey (see www.gap.uidaho.edu/). Figure 2.24 shows the basic gap process. Imagery from the Landsat Thematic Mapper (TM) satellite, which has a spatial resolution of about 30 m, is combined with data from a digital elevation model to produce a map of vegetation cover for each state, using a standard classification system. Figure 2.25 shows the gap map for California, constructed by the Biogeography Lab at the

University of California, Santa Barbara. Data on the geographic ranges of various species is compiled into a series of species range maps, using information from field biologists, museum collections, and other sources. These range maps are compared with the vegetation cover map, to derive a series of models that identify the specific vegetation cover types favored by each species. For example, the Northern Spotted Owl is known to favor forests in the Pacific Northwest that have not been disturbed by logging.

Each individual species map is next overlaid, to obtain a map of species richness, in order to identify areas that harbor particularly large numbers of species (wetlands are often associated with high species richness). Species that are known to be of particular concern are identified. A map of protection status is constructed, to identify all protected areas, and the various laws, and regulations that apply to each area. Finally, a plan is developed for additional protection, in areas where the lack of protection appears to be leading to endangerment, and extinction. Methods have been developed for automating this final step in a GIS optimum search process (see Section 14.3).

■ Predicted range of the Gila monster

::: Protected areas

Figure 2.23 Comparison of the actual range and extent of protection for the Gila monster lizard in southern Utah, USA (Source: USGS, Dept of the Interior)

2.3.5.3 Method

Gap analysis requires many different kinds of input, and a substantial amount of human interaction in the various stages. The classification of imagery requires humans to identify training areas on the ground where vegetation is known to belong to a certain class. The imagery for each training area is then analyzed, to identify the defining characteristics of the class in the imagery, and these characteristics are then used to find other areas of the same class in the imagery. Humans are also needed to build species range maps, and to develop the predictive models that associate species with particular vegetation cover types. Finally, humans use the GIS to develop conservation plans. All of this would be impossible to do by hand, and gap analysis is an excellent example of humans working in collaboration with computers, dividing the work so that the GIS can do what it does best, and the human can perform those tasks that require human intelligence, and experience.

At the core of gap analysis are the operations of overlay, when various layers of information are combined, and the measurement of area. These techniques are discussed in detail in Sections 13.4.3, and 13.3 respectively.

2.3.5.4 Scientific foundations, geographic principles, and techniques

Scientific foundations

Gap analysis is founded on the notion that the presence and abundance of a species in an area can be predicted from knowledge of the area's

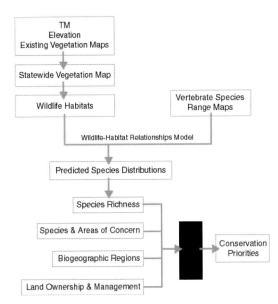

Figure 2.24 A flowchart of the gap analysis process (Source: USGS, Dept of the Interior)

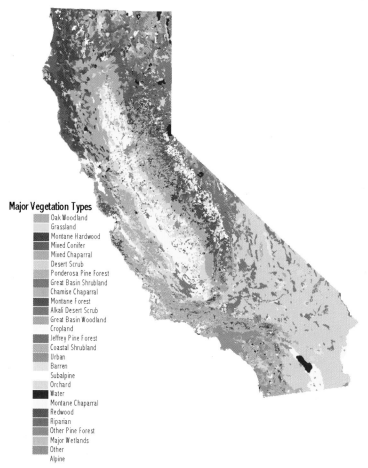

Major Vegetation Types

- Oak Woodland
- Grassland
- Montane Hardwood
- Mixed Conifer
- Mixed Chaparral
- Desert Scrub
- Ponderosa Pine Forest
- Great Basin Shrubland
- Chamise Chaparral
- Montane Forest
- Alkali Desert Scrub
- Great Basin Woodland
- Cropland
- Jeffrey Pine Forest
- Coastal Shrubland
- Urban
- Barren
- Subalpine
- Orchard
- Water
- Montane Chaparral
- Redwood
- Riparian
- Other Pine Forest
- Major Wetlands
- Other
- Alpine

Figure 2.25 Land cover and vegetation map produced by the California Gap Project (Source: California Department of Fish and Game)

vegetation. That knowledge in turn comes from images from space, which (radar images aside) must be obtained at times when the ground is not obscured by cloud. The season when the image was obtained is crucial, because many vegetation types change their appearance dramatically from one season to another.

The knowledge of vegetation used in gap analysis is of a particular type, identified in Box 3.3 as *nominal* data, meaning that variation is described through changes in *class*. By implication, every location assigned to a given class appears the same to the species that range there – and sharp changes in species should be observed at class boundaries. The problems associated with this approach to mapping are discussed at length in Chapter 6. Gap analysis takes a fairly coarse approach to mapping, typically ignoring areas of a specific class if they are of less than 100 hectares, and this will have an effect on the

accuracy of the model's predictions, particularly for species with very small ranges.

Geographic principles

Gap analysis takes a very *geographic* approach to the problem of species protection, and its fundamental driving force is the geographic mismatch between areas that need protection, and areas that supply it. It makes use of the core GIS idea that the world can be understood as a series of layers of different types of information, and overlays them to arrive at its conclusions. This idea, and alternatives to it, are discussed at length in Chapter 3.

One of the most important concepts in the final stage of gap analysis, the design, and development of conservation plans, is the *corridor*, a link between ranges that allows individuals of a species to mix, and breed. In California, for example, the Sierra, and Coastal Ranges are both good habitat for black bear and mountain lion, but these two areas are separated by the Central Valley, an area of intensive agriculture and human settlement that is generally inhospitable to both species. If the barrier is impenetrable, reproduction can occur only within the two subpopulations. If a corridor can be established, perhaps along less-settled river valleys and ridges, making movement between the two areas possible, then the two subpopulations can interbreed. A larger gene pool is more likely to lead to long-term success of the species.

Techniques

The techniques needed for gap analysis are widely available in GIS. This application requires extensive interaction between GIS and user, using techniques such as on-screen digitizing, which allow users to work directly with digital maps displayed on the screen (Section 10.3.2.2), and trained classification of imagery.

2.3.5.5 Generic scientific questions arising from the application

Many environmental applications of GIS are based on sound science. The gap analysis example requires knowledge of the ranges of species, which is typically obtained by biologists working in the field, making detailed observations over long periods of time. The dynamic simulation models used to predict groundwater movement are based on sound theory, such as the flow equations first formulated by Henry Darcy. Uncertainty always creeps into any prediction, however, for a number of reasons. Data are never perfect, being subject to measurement error (Chapter 15), and uncertainty due to the need to generalize, abstract, and approximate (Chapter 7). Equations such as Darcy's predict only under ideal circumstances, and such circumstances are rarely approximated in the real world. So predictions from any environmental analysis are always subject to some degree of uncertainty. GIS users need to be aware of this, because the GIS itself will always appear to be precise, providing abundant significant digits with all its numerical outputs, and crisp lines, and clear colors in its maps.

Besides uncertainty, users of GIS also need to be wary of the tendency to think of the system as a *black box*, containing a model or analysis procedure developed by someone else, perhaps a private company, and not completely documented. GIS documentation does not always specify precisely what the GIS does to data, and it can sometimes be impossible to find out from the

software developer. This impenetrability is sharply in conflict with an old principle of scientific reporting, that results of an analysis *should always be reported in sufficient detail to allow someone else to replicate them*. On the other hand science is complex these days, and all of us from time to time must use tools developed by others that we do not fully understand. It is up to all of us to demand to know as many of the details of GIS analysis as is reasonably possible.

> Users of GIS should always strive to know exactly what the system is doing to their data.

2.3.5.6 Practice

GIS is now widely used in all areas of environmental science, from ecology to geology, and from oceanography to alpine geomorphology. It is used to capture and store data in the field, for analysis on a battery-powered laptop in a field tent, and for uploading via a satellite link to a home institution. It is also used in the office, working with data collected by satellites and other field scientists, and using complex simulation models that allow us to evaluate future scenarios. Some applications amount to little more than the use of a GIS as a mapping tool, to assemble and present data. But even here the design of the map, and the selection of data, raise issues that can be quite complex and sophisticated. At the other end of the spectrum, GIS is used as a platform for intensive numerical and statistical analysis.

2.4 Concluding Comments

This chapter has presented a selection of GIS application areas, and specific instances within each of the selected areas. The emphasis throughout has been on the range of contexts, from day-to-day problem-solving to curiosity-driven science. The principles of the scientific method have been stressed throughout – the need to maintain an enquiring mind, constantly asking questions about what is going on, and what it means; the need to use terms that are well-defined, and understood by others, so that knowledge can be communicated; the need to describe procedures in sufficient detail that they can be replicated by others; and the need for accuracy, in observations, measurements, and predictions. These principles are valid whether the context is a simple inventory of the assets of a utility company, or the simulation of complex biological systems.

Questions for Further Study

1. Look at one of the applications areas in the Reference Links. To what extent do you believe that the author of your chapter has demonstrated that GIS has been "successful" in application? Suggest some of the implicit and explicit assumptions that are made in order to achieve a "successful" outcome.
2. Look at one of the applications areas in the Reference Links, and re-examine the list of critiques of GIS at the end of Section 1.7. To what extent do you think that the critiques are relevant to the applications that you have studied?
3. Compare and contrast the views of Richard Webber (Box 2.1) and **either** David Mark (Box 1.7) **or** Brian J L Berry (Box 1.8) on the interrelationship between geography and GIS.

Online Resources

NCGIA Core Curricula (www.ncgia.ucsb.edu/pubs/core.html):

Core Curriculum in GIScience, Section 4.1 (Land Information Systems and Cadastral Applications, Steven J Ventura) and 4.2 (links to material from PrecisionAg.org)

Core Curriculum in GIS, 1990, Units 51 (GIS Application Areas), 52 (Resource Management Applications), 53 (Urban Planning and Management Applications), 54 (Cadastral Records and LIS), 55 (Facilities Management), and 56 (Commercial Applications)

ESRI Virtual Campus courses (campus.esri.com):

GIS Applications for Tax Assessors, by Feng Yang

Introduction to Urban and Regional Planning using ArcView GIS, by Chris Pettit and David Pullar

Characterizing Forests using ArcView GIS, by Glen Jordan

Spatial Hydrology using ArcView GIS, by David R Maidment

Spatial Analysis in Agriculture: A GIS Approach, by Terry Brase

Introduction to Successful Marketing using ArcView GIS, by Christine M Koontz and Dean K Jue

Mapping for Health Care Professionals using ArvView GIS, by Zvia Segal Naphtali

Integrating Marine Science GIS into a K-12 Classroom, by Genevieve F Healy

Conservation GIS using ArcView, by Conservation GIS Consortium

Case studies on the use of GIS: www.agi.org.uk/pag-es/case-stu/case-stu.htm

Reference Links

Maguire D J, Goodchild M F, and Rhind D W (eds) 1991 *Geographical Information Systems: Principles, and applications*. Harlow, UK: Longman. See the selected chapters from Section III Applications (Chapter 35 onwards) (Text available online from 'Links to Big Book 1' at www.wiley.com/gis and www.wiley.co.uk/gis).

Longley P A, Goodchild M F, Maguire D J, and Rhind D W (eds) 1999 *Geographical Information Systems: Principles, techniques, management, and applications*. New York: John Wiley. Part 4: Applications (Chapters 57–71). Including: Chapter 51, GIS for business and service planning (Birkin M, Clarke, G P, Clarke M) Chapter 60, GIS in emergency management (Cova T)

References

Bonham-Carter G F 1994 *Geographic Information Systems for Geoscientists: Modeling with GIS*. New York: Pergamon.

Bromley R and Coulson M 1989 The value of corporate GIS to local authorities: evidence of a needs study in Swansea City Council. *Mapping Awareness* 3(5): 32–5.

De Wit M J M 1999 Nutrient fluxes in the Rhine and Elbe basins. *Nederlandse Geografische Studies* 259, Utrecht, The Netherlands: Koninklijk Nederlands Aardrijkskundig Genootschap (Utrecht University).

Goodchild M F, Parks B O, and Steyaert L T 1993 *Environmental Modeling with GIS*. New York: Oxford University Press.

Greene R W 2000 *GIS in Public Policy*. Redlands, California: ESRI Press.

Haines-Young R, Green D R, and Cousins S 1993 *Landscape Ecology, and Geographic Information Systems*. London: Taylor and Francis.

Johnston C A 1998 *Geographic Information Systems in Ecology*. Oxford: Blackwell.

Martin D J 1996 *Geographic Information Systems: Socioeconomic applications* (2nd edn). London: Routledge.

Martin D J and Longley P A 1995 Data sources, and their geographical integration. In P A Longley and G P Clarke (eds) *GIS for Business, and Service Planning*. New York: John Wiley 15–32.

O'Looney J 2000 *Beyond Maps: GIS and decision making in local government*. Redlands, California: ESRI Press.

Rogers E M 1995 *Diffusion of Innovations* (4th edn). New York: Free Press.

REPRESENTING GEOGRAPHY

3

This chapter introduces the concept of representation, or the construction of a digital model of some aspect of the Earth's surface. Representations have many uses, because they allow us to learn, think, and reason about places and times that are outside our immediate experience. This is the basis of scientific research, planning, and many forms of day-to-day problem-solving.

The geographic world is extremely complex, revealing more detail the closer one looks, almost ad infinitum. So in order to build a representation of any part of it, it is necessary to make choices about what to represent, at what level of detail, over what time period, etc. The large number of possible choices creates many opportunities for designers of GIS software.

Learning objectives

After reading this chapter you will know:

- The importance of understanding representation in GIS;

- The concepts of fields and objects and their fundamental significance;

- Raster and vector representation and how they affect many GIS principles, techniques, and applications;

- The paper map and its position as a GIS product;

- The art and science of representing real-world phenomena in GIS.

3.1 Introduction

We live on the surface of the Earth, and spend most of our lives in a relatively small fraction of that space. Of the approximately 500 million square kilometers of surface, only one third is land, and only a fraction of that is occupied by the cities and towns in which most of us live. The rest of the Earth, including the parts we never visit, the atmosphere, and the solid under our feet, remains unknown to us except through the information that is communicated to us through books, newspapers, television, the Web, or the spoken word. We live lives that are almost infinitesimal in comparison with the 4.5 billion years of Earth history, or the over 10 billion years since the universe began, and know about the Earth before we were born only through the evidence compiled by geologists, archaeologists, historians, etc. Similarly, we know nothing about the world that is to come, where we have only predictions to guide us.

Because we can observe so little of the Earth directly, we rely on a host of methods for learning about its other parts, for deciding where to go as tourists or shoppers, choosing where to live, running the operations of corporations, agencies, and governments, and many other activities. Almost all human activities at some time require knowledge about parts of the Earth that are outside our direct experience, because they occur either elsewhere in space, or elsewhere in time.

All human activities require knowledge about the Earth – past, present, or future.

One way to visualize the space occupied by a human through time is as a three-dimensional diagram, in which the two horizontal axes denote location on the Earth's surface, and the vertical axis denotes time. In Figure 3.1, the lives of three people are shown as they move through space and time, from one residential location to another. Note how large areas of the diagram are unvisited by any of the three people. The diagram is crude, and if we examined each track or trajectory in more detail we would see the movements that take people on annual vacations and business trips. If we looked even closer we would see daily movements, as people go to work, or to shop. Each closer perspective would display more information, and a vast storehouse would be required to capture the precise trajectories of all humans throughout their lifetimes.

The real trajectories of the three individuals shown in Figure 3.1 are complex, and the figure is only a representation of them – a model on a piece of paper, generated by a computer from a database. We use the terms representation and model because they imply a simple relationship between the contents of the figure and the database, and the real-world trajectories of the three individuals. Such representations or models serve many useful purposes, and occur in many

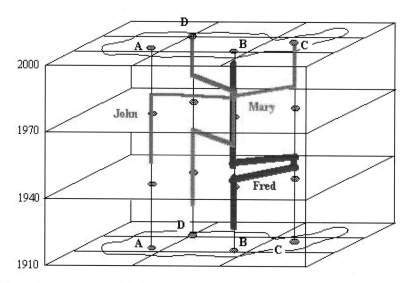

Figure 3.1 Schematic representation of the lives of three US citizens in space (two horizontal axes) and time (vertical axis). Fred, John, and Mary were born in three different places (B, A, and D respectively) but lived for a time in the same place (B). Fred has lived in B all his life, except for a few years in C during World War II. Note how the diagram simplifies what is actually a very complex pattern (e.g. journeys to work), much of which is not visible at this level of generalization

different forms. For example, representations occur:

- in the human mind, when our senses capture information about our surroundings (such as the images captured by the eye, or the sounds captured by the ear), and memory preserves such representations for future use;

- in photographs, which are two-dimensional models of the light emitted or reflected by objects in the world into the lens of a camera;

- in spoken descriptions and written text, in which people describe some aspect of the world in language, in the form of travel accounts or diaries; or

- in the numbers that result when aspects of the world are measured, using such devices as thermometers, rulers, or speedometers.

By building representations, we humans can assemble far more knowledge about our planet than we ever could as individuals. We can build representations that serve such purposes as planning, resource management and conservation, travel, or the day-to-day operations of a parcel delivery service.

> Representations help us assemble far more knowledge about the Earth than is possible on our own.

Representations are reinforced by the rules and laws that we humans have learned to apply to the unobserved world around us. When we encounter a fallen log in a forest we are willing to assert that it once stood upright, and once grew from a small shoot, even though no-one actually observed or reported either of these stages. We predict the future occurrence of eclipses based on the laws we have discovered about the motions of the Solar System. In GIS applications, we often rely on methods of spatial interpolation to guess the conditions that exist in places where no observations were made, based on the rule (often elevated to the status of a First Law of Geography and attributed to Waldo Tobler, see Box 5.1) that all places are similar, but nearby places are more similar than distant places.

> Tobler's First Law of Geography: Everything is related to everything else, but near things are more related than those far apart.

3.2 Digital Representation

This book is about one particular form of representation that is becoming increasingly

important in our society – representation in digital form. Today, almost all communication between people through such media as the telephone, FAX, music, television, newspapers and magazines, or e-mail is at some time in its life in digital form, and information technology based on digital representation is moving into all aspects of our lives, from science to commerce to daily existence. Almost half of all households in some industrial societies now own at least one powerful digital information processing device (a computer); a large proportion of all work in offices now occurs using digital computing technology; and digital technology has invaded many devices that we use every day, from the microwave oven to the automobile.

One interesting characteristic of digital technology is that the representation itself is rarely if ever seen by the user, because only a few technical experts ever see the individual elements of a digital representation. What we see instead are *views*, designed to present the contents of the representation in a form that is meaningful to us.

The term *digital* derives from *digits*, or the fingers, and our system of counting based on the 10 digits of the human hand. But while the counting system has 10 symbols (0 through 9), the representation system in digital computers uses only two (0 and 1). In a sense, then, the term *digital* is a misnomer for a system that represents all information using some combination of the two symbols 0 and 1, and the more exact term *binary* is more appropriate. In this book we follow the convention of using *digital* to refer to technology based on binary representations.

> Computers represent phenomena as binary digits. Every item of useful information about the Earth's surface is ultimately reduced by a GIS to some combination of 0s and 1s.

Over the years many standards have been developed for converting information into digital form. Box 3.1 shows the system used for representing whole numbers, or integers. Text is most often converted using the ASCII coding system, which assigns seven or eight binary digits (*bits*) to each alphabetic character, distinguishing between upper and lower case, and has additional codes for standard typographic symbols. There are many competing coding standards for images and photographs (GIF, JPEG, TIFF, etc.) and for movies (e.g. MPEG) and sound (e.g. MIDI, MP3). Much of this book is about the coding systems used to represent geographic data, especially Chapter 9 and, as you might guess, that turns out to be comparatively complicated.

Box 3.1   The binary counting system

The binary counting system uses only two symbols, 0 and 1, to represent numerical information. There are only two options for a single digit, but there are four possible combinations for two digits (00, 01, 10, and 11), and eight possible combinations for three digits (000, 001, 010, 011, 100, 101, 110, 111). Digits in the binary system (known as binary digits, or *bits*) behave like digits in the decimal system but using powers of two. The rightmost digit denotes units, the next digit to the left denotes twos, the next to the left denotes fours, etc. For example, the binary number 11001 denotes one unit, no twos, no fours, one eight, and one sixteen, and is equivalent to 25 in the normal (decimal) counting system. We call this the *integer* digital representation of 25, because it represents 25 as a whole number, and is readily amenable to arithmetic operations.

The seven-bit ASCII system assigns codes to each symbol of text, including letters, numbers, and common symbols. The number 2 is assigned ASCII code 50 (0110010 in binary), and the number 5 is 53 (0110101), so if 25 were coded as two characters using seven-bit ASCII its digital representation would be 14 bits long (01100100110101). The three characters 2 = 2 would be coded as 50, 61, 50 (011001001111010110010). ASCII is used for coding text, but if numbers are to be processed using arithmetic then systems such as the integer coding above are used instead.

Digital technology is successful for many reasons, not the least of which is that all kinds of information share a common basic format (0s and 1s), and can be handled in ways that are largely independent of their actual meaning. The Internet, for example, operates on the basis of packets of information, consisting of strings of 0s and 1s, which are sent through the network based on the information contained in the packet's header. The network needs to know only what the header means, and how to read the instructions it contains regarding the packet's destination. The rest of the contents are no more than a collection of bits, representing anything from an e-mail message to a short burst of music or highly secret information on its way from one military installation to another, and are never normally examined or interpreted during transmission. This allows one digital communications network to serve every need, from electronic commerce to chatrooms, and it allows manufacturers to build processing and storage technology for vast numbers of users who have very different applications in mind. Compare this with earlier ways of communicating, which required printing presses and delivery trucks for one application (newspapers) and networks of copper wires for another (telephone).

Digital representations of geography hold enormous advantages over previous types – paper maps, written reports from explorers, or spoken accounts. We can use the same cheap digital devices – the components of PCs, the Internet, or mass storage devices – to handle every type of information, independent of its meaning. Digital data are easy to copy, they can be transmitted at close to the speed of light, they can be stored at high density in very small spaces, and they are less subject to the physical deterioration that affects paper and other physical media. Perhaps more importantly, data in digital form are easy to transform, process, and analyze. Geographic information systems allow us to do things with digital representations that we were never able to do with paper maps: to measure accurately and quickly, to overlay and combine, and to change scale, zoom, and pan without respect to map sheet boundaries. The vast array of possibilities for processing that digital representation opens up is reviewed in Chapters 13 and 14, and also covered in the applications that are distributed throughout the book.

Digital representation has many uses because of its simplicity and low cost.

3.3   Representation of What and for Whom?

Thus far we have seen how humans are able to build representations of the world around them, but we have not yet discussed why representations are useful, and why humans have become so ingenious at creating and sharing them. The emphasis here and throughout the book is on one type of representation, termed *geographic*, and defined as a representation of some part of the Earth's surface or near surface.

Geographic representation is concerned with the Earth's surface or near surface.

Geographic representations are among the most ancient, having their roots in the needs of very

Box 3.2 Prince Henry the Navigator

Prince Henry of Portugal, who died in 1460, was known as Henry the Navigator because of his keen interest in exploration. In 1433 Prince Henry sent a ship from Portugal to explore the west coast of Africa in an attempt to find a sea route to the Spice Islands. This ship was the first to travel south of Cape Bojador (latitude 26 degrees 20 minutes N). To make this and other voyages Prince Henry assembled a team of mapmakers, sea captains, geographers, ship builders, and many other skilled craftsmen. Prince Henry showed the way for Vasco de Gama and other famous 15th century explorers. His management skills could be applied in much the same way in today's GIS projects.

Figure 3.2 Prince Henry the Navigator, originator of the Age of Discovery in the 15th century, and promoter of the concept of geographic knowledge as the common property of humanity

early societies. The tasks of hunting and gathering can be much more efficient if hunters are able to communicate the details of their successes to other members of their band – the locations of edible roots or game, for example. Maps must have originated in the sketches early people made in the dirt of campgrounds or on cave walls, long before language became sufficiently sophisticated to convey equivalent information through speech. We know that the peoples of the Pacific built representations of the locations of islands out of simple materials to guide each other, and that very simple forms of representation are used by social insects such as bees to communicate the locations of food resources.

Hand-drawn maps and speech are effective media for communication between members of a small band, but communication at much larger scales became possible with the invention of the printing press in the 15th century. Now large numbers of copies of a representation could be made and distributed, and for the first time it became possible to imagine that something could be known by every human being – that knowledge could be the common property of humanity. Only one major restriction affected what could be distributed using this new mechanism: the representation had to be flat. If one were willing to accept that constraint, however, paper proved to be enormously effective; it was cheap, light and thus easily transported, and durable. Only fire proved to be disastrous for paper, and human history is replete with instances of the loss of vital information through fire, from the burning of

the Alexandria Library in the 7th century that destroyed much of the accumulated knowledge of classical times to the major conflagrations of London in 1666, San Francisco in 1906, or Tokyo in 1945.

One of the most important periods for geographic representation began in the early 15th century in Portugal. Henry the Navigator (Box 3.2) is often credited with originating the Age of Discovery, the period of European history that led to the accumulation of large amounts of information about other parts of the world through sea voyages and land explorations. Maps became the medium for sharing information about new discoveries, and for administering vast colonial empires, and their value was quickly recognized. Although detailed representations now exist of all parts of the world, including Antarctica, in a sense the spirit of the Age of Discovery continues in the explorations of the oceans, caves, and outer space.

It was the creation, dissemination, and sharing of accurate representations that distinguished the Age of Discovery from all previous periods in human history (and it would be unfair to ignore its distinctive negative consequences, notably the spread of European diseases and the growth of the slave trade). Information about other parts of the world was assembled in the form of maps and journals, reproduced in large numbers using the recently invented printing press, and distributed on paper. Even the modest costs associated with buying copies were eventually addressed through the development of free public lending libraries in the 19th century, which gave access to virtually everyone. Today, we benefit from what is now a longstanding tradition of free and open access to much of humanity's accumulated store of knowledge about the geographic world, in the form of paper-based representations, through the institution of libraries and the copyright doctrine that gives people rights to material for personal use (see Section 17.1 for a complete account of laws affecting ownership and access). The Internet has already become the delivery mechanism for providing distributed access to geographic information.

In the Age of Discovery maps became extremely valuable representations of the state of geographic knowledge.

It is not by accident that the list of important applications for geographic representations closely follows the list of applications of GIS (see Section 1.1 and Chapter 2), since representation is at the heart of our ability to solve problems using digital tools. Any application of GIS requires clear attention to questions of *what* should be represented, and *how*. There is a multitude of possible ways of representing the geographic world in digital form, none of which is perfect, and none of which is ideal for all applications.

The key GIS representation issues are what to represent and how to represent it.

One of the most important criteria for the usefulness of a representation is its *accuracy*. Because the geographic world is almost infinitely complex, there are always choices to be made in building any representation – what to include, and what to leave out. When US President Thomas Jefferson dispatched Meriwether Lewis to explore and report on the nature of the lands from the upper Missouri to the Pacific, he said Lewis possessed "a fidelity to the truth so scrupulous that whatever he should report would be as certain as if seen by ourselves". But he clearly didn't expect Lewis to report *everything* he saw in complete detail: Lewis exercised a large amount of judgment about what to report, and what to omit. The question of accuracy is taken up at length in two subsequent chapters: in Chapter 6, which deals with general aspects of the problem, and in Chapter 15, which addresses the technical and practical aspects.

Many plans for the real world can be tried out first on models or representations.

One more vital interest drives our need for representations of the geographic world, and also the need for representations in many other human activities. When a pilot must train to fly a new type of aircraft, it is much cheaper and less risky for him or her to work with a flight simulator than with the real aircraft. Flight simulators can represent a much wider range of conditions than a pilot will normally experience in flying. Similarly, when decisions have to be made about the geographic world, it is effective to experiment first on models or representations, exploring different scenarios. Of course this works only if the representation behaves as the real aircraft or world does, and a great deal of knowledge must be acquired about the world before an accurate representation can be built that permits such simulations. But the use of representations for training, exploring future scenarios, and recreating the past is now common in many fields, including surgery, chemistry, and engineering, and with technologies like GIS is becoming increasingly common in dealing with the geographic world.

3.4 The Fundamental Problem

Geographic data are built up from atomic elements, or facts about the geographic world. At its most primitive, an atom of geographic data (strictly, a datum) links a place, often a time, and some descriptive property. The first of these, place, is specified in one of several ways that are discussed at length in Chapter 4, and there are also many ways of specifying the second, time. We often use the term *attribute* to refer to the last of these three. For example, consider the statement "The temperature at local noon on December 2nd 1999 at latitude 34 degrees 45 minutes north, longitude 120 degrees 0 minutes west, was 18 degrees Celsius". It ties location and time to the property or attribute of atmospheric temperature.

Geographic data link place, time, and attributes.

Other facts can be broken down into their primitive atoms. For example, the statement "Mount Everest is 8848 m high" can be derived from two atomic geographic facts, one giving the location of Mt Everest in latitude and longitude, and the other giving the elevation of that latitude and longitude. Note, however, that the statement would not be a geographic fact to a community that had no way of knowing where Mt Everest is located.

Many aspects of the Earth's surface are comparatively static and slow to change. Height above sea level changes slowly because of erosion and movements of the Earth's crust, but these processes operate on scales of hundreds or thousands of years, and for most applications except geophysics we can safely omit time from the representation of elevation. On the other hand atmospheric temperature changes daily, and dramatic changes sometimes occur in minutes with the passage of a cold front or thunderstorm, so time is distinctly important, though such climatic variables as mean annual temperature can be represented as static.

The range of attributes in geographic information is vast. We have already seen that some vary slowly and some rapidly. Some attributes are physical or environmental in nature, while others are social or economic. Some attributes simply identify a place or an individual – examples include street addresses, social security numbers, or the parcel identifiers used for recording land ownership. Other attributes measure something at a location and perhaps at a time (e.g. atmospheric temperature or elevation), while others classify into categories (e.g. the class of land use, differentiating between agriculture, industry, or residential land). Because attributes are important outside the domain of GIS there are standard terms for the different types (see Box 3.3).

Geographic attributes are classified as nominal, ordinal, interval, ratio, and cyclic.

In principle if we collected enough atoms of geographic information we would be able to build a complete representation of the world. The idea of integrating all available geographic information into a single digital representation underlies the idea of *Digital Earth*, a concept that originated with (then) US Vice President Al Gore in his book *Earth in the Balance* (Gore 1992), and was explored further in one of his speeches (see Box 3.4). Although the information would presumably be distributed among many different archives, the user would see Digital Earth as a single entity, and would be able to explore the world by interacting with its digital representation. Digital Earth could be a powerful tool in education, allowing students to learn by exploring the digital representation, a point made forcefully by Gore in the excerpt shown in Box 3.4. In a sense it merely updates the longstanding tradition of geographic education, which has allowed students to explore the world through older representations, such as books and photographs, and through spoken communication from teachers, but it adds the very powerful notion of being able to experiment with the world by manipulating its digital model. In a sense, Digital Earth would make everything we know about the geographic world accessible through a single portal, and would allow us to use that knowledge to explore distant places and also distant times. It would be much more powerful than its predecessors, which we might term *Written Earth*, and *Spoken Earth*. Other aspects of Digital Earth are explored in Section 19.3.3.1.

But this scenario misses a fundamental problem, which is that the world is in effect infinitely complex, and the number of atoms required for a complete representation is similarly infinite. The closer we look at the world, the more detail it reveals – and it seems that this process extends *ad infinitum*. The shoreline of Maine appears complex on a map, but even more complex when examined in greater detail, and as more detail is revealed the shoreline appears to get longer and longer, and more and more convoluted. To characterize the world completely we would have to specify the location of every person, every blade of grass, and every grain of sand – in fact, every subatomic particle, clearly an impossible task. So in practice any representation must be partial – it must limit the level of detail provided, or ignore change through time, or ignore certain attributes, or simplify in some other way.

The world is infinitely complex, but computer systems are finite. Representation is all about the choices that are made in capturing knowledge about the world.

One very common way of limiting detail is by throwing away or ignoring information that applies only to small areas, in other words not looking too closely. The image you see on a computer screen is composed of a million or so basic elements or *pixels*, and if the whole Earth were displayed at once each pixel would cover an area roughly 10 km on a side, or about 100 sq km. At this level of detail the island of Manhattan occupies roughly 10 pixels, and virtually everything on it except Central Park is a blur. We would say that such an image has a *spatial resolution* of about 10 km, and know that anything much less than 10 km across is virtually invisible.

It is easy to see how this helps with the problem of too much information. The Earth's surface covers about 500 million sq km, so if this level of detail is sufficient for an application, a property of the surface such as elevation can be described with only 5 million pieces of information, instead of the 500 million it would take to describe elevation with a resolution of 1 km, and the 500 trillion (500 000 000 000 000) it would take to describe elevation with 1m resolution.

Another strategy for limiting detail is to observe that many properties remain constant over large areas. For example, in describing the elevation

Box 3.3 Types of attributes

The simplest type of attribute, termed *nominal*, is one that serves only to identify or distinguish one entity from another. Placenames are a good example, as are names of houses, or the numbers on a driver's license – each serves only to identify the particular instance of a class of entities from other members of the same class. Nominal attributes include numbers, letters, and even colors. Even though a nominal attribute can be numeric it makes no sense to apply arithmetic operations to it: adding two nominal attributes, such as two drivers' license numbers, creates nonsense.

Attributes are *ordinal* if their values have a natural order. For example, Canada rates its agricultural land by classes of soil quality, with Class 1 being the best, Class 2 not so good, etc. Adding or taking ratios of such numbers makes little sense, since 2 is not twice as much of anything as 1, but at least ordinal attributes have inherent order. Averaging makes no sense either, but the *median*, or the value such that half of the attributes are higher-ranked and half are lower-ranked, is an effective substitute for the average for ordinal data as it gives a useful central value.

Attributes are *interval* if the differences between values make sense. The scale of Celsius temperature is interval, because it makes sense to say that 30 and 20 are as different as 20 and 10. Attributes are *ratio* if the ratios between values makes sense. Weight is ratio, because it makes sense to say that a person of 100 kg is twice as heavy as a person of 50 kg; but Celsius temperature is only interval, because 20 is not twice as hot as 10 (and this argument applies to all scales that are based on similarly arbitrary zero points, including longitude).

In GIS it is sometimes necessary to deal with data that fall into categories beyond these four, and Chrisman (1997) has written about the special problems of classifying GIS data. For example, data can be directional or *cyclic*, including flow direction on a map, or compass direction, or longitude. The special problem here is that the number following 359 is 0. Averaging two directions such as 359 and 1 yields 180, so the average of two directions close to North can appear to be South. This is somewhat analogous to the famous Y2K bug, which originated because the next year after (19)99 was (20)00, not (19)00, a problem for early systems that did not record the first two digits of the year. There are excellent texts on dealing with directional data (e.g. Mardia and Jupp 2000). Because they occur sometimes in GIS, and few designers of GIS software have made special arrangements for them, it is important to be alert to the possibility. Another set of problems arise because latitude and longitude are often written in the form of degrees, minutes, and seconds (DMS), and computers are not normally able to deal with the fact that adding one minute to a latitude of 30 degrees, 59 minutes produces 31 degrees and no minutes, not 30 degrees and 60 minutes. The normal way of dealing with this problem in GIS is to express latitude and longitude in decimal degrees, not DMS, so 30 degrees 59 minutes would normally be stored as 30.98333 . . .

of the Earth's surface we could take advantage of the fact that roughly two-thirds of the surface is covered by water, with its surface at sea level. Of the 5 million pieces of information needed to describe elevation at 10 km resolution, approximately 3.4 million will be recorded as zero, a colossal waste. If we could find an efficient way of identifying the area covered by water, then we would need only 1.6 million real pieces of information.

Humans have found many ingenious ways of describing the Earth's surface efficiently, because the problem we are addressing is as old as representation itself, and as important for paper-based representations as it is for binary representations in computers. But this ingenuity is itself the source of a substantial problem for GIS: there are many ways of representing the Earth's surface, and users of GIS thus face difficult and at times confusing choices. This chapter discusses some of those choices, and the issues are pursued further in subsequent chapters on uncertainty (Chapter 6), generalization (Chapter 7), and data modeling (Chapter 9).

3.5 Discrete Objects and Fields

3.5.1 Discrete objects

Mention has already been made of level of detail as a fundamental choice in representation. Another, perhaps even more fundamental choice, is between two conceptual schemes. There is good evidence that we as humans like to simplify the world around us by naming things, and seeing individual things as instances of broader categories. We prefer a world of black and white,

Box 3.4 Al Gore and Digital Earth

Al Gore was Vice-President of the United States from 1993 to 2001. In a speech written for presentation at the opening of the California Science Museum, Los Angeles, January 1998 he discussed his vision for a Digital Earth:

"Imagine, for example, a young child going to a Digital Earth exhibit at a local museum. After donning a head-mounted display, she sees Earth as it appears from space. Using a data glove, she zooms in, using higher and higher levels of resolution, to see continents, then regions, countries, cities, and finally individual houses, trees, and other natural and man-made objects. Having found an area of the planet she is interested in exploring, she takes the equivalent of a 'magic carpet ride' through a 3D visualization of the terrain. Of course, terrain is only one of the numerous kinds of data with which she can interact. Using the system's voice recognition capabilities, she is able to request information on land cover, distribution of plant and animal species, real-time weather, roads, political boundaries, and population. She can also visualize the environmental information that she and other students all over the world have collected as part of the GLOBE project. This information can be seamlessly fused with the digital map or terrain data. She can get more information on many of the objects she sees by using her data glove to click on a hyperlink. To prepare for her family's vacation to Yellowstone National Park, for example, she plans the perfect hike to the geysers, bison, and bighorn sheep that she has just read about. In fact, she can follow the trail visually from start to finish before she ever leaves the museum in her hometown.

She is not limited to moving through space, but can also travel through time. After taking a virtual field-trip to Paris to visit the Louvre, she moves backward in time to learn about French history, perusing digitized maps overlaid on the surface of the Digital Earth, newsreel footage, oral history, newspapers, and other primary sources. She sends some of this information to her personal e-mail address to study later. The time-line, which stretches off in the distance, can be set for days, years, centuries, or even geological epochs, for those occasions when she wants to learn more about dinosaurs".

of good guys and bad guys, to the real world of shades of gray.

The two fundamental ways of representing geography are discrete objects and fields.

This preference is reflected in one way of viewing the geographic world, known as the *discrete object* view. In this view, the world is empty, except where it is occupied by objects with well-defined boundaries that are instances of generally recognized categories. Just as the desktop is littered with books, pencils, or computers, the geographic world is littered with cars, houses, fields, and other discrete objects. Thus the landscape of Minnesota is littered with lakes, and the landscape of Scotland is littered with mountains. One characteristic of the discrete object view is that objects can be counted, so license plates issued by the State of Minnesota carry the legend "10,000 lakes", and climbers know that there are exactly 284 mountains in Scotland over 3,000 ft (the so-called Munros, from Sir Hugh Munro who originally listed 277 of them in 1891 – the count was expanded to 284 in 1997).

The discrete object view represents the world as objects with well-defined boundaries in empty space.

Biological organisms fit this model well, and this allows us to count the number of residents in an area of a city, or to describe the behavior of individual Grizzly Bears. Manufactured objects also fit the model, and we have little difficulty counting the number of cars produced in a year, or the number of airplanes owned by an airline. But other phenomena are messier. It is not at all clear what constitutes a mountain, for example, and the problems of defining one have been parodied in a movie, "The Englishman Who Went up a Hill but Came Down a Mountain" (1995, directed by Christopher Monger), which deals with the nebulous difference between a mountain and a large hill.

Geographic objects are identified by their dimensionality. Objects that occupy area are termed two-dimensional, and generally referred to as areas. The term *polygon* is also common for technical reasons explained later. Other objects are more like one-dimensional lines, including roads, railways, or rivers, and are often represented as one-dimensional objects and generally referred to as lines. Other objects are more like zero-dimensional points, such as individual animals or buildings, and are referred to as points.

Of course in reality all objects that are perceptible to humans are three dimensional, and their representation in fewer dimensions can be at best an approximation. But the ability of GIS to handle truly three-dimensional objects as volumes with associated surfaces is very limited. Some GIS allow for a third (vertical) coordinate to be specified for all point locations. Buildings are sometimes represented by assigning height as an attribute, though if this option is used it is impossible to distinguish flat roofs from any other kind. Various strategies have been used for representing overpasses and underpasses in transportation networks, because this information is vital for navigation but not normally represented in strictly two-dimensional network representations. One common strategy is to represent turning options at every intersection – so an overpass appears in the database as an intersection with no turns (Figure 3.3).

The discrete object view leads to a powerful way of representing geographic information about objects. Think of a class of objects of the same dimensionality – for example, all of the Grizzly Bears (Figure 3.4) in the Kenai Peninsula of Alaska. We would naturally think of these objects as points. We might want to know the sex of each bear, and its date of birth, if our interests were in monitoring the bear population. We might also have a collar on each bear that transmitted the bear's location at regular intervals. All of this information could be expressed in a table, such

as the one shown in Table 3.1, with each row corresponding to a different discrete object, and each column to an attribute of the object. To reinforce a point made earlier, this is a very efficient way of capturing raw geographic information on Grizzly Bears.

But it is not perfect as a representation for all geographic phenomena. Imagine visiting the Earth from another planet, and asking the humans what they chose as a representation for the infinitely complex and beautiful environment around them. The visitor would hardly be impressed to learn that they chose tables, especially when the phenomena represented were natural phenomena such as rivers, landscapes, or oceans. Nothing on the natural Earth looks remotely like a table. It is not at all clear how the properties of a river should be represented as a table, or the properties of an ocean. So while the discrete object view works well for some kinds of phenomena, it misses the mark badly for others.

3.5.2 Fields

While we might think of terrain as composed of discrete mountain peaks, valleys, ridges, slopes, etc., and think of listing them in tables and counting them, there are unresolvable problems of definition for all of these objects. Instead, it is much more useful to think of terrain as a continuous surface, in which elevation can be defined rigorously at every point. Such continuous surfaces form the basis of the other common view of geographic phenomena, known

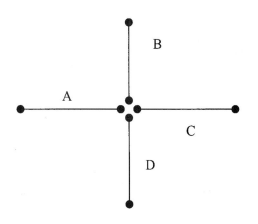

From link	To link	Turn?
A	B	No
A	C	Yes
A	D	No
B	C	No
B	D	Yes
B	A	No
C	D	No
C	A	Yes
C	B	No
D	A	No
D	B	Yes
D	C	No

Figure 3.3 The problems of representing a three-dimensional world using a two-dimensional technology. The intersection of links A, B, C, and D is an overpass, so no turns are possible between such pairs as A and B

Figure 3.4 Bears are easily conceived as discrete objects, maintaining their identity as objects through time and surrounded by empty space

as the *field* view (and not to be confused with other meanings of the word field). In this view the geographic world can be described by a number of *variables*, each measurable at any point on the Earth's surface, and changing in value across the surface.

> The field view represents the real world as a finite number of variables, each one defined at every possible position.

Objects are distinguished by their dimensions, and naturally fall into categories of points, lines, or areas. Fields, on the other hand, can be distinguished by what varies, and how smoothly. A field of elevation, for example, varies much more smoothly in a landscape that has been worn down by glaciation or flattened by blowing sand than one recently created by cooling lava. Cliffs are places in fields where elevation changes suddenly, rather than smoothly. Population density is a kind of field, defined everywhere as the number of people per unit area, though the definition breaks down if the field is examined so closely that the individual people become visible. Fields can also be created from classifications of land, into categories of land use, or soil type. Such fields change suddenly at the boundaries between different classes. Other types of fields can be defined by continuous variation along lines, rather than across space. Traffic density, for example, can be defined everywhere on a road network, and flow volume can be defined everywhere on a river. Figure 3.5 shows some examples of field-like phenomena (see also Box 3.5).

Here is a simple example illustrating the difference between the discrete object and field conceptualizations. Suppose you were hired for the summer to count the number of lakes in Minnesota, and promised that your answer would appear on every license plate issued by the state. The task sounds simple, and you were happy to get the job. But on the first day you started to run into difficulty. What about small ponds, do they count as lakes? What about wide stretches of rivers? What about swamps that dry up in the summer? What about a lake with a narrow section connecting two wider parts, is it one lake or two? Your biggest dilemma concerns the scale of mapping, since the number of lakes shown on a map clearly depends on the map's level of detail – a more detailed map almost certainly will show more lakes.

Your task clearly reflects a discrete object view of the phenomenon. The action of counting implies that lakes are discrete, two-dimensional

Table 3.1 An example of the representation of geographic information as a table: the locations and attributes of each of four Grizzly Bears in the Kenai Peninsula of Alaska. Locations have been obtained from radio collars. Only one location is shown for each bear, at noon on July 31 2000

Bear ID	Sex	Estimated year of birth	Date of collar installation	Location, noon on 31 July 2000
001	M	1996	02241999	−150.6432, 60.0567
002	F	1994	03311999	−149.9979, 59.9665
003	F	1991	04211999	−150.4639, 60.1245
004	F	1992	04211999	−150.4692, 60.1152

(A)

(B)

Figure 3.5 Examples of field-like phenomena. (A) Image of part of the lower Colorado River in the Southwestern USA. The lightness of the image at any point measures the amount of radiation captured by the satellite's imaging system. (B) A simulated image derived from the Shuttle Radar Topography Mission, a new source of high-quality elevation data. The image shows the Carrizo Plain area of Southern California, USA, with a simulated sky and with land cover obtained from other satellite sources (Courtesy: NASA/JPL/Caltech)

Figure 3.6 Lakes are difficult to conceptualize as discrete objects, because it is often difficult to tell where a lake begins and ends, or what distinguishes a wide river from a lake

Table 3.2 A scale of lakeness suitable for defining lakes as a field

Lakeness	Definition
1	Location is always dry under all circumstances
2	Location is sometimes flooded in Spring
3	Location supports marshy vegetation
4	Water is always present to a depth of less than 1 m
5	Water is always present to a depth of more than 1 m

objects littering an otherwise empty geographic landscape. In a field view, on the other hand, all points are either lake or non-lake. Moreover, we could refine the scale a little to take account of marginal cases; for example, we might define the scale shown in Table 3.2, which has five degrees of lakeness. The complexity of the view would depend on how closely we looked, of course, and so the scale of mapping would still be important. But all of the problems of defining a lake as an object would disappear (though there would still be problems in defining the levels of the scale). Instead of counting, our strategy would be to lay a grid over the map, and assign each grid cell a score on the lakeness scale. The size of the grid cell would determine how accurately the result approximated the value we could theoretically obtain by visiting every one of the infinite number of points in the state. At the end, we would tabulate the resulting scores, counting the number of cells having each value of lakeness, or averaging the lakeness score. We could even design a new and scientifically more reasonable license plate – "Minnesota, 12% lake" or "Minnesota, average lakeness 2.02".

The difference between objects and fields is also illustrated well by photographs (e.g. Figure 3.5A). The image in a photograph is created by variation in the chemical state of the material in the photographic film – in early photography, minute particles of silver were released from molecules of silver nitrate to darken the image when the unstable molecules were exposed to light. We think of the image as a field of continuous variation in color or darkness. But when we look at the image, the eye and brain begin to infer the presence of discrete objects, such as people, rivers, fields, cars, or houses, as they interpret the content of the image.

Box 3.5   2.5 dimensions

Areas are two-dimensional objects, and volumes are three dimensional, but GIS users sometimes talk about "2.5D". Almost without exception the elevation of the Earth's surface has a single value at any location (exceptions include overhanging cliffs). So elevation is conveniently thought of as a field, a variable with a value everywhere in two dimensions, and a full 3D representation is only necessary in areas with an abundance of over-hanging cliffs, if these are important features. The idea of dealing with a three-dimensional phenomenon by treating it as a single-valued function of two horizontal variables gives rise to the term "2.5D". Figure 3.5(B) shows an example of an elevation surface.

3.6 Rasters and Vectors

Fields and discrete objects define two conceptual views of geographic phenomena, but they do not solve the problem of digital representation. A field view still potentially contains an infinite amount of information if it defines the value of the variable at every point, since there is an infinite number of points in any defined geographic area. Discrete objects can also require an infinite amount of information for full description – for example, a coastline contains an infinite amount of information if it is mapped in infinite detail. Thus fields and objects are no more than conceptualizations, or ways in which we think about geographic phenomena; they are not designed to deal with the limitations of computers.

> Raster and vector are two methods of representing geographic data in digital computers.

Two methods are used to reduce geographic phenomena to forms that can be coded in computer databases, and we call these raster and vector. In principle both can be used to code both fields and discrete objects, but in practice there is a strong association between raster and fields, and between vector and discrete objects.

3.6.1 Raster data

In a raster representation geographic space is divided into an array of cells that are usually square, but sometimes rectangular (Figure 3.7). All geographic variation is then expressed by assigning properties or attributes to these cells. The cells are sometimes called pixels (short for *picture elements*), which is the preferred term in the area of visualization.

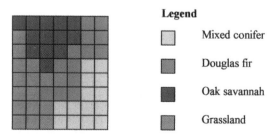

Legend

Mixed conifer

Douglas fir

Oak savannah

Grassland

Figure 3.7 Raster representation. Each color represents a different value of a nominal-scale field variable denoting land cover class

> Raster representations divide the world into arrays of cells and assign attributes to the cells.

One of the commonest forms of raster data comes from remote sensing satellites, which capture information in this form and send it to ground to be distributed and analyzed. Data from the Landsat satellite, for example, which is commonly used in GIS applications, come in cells that are 30 m on a side on the ground, or approximately 1/10 of a hectare in area. Other similar data can be obtained from sensors mounted on aircraft. Imagery varies according to the spatial resolution (expressed as the length of a cell side as measured on the ground), and also according to the timetable of image capture by the sensor. Some satellites are in *geostationary* orbit over a fixed point on the Earth, and capture images constantly. Others pass over a fixed point at regular intervals (e.g. every 12 days). Finally, sensors vary according to the part or parts of the spectrum that they sense. The *electromagnetic* spectrum includes the visible portions picked up by the human eye, and also portions that are invisible to the human eye but detectable by a sensor. Figure 3.8 shows the spectrum laid out according to *frequency* (number of oscillations per second) and *wavelength* (the length of a single wave when radiation is traveling at the speed of light). The visible parts of the spectrum are most important for remote sensing, but some invisible parts of the spectrum are particularly useful in detecting heat, and the phenomena that produce heat, such as volcanic activities. Many sensors capture images in several areas of the spectrum, or *bands*, simultaneously, because the relative amounts of radiation in different parts of the spectrum are often useful indicators of certain phenomena, such as green leaves, or water, on the Earth's surface. The AVIRIS (Airborne Visible InfraRed Imaging Spectrometer) captures no fewer than 224 different parts of the spectrum, and is being used to detect particular minerals in the soil, among other applications. Remote sensing is a complex topic, and further details are available in Section 10.2.1, texts such as Ryerson's (1998), and at Web sites such as www.nasa.gov.

Square cells fit together nicely on a flat table or a sheet of paper, but they will not fit together neatly on the curved surface of the Earth. So just as representations on paper require that the Earth be flattened, or projected, so too do rasters (because of the distortions associated with flattening, the cells in a raster can never be perfectly equal in size on the Earth's surface). Projections, or ways of flattening the Earth, are described in Section

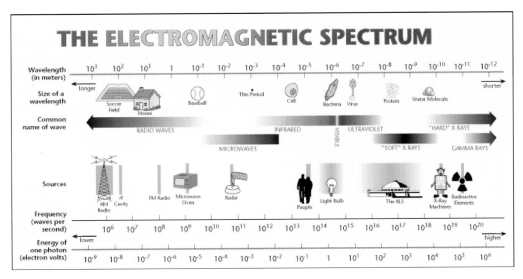

Figure 3.8 The electromagnetic spectrum. The most important wavelengths (or frequencies) for remote sensing are the visible portion (note how narrow it is in comparison with the entire spectrum), and the adjacent part of the infrared. The second row of the diagram identifies familiar objects whose size is similar to that wavelength (occupying roughly the same length as a single wave of radiation traveling at the speed of light). Other uses of the spectrum besides remote sensing are also shown (Source: Center for Beam Physics, Lawrence Berkeley National Laboratory)

4.7. Many of the terms that describe rasters suggest the laying of a tile floor on a flat surface – we talk of raster cells *tiling* an area, and a raster is said to be an instance of a *tesselation*. The mirrored ball hanging above a dance floor recalls the impossibility of covering a spherical object like the Earth perfectly with flat, square pieces.

When information is represented in raster form all detail about variation within cells is lost, and instead the cell is given a single value. Suppose we wanted to represent the map of the counties of Texas as a raster. Each cell would be given a single value to identify a county, and we would have to decide the rule to apply when a cell falls in more than one county. Often the rule is a simple plurality: the county with the *largest share* of the cell's area gets the cell. Sometimes the rule is based on the *central point* of the cell, and the county at that point is assigned to the whole cell. Figure 3.9 shows these two rules in operation. The largest share rule is almost always preferred, but the central point rule is sometimes used in the interests of faster computing.

Table 3.3 shows some common instances of raster data, the approximate cell size on the ground, and the rule used to assign values, and includes instances where the value assigned is the average or mean over the cell's area. Digital orthophotos are raster photographic images, mostly acquired from aircraft, and processed to remove geometric distortions due to the position

of the sensor or camera and the varying elevation of the Earth's surface.

3.6.2 Vector data

In a vector representation, all lines are captured as points connected by precisely straight lines (some GIS software allows points to be connected by curves rather than straight lines, but in most cases curves have to be approximated by increasing the density of points). An area is captured as a series of points or *vertices* connected by straight lines as shown in Figure 3.10. The straight edges between vertices explain why areas in vector representation are often called *polygons*, and in GIS-speak the terms polygon and area are often used interchangeably. Lines are captured in the same way, and the term *polyline* has been coined to describe a curved line represented by a series of straight segments connecting vertices.

To capture an area object in vector form, we need only specify the locations of the points that form the vertices of a polygon. This seems simple, and also much more efficient than a raster representation, which would require us to list all of the cells that form the area. These ideas are captured succinctly in the comment "Raster is vaster, and vector is correcter". To create a precise approximation to an area in raster, it

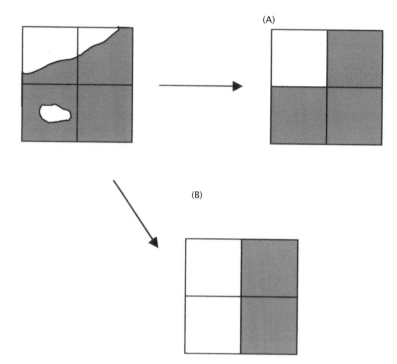

Figure 3.9 Effect of a raster representation using (A) the largest share rule and (B) the central point rule

Table 3.3 The representation used by some common sources of raster data

Dataset	Approximate ground cell size	Assignment rule
US Geological Survey digital elevation data	30 m or 90 m	Central point
Landsat Thematic Mapper imagery	30 m	Mean value
US Geological Survey Digital Orthophoto	1 m	Mean value

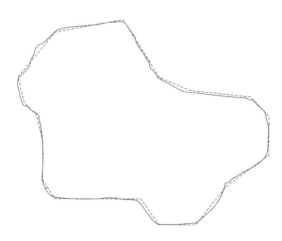

Figure 3.10 An area (red line) and its approximation by a polygon (blue line)

would be necessary to resort to using very small cells, and the number of cells would rise proportionally (in fact, every halving of the width and height of each cell would result in a quadrupling of the number of cells). But things are not quite as simple as they seem. The apparent precision of vector is often unreasonable, since many geographic phenomena simply cannot be located with high accuracy. So although raster data may look less attractive, they may be more honest to the inherent quality of the data. Also, various methods exist for compressing raster data that can greatly reduce the capacity needed to store a given dataset. They include specialized methods such as run-length encoding (Figure 3.11), as well as generic methods used in the computing industry, such as JPEG and MrSID (see Section 9.2.2). So the choice between raster and vector is often complex, as shown in Table 3.4.

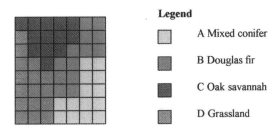

Legend

☐ A Mixed conifer

▨ B Douglas fir

■ C Oak savannah

▨ D Grassland

Normal row-by-row coding

CCCCCBBDCCCCBBDCCCBBBDDCBBAADDDDBAADDBBBAADDDAAAADDD
AAAA

Run-length encoding

5C 2B 1D 4C 2B 1D 3C 3B 2D 1C 2B 2A 4D 1B 2A 2D 3B 2A 3D 4A 3D 4A

Figure 3.11 A simple method of compressing a raster. In normal coding this raster would require 56 entries in the database, one for every cell in the 7 by 8 array. Run length encoding records cells in the same order, but uses a combination of multipliers and classes (e.g. the first five Cs are replaced by two entries 5 and C). In this form the array occupies only 22 pairs, or 44 entries in storage)

Table 3.4 Relative advantages of raster and vector representation

Issue	Raster	Vector
Volume of data	Depends on cell size	Depends on density of vertices
Sources of data	Remote sensing, imagery	Social and environmental data
Applications	Resources, environmental	Social, economic, administrative
Software	Raster GIS, image processing	Vector GIS, automated cartography
Resolution	Fixed	Variable

3.7 The Paper Map

The paper map has long been a powerful and effective means of communicating geographic information. It is an instance of an *analog* representation, or a physical model in which the real world is scaled – in the case of the paper map, part of the world is scaled to fit the size of the paper. A key property of a paper map is its *scale* or *representative fraction*, defined as the ratio of distance on the map to distance on the Earth's surface. For example, a map with a scale of 1:24,000 reduces everything on the Earth to one 24,000th of its real size. This is a bit misleading, because the Earth's surface is curved but a paper map is flat, so scale cannot be exactly constant.

A paper map is: a source of data for geographic databases; an analog product from a GIS; and an effective communication tool.

Maps have been so important, particularly prior to the development of digital technology, that many of the ideas associated with GIS are actually inherited directly from paper maps. For example, scale is often cited as a property of a digital database, even though the definition of scale makes no sense for digital data – ratio of distance *in the computer* to distance on the ground; how can there be distances in a computer? What is meant is a little more complicated: when a scale is quoted for a digital database it is usually the scale of the map that formed the source of the data. So if a database is said to be at a scale of 1:24,000 one can safely assume that it was created from a paper map at that scale. Further discussion of scale can be found in Box 5.3; in Chapter 6, where it is important to the concept of uncertainty, and in Chapter 7 on generalization.

There is a close relationship between the contents of a map and the raster and vector representations discussed in the previous section. The US Geological Survey, for example, distributes two digital versions of its topographic maps, one in raster form and one in vector form, and both attempt to capture the contents of the map as closely as possible. In the raster form, or *digital raster graphic* (DRG), the map is scanned at a very high density, using very small pixels, so that the raster looks very much like the original (Figure 3.12, and see also Figure 7.2). The coding of each pixel simply records the color of the map picked up by the scanner, and the dataset includes all of the textual information surrounding the actual map.

In the vector form, or *digital line graph* (DLG), every geographic feature shown on the map is represented as a point, polyline, or polygon. The symbols used to represent point features on the map, such as the symbol for a windmill, are replaced in the digital data by points with associated attributes. Contours, which are shown on the map as lines of definite width, are replaced by polylines of no width, and given attributes that record their elevations.

In both cases, and especially in the vector case, there is a significant difference between the analog representation of the map and its digital equivalent. So it is quite misleading to think of the contents of a digital representation as a map, and to think of a GIS as a container of digital maps. Digital representations can include information that would be very difficult to show on maps. For example, they can represent the curved surface of the Earth, without the need for the distortions associated with flattening. They can represent changes, whereas maps must be static because it is very difficult to change their contents once they have been printed or drawn. Digital databases can represent all three spatial dimensions, including the vertical, whereas maps must always show two-dimensional views or limited three-dimensional perspectives. So while the paper map is a useful metaphor for the contents of a geographic database, we must be careful not to let it limit our thinking about what is possible in

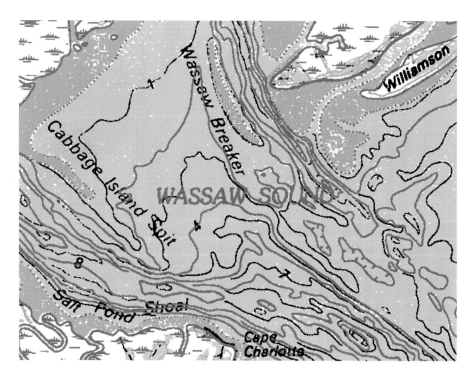

Figure 3.12 Example of a Digital Raster Graphic, a raster scan of a small fragment of a US Geological Survey 1:100 000 map sheet (Source: USGS)

the way of representation. This issue is pursued at greater length in Chapter 9.

Questions for Further Study

1. What fraction of the Earth's surface have you experienced in your lifetime? Make a diagram like that shown in Figure 3.1, at appropriate levels of detail, to show (a) where you have lived in your lifetime, (b) how you spent yesterday. How would you describe what is missing from each of these diagrams?
2. Identify the limits of your own neighborhood, and start making a list of the discrete objects you are familiar with in the area. What features are hard to think of as discrete objects? For example, how will you divide up the various roadways in the neighborhood into discrete objects – where do they begin and end?
3. Make a list of the field variables that could be measured or mapped in your neighborhood.
4. There are many geographic data clearinghouses on the Web – an example is the US Federal Geographic Data Committee's National Geospatial Data Clearinghouse at www.fgdc.gov. Explore the contents of a clearinghouse, and summarize the types of data available using the terms introduced in this chapter – discrete objects, fields, raster, vector, and the attribute data types (nominal, ordinal, interval, ratio, cyclic).
5. The early explorers had limited ways of communicating what they saw, but many were very effective at it. Examine the published diaries, notebooks, or dispatches of one or two early explorers, and look at the methods they used to communicate with others. What words did they use to describe unfamiliar landscapes, and how did they mix words with sketches?

Online Resources

NCGIA Core Curricula (www.ncgia.ucsb.edu/pubs/core.html):
Core Curriculum in GIScience, Section 1.1 (The World in Spatial Terms, edited by Reg Golledge), 1.1.1 (Human Cognition of the Spatial World, Dan Montello) and 1.4.2 (Maps as Representations of the World, Judy Olson)

Core Curriculum in GIS, 1990, Units 2 (Maps and Map Analysis), 10 (Spatial Databases as Models of Reality), 22 (The Object/Layer Debate), and 73 (GIS and Spatial Cognition)
ESRI Virtual Campus courses (campus.esri.com): Turning Data into Information, by Paul Longley, Michael Goodchild, David Maguire, and David Rhind (Module "Introduction")

Reference Links

Maguire D J, Goodchild M F, and Rhind D W (eds) 1991 *Geographical Information Systems: Principles and applications*. Harlow, UK: Longman (Text available online from 'Links to Big Book 1' at www.wiley.com/gis and www.wiley.co.uk/gis).
Chapter 9, Concepts of space and geographical data (Gatrell A C)
Chapter 14, GIS and remote sensing (Simonett D S)
Chapter 16, High-level spatial data structures for GIS (Egenhofer M J, Herring J)
Longley P A, Goodchild M F, Maguire D J, and Rhind D W (eds) 1999 *Geographical Information Systems: Principles, techniques, management and applications*. New York: John Wiley.
Chapter 5, Spatial representation: the scientist's perspective (Raper J F)
Chapter 6, Spatial representation: the social scientist's perspective (Martin D J)
Chapter 7, Spatial representation: a cognitive view (Mark D M)
Chapter 8, Time in GIS and geographical databases (Peuquet D J)
Chapter 9, Representation of terrain (Hutchinson M F, Gallant J C)
Chapter 32, Digital remotely sensed data and their characteristics (Barnsley M)

References

Chrisman N R 1997 *Exploring Geographic Information Systems*. New York: John Wiley.
Gore A 1992 *Earth in the Balance: Ecology and the human spirit*. Boston: Houghton Mifflin.
Mardia K V and Jupp P E 2000 *Directional Statistics*. New York: John Wiley.
Ryerson R A (Ed.) 1998 *Manual of Remote Sensing*. New York: John Wiley.

GEOREFERENCING

Geographic location is the element that distinguishes geographic information from all other types, so methods for specifying location on the Earth's surface are essential to the creation of useful geographic information. Humanity has developed many such techniques over the centuries, and this chapter provides a basic guide for GIS students – what you need to know about georeferencing to succeed in GIS. The first section lays out the principles of georeferencing, including the requirements that any effective system must satisfy. Subsequent sections discuss commonly used systems, starting with the ones closest to everyday human experience, including placenames and street addresses, and moving to the more accurate scientific methods that form the basis of geodesy and surveying. The final section deals with issues that arise over conversions between georeferencing systems, and with the concept of a gazetteer.

Learning objectives

By the end of this chapter you will:

● know the requirements for an effective system of georeferencing;

● be familiar with the problems associated with placenames, street addresses, and other systems used every day by humans;

● know how the Earth is measured and modeled for the purposes of positioning;

● understand the basic principles of map projections, and the details of some commonly used projections.

4.1 Introduction

Chapter 3 introduced the idea of an atomic element of geographic information: a triple of location, optionally time, and attribute. To make geographic information systems work there must be techniques for assigning values to all three of these, in ways that are commonly understood by people who wish to communicate. All the world agrees on the basic calendar and time system, so there are only minor problems associated with communicating that element of the atom when it is needed (although different time zones, different names of the months in different languages, and systems such as the classical Japanese convention of dating by the year of the Emperor's reign all sometimes manage to confuse us).

Time is optional in a geographic information system, but location is not, so this chapter focuses on techniques for specifying location, and the problems and issues that arise. Locations are the basis for many of the benefits of geographic information systems: the ability to map, to tie different kinds of information together because they refer to the same place, or to measure distances and areas. Without locations, data are said to be non-spatial or *aspatial* and would have little value within a geographic information system.

Time is an optional element in geographic information, but location is essential.

Several terms are commonly used to describe the act of assigning locations to atoms of information. We use the verbs *to georeference*, *to geolocate*, and *to geocode*, and say that facts have been *georeferenced* or *geocoded*. We talk about *tagging* records with geographic locations, or about *locating* them. The term georeference will be used throughout this chapter.

The primary requirements of a georeference are that it be *unique*, so that there is only one location associated with a given georeference, and therefore no confusion about the location that is referenced; and that its meaning be *shared* among all of the people who wish to work with the information, including their geographic information systems. For example, the georeference 909 West Campus Lane, Goleta, California, USA points to a single house – there is no other house with that address – and its meaning is shared sufficiently widely to allow mail to be delivered to the address from virtually anywhere on the planet. The address will not be meaningful to most people living in China because they are not familiar with the Roman alphabet, but it will be meaningful to a sufficient number of people within China's postal service, so a letter mailed

from China to that address will likely be delivered successfully. Uniqueness and shared meaning are sufficient also to allow people to link different kinds of information based on common location: for example, a driving record that is georeferenced by street address can be linked to a record of purchasing. The negative implications of this kind of record linking for human privacy are discussed further in Section 17.3.1.

To be as useful as possible a georeference must be *persistent through time*, because it would be very confusing if georeferences changed frequently, and very expensive to update all of the records that depend on them. This can be problematic when a georeferencing system serves more than one purpose, or is used by more than one agency with different priorities. For example, a municipality may expand by incorporating more land, creating problems for mapping agencies, and for researchers who wish to study the municipality through time. Street names sometimes change, and postal agencies sometimes revise postal codes. Changes even occur in the names of cities (Saigon to Ho Chi Minh City), or in their conventional transcriptions into the Roman alphabet (Peking to Beijing).

To be most useful, georeferences should stay constant through time.

Every georeference has an associated spatial resolution (see Box 5.3), equal to the size of the area that is assigned that georeference. A mailing address could be said to have a spatial resolution equal to the size of the mailbox, or perhaps to the area of the parcel of land or structure assigned that address. A US state has a spatial resolution that varies from the size of Rhode Island to that of Alaska, and many other systems of georeferencing have similarly wide-ranging spatial resolutions.

Many systems of georeferencing are unique only within an area or *domain* of the Earth's surface. For example, there are many cities with the name Springfield in the USA (18 according to a recent edition of the Rand McNally Road Atlas; similarly there are nine Whitchurches in the 2000 AA Road Atlas of the United Kingdom). City name is unique within the domain of a state, however, a property that was engineered with the advent of the postal system in the 19th century. Today there is no danger of there being two Springfields in Massachusetts, and a driver can confidently ask for directions to ''Springfield, Massachusetts'' in the knowledge that there is no danger of being sent to the wrong Springfield. But people living in London, Ontario, Canada are well aware of the dangers of talking about ''London'' without specifying the appropriate domain. Even in

Figure 4.1 Placenames are not necessarily unique at the global level – there are many Londons, for example, besides the largest and most prominent one in the UK. People living in other Londons must often add additional information (e.g. London, *Ontario, Canada*) to resolve ambiguity

Toronto, Ontario a reference to ''London'' may be misinterpreted as a reference to the older (UK)

London on a different continent, rather than to the one 200 km away in the same province (Figure 4.1). Street name is unique in the USA within municipal domains, but not within larger domains such as county or state. The six digits of a UK National Grid reference repeat every 100 km, so additional letters are needed to achieve uniqueness within the national domain (see Box 4.1). Similarly there are 120 places on the Earth's surface with the same Universal Transverse Mercator georeference (see Section 4.7.2), and a zone number and hemisphere must be added to make a reference unique in the global domain.

While some georeferences are based on simple names, others are based on various kinds of *measurements*, and are called *metric* georeferences. They include latitude and longitude, and various kinds of coordinate systems, all of which are discussed in more detail below. One enormous advantage of such systems is that they provide the potential for infinitely fine spatial resolution: provided we have sufficiently accurate measuring devices, and use enough decimal places, it is possible with such systems to locate

Box 4.1 A national system of georeferencing: the National Grid of Great Britain

The National Grid is administered by the Ordnance Survey of Great Britain, and provides a unique georeference for every point in England, Scotland, and Wales. The first designating letter defines a 500 km square, and the second defines a 100 km square (see Figure 4.2). Within each square, two measurements, called easting and northing, define a location with respect to the lower left corner of the square. The number of digits defines the precision – three digits for easting and three for northing (a total of six) define location to the nearest 100 m.

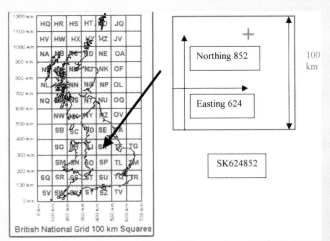

Figure 4.2 The National Grid of Great Britain, illustrating how a point is assigned a grid reference that locates it uniquely to the nearest 100 m (Source: Peter H. Dana)

information to any level of accuracy. Another is that from measurements of two or more locations it is possible to compute distances, a very important requirement of georeferencing in GIS.

Metric georeferences are much more useful, because they allow distances to be calculated.

Other systems simply *order* locations. In most countries mailing addresses are ordered along streets, often using the odd integers for addresses on one side, and the even integers for addresses on the other. This means that it is possible to say that 3000 State Street and 100 State Street are further apart than 200 State Street and 100 State Street, and allows postal services to sort mail for easy delivery. In the Western United States, it is often possible to infer estimates of the distance between two addresses on the same street by knowing that 100 addresses are assigned to each city block, and that blocks are typically between 120 and 160 m long.

This section has reviewed some of the general properties of georeferencing systems, and Table 4.1 summarizes some commonly used systems. The following sections discuss the specific properties of the ones that are most important in GIS applications.

4.2 Placenames

Giving places names is the simplest form of georeferencing, and was most likely the one first developed by early hunter-gatherer societies. Any distinctive feature on the landscape, such as a particularly old tree, can serve as a point of reference for two people who wish to share information, such as the existence of good game in the tree's vicinity. Human landscapes rapidly became littered with names, as people sought distinguishing labels to use in describing aspects of their surroundings, and other people adopted them. Today, of course, we have a complex system of naming oceans, continents, cities, mountains, rivers, and other prominent features that is almost universally shared, and each country maintains a system of authorized naming, often through national or state committees assigned with the task of standardizing geographic names.

Language extends the power of placenames through words such as "between", which serve to refine references to location, or "near", which serve to broaden them (Figure 4.3). "Where State Street crosses Mission Creek" is an instance of combining two placenames to achieve greater refinement of location than either name could

Figure 4.3 There are many different ways of defining London and determining its boundaries. Features in the center, such as the Palace of Westminster and Big Ben, are more likely to be regarded as in London than features on the periphery, such as Heathrow Airport

achieve individually. Even more powerful extensions come from combining placenames with directions and distances, as in "200 m north of the old tree" or "50 km west of Springfield".

Many commonly used placenames have meanings that vary between people, and with the context in which they are used

But placenames are of limited use as georeferences. First, they often have very coarse spatial resolution. "Asia" covers over 43 million sq km, so the information that something is located "in Asia" is not very helpful in pinning down its location. In a similar vein, even Rhode Island, the smallest state of the USA, has a land area of over 2700 sq km. Second, only certain placenames are officially authorized by national or subnational government agencies. Many more are recognized only locally, so their use is limited

Table 4.1. Some commonly used systems of georeferencing

System	Domain of uniqueness	Metric?	Example	Spatial resolution
Placename	varies	no	London, Ontario, Canada	varies by feature type
Postal address	global	no, but ordered along streets in most countries	909 West Campus Lane, Goleta, California, USA	size of one mailbox
Postal code	country	no	93117 (US ZIP code); WC1E 6BT (UK Unit Postcode)	area occupied by a defined number of mailboxes
Telephone calling area	country	no	805	varies
Cadastral system	local authority	no	Parcel 01452954, City of Springfield, MA, USA	area occupied by a single parcel of land
Public Land Survey System	Western USA only, unique to Prime Meridian	yes	Sec 5, Township 6E, Range 4N	defined by level of subdivision
Latitude/ longitude	global	yes	119 degrees 45 minutes West, 34 degrees 40 minutes North	infinitely fine
Universal Transverse Mercator	zones six degrees of longitude wide, and N or S hemisphere	yes	563146E, 4356732N	infinitely fine
State Plane Coordinates	USA only, unique to state and to zone within state	yes	55086.34E, 75210.76N	infinitely fine

to communication between people in the local community. Placenames may even be lost through time: although there are many contenders, we do not know with certainty where the "Camelot" described in the English legends of King Arthur was located.

The meaning of placenames can become lost through time.

4.3 Postal Addresses and Postal Codes

Postal addresses were introduced after the development of mail delivery in the 19th century. They rely on several assumptions:

● Every dwelling and office is a potential destination for mail.

● Dwellings and offices are arrayed along paths, roads, or streets, and numbered accordingly.

● Paths, roads, and streets have names that are unique within local areas.

● Local areas have names that are unique within larger regions.

● Regions have names that are unique within countries.

If the assumptions are true, then mail address provides a unique identification for every dwelling on Earth.

Today, postal addresses are an almost universal means of locating many kinds of human activity: delivery of mail, place of residence, or place of business. They fail, of course, in locating anything that is not a potential destination for mail, including almost all kinds of natural features (Mt Everest does not have a postal address, and neither does Manzana Creek in Los Padres National Forest in California, USA). They are not as useful when dwellings are not numbered consecutively along streets, as happens in some cultures (notably in Japan, where street numbering reflects date of construction, not sequence along the street – it is temporal, rather than spatial) and in large building complexes like condominiums. Many GIS applications rely on the ability to locate activities by postal address, and to convert addresses to some more universal system of georeferencing, such as latitude and longitude, for mapping and analysis.

> Postal addresses work well to georeference dwellings and offices, but not natural features.

Postal codes were introduced in many countries in the late 20th century in order to simplify the sorting of mail. In the Canadian system, for example, the first three characters of the six-character code identify a Forward Sortation Area, and mail is initially sorted so that all mail directed to a single FSA is together. Each FSA's incoming mail is accumulated in a local sorting station, and sorted a second time by the last three characters of the code, to allow it to be delivered easily. Figure 4.4 shows a map of the FSAs for an area of the Toronto metropolitan region. The full six characters are unique to roughly 10 houses, a single large business, or a single building. Much effort went into ensuring widespread adoption of the coding system by the general public and businesses, and computer programs were developed to assign codes automatically to addresses for large-volume mailers.

Postal codes have proven very useful for many purposes besides the sorting and delivery of mail. Although the area covered by a Canadian FSA or a

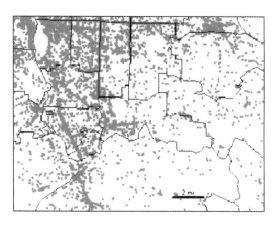

  ⁕ Business
  —— Zip Code Boundary

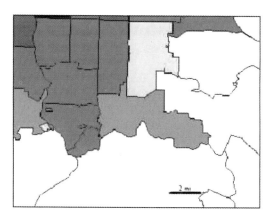

  0–12
  13–36
  37–67
  68–116
  117–176
  177–307
  308–558

Figure 4.5 The use of ZIP code boundaries as a convenient basis for summarizing data. In this instance each business has been allocated to its ZIP code, and the ZIP code areas have been shaded according to the density of businesses per square mile (Source: ESRI)

US ZIP code varies, and can be changed whenever the postal authorities want, it is sufficiently constant to be useful for mapping purposes, and many businesses routinely make maps of their customers by counting the numbers present in each postal code area, and dividing by total population to gain a picture of market penetration. Figure 4.5 shows an example of summarizing data by ZIP code. Most people know the

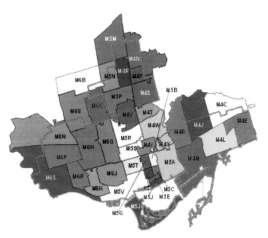

Figure 4.4 Forward Sortation Areas (FSAs) of the central part of the Toronto metropolitan region. FSAs form the first three characters of the six-character Canadian postal code

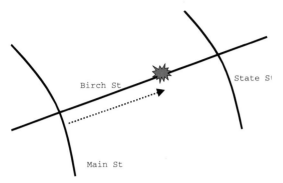

Figure 4.6 Linear referencing – an incident's position is determined by measuring its distance (87 m) along one road (Birch St) from a well-defined point (its intersection with Main St)

postal code of their home, and in some instances postal codes have developed popular images (the ZIP code for Beverly Hills, California, 90210, became the title of a successful television series).

4.4 Linear Referencing Systems

A linear referencing system defines location on a network by measuring distance from a defined point of reference along a defined path in the network. Figure 4.6 shows an example, an accident whose location is reported as being a measured distance from a street intersection, along a named street. Linear referencing is closely related to the use of street addresses, but uses an explicit measurement of distance rather than the much less reliable surrogate of street address number.

Linear referencing is widely used in applications that center on a linear network. This includes highways (e.g. Mile 1240 of the Alaska Highway), railroads (e.g. 24.9 miles from Paddington Station in London on the main line to Bristol, England), electrical transmission, pipelines, and canals. Linear references are used by highway agencies to define the locations of bridges, signs, potholes, and accidents, and to record pavement condition.

Linear referencing systems are widely used in managing transportation infrastructure, and in dealing with emergencies.

Linear referencing provides a sufficient basis for georeferencing for some applications. Highway departments often base their records of accident locations on linear references, as well as their inventories of signs and bridges (GIS has many

applications in transportation that are known collectively as GIS-T, and in the developing field of intelligent transportation systems or ITS). But for other applications it is important to be able to convert between linear references and other forms, such as latitude and longitude. For example, the Onstar system that is installed in many Cadillacs sold in the USA is designed to radio the position of a vehicle automatically as soon as it is involved in an accident. When the airbags deploy, a GPS (Global Positioning System; see Section 10.2.2.2) receiver determines position, which is then relayed to a central dispatch office. Emergency response centers often use street addresses and linear referencing to define the locations of accidents, so the latitude and longitude received from the vehicle must be converted before an emergency team can be sent to the accident.

Linear referencing systems are often difficult to implement in practice in ways that are robust in all situations. In an urban area where intersections occur frequently it is relatively easy to measure distance from the nearest one (e.g. on Birch St. 87 m west of the intersection with Main St.). But in rural areas it may be a long way from the nearest intersection. Even in urban areas it is not uncommon for two streets to intersect more than once (e.g. Birch may have two intersections with Columbia Crescent). There may also be difficulties in defining distance accurately, especially if roads include steep sections where the distance driven is significantly longer than the distance evaluated on a two-dimensional digital representation.

4.5 Cadasters and the US Public Land Survey System

The *cadaster* is defined as the map of land ownership in an area, maintained for the purposes of taxing land, or of creating a public record of ownership. The process of *subdivision* creates new parcels by legally subdividing existing ones.

Parcels of land in a cadaster are often uniquely identified, by number or by code, and are also reasonably persistent through time, and thus satisfy the requirements of a georeferencing system. But very few people know the identification code of their home parcel, and use of the cadaster as a georeferencing system is thus limited largely to local officials.

The US Public Land Survey System (PLSS) evolved out of the need to survey and distribute the vast land resource of the Western USA, starting in the early 19th century, and expanded to become the dominant system of cadaster for all of the USA west of Ohio, and all of Western Canada. Its essential simplicity and regularity make it useful for many purposes, and understandable by the general public. Its geometric regularity also allows it to satisfy the requirement of a metric system of georeferencing, because each georeference is defined by measured distances.

> The Public Land Survey System defines land ownership over much of western North America, and is a useful system of georeferencing.

To survey an area using the PLSS, a surveyor first lays out an accurate north–south line or *prime meridian*. Rows are laid out six miles apart and perpendicular to this line, to become the *ranges* of the system. Then blocks or *townships* are laid out in six mile by six mile squares on either side of the prime meridian (see Figure 4.7). Each square is referenced by township number, range number, whether it is to the east or to the west, and the name of the prime meridian. Thirty-six *sections* of one mile by one mile are laid out inside each township, and numbered using a standard system (note how the numbers reverse in every other row). Each section is divided into four quarter-sections of a quarter of a square mile, or 160 acres, the size of the nominal family farm orhomestead in the original conception of the PLSS. The process can be continued by

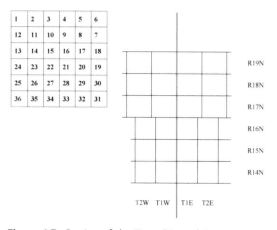

Figure 4.7 Portion of the Township and Range system (Public Land Survey System) widely used in the Western USA as the basis of land ownership. Townships are laid out in six mile squares on either side of an accurately surveyed Principal Meridian (as shown on the right). The offset shown between ranges 16N and 17N is needed to accommodate the Earth's curvature (shown much exaggerated). The square mile sections within each township are numbered as shown at the upper left

subdividing into four to obtain any level of spatial resolution.

The PLSS would be a wonderful system if the Earth were flat. To account for its curvature the squares are not perfectly six miles by six miles, and the rows must be offset frequently. Figure 4.7 shows this offsetting for a small area. Nevertheless, it remains an efficient system, and one with which many people in the Western USA and Western Canada are familiar. It is often used to specify location, particularly in managing natural resources in the oil and gas industry and in mining, and in agriculture. Systems have been built to convert PLSS locations automatically to latitude and longitude.

4.6 Measuring the Earth: Latitude and Longitude

The most powerful systems of georeferencing are those that provide the potential for very fine spatial resolution, that allow distance to be computed between pairs of locations, and that support other forms of spatial analysis. The system of latitude and longitude is in many ways the most comprehensive, and is often called the *geographic* system of coordinates, based on the Earth's rotation about its center of mass.

To define latitude and longitude we first identify the *axis* of the Earth's rotation. The Earth's center of mass lies on the axis, and the plane through the center of mass perpendicular to the axis defines the *Equator*. Slices through the Earth parallel to the axis, and perpendicular to the plane of the Equator, define lines of constant longitude (Figure 4.8), rather like the segments of an orange. A slice through a line marked on the ground at the Royal Observatory in Greenwich, England defines zero longitude, and the angle between this slice and any other slice defines the latter's measure of longitude. Each of the 360 degrees of longitude is divided into 60 minutes, and each minute into 60 seconds. But it is more conventional to refer to longitude by degrees East or West, so longitude ranges from 180 degrees West to 180 degrees East. Finally, because computers are designed to handle numbers ranging from very large and negative to very large and positive, we normally store longitude in computers as if West was negative and East was positive; and we store parts of degrees using decimals rather than minutes and seconds. A line of constant longitude is termed a *meridian*.

Longitude can be defined in this way for any rotating solid, no matter what its shape, because

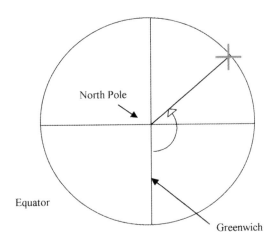

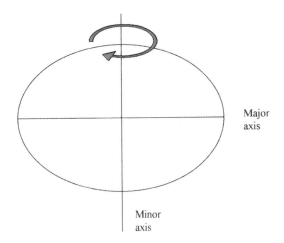

Figure 4.8 Definition of longitude. The Earth is seen here from above the North Pole, looking along the Axis, with the Equator forming the outer circle. The location of Greenwich defines the Prime Meridian. The longitude of the point at the center of the red cross is determined by drawing a plane through it and the axis, and measuring the angle between this plane and the Prime Meridian

Figure 4.9 Definition of the ellipsoid, formed by rotating an ellipse about its minor axis (corresponding to the axis of the Earth's rotation)

the axis of rotation and the center of mass are always defined. But the definition of latitude requires that we know something about the shape. The Earth is a complex shape that is only approximately spherical. A much better approximation or *figure of the Earth* is the *ellipsoid of rotation*, the figure formed by taking a mathematical ellipse and rotating it about its shorter axis (Figure 4.9). The term *spheroid* is also commonly used.

The difference between the ellipsoid and the sphere is measured by its *flattening*, or the reduction in the minor axis relative to the major axis. Flattening is defined as:

$$f = (a - b)/a$$

where *a* and *b* are the lengths of the major and minor axes respectively (we usually refer to the *semi*-axes, or half the lengths of the axes, because these are comparable to radii). The actual flattening is about 1 part in 300.

> The Earth is slightly flattened, such that the distance between the Poles is about 1 part in 300 less than the diameter at the Equator.

Much effort was expended over the past 200 years in finding ellipsoids that best approximated the shape of the Earth in particular countries, so that national mapping agencies could measure position and produce accurate maps. Early ellipsoids varied significantly in their basic parameters, and were

generally not centered on the Earth's center of mass. But the development of intercontinental ballistic missiles in the 1950s and the need to target them accurately, as well as new data available from satellites, drove the push to a single international standard. Without a single standard, the maps produced by different countries using different ellipsoids could never be made to fit together along their edges, and artificial steps and offsets were often necessary in moving from one country to another (navigation systems in aircraft would have to be corrected, for example).

The ellipsoid known as WGS84 (the World Geodetic System of 1984) is now widely accepted, and North American mapping is being brought into conformity with it through the adoption of the virtually identical North American Datum of 1983 (NAD83). But many other ellipsoids remain in use in other parts of the world, and much older data still adhere to earlier standards, such as the North American Datum of 1927 (NAD27). Table 4.2 gives details of these standards. A quick scan of the second column of this table shows that 6378 km is a reasonable approximation of the Earth's radius for purposes that only require accuracy to a km or so.

We can now define latitude. Figure 4.10 shows a line drawn through a point of interest perpendicular to the ellipsoid at that location. The angle made by this line with the plane of the Equator is defined as the point's latitude, and varies from 90 South to 90 North. Again, south latitudes are usually stored as negative numbers, and north latitudes as positive. Latitude is often symbolized by the Greek letter phi (ϕ) and longitude by the

Table 4.2. Some important ellipsoids. WGS84 has become the most important ellipsoid for international use

Ellipsoid	Semi-major axis (m)	1/flattening NB
Airy 1830	6377563.396	299.3249646
Modified Airy	6377340.189	299.3249646
Australian National	6378160	298.25
Bessel 1841 (Namibia)	6377483.865	299.1528128
Bessel 1841	6377397.155	299.1528128
Clarke 1866	6378206.4	294.9786982
Clarke 1880	6378249.145	293.465
Everest (India 1830)	6377276.345	300.8017
Everest (Sabah and Sarawak)	6377298.556	300.8017
Everest (India 1956)	6377301.243	300.8017
Everest (Malaysia 1969)	6377295.664	300.8017
Everest (Malaysia and Singapore)	6377304.063	300.8017
Everest (Pakistan)	6377309.613	300.8017
Modified Fischer 1960	6378155	298.3
Helmert 1906	6378200	298.3
Hough 1960	6378270	297
Indonesian 1974	6378160	298.247
International 1924	6378388	297
Krassovsky 1940	6378245	298.3
GRS 80	6378137	298.257222101
South American 1969	6378160	298.25
WGS 72	6378135	298.26
WGS 84	6378137	298.257223563

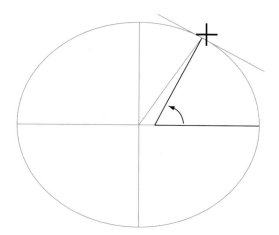

Figure 4.10 Definition of the latitude of the blue point, as the angle between the Equator and a line drawn perpendicular to the ellipsoid

surface. Two points on the same north–south line of longitude, and separated by one degree of latitude are 1/360 of the circumference of the Earth apart, or about 111 km apart. One minute of latitude corresponds to 1.86 km, and also defines one nautical mile, a unit of distance that is still commonly used in navigation. One second of latitude corresponds to about 30 m. But things are more complicated in the east–west direction, and these figures only apply to east–west distances along the Equator, where lines of longitude are furthest apart. Away from the Equator the length of a line of latitude gets shorter and shorter, until it vanishes altogether at the poles. The degree of shortening is approximately equal to the cosine of latitude, or $\cos \phi$, which is 0.866 at 30 degrees North or South, 0.707 at 45 degrees, and 0.500 at 60 degrees. So a degree of longitude is only 55 km along the northern boundary of the Canadian province of Alberta (exactly 60 degrees North).

Greek letter lambda (λ), so the respective ranges can be expressed in mathematical shorthand as: $-180 \leq \lambda \leq 180$; $-90 \leq \phi \leq 90$. A line of constant latitude is termed a *parallel*.

It is important to have a sense of what latitude and longitude mean in terms of distances on the

Lines of latitude and longitude are equally far apart only at the Equator; towards the Poles lines of longitude converge.

Given latitude and longitude it is possible to determine distance between any pair of points,

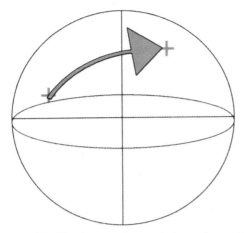

Figure 4.11 The shortest distance between two points on the sphere is an arc of a great circle, defined by slicing the sphere through the two points and the center (all lines of longitude, and the Equator, are great circles). The circle formed by a slice that does not pass through the center is a small circle (all lines of latitude except the Equator are small circles)

not just pairs along lines of longitude or latitude. It is easiest to pretend for a moment that the Earth is spherical, because the flattening of the ellipsoid makes the equations much more complex. But on a spherical Earth the shortest path between two points is a *great circle*, or the arc formed if the Earth is sliced through the two points and through its center (Figure 4.11; an off-center slice creates a *small circle*). The length of this arc on a spherical Earth of radius R is given by:

$$R\cos^{-1}[\sin\phi_1\sin\phi_2 + \cos\phi_1\cos\phi_2\cos(\lambda_1 - \lambda_2)]$$

where the subscripts denote the two points (and see the discussion of Measurement in Section 13.3.1). For example, the distance from a point on the Equator at longitude 90 East (in the Indian Ocean between Sri Lanka and the Indonesian island of Sumatra) and the North Pole is found by evaluating the equation for $\phi_1 = 0$, $\lambda_1 = 90$, $\phi_2 = 90$, $\lambda_2 = 90$. It is best to work in radians (1 radian is 57.30 degrees, and 90 degrees is $\pi/2$ radians). The equation evaluates to $R\cos^{-1}0$, or $R\,\pi/2$, or one quarter of the circumference of the Earth. Using a radius of 6378 km this comes to 10,018 km, or close to 10,000 km (not surprising, since the French originally defined the meter in the late 18th century as one ten millionth of the distance from the Equator to the Pole).

4.7 Projections and Coordinates

Latitude and longitude define location on the Earth's surface in terms of angles with respect to well-defined references: the Royal Observatory at Greenwich, UK, the center of mass, and the axis of rotation. As such, they constitute the most comprehensive system of georeferencing, and support a range of forms of analysis, including the calculation of distance between points, on the curved surface of the Earth. But many technologies for working with geographic data are inherently flat, including paper and printing, which evolved over many centuries long before the advent of digital geographic data and GIS. For various reasons, therefore, much work in GIS deals with a flattened or *projected* Earth, despite the price we pay in the distortions that are an inevitable consequence of flattening. Specifically, the Earth is often flattened because:

- paper is flat, and paper is still used as a medium for inputting data to GIS by scanning or digitizing (see Section 10.3.2), and for outputting data in map or image form;

- rasters are inherently flat, since it is impossible to cover a curved surface with equal squares without gaps or overlaps;

- photographic film is flat, and film cameras are still used widely to take images of the Earth from aircraft to use in GIS;

- when the Earth is seen from space, the part in the center of the image has the most detail, and detail drops off rapidly, the back of the Earth being invisible; in order to see the whole Earth with approximately equal detail it must be distorted in some way, and it is most convenient to make it flat.

The Cartesian coordinate system (Figure 4.12) assigns two coordinates to every point on a flat surface, by measuring distances from an origin

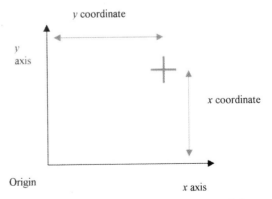

Figure 4.12 A Cartesian coordinate system, defining the location of the red point in terms of two measured distances from the Origin, parallel to the two axes

parallel to two axes drawn at right angles. We often talk of the two axes as x and y, and of the associated coordinates as the x and y coordinate, respectively. Because it is common to align the y axis with North in geographic applications, the coordinates of a projection on a flat sheet are often termed *easting* and *northing*.

> Although nothing in a digital computer is flat, there are several good reasons for using projections in GIS to flatten the Earth.

One way to think of a map projection, therefore, is that it transforms a position on the Earth's surface identified by latitude and longitude (ϕ, λ) into a position in Cartesian coordinates, (x, y). Every recognized map projection, of which there are many, can be represented as a pair of mathematical functions:

$$x = f(\phi, \lambda)$$
$$y = g(\phi, \lambda)$$

For example, the famous Mercator projection uses the functions:

$$x = \lambda$$
$$y = \ln \tan[\phi/2 + \pi/4]$$

where ln is the natural log function. The inverse transformations that map Cartesian coordinates back to latitude and longitude are also expressible as mathematical functions; in the Mercator case they are:

$$\lambda = x$$
$$\phi = 2 \tan^{-1} e^{y} - \pi/2$$

where e denotes the constant 2.71828. Many of these functions have been implemented in GIS, allowing users to work with virtually any recognized projection and datum, and to convert easily between them.

> Two datasets can differ in both the projection and the datum, so it is important to know both for every dataset.

Projections necessarily distort the Earth, so it is impossible in principle for the scale (distance on the map compared with distance on the Earth, see Box 5.3) of any flat map to be perfectly uniform, or for the pixel size of any raster to be perfectly constant. But projections can preserve certain properties, and two such properties are particularly important, although any projection can achieve at most one of them, not both:

- the *conformal* property, which ensures that the shapes of small features on the Earth's surface are preserved on the projection: in other words, that the scales of the projection in the x and y directions are always equal;
- the *equal area* property, which ensures that areas measured on the map are always in the same proportion to areas measured on the Earth's surface.

The conformal property is useful for navigation, because a straight line drawn on the map has a constant bearing (the technical term for such a line is a *loxodrome*). The equal area property is useful for various kinds of analysis involving areas, such as the computation of the area of someone's property.

Besides their distortion properties, another common way to classify map projections is by analogy to a physical model of how positions on the map's flat surface are related to positions on the curved Earth. There are three major classes (Figure 4.13):

- *cylindrical* projections, which are analogous to wrapping a cylinder of paper around the Earth, projecting the Earth's features onto it, and then unwrapping the cylinder;
- *azimuthal* or *planar* projections, which are analogous to touching the Earth with a sheet of flat paper; and
- *conic* projections, which are analogous to wrapping a sheet of paper around the Earth in a cone.

In each case, the projection's *aspect* defines the specific relationship, e.g. whether the paper is wrapped around the Equator, or touches at a pole. Where the paper coincides with the surface the scale of the projection is 1, and where the paper is some distance outside the surface the projected feature will be larger than it is on the Earth. *Secant* projections attempt to minimize distortion by allowing the paper to cut through the surface, so that scale can be both greater, and less than 1 (Figure 4.13; projections in which scale is always 1 or greater are called *tangent*).

All three types can have either conformal or equal area properties, but, as stated above, not both. Figure 4.14 shows examples of several common projections, and shows how the lines of latitude and longitude map onto the projection, in a (distorted) grid known as a *graticule*.

The next sections describe several particularly important projections in detail, and the coordinate systems that they produce. Each is important to GIS, and users are likely to come across them frequently. The map projection (and datum) used to make a dataset is sometimes not known, so it

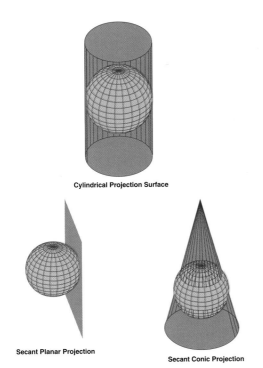

Cylindrical Projection Surface

Secant Planar Projection

Secant Conic Projection

Figure 4.13 The basis for three types of map projections – cylindrical, planar, and conic. In each case a sheet of paper is wrapped around the Earth, and positions of objects on the Earth's surface are projected onto the paper. The cylindrical projection is shown in the *tangent* case, with the paper touching the surface, but the planar and conic projections are shown in the *secant* case, where the paper cuts into the surface (Source: Peter H. Dana)

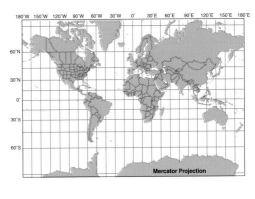

Mercator Projection

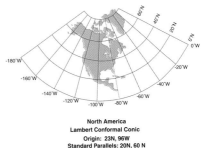

North America
Lambert Conformal Conic
Origin: 23N, 96W
Standard Parallels: 20N, 60 N

Figure 4.14 Examples of some common map projections. The Mercator projection is a tangent cylindrical type, shown here in its familiar Equatorial aspect (cylinder wrapped around the Equator). The Lambert Conformal Conic projection is a secant conic type. In this instance the cone onto which the surface was projected intersected the Earth along two lines of latitude: 20 North and 60 North (Source: Peter H. Dana)

is helpful to know enough about map projections, and coordinate systems to make intelligent guesses when trying to combine such a dataset with other data. Several excellent books on map projections are listed in the References.

4.7.1 The Plate Carrée or Cylindrical Equidistant projection

The simplest of all projections simply maps longitude as *x* and latitude as *y*, and for that reason is also known informally as the *unprojected* projection. The result is a heavily distorted image of the Earth, with the poles smeared along the entire top and bottom edges of the map, and a very strangely shaped Antarctica. Nevertheless, it is the view that we most often see when images are created of the entire Earth from satellite data (for example in illustrations of sea surface temperature that show the El Niño or La Niña effects). The projection is not conformal (small shapes are distorted) and not

equal area, though it does maintain the correct distance between every point and the Equator. It is normally used only for the whole Earth, and maps of large parts of the Earth, such as the USA or Canada, look distinctly odd in this projection. Figure 4.15 shows the projection applied to North America, and also shows a comparison of three familiar projections of the United States: the Plate Carrée, Mercator, and Lambert Conformal Conic.

> When longitude is assigned to *x* and latitude to *y* a very odd-looking Earth results.

Serious problems can occur when doing analysis using this projection. Moreover, since most methods of analysis in GIS are designed to work with Cartesian coordinates rather than latitude and longitude, the same problems can arise in using analysis when a dataset uses latitude and longitude, or so-called geographic coordinates. For example, a command to generate a circle of radius one unit in this projection will create a figure that is two degrees of latitude across in

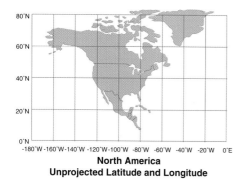

North America
Unprojected Latitude and Longitude

Three Map Projections Centered at 39 N and 96 W

Mercator

Lambert Conformal Conic

Un-Projected Latitude and Longitude

Figure 4.15 (A) The so-called *unprojected* or Plate Carrée projection, a tangent cylindrical projection formed by using longitude as *x* and latitude as *y*. (B) A comparison of three familiar projections of the USA. The Lambert Conformal Conic is the one most often encountered when the USA is projected alone, and is the only one of the three to curve the parallels of latitude, including the northern border on the 49th Parallel (Source: Peter H. Dana)

the north–south direction, and two degrees of longitude across in the east–west direction. On the Earth's surface this figure is not a circle at all, and at high latitudes it is a very squashed ellipse. What happens if you ask your favorite GIS to

generate a circle, and add it to a dataset that is in geographic coordinates? Does it recognize that you are using geographic coordinates and automatically compensate for the differences in distances east–west and north–south away from the Equator, or does it in effect operate on a Plate Carrée projection and create a figure that is an ellipse on the Earth's surface? If you ask it to compute distance between two points defined by latitude and longitude, does it use the true shortest (great circle) distance based on the equation in Section 4.6, or the formula for distance in a Cartesian coordinate system on a distorted plane?

It is wise to be careful when using a GIS to analyze data in latitude and longitude rather than in projected coordinates, because serious distortions of distance, area, and other properties may result.

4.7.2 The Universal Transverse Mercator projection

The UTM system is often found in military applications, and in datasets with global or national coverage. It is based on the Mercator projection, but in *transverse* rather than Equatorial aspect, meaning that the projection is analogous to wrapping a cylinder around the Poles, rather than around the Equator. There are 60 zones in the system, and each zone corresponds to a half cylinder wrapped along a particular line of longitude, each zone being 6 degrees wide. Thus Zone 1 applies to longitudes from 180W to 174W, with the half cylinder wrapped along 177W; Zone 10 applies to longitudes from 126W to 120W centered on 123W, etc. (Figure 4.16).

The UTM system is secant, with lines of scale 1 located on both sides of the central meridian. The

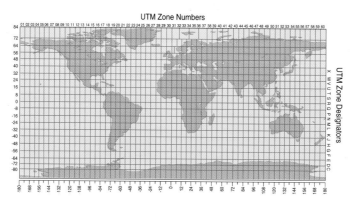

UTM Zone Numbers

Figure 4.16 The system of zones of the Universal Transverse Mercator system. The zones are identified at the top. Each zone is six degrees of longitude in width (Source: Peter H. Dana)

projection is conformal, so small features appear with the correct shape and scale is the same in all directions. Scale is 0.9996 at the central meridian and at most 1.0004 at the edges of the zone. Both parallels and meridians are curved on the projection, with the exception of the zone's central meridian and the Equator. Figure 4.17 shows the major features of one zone.

The coordinates of a UTM zone are defined in meters, and set up such that the central meridian's Easting is always 500,000 m, so Easting varies from near zero to near 1,000,000 m. In the Northern Hemisphere the Equator is the origin of Northing, so a point at Northing 5,000,000 m is approximately 5000 km from the Equator. In the

Southern Hemisphere the Equator is given a Northing of 10,000,000 m, and all other Northings are less than this.

> UTM coordinates are in meters, making it easy to make accurate calculations of short distances between points.

Because there are effectively 60 different projections in the UTM system, maps will not fit together across a zone boundary. Zones become so much of a problem at high latitudes that the UTM system is normally replaced with azimuthal projections centered on each Pole (known as the UPS or Universal Polar Stereographic system) above 80 degrees latitude. The problem is especially critical for cities that cross zone boundaries, such as Calgary, Alberta, Canada (crosses the boundary at 114W between Zone 11 and Zone 12). In such situations one zone can be extended to cover the entire city, but this results in distortions that are larger than normal. Another option is to define a special zone, with its own central meridian selected to pass directly through the city's center. Italy is split between Zones 32 and 33, and Italian maps carry both sets of eastings and northings.

UTM coordinates are easy to recognize, because they commonly consist of a six-digit integer followed by a seven-digit integer (and decimal places if precision is greater than a meter), and sometimes include zone numbers, and hemisphere codes. They are a good basis for analysis, because distances can be calculated from them for points within the same zone with no more than 0.04% error. But they are complicated enough that their use is effectively limited to professionals (the so-called "spatially aware professionals" or SAPs defined in Section 1.4) except in applications where they can be hidden from the user. UTM grids are marked on many topographic maps, and many countries project their topographic maps using UTM, so it is easy to obtain UTM coordinates from maps for input to digital datasets, either by hand or automatically using scanning or digitizing (Section 10.3.2).

4.7.3 State Plane Coordinates and other local systems

Although the distortions of the UTM system are small, they are nevertheless too great for some purposes, particularly in accurate surveying. Zone boundaries are also a problem in many applications, because they follow arbitrary lines of longitude rather than boundaries between jurisdictions. In the 1930s each US state agreed to adopt its own projection and coordinate system,

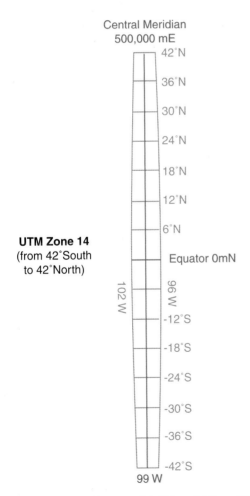

Figure 4.17 Major features of UTM Zone 14 (from 102W to 96W). The central meridian is at 99W. Scale factors vary from 0.9996 at the central meridian to 1.0004 at the zone boundaries. See text for details of the coordinate system (Source: Peter H. Dana)

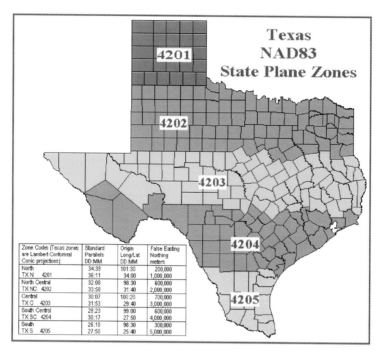

Texas
NAD83
State Plane Zones

4201

4202

4203

4204

4205

Zone Codes (Texas zones are Lambert Conformal Conic projections)	Standard Parallels DD:MM	Origin Long/Lat DD:MM	False Easting Northing meters
North TX N 4201	34:39 36:11	101:30 34:00	200,000 1,000,000
North Central TX NC 4202	32:08 33:58	98:30 31:40	600,000 2,000,000
Central TX C 4203	30:07 31:53	100:20 29:40	700,000 3,000,000
South Central TX SC 4204	28:23 30:17	99:00 27:50	600,000 4,000,000
South TX S 4205	26:10 27:50	98:30 25:40	300,000 5,000,000

Figure 4.18 The five State Plane Coordinate zones of Texas. Note that the zone boundaries are defined by counties, rather than parallels, for administrative simplicity (Source: Peter H. Dana)

generally known as State Plane Coordinates (SPC), in order to support these high-accuracy applications. Projections were chosen to minimize distortion over the area of the state, so choices were often based on the state's shape. Some large states decided that distortions were still too great, and designed their SPCs with internal zones (for example, Texas has five zones based on the Lambert Conformal Conic projection, Figure 4.18, while Hawaii has five zones based on the Transverse Mercator projection). Many GIS have details of SPCs already stored, so it is easy to transform between them and UTM, or latitude and longitude. The system was revised in 1983 to accommodate the shift to the new North American Datum (NAD83).

All US states have adopted their own specialized coordinate systems for applications such as surveying that require very high accuracy.

Many other countries have adopted coordinate systems of their own. For example, the UK uses a single projection and coordinate system known as the National Grid that is based on the Transverse Mercator projection (see Box 4.1) and is marked on all topographic maps. Canada uses a uniform coordinate system based on the Lambert Conformal Conic projection, which has properties that are useful at mid to high latitudes, for applications where the multiple zones of the UTM system would be problematic.

4.8 Converting Georeferences

GIS are particularly powerful tools for converting between projections and coordinate systems, because these transformations can be expressed as numerical operations. In fact this ability was one of the most attractive features of early systems for handling digital geographic data, and drove many early applications. But other conversions, e.g. between placenames and geographic coordinates, are much more problematic. Yet they are essential operations. Almost everyone knows their mailing address, and can identify travel destinations by name, but few are able to specify these locations in coordinates, or to interact with geographic information systems on that basis. GPS (Section 10.2.2.2) is attractive precisely because it allows its users to determine their latitude and longitude, or UTM coordinates, directly at the touch of a button.

Methods of converting between georeferences are important for:

● converting lists of customer addresses to co-ordinates for mapping or analysis;

● combining datasets that use different systems of georeferencing;

● converting to projections that have desirable properties, e.g. no distortion of area, for analysis;

● searching the Internet or other distributed data resources for data about specific locations;

● positioning GIS map displays by recentering them on places of interest that are known by name (these last two are sometimes called *locator* services).

The oldest method of converting georeferences is the *gazetteer*, the name commonly given to the index in an atlas that relates placenames to latitude and longitude, and to relevant pages in the atlas where information about that place can be found. In this form the gazetteer is a useful locator service, but it works only in one direction as a conversion between georeferences (from placename to latitude and longitude). Gazetteers have evolved substantially in the digital era, and it is now possible to obtain large databases of placenames and associated coordinates and to access services that allow such databases to be queried over the Internet (e.g. the Alexandria Digital Library gazetteer, alexandria.ucsb.edu; the US Geographic Names Information System, mapping.usgs.gov/www.gnis).

4.9 Summary

This chapter has looked in detail at the complex ways in which humans refer to specific locations on the planet. Any form of geographic informa-tion must involve some kind of georeference, and so it is important to understand the common methods, and their advantages and disadvan-tages. Many of the benefits of GIS rely on accurate georeferencing – the ability to link different items of information together through common geographic location; the ability to measure distances and areas on the Earth's surface, and to perform more complex forms of analysis; and the ability to communicate geographic information in forms that can be understood by others.

Georeferencing began in early societies, to deal with the need to describe locations. As humanity has progressed, we have found it more and more necessary to describe locations accurately, and

over wider and wider domains, so that today our methods of georeferencing are able to locate phenomena unambiguously and to high accuracy anywhere on the Earth's surface. Today, with modern methods of measurement, it is possible to direct another person to a point on the other side of the Earth to an accuracy of a few centimeters, and this level of accuracy and referencing is regularly achieved in such areas as geophysics and civil engineering.

But georeferences can never be perfectly accurate, and it is always important to know something about spatial resolution. Questions of measurement accuracy are discussed in Section 6.3.2 while Chapter 15 deals with techniques for representation of phenomena that are inherently fuzzy, such that it is impossible to say with certainty whether a given point is inside or outside the georeference.

Questions for Further Study

1. Visit your local map library, and determine: (1) the projections and datums used by selected maps; and (2) the coordinates of your house in several common georeferencing systems.
2. Summarize the arguments for and against a single global figure of the Earth, such as WGS84.
3. How would you go about identifying the projection used by a common map source, such as the weather maps shown by a TV station or in a newspaper?
4. What would be the best type of map projection to use in GIS projects that (1) used the GIS to measure the areas of land available for certain types of use, or (2) used the GIS to create maps for ocean navigation?
5. Calculate the distance from your home city to a city on another continent.
6. Transform the latitude and longitude of your home into its Mercator projection equivalent, using the equations given in this chapter.
7. Compare and contrast the advances that have been made in increasing georeferencing resolution for (1) socioeconomic, and (2) environmental data.

Online Resources

NCGIA Core Curricula (www.ncgia.ucsb.edu/pubs/core.html):
Core Curriculum in GIScience, Sections 1.3 (Position on the Earth, edited by Ken Foote), 1.3.1 (Coordinate Systems Overview, Peter Dana), 1.3.2 (Latitude and Longitude, Anthony

Kirvan), 1.3.3 (The Shape of the Earth, Peter Dana), 1.3.4 (Discrete Georeferencing, David Cowen), 1.3.5 (Global Positioning Systems Overview, Peter Dana), 1.4.1 (Projections and Transformations), and 1.4.2 (Maps as Representations of the World, Judy Olson)

Core Curriculum in GIS, 1990, Units 26 (General Coordinate Systems), 27 (Map Projections), and 29 (Discrete Georeferencing)

ESRI Virtual Campus courses (campus.esri.com):
Understanding Map Projections and Coordinate Systems, by ESRI
Understanding Geographic Data, by David DiBiase (see the lesson on ''Map Scale and Map Projections'' in the Module ''What is Geographic Data?'')

Reference Links

Maguire D J, Goodchild M F, and Rhind D W (eds) 1991 *Geographical Information Systems: Principles and applications*. Harlow, UK: Longman (Text available online from 'Links to Big Book 1' at www.wiley.com/gis and www.wiley.co.uk/gis).
Chapter 10, Coordinate systems and map projections for GIS (Maling D)

Longley P A, Goodchild M F, Maguire D J, and Rhind D W (eds) 1999 *Geographical Information Systems: Principles, techniques, management and applications*. New York: John Wiley.
Chapter 30, Spatial referencing and coordinate systems (Seeger H).

References

Bugayevskiy L M and Snyder J P 1995 *Map Projections: A reference manual*. London: Taylor, and Francis.
Maling D H 1992 *Coordinate Systems and Map Projections* (2nd edn). Oxford: Pergamon.
Snyder J P 1997 *Flattening the Earth: Two thousand years of map projections*. Chicago: University of Chicago Press.

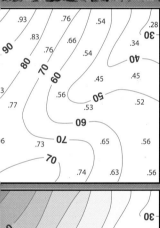

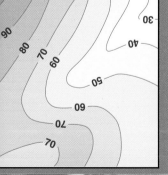

THE NATURE OF GEOGRAPHIC DATA

5

Geographic information is sufficiently special to warrant its own type of information system. This chapter elaborates on the spatial is special theme by examining the nature of geographic data. It sets out what it is about geographic data that distinguishes them from all other kinds, and what special tools are needed to analyze and work with them. Most geographic data come to us in the form of samples, selected from the more complex reality we are trying to represent, so it is important to understand the nature of samples, and the basis on which we can infer information about the gaps between samples. Another key property of geographic information is its level of detail. This chapter introduces the key concept of spatial autocorrelation, and shows how it can be measured. The concept of fractals provides a solid theoretical foundation for geographic representations, especially in making it possible to predict the effects of changing levels of detail.

Learning objectives

After reading this chapter you will understand:

- How Tobler's First Law of Geography is formalized through the concept of spatial autocorrelation;

- The relationship between scale and the level of geographic detail in a representation;

- The principles of building representations around geographic samples;

- How the property of smoothness and continuous variation can be used to characterize geographic variation;

- How fractals can be used to measure and simulate surface roughness.

5.1 Introduction

Our lives may be infinitesimally small compared with the geographic extent and history of the world, but they are nevertheless very intricate in detail. Human behavior exhibits structure in geographic space, both in our day-to-day (often repetitive) decisions about where to go, what to do, how much time to spend doing it, and in our longer-term (one-off) decisions about where to live, how to achieve career objectives, and how to balance work, leisure, and family pursuits. In terms of GIS applications, these are respectively examples of the *operational* and *strategic* decisions discussed in Section 1.1. A principal objective of geographic analysis is to understand how both types of decisions are structured over space. For example, how do people structure the searches they make for new housing when they move, and how can that information help us to design better information systems to help them? Decisions are also structured over time: "the past is the key to the present" aptly summarizes the way in which past actions and thinking condition future actions.

Together, space and time define the geographic context of our past actions, and set the geographic limits of new operational and strategic decisions – they condition what we know, what we perceive to be our options, and how we choose among them. If we try to explain our activity patterns over either the short or the long term, we need to specify how and why spatial and temporal context affects what we do. In a day-to-day operational sense it is obvious that the locus of our daily activities is very much determined by where we live and work. In a strategic sense, where we were born, grew up, and went to college can all affect where we choose to work or live in later life.

Our behavior in geographic space often reflects past patterns of behavior.

In Chapter 3, Figure 3.1 was derived from one of the very few theories generated wholly within the discipline of geography. It derived from Torsten Hägerstrand's time geography treatise, conceived in the 1950s, and illustrates the trajectory of an individual's day-to-day activities across space and time. We think of much of our own spatial behavior (such as the daily commute to work, or shopping trips) as routine, even perfectly repetitive. Yet when we come to represent the spatial and temporal activity patterns of many individuals, the task becomes difficult and error prone. In the terminology of Section 1.3, generalized *laws* of the behavior of aggregations of humans are most unlikely to work perfectly.

This is also true of spatial and temporal representations in general – whether our interest be in the representation of landscape evolution, or urban growth, or disease diffusion, for example. We saw in Chapter 3 that the fundamental problem of GIS lies in identifying what to leave in and what to take out of digital representations. We wish to record all spatial and temporal events (or occurrences) of significance, without becoming mired in irrelevant detail, and need to acknowledge that they may be unevenly distributed over space and time.

Thus far, we have thought of good GIS data as abstracted from continuous spatial distributions and from sequences of events over continuous time. Some events, such as the daily rhythm of the journey to work, are clearly incremental extensions of past practice, while others, such as residential relocation, constitute sudden breaks with the past. Similarly, landscapes of gently undulating terrain are best thought of as smooth and continuous, while others (such as the landscapes developed about fault systems, or mountain ranges) are best conceived as discretely bounded, jagged, and irregular. Smoothness and irregularity turn out to be among the most important distinguishing characteristics of geographic data.

Some geographic phenomena vary smoothly across space, while others can exhibit extreme irregularity, in violation of Tobler's Law.

The scale or level of detail at which we seek to represent reality often determines whether spatial and temporal phenomena appear regular or irregular. But so too does the degree and nature of *spatial heterogeneity*, a term that refers to the tendency of geographic places and regions to be different from each other. Everyone would recognize the extreme difference of landscapes between such regions as the Antarctic, the Nile delta, the Sahara desert, or the Amazon basin, and many would recognize the more subtle differences between the Central Valley of California, the Northern Plain of China, and the valley of the Ganges in India. Heterogeneity occurs both in the way the landscape looks, and in the way processes act on the landscape (the form/process distinction of Section 1.3). With respect to Hägerstrand's time geography, it is clear that some processes, such as the journey to work, simply oscillate while the daily activity pattern maintains a constant, controlled variation. But other patterns of behavior are uncontrolled over time, as in the distance a child travels from home on unaccompanied trips as he or she grows up, or in the patterns people trace when they move from one place of residence to another. While the spatial variation in some processes simply oscillates about an average, other processes vary ever more the longer they are observed. As a general rule, spatial data exhibit an increasing range of values, or increased heterogeneity, with increased distance. In this chapter we focus on the way that phenomena vary across space, and the general nature of geographic variation. Later, we return to the techniques for measuring spatial variation – we consider descriptive measures in Section 14.2 and variograms in Section 13.4.4.3.

The real-world GIS applications described in this book variously share practical goals of operational and strategic problem-solving. What, then, is an appropriate scale or level of detail at which to build a representation for a particular application? What is the most appropriate snapshot interval for recording events and occurrences? How might we generalize from our measurements in order to identify the spatial structure of a given application in a GIS? And what formal methods and techniques can we use to relate key spatial events and outcomes to one another?

of GIS, of selecting what to leave in and what to take out of our digital representations of the real world? There are some clues that we can develop from Hägerstrand's time geography, since unevenness in the spatio-temporal outcomes of human activities presents a special case of spatial and temporal processes in natural and artificial (human-made) environments. In Section 3.1 we presented Waldo Tobler's (Box 5.1) formalization of our most general understanding of the spatial distribution of events and occurrences. This First Law of Geography states that everything is related to everything else, but near things are more related than distant things. Formally, this property is known as *spatial autocorrelation*, measures of which quantify the degree to which near and more distant things are interrelated. The relationship between consecutive events in *time* is formalized in the analogous concept of *temporal autocorrelation*. The analysis of time series data is in some senses more straightforward, since the direction of causality is only one way – past events are sequentially related to the present and to the future. This chapter (and book) is principally concerned with spatial, rather than temporal, autocorrelation.

5.2 The Fundamental Problem Revisited

How can we respond to the fundamental problem

Explanation in time need only look to the past, but explanation in space must look in all directions simultaneously.

Box 5.1 *Waldo Tobler and the First Law of Geography*

Waldo Tobler (Figure 5.1) was born in Switzerland. He received his degrees in Geography from the University of Washington in Seattle, spent several years at the University of Michigan, and is currently Professor Emeritus at the University of California in Santa Barbara. Very unusually for a geographer, until his retirement he held the positions of Professor of Geography and Professor of Statistics at the same institution. He has used computers in geographic research for over forty years, with emphasis on mathematical modeling and graphic interpretations. He formulated the "First Law of Geography" in 1970 while producing a computer movie. He is also the inventor of novel and unusual map projections, among which was the first derivation of the partial differential equations for cartograms (Section 12.2.4). He also invented a method for interpolating spatial data (Section 5.5), which results in smooth two-dimensional mass-preserving surfaces. Waldo was one of the principal investigators and a Senior Scientist in the National Science Foundation sponsored National Center for Geographic Information and Analysis, and is a member of the National Academy of Sciences of the United States.

Today Waldo remains an active researcher. His current concerns relate to ideas in computational geography including the analysis of geographical vector fields and the development of global trade models. His university teaching has centered upon courses on the History of Cartography, Geographic Transformations, and Migration.

Figure 5.1 Waldo Tobler, geographer, and statistician

In the analysis of spatial data, the processes that give rise to spatial autocorrelation can be two- and even three-dimensional. An understanding of these processes and their outcomes has very important implications for the way in which we abstract and collect data to hold in representations, and for the ways in which we seek to draw inferences between events and occurrences. First, understanding of the nature of spatial autocorrelation can help us to generalize from sample observations in order to build spatial representations – in fact, if spatial autocorrelation did not exist the entire process of trying to simplify the world in a representation would be impossible. Second, the property of spatial autocorrelation strains or can even negate the validity of many of the conventional methods and techniques that have been developed to tell us about the relatedness of occurrences and events. Thus the property of autocorrelation can both aid and frustrate the representation of spatial distributions. On the downside, it diminishes the power of conventional statistical analysis. Yet, more positively, it is a property which can be exploited, since if we can measure and understand the nature of autocorrelation, it becomes easier to build well-generalized representations of spatial distributions.

Recognition of the importance of geographic *scale* or level of detail is fundamental to an understanding of the likely strength and nature of autocorrelation in any given application.

Together, the scale and spatial structure of a particular application suggest ways in which we should *sample* geographic reality, and the ways in which we should *interpolate* between sample observations in order to build our representation. The concepts of scale, sampling, and interpolation are key to geographic representation, and we will return to them throughout much of this book. We will focus here upon the rudiments of interpolation, although we will return to consider it in more detail in Chapter 14.

5.3 Spatial Autocorrelation and Scale

In Chapter 3 (Box 3.3) we classified attribute data into the nominal, ordinal, interval, ratio, and cyclic scales of measurement. Objects existing in space are described by locational (spatial) descriptors, and are conventionally classified using the taxonomy shown in Box 5.2.

Spatial autocorrelation measures attempt to deal simultaneously with similarities in the location of spatial objects (Box 5.2) and their attributes (Box 3.3). If features that are similar in location are also similar in attributes, then the pattern as a whole is said to exhibit *positive spatial autocorrelation*. Conversely, *negative spatial autocorrelation* is said to exist when features which are close together in space tend to be more dissimilar in attributes than features

---

Box 5.2 Types of spatial object

We saw in Section 3.4.1 that geographic objects are classified according to their *topologic dimension*, which provides a measure of the way they fill space. For present purposes we assume that dimensions are restricted to *integer* (whole number) values, though in later sections (Sections 5.8, and 14.2.6) we relax this constraint and consider geographic objects of non-integer (fractional, or *fractal*) dimension. All geometric objects can be used to represent occurrences at absolute locations (*natural* objects), or they may be used to summarize spatial distributions (*artificial* objects).

A *point* has neither length nor breadth nor depth, and hence is said to be of dimension 0. Points may be used to indicate spatial occurrences or events, and their spatial patterning. *Point pattern analysis* is used to identify whether occurrences or events are inter-related – as in the analysis of the incidence of crime, or in identifying whether patterns of disease infection might be related to environmental or social factors. The *centroid* of an area object is an artificial point reference, which is located so as to provide a summary measure of the location of the object (Section 14.2.1).

Lines have length, but not breadth or depth, and hence are of dimension 1. They are used to represent linear entities such as roads, pipelines, and cables, which frequently build together into networks. They can also be used to measure distances between spatial objects, as in the measurement of inter-centroid distance. In order to reduce the burden of data capture and storage, lines are often held in GIS in *generalized* form (see Chapter 6).

Area objects have the two dimensions of length and breadth, but not depth. They may be used to represent natural objects, such as agricultural fields, but are also commonly used to represent artificial aggregations, such as census tracts (see below). Areas may bound linear features and enclose points, and GIS functions can be used to identify whether a given area encloses a given point (Section 13.4.2).

Box 5.2 *Continued*

Surface or *volume* objects have length, breadth, and depth, and hence are of dimension 3. They are used to represent natural objects such as river basins, or artificial phenomena such as the population potential of shopping centers (Section 13.4.5). Surfaces are frequently derived by *interpolation* between lower-dimension entities such as spot (point) heights, and contour lines (Chapter 14), where the depth dimension is actually an attribute value attached to a particular point in two-dimensional space (see the discussion of "2.5D" in Box 3.5).

Time is often considered to be the fourth dimension of spatial objects, although GIS remains poorly adapted to the modeling of temporal change.

The relationship between higher- and lower-dimension spatial objects is analogous to that between higher- and lower-order attribute data, in that lower-dimension objects can be derived from those of higher dimension but not vice versa. Certain phenomena, such as population, may be held as natural or artificially imposed spatial object types. The chosen way of representing phenomena in GIS not only defines the apparent nature of geographic variation, but also the way in which geographic variation may be analyzed. Some objects, such as agricultural fields or digital terrain models, are represented in their natural state. Others are transformed from one spatial object class to another, as in the transformation of population data from individual points to census tract areas, for reasons of confidentiality, convenience, or convention. Some high-order representations are created by interpolation between lower-order objects, as in the creation of digital terrain models (DTMs) from spot height data.

The classification of spatial phenomena into object types is fundamentally dependent upon scale. For example, on a less-detailed map of the world, New York is represented as a zero-dimensional point. On a more-detailed map such as a road atlas it will be represented as a two-dimensional area. Yet if we visit the city, it is very much experienced as a three-dimensional entity, and virtual reality systems seek to represent it as such (see Section 12.3).

which are further apart (in opposition to Tobler's Law). Zero autocorrelation occurs when attributes are independent of location. Figure 5.2 presents some simple field representations of a geographic variable in 64 cells that can each take one of two values, coded blue and white (see Goodchild 1986 for a more extended discussion of spatial autocorrelation and this example). Each of the five illustrations contains the same set of attributes, 32 white cells and 32 blue cells, yet the spatial arrangements are very different. Figure 5.2(A) presents the familiar chess board, and illustrates extreme negative spatial auto-correlation between neighboring cells. Figure 5.2(E) presents the opposite extreme of positive autocorrelation, when blue and white cells cluster together in homogeneous regions. The other illustrations show arrangements which exhibit intermediate levels of autocorrelation. Figure 5.2(C) corresponds to spatial inde-pendence, or no autocorrelation, Figure 5.2(B) shows a relatively dispersed arrangement, and Figure 5.2(D) a relatively clustered one.

Spatial autocorrelation is determined both by similarities in position, and by similarities in attributes.

The patterns shown in Figure 5.2 are examples of a particular case of spatial autocorrelation. In terms of the classification developed in Chapter 3 (Box 3.3) the attribute data are *nominal* (blue and white simply identify two different possibilities, with no implied order and no possibility of difference, or ratio) and their spatial distribution is conceived as a field, with a single value everywhere. The figure gives no clue as to the true dimensions of the area being represented. Usually, similarities in attribute values may be more precisely measured on higher-order mea-surement scales, enabling continuous measures of spatial variation (see Section 15.1.2 for a discussion of precision). As we see below, the way in which we define what we mean by *neighboring* in investigating spatial arrangements may be more or less sophisticated. In considering the various arrangements shown in Figure 5.2, we have only considered the relationship between the attributes of a cell and those of its four *immediate* neighbors. But we could include a cell's four diagonal neighbors in the comparison (compare Figure 14.18), and more generally there is no reason why we should not interpret Tobler's Law in terms of a gradual incremental attenuating effect of distance as we traverse successive cells.

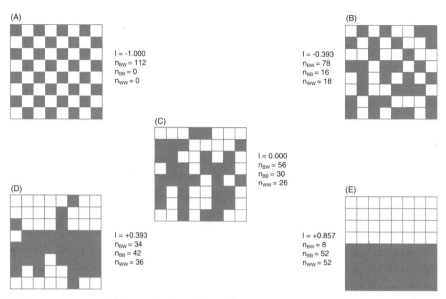

(A)

I = -1.000
n_{BW} = 112
n_{BB} = 0
n_{WW} = 0

(B)

I = -0.393
n_{BW} = 78
n_{BB} = 16
n_{WW} = 18

(C)

I = 0.000
n_{BW} = 56
n_{BB} = 30
n_{WW} = 26

(D)

I = +0.393
n_{BW} = 34
n_{BB} = 42
n_{WW} = 36

(E)

I = +0.857
n_{BW} = 8
n_{BB} = 52
n_{WW} = 52

Figure 5.2 Field arrangements of blue and white cells exhibiting: (A) extreme negative spatial autocorrelation; (B) a dispersed arrangement; (C) spatial independence; (D) spatial clustering; and (E) extreme positive spatial autocorrelation. The values of the *I* statistic are calculated using the equation in Section 5.6 n_{BW}, n_{BB} and n_{WW} denote numbers of blue–white, blue–blue, and white–white joins respectively, and I refers to the Moran statistic (Section 5.6.2) (Source: Goodchild M F 1986 *Spatial Autocorrelation* CATMOG, Norwich: GeoBooks)

We began this chapter by referring to Hägerstrand's time-space geography in order to illustrate the way in which the rhythm of day-to-day activities can be highly, even perfectly, repetitive in the short term. Activity patterns often exhibit strong positive temporal autocorrelation (where you were at this time last week, or this time yesterday is likely to affect where you are now), but

Box 5.3 *The many meanings of scale*

Unfortunately the word *scale* has acquired too many meanings in the course of time. Because they are to some extent contradictory, it is best to use other terms that have clearer meaning where possible.

Scale is in the details. Many scientists use scale in the sense of spatial resolution, or the level of spatial detail in data. Data are fine-scaled if they include records of small objects, and coarse-scaled if they do not.

Scale is about extent. Scale is also used by scientists to talk about the geographic extent or scope of a project: a large-scale project covers a large area, and a small-scale project covers a small area. Scale can also refer to other aspects of the project's scope, including the cost, or the number of people involved.

The scale of a map. Geographic data are often obtained from maps, and often displayed in map form. Cartographers use the term scale to refer to a map's *representative fraction* (the ratio of distance on the map to distance on the ground – see Section 3.7). Unfortunately this leads to confusion (and often bemusement) over the meaning of *large* and *small* with respect to scale. To a cartographer a large scale corresponds to a large representative fraction, in other words to plenty of geographic detail. This is exactly the opposite of what most other scientists understand by a large-scale study. In this book we have tried to avoid this problem by using *coarse* and *fine* instead.

Scale is often integral to the trade off between the level of spatial resolution and the degree of attribute detail that can be stored in a given application – as in the trade off between spatial and spectral resolution in remote sensing (Section 10.2.1).

For an extensive discussion of these and other meanings of scale (e.g. the degree of spectral or temporal coarseness), and their implications, see Quattrochi and Goodchild (1996). The scale at which data from different sources are usually made available is discussed in Section 10.4.

only if measures are made at the same time every day, that is, at the temporal scale of the daily interval. If, say, sample measurements were taken every 17 hours, measures of the temporal auto-correlation of your activity patterns would likely be much lower. This issue of *sampling interval* is also of direct importance in the measurement of spatial autocorrelation, because spatial events and occurrences can conform to spatial structure. The hierarchical hexagonal market areas of Christaller's Central Place Theory shown in Figure 5.3 provide one of the best-known examples in geography. Just as temporal periodicity may be missed if measures are made at an inappropriate time interval (or temporal scale), so spatial periodicity may only be detected if the spatial scale of measurement is appropriate to the distribution of the spatial phenomenon. In short, measures of spatial and temporal autocorrelation are both *scale dependent* (this point is explored in more detail in Section 14.2.5).

The nested hexagonal market areas shown in Figure 5.3 also introduce the notion of hierarchy, and the possibility that spatial patterning may be repetitive across a range of scales. This is illustrated using a mosaic of squares, rather than hexagons, in Figure 5.4. Figure 5.4(A) presents a coarse-scale representation of attributes in nine

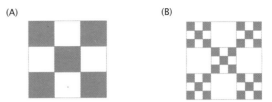

Figure 5.4 A Sierpinski carpet at two levels of resolution: (A) coarse scale; and (B) finer scale

Figure 5.5 Individual rocks may resemble larger scale structures, such as the mountains from which they are broken, in form

squares, and a pattern of negative spatial auto-correlation. In Figure 5.4(B), a finer-scale representation reveals that the smallest black cells replicate the pattern of the whole area in a recursive manner. The pattern of spatial auto-correlation at the coarser scale is replicated at the finer scale, and the overall pattern is said to exhibit the property of *self-similarity*, much like the idealized landscapes of Central Place Theory. Self-similar structure is characteristic of natural as well as social systems: for example, a rock may resemble the physical form of the mountain from which it was broken (Figure 5.5), or small coastal features may resemble larger bays and inlets in structure and form. Self-similarity is a core concept of fractals, a topic introduced in Section 5.8.

5.4 Spatial Sampling

The quest to represent the myriad complexity of the real world requires us to abstract, or sample, events and occurrences from a *sample frame*, defined as the universe of eligible elements of interest. Thus the process of sampling elements

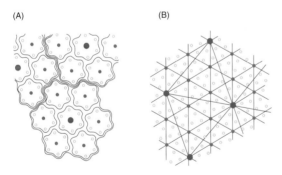

Figure 5.3 The theoretical size and spacing of central places in an urban system according to Christaller's administrative principle, in which hinterlands are nested hierarchically. This is one variant of Central Place Theory, which can be used to identify the hinterlands (A) and routes (B) associated with settlements at different levels in the system. Lowest order centers (open yellow circles) have the smallest thresholds and ranges, and only grocery (convenience) store functions; second order centers (small green circles) have butchers' shops; and third order centers (large red circles) have hardware stores. This normative landscape exhibits clear spatial structure, and a systematic sample design would likely pick up particular levels in the hierarchy. (Adapted from Johnston R J, Gregory D, Pratt G, Watts M (eds) 2000 *The Dictionary of Human Geography* (4th Edition). Oxford: Blackwell Publishers Ltd, 73)

from a sample frame can very much determine the apparent nature of geographic data. A spatial sampling frame might be bounded by the extent of a field of interest, or by the combined extent of a set of areal objects. We can think of sampling as the process of selecting points from a continuous field or, if the field has been represented as a mosaic of areal objects, of selecting some of these objects while discarding others. Scientific sampling requires that each element in the sample frame has a known and prespecified chance of selection.

In some important senses, we can think of any geographic representation as a kind of sample, in that the elements of reality that are retained are abstracted from the real world in accordance with some overall design. This is the case in remote sensing, for example (Section 10.2.1), in which each pixel value is a spatially averaged reflectance value calculated at the spatial resolution characteristic of the sensor. In many situations, we will need consciously to select some observations, and not others, in order to create a generalizable abstraction. This is because, as a general rule, the resources available to any given project do not stretch to measuring every single one of the elements (e.g. soil profiles, migrating animals, shoppers) that we know to make up our population of interest. And even if resources were available, science tells us that this would be wasteful, since procedures of *statistical inference* allow us to infer from samples to the populations from which they were drawn. We will return to the process of statistical inference in Sections 5.7 and 14.4. Here, we will confine ourselves to the question, how do we ensure a good sample?

Geographic data are only as good as the sampling scheme used to create them.

Classical statistics often emphasize the importance of randomness in sound sample design. The purest form, simple random sampling, is well known: each element in the sample frame is assigned a unique number, and a prespecified number of elements are selected using a random number generator. In the case of a spatial sample from continuous space, x, y coordinate pairs might be randomly sampled within the range of x and y values (see Chapter 4 for information on coordinate systems). Since each randomly selected element has a known and prespecified probability of selection, it is possible to make robust and defensible generalizations to the population from which the sample was drawn. A spatially random sample is shown in Figure 5.6(A). Random sampling is integral to probability theory, and this enables us to use the distribution of values in

our sample to tell us something about the likely distribution of values in the parent population from which the sample was drawn.

However, sheer bad luck can mean that randomly drawn elements are disproportionately concentrated amongst some parts of the population at the expense of others, particularly when the size of the sample is small relative to the population from which it was drawn. For example, a survey of household incomes might happen to select households with unusually low incomes. Spatially systematic sampling aims to circumvent this problem, and ensure greater evenness of coverage across the sample frame. This is achieved by identifying a regular sampling interval k (equal to the reciprocal of the sampling fraction N/n, where n is the required sample size and N is the size of the population), and proceeding to select every kth element. In spatial terms, the sampling interval of spatially systematic samples maps into a regularly spaced grid, as shown in Figure 5.6(B). This advantage over simple random sampling may be two-edged, however, if the sampling interval and the spatial structure of the study area coincide, that is, the sample frame exhibits *periodicity* – of the kind shown in the central place landscape (Figure 5.3), for example. In such instances, there may be a consequent failure to detect the true extent of heterogeneity of population attributes. Imagine, for example, that the grid pattern of Figure 5.6(B) were to coincide with the grid plan of a city. In a survey of urban land use, it is extremely unlikely that the attributes of street intersection locations would be representative of land uses elsewhere in the block structure. A number of hybrid sample designs have been devised to get around the vulnerability of spatially systematic sample designs to periodicity, and the danger that simple random sampling may generate freak samples. These include stratified random sampling to ensure evenness of coverage (Figure 5.6(C)) and periodic random changes in the grid width of a spatially systematic sample (Figure 5.6(D)), perhaps subject to minimum spacing intervals.

In certain circumstances, it may be more efficient to restrict measurement to a smaller range of sites – because of the prohibitive costs of transport over large areas, for example. Clustered sample designs, such as that shown in Figure 5.6(E), may be used to generalize about attributes if the cluster presents a microcosm of surrounding conditions. For example, political opinion polls are often taken in shopping centers where shoppers can be deemed broadly representative of the population at large. However, they are unlikely to provide a comprehensive detailed picture of spatial structure.

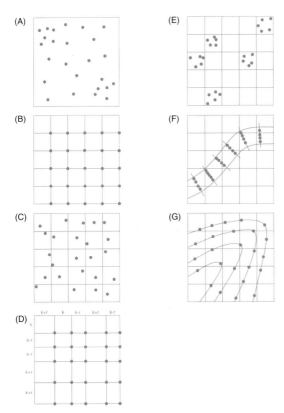

Figure 5.6 Spatial sample designs: (A) simple random sampling; (B) systematic sampling; (C) systematic sampling with local random allocation; (D) systematic sampling with random variation in grid spacing; (E) clustered sampling; (F) transect sampling; and (G) contour sampling

Use of either simple random or spatially systematic sampling presumes that each observation is of equal importance, and hence of equal weight, in building a representation. As such, these sample designs are suitable for circumstances in which spatial structure is weak or non-existent, or where (as in circumstances fully described by Tobler's Law) the attenuating effect of distance is constant in all directions. They are also suitable in circumstances where spatial structure is unknown. Yet in most practical applications, spatial structure is (to some extent at least) known, even if it cannot be wholly explained by Tobler's Law. These conditions make it both more efficient and necessary to devise application-specific sample designs. This makes for improved quality of representation, with minimum resource costs of collecting data. Relevant sample designs include sampling along a transect, such as a soil profile (Figure 5.6(F)) or along a contour line (Figure 5.6(G)).

Consider the area of Leicestershire, UK, illustrated in Figure 5.7, taken from a widely-used series of computer exercises in GIS (Langford 1993). It depicts a landscape in which the hilly relief of an upland area falls away sharply towards a river's flood plain. In identifying the sample spot heights that we might measure and hold in a GIS to create a representation of this area, we would be advised to sample a disproportionate number of observations in and immediately around the upland area of the study area where the local variability of heights is greatest.

In a socio-economic context, imagine that you are required to identify the total repair cost of bringing all housing in a city up to a specified standard. (Such applications are common in, for example, forming bids for Federal or Central Government funding.) A GIS that showed the time period in which different neighborhoods were developed (such as that for the Baltimore-Washington area shown in Figure 2.20) would provide a useful guide to effective use of sampling resources. Newer houses are all likely to be in more or less the same condition, while the repair costs of the older houses are likely to be much more variable, and dependent upon the attention that the occupants have lavished upon them. As a general rule, the older neighborhoods warrant a higher sampling interval than the newer ones, but other considerations may also be accommodated into the sampling design as well – such as construction type (duplex versus apartment, etc.) and local geology (as an indicator of risk of subsidence).

In any application, where the events or phenomena that we are studying are spatially heterogeneous, we will require a large sample to capture the full variability of attribute values at all possible locations. Other parts of the study area may be much more homogeneous in attributes, and a sparser sampling interval may thus be more appropriate. Both simple random and systematic sample designs (and their variants) may be adapted in order to allow a differential sampling interval over a given study area (see Section 15.1.1 for more on this issue with respect to sampling vegetation cover). Thus it may be sensible to partition the sample frame into sub-areas, based on our knowledge of spatial structure – specifically our knowledge of the likely variability of the attributes that we are measuring. Other application-specific special circumstances include:

● Whether source data are ubiquitous or must be specially collected.

● The resources available for any survey undertaking.

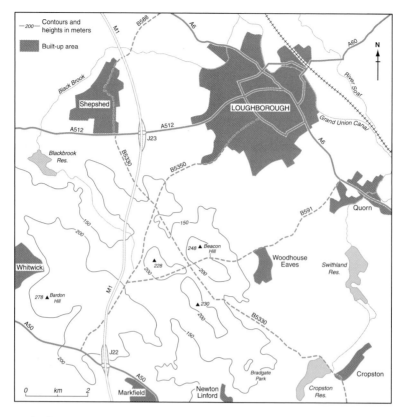

Figure 5.7 An example of physical terrain in which differential sampling would be advisable to construct a representation of elevation (Source: M. Langford, MRRL, University of Leicester)

● The accessibility of all parts of the study area to field observation (still difficult even in the era of widespread availability of Global Positioning System receivers: Section 10.2.2.2).

Stratified sampling designs attempt to allow for the unequal abundance of different phenomena on the Earth's surface.

5.5 Spatial Interpolation

In selectively abstracting, or sampling, part of reality to hold within our representation, it follows that judgment is required to fill in the gaps between the observations that make up a representation. This requires understanding of the likely attenuating effect of distance between the sample observations, and thus of the nature of geographic data (Figures 5.8 and 5.9). That is to say, we need to make an informed judgment

Figure 5.8 We require different ways of interpolating between points, as well as different sample designs, for representing mountains, forested hillsides, and lakeland areas

(A)

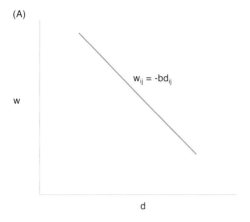

(B)

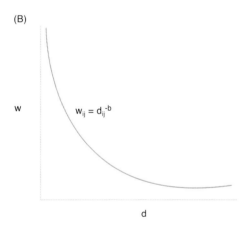

(C)

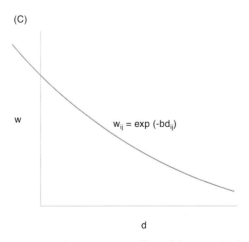

Figure 5.9 The attenuating effect of distance: (A) linear distance decay, $w_{ij} = -b\,d_{ij}$; (B) negative power distance decay, $w_{ij} = d_{ij}^{-b}$; and (C) negative exponential distance decay, $w_{ij} = \exp(-b\,d_{ij})$

about an appropriate *interpolation* function. A literal interpretation of Tobler's Law implies a continuous, smooth, attenuating effect of distance upon the attribute values of adjacent or contiguous spatial objects, or incremental variation in attribute values as we traverse a field. The polluting effect of chemical or oil spillage decreases in a predictable (and in still waters, uniform) fashion with distance from the point source; aircraft noise pollution decreases with distance from the linear flight path, and the number of visits to a National Park decreases at a regular rate as we traverse the counties that adjoin it. This section focuses on principles, and introduces some of the functions that are used to describe effects over distance, or the nature of geographic variation, while Section 13.4.4 discusses the process of spatial interpolation from a technical perspective.

The precise nature of the function used to represent the effects of distance is likely to vary between applications, and Figure 5.9 illustrates several hypothetical types. In mathematical terms, we take b as a parameter which affects the rate at which the weight w_{ij} declines with distance: a small b produces a slow decrease, and a large b a more rapid one. In most applications, the choice of distance attenuation function is the outcome of past experience, the fit of a particular application dataset, and convention. Figure 5.9(A) presents the simple case of linear distance decay, given by the expression:

$$w_{ij} = -b\,d_{ij},$$

as might reflect the noise levels experienced across a transect perpendicular to an aircraft flight path. Figure 5.9(B) presents a negative power distance decay function, given by the expression:

$$w_{ij} = d_{ij}^{-b},$$

which has been used by some researchers to describe the decline in the density of resident population with distance from historic central business district (CBD) areas. Figure 5.9(C) illustrates a negative exponential statistical fit, given by the expression:

$$w_{ij} = e^{-b\,d_{ij}},$$

conventionally used in human geography to represent the decrease in retail store patronage with distance from it.

Each of the attenuation functions illustrated in Figure 5.9 is idealized, in that the effects of distance are presumed to be both regular, continuous, and *isotropic*, that is, uniform in every direction. This may be appropriate for many applications. The notion of smooth, and

Figure 5.10 An early computer graphics simulation of landscape using continuous mathematical functions (Source: Batty and Longley 1994, *Fractal Cities*, Academic Press Ltd, London, UK)

continuous variation underpins many of the representational traditions in cartography, as in the creation of *isopleth* (or isoline) maps. This is described in Box 5.4. To some extent at least, high school math also conditions us to think of spatial variation as continuous, and as best represented by interpolating smooth curves between everything. Yet our understanding of spatial structure often tells us that variation is not smooth and continuous – why else would we adopt the application-specific sample designs described in the spatial structures in the applications that we have already illustrated in Section 5.4? This is apparent from the early computer graphics work of the geographer Michael Batty, an example of which is shown in Figure 5.10. In this illustration, the rolling landscape is made up of continuous sine waves, while the clouds are created using polar functions. Yet the circumstances in which such continuous functions are appropriate are the exception rather than the rule, and the true nature of geographic data might be better represented using other interpolation methods and functions. Thus we need to adapt our thinking to ac-commodate discontinuities in physical (e.g. fault lines or cliffs) and socio-economic distributions (e.g. shifts in household income distributions on crossing the US – Mexico border). These physical, and social issues are illustrated in Figure 5.11, which shows the location of a hypothetical retail store. It is tempting to suggest that the customer catchment of the store is bounded by an approximately circular area, the radius of which is determined by a 10 minute drive time at average speed. In

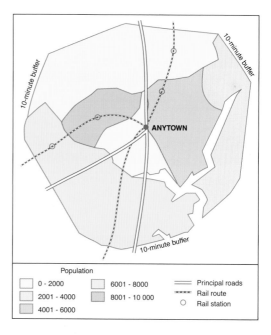

Population

☐ 0 - 2000	☐ 6001 - 8000	▬ Principal roads
☐ 2001 - 4000	▨ 8001 - 10 000	▭ Rail route
☐ 4001 - 6000		○ Rail station

Figure 5.11 Discontinuities in a retail catchment (Source: Adapted from Birkin M, Clarke G P, Clarke M, and Wilson A G 1996 *Intelligent GIS*, Wiley, New York)

practice, the circle is modified slightly because census counts of population are only available for administrative zones. But what is more important is that the catchment depends also on:

- Physical factors, such as rivers and (especially for walking trips) topographic relief.
- Road and rail infrastructure and associated capacity, congestion and access (e.g. rail stations and road access ramps).
- Socio-economic factors, such as the economic divides that abruptly differentiate many communities and have clear implications for shopping behavior.

The population base to Figure 5.11 raises an important issue of the representation of spatial structure. Remember that we said that the circular retail catchment had to be adapted to fit the administrative geography of population enumeration. The distribution of population is shown using choropleth mapping (Box 5.4), which implicitly assumes that population is uniformly distributed within zones and that the only important changes in distribution take place at zone boundaries. Such representations can obscure continuous variations and mask the true pattern of distance attenuation.

Box 5.4 Isopleth and choropleth maps

Isopleth maps are used to visualize phenomena that are conceptualized as fields, and measured on interval or ratio scales. An *isoline* connects points with equal attribute values, such as contour lines (equal height above sea level), *isohyets* (points of equal precipitation), *isochrones* (points of equal travel time), or *isodapanes* (points of equal transport cost). Figure 5.12 illustrates the procedures that are used to create a surface about a set of point measurements (Figure 5.12(A)), such as might be collected from rain gauges across a study region (and see Section 13.4.4 for more technical detail on the process of spatial interpolation). A parsimonious number of user-defined values is identified to define the contour intervals (Figure 5.12(B)). The cartographer or GIS then interpolates a contour between point observations of greater and lesser value (Figure 5.12(C)) using standard procedures of inference, and the other contours are then interpolated using the same procedure (Figure 5.12(D)). Hue or shading can be added to improve user interpretability (Figure 5.12(E)).

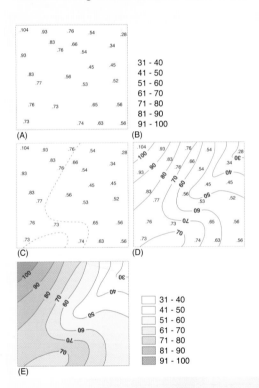

Choropleth maps are constructed from values describing the properties of non-overlapping areas, such as counties or census tracts. Each area is colored, shaded, or cross-hatched to symbolize the value of a specific variable, as in Figure 5.13. Two types of variables can be used, termed *spatially extensive* and *spatially intensive*. The former are variables whose values are true only of entire areas, such as total population, or total number of children under 5 years of age. The latter are variables that could potentially be true of every part of an area, if the area were homogeneous – examples include densities, rates, or proportions. Conceptually, a spatially intensive variable is a field, averaged over each area, whereas a spatially extensive variable is a field of density whose values are summed or integrated to obtain each area's value.

Geographic rules define what happens to the properties of objects when they are split or merged (see Section 9.2.4). Figure 5.13 compares a map of total population (spatially extensive) with a map of population density (spatially intensive). The former is highly misleading – the same color is applied uniformly to each part of an area, yet we know that the mapped property cannot be true of each part of the area (see also Section 12.4).

Figure 5.12 The creation of isopleth maps: (A) point attribute values; (B) user defined classes; (C) interpolation of class boundary between points; (D) addition and labelling of other class boundaries; and (E) use of hue to enhance perception of trends (Source: after Kraak and Ormeling 1996, 161)

5.6 Measuring Distance Effects as Spatial Autocorrelation

An understanding of spatial structure helps us to deduce a good sampling strategy, to use an appropriate means of interpolating between sampled points, and hence to build a spatial representation that is fit for purpose. In effect, this approach amounts to anticipating the likely pattern of spatial autocorrelation in order to build a spatial representation of the world. However, in many applications we do not understand enough about geographic variability, distance effects, and spatial structure to proceed in this clear, essentially *deductive* manner. A further branch of spatial analysis thus emphasizes the *measurement*

Box 5.4 Continued

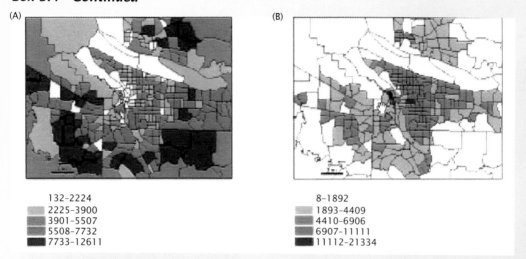

(A)

132–2224
2225–3900
3901–5507
5508–7732
7733–12611

(B)

8–1892
1893–4409
4410–6906
6907–11111
11112–21334

Figure 5.13 Choropleth maps of (A) a spatially extensive variable, total population, and (B) a related but spatially intensive variable, population density. Many cartographers would argue that (A) is misleading, and that spatially extensive variables should always be converted to spatially intensive form (as densities, ratios, or proportions) before being displayed as choropleth maps. (Source: Mitchell A 1999 *The ESRI Guide to GIS Analysis*, 70)

of spatial autocorrelation as an end in itself. While the techniques of spatial sampling and interpolation that have been discussed thus far present an essentially deductive means of representing space, the calculation of measures of spatial autocorrelation amounts to a more *inductive* approach to building up an understanding of the nature of geographic data.

> Induction reasons from data to build up understanding, while deduction begins with theory and principle as a basis for looking at data.

Methods of measuring spatial autocorrelation depend on the types of objects used as the basis of a representation and, as we saw in Section 5.3, the scale of attribute measurement is important too. Interpretation depends on how the objects relate to our conceptualization of the phenomena they represent. If the phenomenon of interest is conceived as a field, then spatial autocorrelation measures the smoothness of the field using data from the sample points, lines, or areas that represent the field. If the phenomena of interest are conceived as discrete objects, then spatial autocorrelation measures how the attribute values are distributed among the objects, distinguishing between arrangements that are clustered, random, and locally contrasting. Figure

5.14 shows examples of each of the four object types, with associated attributes, chosen to represent situations in which a scientist might wish to measure spatial autocorrelation. The point data in Figure 5.14(A) comprise data on well bores over an area of 30 km^2, and together provide information on the depth of an aquifer beneath the surface (the blue shading identifies those within a predetermined threshold). We would expect values to exhibit strong spatial autocorrelation, with departures from this norm indicative of changes in bedrock structure or form. The line data in Figure 5.14(B) present accidents in the Southwestern Ontario provincial highway network. Low spatial autocorrelation in these statistics implies that local causative factors (such as badly laid out junctions) account for most accidents, whereas strong spatial autocorrelation would imply a more regional scale of variation, and a link between accident rates and lifestyles, climate, or population density. The area data in Figure 5.14(C) illustrate the socio-economic patterning of an urban area, and beg the question of whether, at a regional scale, there are commonalities in household structure. The volume data in Figure 5.14(D) allow some measure of the spatial autocorrelation of high-rise structures to be made, perhaps as part of a study of the way that the urban core of Seattle functions.

(A)

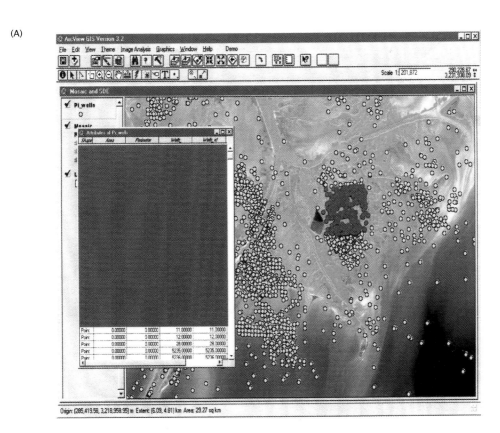

(B)

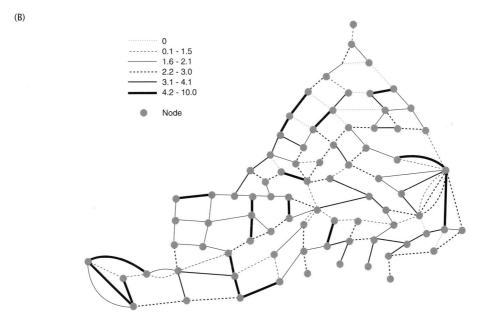

Figure 5.14 Situations in which a scientist might want to measure spatial autocorrelation: (A) point data (wells with attributes stored in a spreadsheet) (Source: ESRI); (B) line data (accident rates in the Southwestern Ontario provincial highway network) (cont.)

(C)

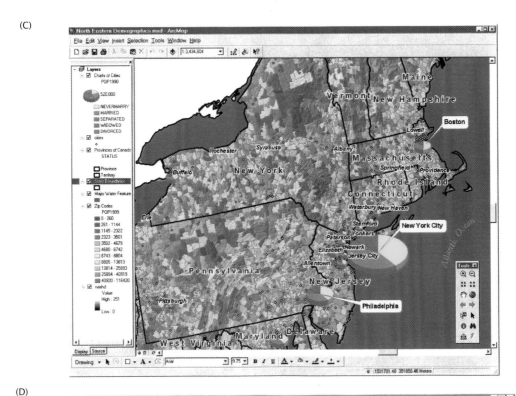

(D)

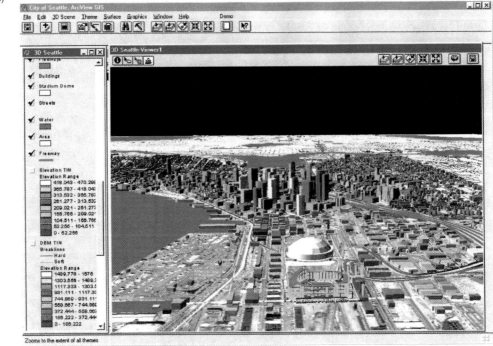

Figure 5.14 Situations in which a scientist might want to measure spatial autocorrelation (cont.): (C) area data (marital status by county in the northeastern USA) (Source: ESRI); and (D) volume data (elevation and volume of buildings in Seattle) (Source: ESRI)

5.6.1 Representing similarity of neighbors

In formal terms, we compare neighboring values of spatial attributes by defining a matrix **W** in which each element w_{ij} measures the similarity of locations i and j (i identifies the row, and j the column of the matrix). w_{ii} is set equal to 0 for all i. How is similarity measured? For areal objects, such as the simple eight-zone system shown in Figure 5.15(A), or the gridded fields shown in Figure 5.2, contiguity (or adjacency) is often taken as the most tangible measure of spatial

Table 5.1 The weights matrix **W** derived from the zoning system shown in Figure 5.15

	1	2	3	4	5	6	7	8
1	0	1	1	1	0	0	0	0
2	1	0	1	0	0	1	1	0
3	1	1	0	1	1	1	0	0
4	1	0	1	0	1	0	0	0
5	0	0	1	1	0	1	0	1
6	0	1	1	0	1	0	1	1
7	0	1	0	0	0	1	0	1
8	0	0	0	0	1	1	1	0

proximity. On this basis, Figure 5.15(A) may be re-expressed as the graph shown in Figure 5.15(B). Coding $w_{ij} = 1$ if regions i and j are contiguous, and $w_{ij} = 0$ otherwise, we may derive a weights matrix (usually denoted **W**) shown in Table 5.1.

This weights matrix provides an example of the simplest way of representing the locational and attribute similarities of a region of contiguous areal objects. Autocorrelation is identified by the presence of neighboring cells or zones that take the same (binary) attribute value. More sophisticated measures of w_{ij} include the straight line distance between points at the centers of zones, or the lengths of common boundaries. A range of different spatial metrics may also be used, such as existence of linkage by air, or travel time by air, road, or rail, or the strength of linkages between individuals or firms in some (non-spatial) network.

The particular application will determine the best way of representing distance effects when measuring spatial autocorrelation.

The type of matrix shown in Table 5.1 allows us to develop measures of spatial autocorrelation, in terms of four of the attribute types (nominal, ordinal, interval/ratio) in Box 3.3 and our dimensioned classes of spatial objects (Box 5.2 above). Attribute measures are compared using a similarity matrix **C**. Here we will restrict ourselves to interval attribute measures of contiguous areal objects. Goodchild (1986) provides general extensions of the measures presented here, and further measures of spatial autocorrelation are reviewed in connection with spatial interpolation in Section 13.4.4.

5.6.2 Measuring spatial autocorrelation

Spatial autocorrelation is concerned with a comparison of two types of information: similarity

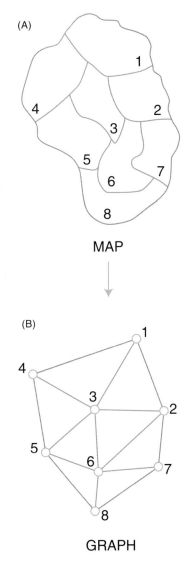

(A)

MAP

(B)

GRAPH

Figure 5.15 (A) A simple mosaic of zones, (B) re-expressed as a graph

among attributes, and similarity of location. The ways in which attributes can be measured depend upon the type of data present, while the calculation of spatial proximity depends upon the type of object. Additionally, there are a number of ways in which different data and object types may be combined into a final index. To simplify things, the following consistent notation will be used:

n number of objects in the sample
i, j any two of the objects
z_i the value of the attribute of interest for object i
c_{ij} the similarity of i's and j's attributes
w_{ij} the similarity of i's and j's locations, $w_{ii} = 0$ for all i ($i \neq j$).

In general, each of the measures of spatial autocorrelation has been created to compare the set of attribute similarities c_{ij} with the set of locational similarities w_{ij}, combining them into a single index of the form of a cross-product, that is:

$$\sum_i \sum_j c_{ij} w_{ij},$$

in other words, the total obtained by multiplying every cell in the **W** matrix with its corresponding entry in the **C** matrix, and summing. Adjustments are made to each index to make it easy to interpret (see below). The results of all autocorrelation tests are conditional upon the specification of the weights matrix **W** of similarity measures.

A variety of ways have been devised to measure the similarity of attributes, c_{ij}, dependent upon their scaling. For nominal data, the usual approach is to set c_{ij} to 1 if i and j take the same attribute value, and zero otherwise. For ordinal data, similarity is usually based on comparing the ranks of i and j, while for interval data both the squared difference $(z_i - z_j)^2$ and the product:

$$(z_i - \bar{z})(z_j - \bar{z})$$

(where $\bar{z}$ denotes the average of the z values) are commonly used. One of the most widely used spatial autocorrelation statistics for the case of area objects and interval scale attributes is the Moran statistic – others include the Geary Index, which has been extensively used in the work of Cliff and Ord (1981).

The Moran Index is positive when nearby areas tend to be similar in attributes, negative when they tend to be more dissimilar than one might expect, and approximately zero when attribute values are arranged randomly and independently in space. It is given by the expression:

$$I = n \sum_i \sum_j w_{ij} c_{ij} \bigg/ \sum_i \sum_j w_{ij} \sum_i (z_i - \bar{z})^2$$

The w_{ij} terms represent the spatial proximity of i and j and can be calculated in any suitable way. c_{ij} again represents the similarity of i's and j's attributes.

5.7   Establishing Dependence in Space

Spatial autocorrelation measures tell us about the inter-relatedness of phenomena across space, one attribute at a time. Another important aspect of the nature of geographic data is the tendency for relationships to exist between different phenomena at the same place – between the values of two different fields, between two attributes of a set of discrete objects, or between the attributes of overlapping discrete objects. This section introduces one of the ways of describing such relationships.

> How the various properties of a location are related is an important aspect of the nature of geographic data.

In a formal statistical sense, regression analysis allows us to identify the *dependence* of one variable upon one or more *independent* variables. For example, we might hypothesize that the value of individual properties in a city is dependent upon a number of variables such as floor area, distance to local facilities such as parks and schools, standard of repair, local pollution levels, and so forth. Formally this may be written:

$$Y = f(X_1, X_2, X_3, \dots, X_K)$$

where Y is the dependent variable and X_1 through X_K are all of the possible independent variables that might impact upon property value. It is important to note that it is the independent variables that together affect the dependent variable, and that the hypothesized causal relationship is one way – that is, that property value is *responsive* to floor area, distance to local facilities, standard of repair, and pollution, and not vice versa. For this reason the dependent variable is termed the *response* variable and the independent variables are termed *predictor* variables in some statistics textbooks.

In practice, of course, we will rarely, if ever, successfully predict the exact values of any sample of properties. We can identify two broad classes of reasons why this might be the case. First, a property price outcome is the response to a huge range of factors, and it is likely that we will have evidence of and be able to measure only a small subset of these. Second, even if we were able to identify and measure every single relevant independent variable, we would in practice only be able to do so to a given level of measurement precision (for a more detailed discussion of precision and its relevance, see Section 15.1.2). Such caveats do not undermine the wider rationale for trying to generalize, since any assessment of the effects of variables we know about is better than no assessment at all. But our conceptual solution to the problems of unknown and imprecisely measured variables is to subsume them all within a statistical error term, and to revise our regression model so that it looks like this:

$$Y = f(X_1, X_2, X_3, \ldots, X_K) + \varepsilon$$

where ε denotes the error term.

We assume that this relationship holds for each case (which we denote using the subscript i) in our population of interest, and thus:

$$Y_i = f(X_{i1}, X_{i2}, X_{i3}, \ldots, X_{iK}) + \varepsilon_i$$

The essential task of regression analysis is to identify the *direction* and *strength* of the association implied by this equation. This becomes apparent if we rewrite it as:

$$Y_i = b_0 + b_1 X_{i1} + b_2 X_{i2} + b_3 X_{i3} + \cdots$$
$$+ b_K X_{iK} + \varepsilon_i$$

where b_1 through b_K are termed regression *parameters*, which measure the direction and strength of the influence of the independent variables X_1 through X_K on Y. b_0 is termed the *constant* or *intercept* term. This is illustrated in simplified form as a scatterplot in Figure 5.16. Here, for reasons of clarity, the values of the

dependent (Y) variable (property value) are regressed and plotted against just one independent (X) variable (floorspace; for more on scatterplots see Section 14.2.4). The scatter of points exhibits an upward trend, suggesting that the response to increased floorspace is a higher property price. A *best fit* line has been drawn through this scatter of points. The gradient of this line is calculated as the b parameter of the regression, and the upward trend of the regression line means that the gradient is positive. The greater the magnitude of the b parameter, the stronger the (in this case positive) effect of marginal increases in the X variable. The value where the regression line intersects the Y axis identifies the property value when floorspace is zero (which can be thought of as the value of the land parcel when no property is built upon it), and gives us the intercept value b_0. The more general multiple regression case works by extension of this principle, and each of the b parameters gauges the marginal effects of its respective X variable.

This kind of effect of floorspace area upon property value is intuitively plausible, and a survey of any sample of individual properties is likely to yield the kind of well-behaved plot illustrated in Figure 5.16. In other cases the overall trend may be more ambiguous. Figure 5.17(A) presents a hypothetical plot of the effect of distance to local school (measured perhaps as straight line distance; see Section 13.3.1 for more on measuring distance in a GIS) upon property value. Here the plot is less well behaved: the

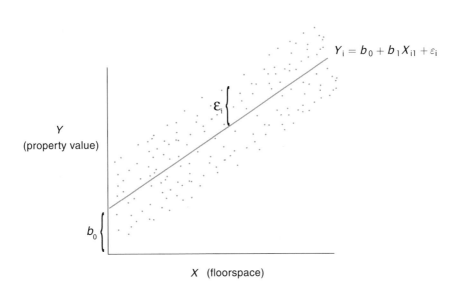

Figure 5.16 The fit of a regression line to a scatter of points, showing intercept, slope, and error terms

overall fit of the regression line is not as good as it might be, and a number of poorly fitting observations (termed high-residual and high-leverage points) present exceptions to a weak general trend. A number of formal statistical measures (notably t statistics and the R^2 measure) as well as less formal diagnostic procedures exist to gauge the statistical fit of the regression model to the data. Details of these can be found in any introductory statistics text, and will not be examined here. Yet an important point (which we also return to in Section 14.4) is that there are many important questions about the applicability of such procedures to spatial data.

It is easiest to assume that a relationship between two variables can be described by a straight line or linear equation, and that assumption has been followed in this discussion. But although a straight line may be a good first approximation to other functional forms (curves, for example), there is no reason to suppose that linear relationships represent the *truth*, in other words, how the world's social and physical variables are actually related. For example, it might be that very close proximity to the school has a negative effect upon property value (because of noise, car parking, and other localized nuisance), and it is properties at intermediate distances that gain the greatest positive neighborhood effect from this amenity. This is shown in Figure 5.17(B): these and other effects might be accommodated by changing the intrinsic *functional form* of the model.

A straight line or linear distance relationship is the easiest assumption to make and analyze, but it may not be the correct one.

Figure 5.16 identifies the discrepancy between one observed property value and the value that is predicted by the regression line (marked ε_i). This difference can be thought of as the error term for individual property i (strictly speaking, it is termed a *residual* when the scatterplot depicts a sample and not a population). The precise slope and intercept of the best fit line is usually identified using the principle of *ordinary least squares* (OLS). OLS regression fits the line through the scatter of points such that the sum of squared residuals across the entire sample is minimized. This procedure is robust and statistically efficient, and yields estimates of the b parameters. But in many situations it is common to try to go further, by *generalizing* results. Suppose the data being analyzed can be considered a representative sample of some larger group. In the field case, sample points might be representative of all of the infinite number of sample points one might

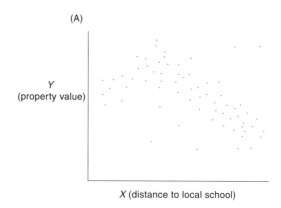

(A)

Y (property value)

X (distance to local school)

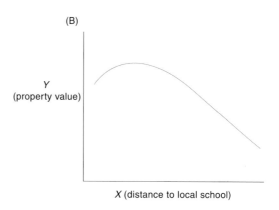

(B)

Y (property value)

X (distance to local school)

Figure 5.17 (A) A scatterplot and (B) hypothetical relationship between distance to local school and domestic property value

select to represent the continuous variation of the field variable (for example, weather stations measuring temperature in new locations, or more soil pits dug to measure soil properties). In the discrete object case, the data analyzed might be only a selection of all of the objects. If this is the case, then statistics provide methods for making accurate and unbiased statements about these larger populations.

Generalization is the process of reasoning from the nature of a sample to the nature of a larger group.

For this to work, several conditions have to hold. First, the sample must be representative, which means for example that every case (element) in the larger group or *population* has a prespecified and independent chance of being selected. The sampling designs discussed in Section 5.4 are one way of ensuring this. But all too often in the analysis of geographic data it turns out to be difficult or impossible to imagine such a population. It is inappropriate, for example, to try to generalize from one study to statements about all of the Earth's surface, if the study was conducted in one area. Generalizations based on samples taken in Antarctica are clearly not representative of all of the Earth's surface. Often GIS provide complete coverage of an area, allowing us to analyze all of the census tracts in a city, or all of the provinces of China. In such cases the apparatus of generalization from samples to populations is unnecessary, and indeed, becomes meaningless.

In addition, the statistical apparatus that allows us to make inferences assumes that there is no *autocorrelation* between errors across space or time. This assumption clearly does not accord with Tobler's Law, where the greater relatedness of near things to one another than distant things is manifest in positive spatial autocorrelation. If strong (positive or negative) spatial autocorrelation is present, the inference apparatus of the ordinary least squares regression procedure rapidly breaks down. The consequence of this is that estimates of the population b parameters become imprecise and the statistical validity of the tests used to confirm the strength and direction of apparent relationships is seriously weakened.

The assumption of zero spatial autocorrelation that is made by many methods of statistical inference is in direct contradiction to Tobler's Law.

The spatial patterning of residuals can provide clues as to whether the structure of space has been correctly specified in the regression equation. Figure 5.17 illustrates the hypothetical spatial distribution of residuals in our property value example – the high clustering of negative residuals around the sewage works suggests that some distance threshold should be added to the specification, or some function that negatively weights property values that are very close to the sewage works. The spatial clustering of residuals can also help to suggest omitted variables that should have been included in the regression specification. Such variables might include the distance to a neighborhood facility that might have a strong positive (e.g. a park) or negative (e.g. a sewage works) effect upon values in our property example (Figure 5.18).

A second assumption of the multiple regression model is that there is no intercorrelation between the independent variables, that is, that no two or more variables essentially measure the same construct. The statistical term for such inter-

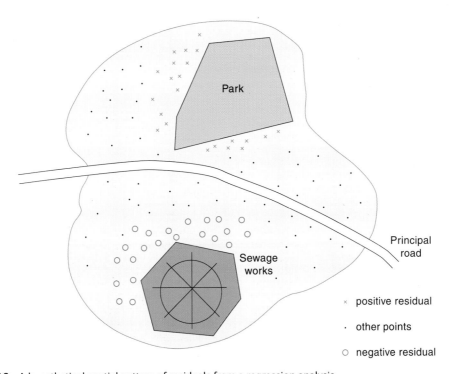

× positive residual

· other points

○ negative residual

Figure 5.18 A hypothetical spatial pattern of residuals from a regression analysis

correlation is *multicollinearity*, and this is a particular problem in GIS applications. GIS is a powerful technology for combining information about a place, and for examining relationships between attributes, whether they be conceptualized as fields, or as attributes of discrete objects. The implication is that each attribute makes a distinct contribution to the total picture of geographic variability. In practice, however, geographic layers are almost always highly correlated. It is very difficult to imagine that two fields representing different variables over the same geographic area would not somehow reveal their common geographic location through similar patterns. For example, a map of rainfall, and a map of population density would clearly have similarities, whether population was dependent on agricultural production and thus rainfall or tended to avoid steep slopes and high elevations where rainfall was also highest.

It is almost impossible to imagine that two maps of different phenomena over the same area would not reveal some similarities.

5.8 Taming Geographic Monsters

Throughout our discussion of the nature of geographic data, we have emphasized that the creation of spatial representations requires understanding of the nature of geographic data, since this informs everything we do, from the design of GIS databases to the decisions we make. We have assumed that, locally at least, spatial variation is smooth, and continuous, with the exceptions of abrupt truncations, and discrete shifts at boundaries. However, much spatial variation is not smooth and continuous, but rather is jagged, and apparently irregular. The processes which give rise to the form of a mountain range produce features that are spatially autocorrelated (for example, the highest peaks tend to be clustered), yet it would be wholly inappropriate to represent a mountainscape using smooth interpolation between peaks and valley troughs.

Jagged irregularity is a property which is also often observed across a range of scales, and detailed irregularity may resemble coarse irregularity in shape, structure, and form. We commented on this in Section 5.3 when we suggested that a rock broken off a mountain may, for reasons of lithology, represent the mountain in form, and this property is often termed *self-similarity*. Batty and Longley (1994) have suggested that cities and city systems are also self-similar in organization

across a range of scales, and this echoes many of the earlier ideas of Christaller (Figure 5.3). They also describe how structures such as the trees in Figure 5.10 can be generated as fractal structures. It is unlikely that idealized smooth curves and conventional mathematical functions will provide useful representations for self-similar, irregular spatial structures. At what scale, if any, does it become appropriate to approximate the San Andreas Fault system by a continuous curve? Urban geographers, for example, have long sought to represent the apparent decline in population density with distance from historic central business districts (CBDs), yet the three-dimensional profiles of cities are characterized by urban canyons between irregularly spaced high-rise buildings (Figure 5.14(D)). Each of these phenomena is characterized by spatial trends (many of the largest faults, the largest mountains, and the largest skyscrapers tend to be close to one another), but they are not contiguous and smoothly joined, and the kinds of surface functions shown in Figure 5.9 present inappropriate generalizations of their structure.

For many years, such features were considered geometrical monsters that defied intuition. More recently, however, a more general geometry of the irregular, termed *fractal geometry* by Benoît Mandelbrot (1983), has come to provide a more appropriate and general means of summarizing the structure and character of spatial objects. Fractals can be thought of as geometric objects that are, literally, between Euclidean dimensions, as described in Box 5.5.

In a self-similar object, each part has the same nature as the whole.

Fractal ideas are important, and for many phenomena a measurement of fractal dimension is as important as measures of spatial autocorrelation, or of medians and modes in standard statistics. An important application of fractal concepts is discussed in Section 14.2.6, and we return again to the issue of length estimation in GIS in Section 13.3.1. Ascertaining the fractal dimension of an object involves identifying the scaling relation between its length or extent and the yardstick (or level of detail) that is used to measure it. Regression analysis, as described in the previous section, provides one (of many) means of establishing this relationship. If we return to the Maine coastline example in Figure 5.20, we obtain scale dependent coast length estimates (L) of 13.6 (4×3.4), 14.1 (2×7.1), and 15.5 (1×16.6) units for the step lengths (r) used in Figures 5.20(B), 5.20(C) and 5.20(D) respectively. (It is arbitrary whether the

Box 5.5 *The strange story of the lengths of geographic objects*

How long is the coastline of Maine (Figure 5.19)? (Benoît Mandelbrot originally posed this question in 1967 with regard to the coastline of Great Britain.)

Figure 5.19 Part of the Maine coastline

We might begin to measure the stretch of coastline shown in Figure 5.20(A). With dividers set to measure radial distances (r) of 100 km step lengths, we would take approximately 3.4 swings and record a length of 340 km (Figure 5.20(B)).

If we then halved the divider span so as to measure 50 km swings, we would take approximately 7.1 swings and the measured length would increase to 355 km (Figure 5.20(C)).

If we halved the divider span once again to measure 25 km swings, we would take approximately 16.6 swings and the measured length would increase still further to 415 km (Figure 5.20(D)).

And so on until the divider span was so small that it picked up all of the detail on this particular (Figure 5.20(A)) representation of the coastline. But that would not be the end of the story.

If we were to resort instead to field measurement, using a tape measure or the Distance Measuring Instruments (DMIs) used by highway departments, the length would increase still further, as we picked up detail that even the most detailed maps do not seek to represent.

If we were to use dividers, or even microscopic measuring devices, to measure every last grain of sand or earth particle, our recorded length measurement would stretch towards infinity, without apparent limit.

In short, the answer to our question is that the length of the Maine coastline is indeterminate. More helpfully, perhaps, any approximation can be thought of as *scale-dependent* – and thus any measurement must also specify scale. The line representation of the coastline also possesses two other properties. First, where small deviations about the overall trend of the coastline resemble larger deviations in form, the coast is said to be *self-similar*. Second, as the path of the coast traverses space, its intricate structure comes to fill up more space than a one-dimensional straight line but less space than a two-dimensional area. As such, it is said to be of *fractional dimension* (and is termed a *fractal*) between 1 (a line) and 2 (an area).

**Box 5.5
Continued**

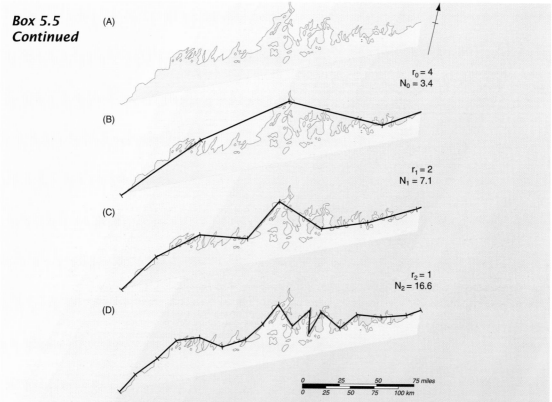

$r_0 = 4$
$N_0 = 3.4$

$r_1 = 2$
$N_1 = 7.1$

$r_2 = 1$
$N_2 = 16.6$

Figure 5.20 The coastline of Maine, at three levels of recursion: (A) the base curve of the coastline; (B) approximation using 100 km steps; (C) 50 km step approximation; and (D) 25 km step approximation

steps are measured in miles or kilometers.) If we then plot the natural log (ln) of L (on the y-axis) against the natural log of r for these and other values, we will build up a scatterplot like that shown in Figure 5.21. If the points lie more or less on a straight line and we fit a regression line through it, the value of the slope (b) parameter is equal to $(1 - D)$, where D is the fractal dimension of the line. This method for analyzing the nature of geographic lines was originally developed by Lewis Fry Richardson (Box 5.6).

5.9 Induction and Deduction, and How It All Comes Together

The abiding message of this chapter is that spatial is special – that geographic data have a unique nature. Tobler's Law presents an elementary general rule about spatial structure, and a starting point for the measurement and simulation of spatially autocorrelated structures. This in turn assists us in devising appropriate spatial sampling schemes, and creating improved representations, which tell us still more about the real world and how we might represent it. A goal of GIS is often to establish causality between different geographically referenced data, and the multiple regression model provides one important means of relating spatial variables to one another, and of inferring from samples to the properties of the populations from which they were drawn. Yet

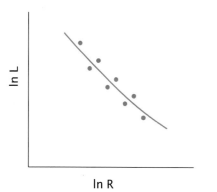

Figure 5.21 The relationship between recorded length (L) and step length (R)

Box 5.6 Lewis Fry Richardson

Lewis Fry Richardson (1881-1953) was one of the founding fathers of the ideas of scaling and fractals. He was brought up a Quaker, and after earning a degree at Cambridge University went to work for the UK Meteorological Office, but his pacifist beliefs forced him to leave in 1920 when the Meteorological Office was militarized under the Air Ministry. His early work on how atmospheric turbulence is related at different scales established his scientific reputation. Later he became interested in the causes of war and human conflict, and in order to pursue one of his investigations found that he needed a rigorous way of defining the length of a boundary between two states. Unfortunately published lengths tended to vary dramatically, a specific instance being the difference between the lengths of the Spanish–Portuguese border as stated by Spain and by Portugal. He developed a method of walking a pair of dividers along a mapped line, and analyzed the relationship between the length estimate and the setting of the dividers, finding remarkable predictability. In the 1960s Benoît Mandelbrot's concept of fractals finally provided the theoretical framework needed to understand this result.

Figure 5.22 Lewis Fry Richardson: a pioneer in the formalization of scale effects

statistical techniques often need to be recast in order to accommodate the special properties of spatial data, and regression analysis is no exception in this regard.

Spatial data provide the foundations to operational and strategic applications of GIS, foundations that must be used creatively yet rigorously if they are to support the spatial analysis superstructure that we wish to erect. Technical competence and understanding are prerequisites to robust and defensible spatial analysis but, as we will see in the Practice Part of this book, we need also to hone spatial analysis to organizational and social context, with humility rather than conviction. In this spirit, induction (reasoning from observations) and deduction (reasoning from principles and theory) can be used alongside each other to develop spatial representation.

Representations need to build upon the general spatial and temporal structures that are known to characterize the real world. Yet general laws (Section 1.3) alone will not tell us enough about spatial and temporal structures to build representations which bear the closest correspondence with geographic reality.

Questions for Further Study

1. The apparatus of inference was developed by statisticians because they wanted to be able to reason from the results of experiments involving small samples to make conclusions about the results of much larger, hypothetical experiments – for example, in using samples to test the effects of drugs. Summarize the problems inherent in using this apparatus for geographic data in your own words.

2. How many definitions and uses of the word *scale* can you identify?

3. Many jurisdictions tout the number of miles of shoreline in their community – for example, Ottawa County, Ohio, USA claims 107 miles of Lake Erie shoreline. What does this mean, and how could you make it more meaningful?

4. What important aspects of the nature of geographic data have not been covered in this chapter?

Online Resources

NCGIA Core Curricula (www.ncgia.ucsb.edu/pubs/core.html):
    Core Curriculum in GIScience, Sections 1.6.1 (Sampling the World)
    Core Curriculum in GIS, 1990, Units 6 (Sampling the World) and 47 (Fractals)
ESRI Virtual Campus courses (campus.esri.com):
    Introduction to ArcView GIS, by ESRI (see "Managing Scale" in the Module "Querying Data in ArcView")

Reference Links

Maguire D J, Goodchild M F, and Rhind D W (eds) 1991 *Geographical Information Systems: Principles and applications.* Harlow, UK: Longman (Text available online from 'Links to Big Book 1' at www.wiley.com/gis and www.wiley.co.uk/gis).
  Chapter 9, Concepts of space and geographic data (Gatrell A C)
Longley P A, Goodchild M F, Maguire D J, and Rhind D W (eds) 1999 *Geographical Information Systems: Principles, techniques, management and applications.* New York: John Wiley.
  Chapter 2, Space, time, geography (Couclelis H)
  Chapter 16, Spatial statistics (Getis A)
  Chapter 17, Interactive techniques and exploratory spatial data analysis (Anselin L)
  Chapter 19, Spatial analysis: retrospect and prospect (Fischer M M)

Cliff A D and Ord J K 1981 *Spatial Processes: Models and applications.* London: Pion.
Goodchild M F 1986 *Spatial Autocorrelation.* Concepts and Techniques in Modern Geography 47. Norwich: GeoBooks.
Kraak M-J and Ormeling F J 1996 *Cartography: Visualization of spatial data.* Harlow, UK: Longman
Langford M 1993 *Getting Started in GIS. A workbook of computer exercises.* Leicester UK: Midlands Regional Research Laboratory.
Mandelbrot B B 1983 *The Fractal Geometry of Nature.* San Francisco: Freeman.
Mitchell A 1999 *The ESRI Guide to GIS Analysis.* Redlands, California: ESRI Press.
Quattrochi D A and Goodchild M F (eds) 1996 *Scale in Remote Sensing and GIS.* Boca Raton, Florida: Lewis Publishers.

References

Batty M and Longley P A 1994 *Fractal Cities: A geometry of form and function.* London: Academic Press.

UNCERTAINTY

6

This chapter introduces the concept of uncertainty, and discusses its principles, dimensions, and causes. It identifies many of the sources of geographic uncertainty and the ways in which they operate in GIS-based representations. This lays the foundations for Chapter 15, which examines techniques for handling uncertainty in GIS. Users of GIS databases face uncertainty about the real-world phenomena represented in databases, because representations can never be perfect replicas of the world. Thus all uses of GIS are subject to uncertainty, and so are all decisions based on GIS. Uncertainty can arise in the form of measurement error and can also be introduced at many other stages in the handling of geographic data, from its acquisition through to its final use.

Learning objectives

By the end of this chapter you will:

- Understand the concept of uncertainty, and the ways that it arises from imperfect representation of geographic phenomena;

- Be aware of the uncertainties introduced in the three stages (conception, measurement, and analysis) of database creation and use;

- Understand the concepts of vagueness, and the uncertainties arising from the definition of key GIS attributes;

- Understand how and why geographic boundaries are often described as indeterminate;

- Be able to define accuracy and measurement error.

6.1 Introduction

GIS can be used to reconcile science with practice, concepts with application, and analytical capability with social context. Yet such reconciliation is almost always imperfect, and in this chapter we will use the term *uncertainty* as an umbrella term to describe the problems that arise out of these imperfections. Thus far we have begun to identify how GIS allows us to represent the apparently endless complexity of reality in machine–readable form, and to develop an understanding of the particular properties of spatial data. Representations may occasionally approach perfect accuracy and precision (see Section 15.1.2) – as might be the case, for example, in the detailed site layout layer of a utility management system, in which strenuous efforts are made to reconcile fine-scale multiple measurements of the built environment. Yet perfect, or nearly perfect, representations of reality are the exception rather than the rule, and such exceptions tend to be of coarse-scale artificial (human-made) phenomena. More usually, the inherent complexity and detail of our world makes it virtually impossible to capture every single facet, at every possible scale, in a digital representation. (Neither is this usually desirable: see Chapter 7.) Furthermore, different individuals see the world in different ways, and in practice no single view is likely to be universally seen as the best or to enjoy uncontested status. In this chapter we discuss how the processes and procedures of abstraction create differences between the contents of our (geographic and attribute) database and real-world phenomena. Later, Chapter 15 reviews some of the techniques that have been developed for handling uncertainty in GIS.

It is impossible to make a perfect representation of the world, so uncertainty about it is inevitable.

Various terms are used to describe these differences, depending on the context. The established scientific notion of measurement *error* focuses on differences between observers or between measuring instruments. It suggests that those differences are measurable, and in Chapter 15 the techniques of measurement of error are discussed in detail. We might be told, for example, that the elevation of Mount Everest is 8850 m, and that the figure was determined by instruments and procedures that create a margin of error of 5 m. Because we do not know whether the error is positive or negative, or even whether the error in this actual instance is 5 m, or 3 m, or 0 m, the published elevation leaves us uncertain about the truth –

and hence error creates a form of uncertainty. It is also likely that the figure of 5 m represents a best estimate of the error and that in practice it could even be a little more than 5 m.

Measurement error is one source of uncertainty.

As we saw in the previous chapter (Section 5.7), the concept of error in multivariate statistics arises from omission of relevant aspects of a phenomenon – as in the failure to specify fully all of the predictor variables in a multiple regression model, for example. Similar problems arise when one or more variables are omitted from the calculation of a composite indicator, as, for example, in omitting road accessibility in an index of land value, or omitting employment status from a measure of social deprivation. The Dutch geographer Gerard Heuvelink has defined *accuracy* as the difference between reality and *our* representation of reality. Although such differences might principally be addressed in formal mathematical terms, the use of the word *our* acknowledges the varying views that are generated by a complex, multi-scale, and inherently uncertain world.

Yet even this established framework is too simple for understanding the defining standards of geographic data or other notions of quality. The terms *ambiguity* and *vagueness* identify further considerations which need to be taken into account in assessing the *quality* of a GIS representation. Error, inaccuracy, ambiguity, and vagueness all contribute to the notion of uncertainty in the broadest sense, which Kate Beard has defined as a measure of the user's understanding of the difference between the contents of a database, and the real phenomena the data are believed to represent. This definition implies that phenomena are real, but includes the possibility that we are unable to describe them exactly. In GIS, the term uncertainty has come to be used as the catch-all term to describe situations in which the digital representation is simply incomplete, and as a measure of the general quality of the representation.

Many geographic representations use inherently vague definitions.

The views outlined in the previous paragraph are themselves controversial, and a rich ground for endless philosophical discussions. Some would argue that uncertainty can be inherent in phenomena themselves, rather than just in their description. Others would argue for distinctions between *vagueness, uncertainty, fuzziness, imprecision, inaccuracy,* and many other terms

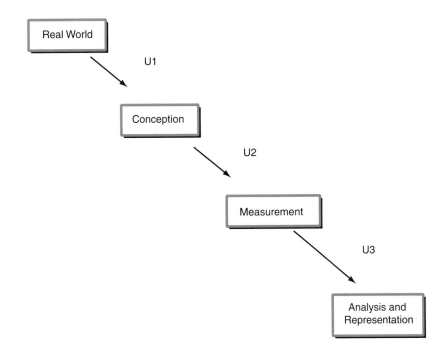

Figure 6.1 A conceptual view of uncertainty (U)

that most people use as if they were essentially synonymous. Fisher (1999) provides a useful and wide-ranging discussion of these terms. We take the catch-all view here, and leave these arguments to further study.

In this chapter, we will discuss some of the principal sources of uncertainty and some of the ways in which uncertainty degrades the quality of a spatial representation. The way in which we conceive of a geographic phenomenon very much prescribes the way in which we are likely to set about measuring it. The measurement procedure, in turn, heavily conditions the ways in which it may be analyzed within a GIS. This chain sequence of events, in which *conception* prescribes *measurement*, which in turn prescribes *analysis* is a succinct way of summarizing much of the content of this chapter (Figure 6.1), which is structured according to the filters U1, U2, and U3.

6.2 U1: Uncertainty in the Conception of Geographic Phenomena

6.2.1 Spatial uncertainty

Our discussion of Tobler's Law (Section 3.1 and Section 5.2) has already suggested one important

property of geographic data, formally measured through the concept of spatial autocorrelation, which makes geographic data handling different from other classes of applications. A further characteristic that sets geography and spatial analysis apart from almost every other science is that only rarely are there *natural* units of analysis. What is the natural unit of measurement for a soil profile? What is the spatial extent of a *pocket* of high unemployment, or a *cluster* of cancer cases? How might we delimit an environmental impact study of spillage from an oil tanker (Figure 6.2)? The questions become still more difficult in bivariate (two variable) and multivariate (more than two variable) studies. At what scale is it appropriate to investigate any relationship between background radiation and the incidence of leukemia? Or to assess any relationship between labor force qualifications and unemployment rates?

In many cases there are no natural units of geographic analysis.

The discrete object view of geographic phenomena is much more reliant upon the idea of natural units of analysis than the field view. Biological organisms are almost always natural units of analysis, as are groupings such as households or families – though

Figure 6.2 How might the spatial impact of an oil tanker spillage be delineated? We can measure the dispersion of the pollutants, but their impacts extend far beyond these narrowly-defined boundaries

Figure 6.3 Seeing the wood for the trees: what absolute or relative incidence rate makes it meaningful to assign the label "oak woodland"?

there are certainly difficult cases, such as the massive networks of fungal strands that are often claimed to be the largest living organisms on Earth. Things we manipulate, such as pencils, books, or screwdrivers, are also obvious natural units. The examples listed in the previous paragraph fall almost entirely into one of two categories – they are either instances of fields, where variation is inherently continuous in space, or they are instances of poorly defined aggregations of discrete objects. In both of these cases it is up to the investigator to make the decisions about units of analysis, rendering the identification of the objects of analysis inherently subjective.

6.2.2 Vagueness

The frequent absence of objective geographic individual units means that, in practice, the labels that we assign to zones are often vague best guesses (Fisher 1999). What absolute or relative incidence of oak trees in a forested zone qualifies it for the label *oak woodland* (Figure 6.3)? Or, in a developing country context in which aerial photography rather than ground enumeration is used to estimate population size, what rate of incidence of dwellings identifies a zone of *dense* population? In each of these instances, it is expedient to transform point-like events (individual trees or individual dwellings) into area objects, and pragmatic decisions must be taken in order to create a working definition of a spatial distribution. These decisions have no *absolute* validity, and raise two important questions:

- Is the defining boundary of a zone crisp and well-defined (Burrough and Frank 1996)?

- Is our assignment of a particular label to a given zone robust and defensible?

Uncertainty can exist both in the positions of the boundaries of a zone and in its attributes.

The questions have statistical implications (can we put numbers on the confidence associated with boundaries or labels?), cartographic implications (how to convey the meaning of vague boundaries and labels through appropriate symbols on maps and GIS displays?), and cognitive implications (do people subconsciously attempt to force things into categories and boundaries to satisfy a deep need to simplify the world?).

6.2.3 Ambiguity

Many objects are assigned different labels by different national or cultural groups (see, for example Mark's (1999) discussion of names given to inland water bodies), and such groups perceive space differently. Geographic prepositions like *across*, *over*, and *in* do not have simple correspondences with terms in other languages. Object names and the topological relations between them may thus be inherently *ambiguous*. Perception, behavior, language, and cognition all play a part in the conception of real-world entities and the relationships between them. GIS cannot present a value-neutral view of the world, yet it can provide a formal framework for the reconciliation of different worldviews.

Many English terms used to convey geographic information are inherently ambiguous.

Ambiguity also arises in the conception and construction of *indicators*. *Direct* indicators are deemed to bear a clear correspondence with a mapped phenomenon. Detailed household income figures, for example, provide a direct indicator of the likely geography of expenditure and demand for goods and services; tree diameter at breast height can be used to estimate stand value; and field nutrient measures can be used to estimate agronomic yield. *Indirect* indicators are used when the best available measure is a perceived surrogate link with the phenomenon of interest. Thus the incidence of central heating amongst households, or rates of multiple car ownership, might provide a surrogate for (unavailable) household income data, while local atmospheric measurements of nitrogen dioxide might provide an indirect indicator of environmental health. Conception of the (direct or indirect) linkage between any indicator and the phenomenon of interest is subjective, hence ambiguous. Such measures will create (possibly systematic) errors of measurement if the correspondence between the two is imperfect. Fisher (1999) illustrates how differences in the conception of deprivation can lead to specification of different composite indicators. With regard to the natural environment, he also describes how conception of critical defining properties of soils can lead to inherent ambiguity in their classification (see Section 15.3).

Ambiguity is introduced when imperfect indicators of phenomena are used instead of the phenomena themselves.

6.2.4 Foundations: the scale of geographic individuals

There is a sense in which vagueness and ambiguity in the conception of *usable* (rather than *natural*) units of analysis undermines the very foundations of GIS. How, in practice, may we create a sufficiently secure base to support geographic analysis? Geographers have long grappled with the problems of defining systems of zones, and have marshaled a range of deductive and inductive approaches (for definitions see Section 5.6) to this end. The long-established regional geography tradition is fundamentally concerned with the delineation of zones characterized by internal homogeneity (with respect to climate, economic development, or agricultural land use, for example), within a zonal scheme which maximizes between-zone heterogeneity, such as

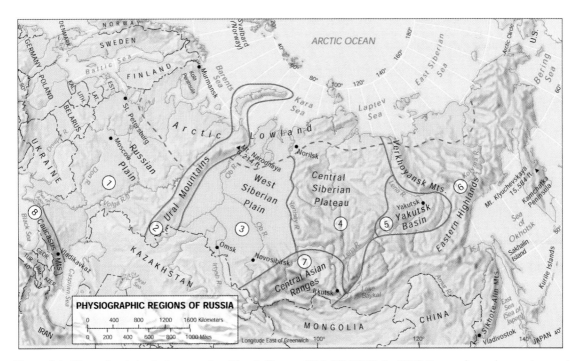

Figure 6.4   The regional physical geography of Russia (Source: H J de Blij, P O Muller 2000 *Geography: realms, regions and concepts* (9th edn.) New York: Wiley, 113)

delineate the breakpoints between the spheres of influence of adjacent facilities or features – as in the definition of travel-to-work areas (Figure 6.5) or the definition of a river catchment. Zones would be defined such that there is maximal interaction within zones, and minimal between zones. The scale at which uniformity or functional integrity is conceived clearly conditions the ways it is measured – in terms of the magnitude of within-zone heterogeneity that must be accommodated in the case of uniform zones, and the degree of leakage between the units of functional zones. This is illustrated in Box 6.1 by the geography of beer consumption in the UK.

Scale has an effect, through the concept of spatial autocorrelation outlined in Section 5.3, upon the outcome of geographic analysis. This was demonstrated more than half a century ago in a classic book by Yule and Kendall (1950), in an analysis which demonstrated that the correlation between wheat and potato yields systematically increased as English county units were amalgamated through a succession of coarser scales (Table 6.1). Fotheringham and Wong (1991) subsequently reaffirmed the existence of scale effects in multivariate analysis, yet drew the discouraging conclusion that in multivariate cases, scale effects did not follow any consistent or predictable trends. This theme of dependence of results on the geographic units of analysis is pursued further in Section 6.4 below.

Relationships typically grow stronger when based on larger geographic units.

GIS appears to trivialize the task of creating composite thematic maps. Yet inappropriate conception of the scale of geographic phenomena can mean that apparent spatial patterning (or the lack of it) in mapped data may be oversimplified,

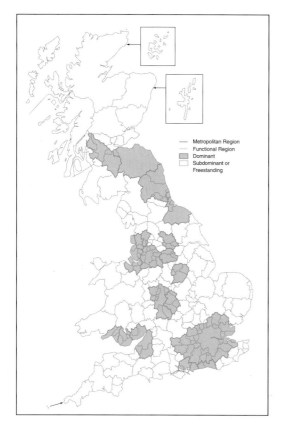

Figure 6.5 A functional regionalization of Great Britain (Source: Champion T and 5 others 1996 *The Population of Britain in the 1990s*. Oxford: Oxford University Press, 9)

the map illustrated in Figure 6.4. Regional geography is fundamentally about delineating *uniform* zones, and many employ multivariate statistical techniques such as cluster analysis. It is possible, for example, to build a database of the 100 most important characteristics of the c.3000 counties of the continental USA, and then to use cluster analysis to determine how the counties might be aggregated into 48 contiguous areas so as to minimize the within-area gross variability across all 100 characteristics. The result would show reasonably homogeneous zones, and would present an interesting alternative to the current state boundaries.

Identification of homogeneous zones and spheres of influence lies at the heart of traditional regional geography as well as contemporary data analysis.

Other geographers have tried to develop *functional* zonal schemes, in which zone boundaries

Table 6.1. In 1950 Yule and Kendall used data for wheat and potato yields from the (then) 48 counties of England to demonstrate that correlation coefficients tend to increase with scale. They aggregated the 48-county data into a succession of coarser zones. The range of their results was from near zero (no correlation) to over 0.99 (almost perfect positive correlation)

No. of geographic zones	Correlation
48	0.2189
24	0.2963
12	0.5757
6	0.7649
3	0.9902

Box 6.1   *Functional and uniform regions for UK beer consumption*

The drinking of beer is often ingrained in the social fabric of a community and its history can span many hundreds of years. The psychology of beer drinking is complex and involves the notion of the local water being particular and the ingredients coming from the local land. Traditional British beers are drunk "live", that is, the yeast that is used to create the alcohol is still present in the beer when drunk, and serves to continuously top up the low levels of carbon dioxide in it. Beer of this type is called "cask conditioned" beer, because the final conditioning of it is done in the wooden cask (its container). Bad treatment of the beer could "kill" it or cause the yeast (sediment) to be poured into the glass. For many, the local beer was the actual *essence* of the local community as it worked on so many mysterious levels and assumed almost reverential qualities. The strong local allegiances of a population to their beer translated into a belief that beer did not "travel well". The net effect of all this was to create natural market areas, each served by a brewery or two with products that had characteristics of the area. The historic geography of these *functional regions* was never fully documented.

By the 1960s the industry was seeking to meet rising demand and achieve economies of scale through consolidation. At the same time came the innovation of keg beer, in which the yeast is killed off by pasteurization and the beer is artificially pressurized with carbon dioxide (like lemonade – rather than the natural build up of carbon dioxide from the yeast fermentation process). It was easily transported over long distances in metal "kegs", required to contain the carbon dioxide. Brewers were thus able to buy into local markets by taking over traditional local breweries and then to close what was viewed as obsolete plant. At the same time, mainstream draught beers were heavily advertised for the first time through the fast-developing and regionally structured medium of television, thereby seeking to mold the geography of beer consumption to that of television advertising regions.

The next development was the introduction of lager to the UK market. Lager is usually a light colored keg beer, like US brands Budweiser or Coors, with less distinctive flavors than traditional British ales and therefore more easy to learn to drink. This new product was generally most attractive to the younger generation, particularly in the regions where the new keg beer produced by the merged breweries was particularly unpalatable. Together these movements had the effect of beginning the break up of the historic functional regions, although residual elements remain. The effects of demography, advertising, and quality of available offerings also came to influence the pattern of consumption – cask versus keg, bitter versus lager, and standard versus premium brands. A rearguard action was also mounted by the Campaign for Real Ale – a pressure group for the wide reintroduction of cask-conditioned ale – particularly in Southern England.

Today's regional geography of beer consumption does not correspond to artificial regions like TV regions, government standard statistical regions, or the functional labor market regions shown in Figure 6.5. It exhibits considerable diversity, even at very local scales. In some

Figure 6.6   Beer regions of the UK (lager in green, premium lager in purple, bitter in yellow, and mild in brown)

Box 6.1   Continued

respects it is very much a legacy regional geography of the natural regions, which were supplied up until the 1960s by local breweries – particularly the areas where drinking of "Mild" bitter is prevalent.

This presents a complicated picture for product marketing, since most goods do not exhibit such stark regional variation. If marketing is carried out over standard statistical regions, such as statistical and TV regions, significant local differences can be lost or obscured. GIS allows analysts in the brewing industry such as Martin Callingham (Box 18.5) to evaluate the drinking habits of the people in the area of a retail outlet (a pub, supermarket, or liquor store), and hence to ascertain the "merchandizing" of the store.

Figure 6.6 is an attempt by Callingham to use cluster analysis to identify today's uniform "beer regions". It is produced at the postal districts scale and identifies the geography of dominance of standard and premium lager (green and purple), bitter (yellow), and mild (brown). The geography of UK beer consumption is thus one in which historic functional regions, regions of uniform marketing activity and a range of demographic and lifestyle factors all exert an influence on beer consumption at a range of scales. GIS helps us to understand this pattern. We can also learn lessons form this about the difficulties inherent in defining appropriate zones for use in GIS analysis.

---

crude, or even illusory. It is also clearly inappropriate to conceive of boundaries as crisp and well-defined if significant leakage occurs across them (as happens, in practice, in the delineation of most functional regions), or if geographic phenomena are by nature fuzzy, vague, or ambiguous. We return to a technical discussion of these topics in Chapter 15.

6.3   U2: Further Uncertainty in the Measurement and Representation of Geographic Phenomena

6.3.1   Representation and measurement

The conceptual models (fields and objects) that were introduced in Chapter 3 impose very different filters upon reality, and the representational models (raster and vector) are characterized by different uncertainties as a consequence. The vector model enables a range of powerful analytical operations to be performed (see Section 9.2.3), yet it also requires *a priori* conceptualization of the nature and extent of geographic individuals and the ways in which they nest together into higher-order zones. The raster model defines individual elements as square cells, with boundaries that bear no relationship at all to natural features, but nevertheless provides a convenient and (usually) efficient structure for data handling within a GIS. However, in the absence of effective automated pattern recognition techniques, human interpretation is usually required to discriminate between real-world spatial entities as they appear in a rasterized image.

Although quite different representations of reality, vector and raster data structures are both attractive in their logical consistency and the ease with which they are able to handle spatial data, and (once the software is written) both are easily implemented in GIS. Yet neither abstraction provides easy measurement fixes and there is no substitute for robust conception of geographic units of analysis (Section 6.2). However the conceptual distinction between fields and discrete objects is often useful in dealing with uncertainty. Figure 6.7 shows a coastline, which is often conceptualized as a discrete line object. But suppose we recognize that its position is uncertain. For example, the coastline shown on a 1:2 000 000 map is a gross generalization, in which major liberties are taken, particularly in areas where the coast is highly indented and irregular. Consequently the 1:2 000 000 version leaves substantial uncertainty about the true location of the shoreline. We might approach this by changing from a line to an area, and mapping the area where the actual coastline lies, as shown in the figure. But another approach would be to reconceptualize the coastline as a field, by mapping a variable whose value represents the probability that a point is land. This is shown in the figure as a raster representation. This would have far more information content, and consequently much more value in many applications. But at the same time it would be difficult to find an appropriate source for the representation — perhaps a fuzzy classification of an air photo, using one of an increasing number of techniques designed to produce representations of the uncertainty associated with objects discovered in images.

Uncertainty can often be conceptualized differently under field and discrete object views.

Indeed, far from offering quick fixes for eliminating or reducing uncertainty, the measure-

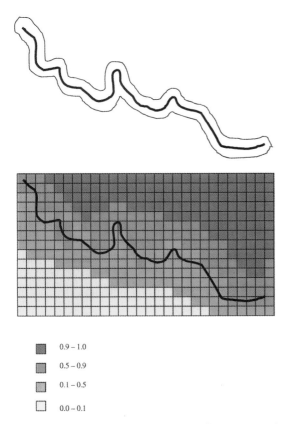

■	0.9 – 1.0
▣	0.5 – 0.9
▨	0.1 – 0.5
□	0.0 – 0.1

Figure 6.7 Contrast between discrete object (top) and field (bottom) conceptualizations of an uncertain coastline. In the discrete object view the line becomes an area delimiting where the true coastline might be. In the field view a continuous surface defines the probability that any point is land

ment process can actually introduce more of it. Given that the vector and raster data models impose quite different filters on reality, it is unsurprising that they can each generate additional uncertainty in rather different ways. In field-based conceptualizations, such as those that underlie remotely sensed images expressed as rasters, spatial objects are not defined *a priori*. Instead, the classification of each cell into one or other category builds together into a representation. In remote sensing, when resolution is insufficient to detect all of the detail in geographic phenomena, the term *mixel* is often used to describe raster cells that contain more than one class of land – in other words, elements in which the outcome of statistical classification suggests the occurrence of multiple land cover categories. The total area of cells classified as mixed should decrease as the resolution of the satellite sensor increases, assuming the number of categories remains

constant, yet a completely mixel-free classification is very unlikely at any level of resolution. Even where the Earth's surface is covered with perfectly homogeneous areas, such as agricultural fields growing uniform crops, the failure of real-world crop boundaries to line up with pixel edges ensures the presence of at least some mixels. Neither does higher resolution imagery solve all problems. Many of today's remote sensing image classifiers are designed to work with medium resolution data (defined as pixel size of between 30 × 30 m and 1000 × 1000 m, and data on between 3 and 7 bands). High resolution data (pixel sizes 10 × 10 m or smaller, with 7 to 256 bands) are characterized by much greater heterogeneity of spectral values and create problems for classification algorithms.

A pixel whose area is divided among more than one class is termed a mixel.

The vector data structure, by contrast, defines spatial entities and specifies explicit topological relations between them. Yet this often entails transformations of the inherent characteristics of spatial objects (Section 13.4). In conceptual terms, for example, while the true individual members of a population might each be defined as point-like objects, they will often appear in a GIS dataset only as aggregate counts for apparently *uniform* zones. Such aggregation can be driven by the need to preserve confidentiality of individual records, or simply by the need to limit data volume. Unlike the field conceptualization of spatial phenomena, this implies that there are good reasons for partitioning space in a particular way. In practice, partitioning of space is often made on grounds that are principally pragmatic, yet are rarely completely random. In much of socio-economic GIS, for example, zones which are designed to preserve the anonymity of survey respondents may be largely *ad hoc* containers. In practice they are likely to have been designed in order to do the job of data collection, and are likely to have no strong integrity as a consequence. They may also reflect the way that a cartographer or GIS interpolates a boundary between sampled points, as in the creation of isopleth maps (Box 5.4). Yet for all this, zones are unlikely to amount to a completely random partitioning of space. Even a census zonal scheme which has been designed principally for ease of data collection will not be devoid of socio-economic structure, since there is some degree of correspondence between social patterning and land-use morphology. As we will see in Section 6.4, this state-of-affairs may be the worst possible outcome for the analysis of spatial distributions.

6.3.2 Accuracy and error

Geography is a science concerned with the examination of events and occurrences at unique locations on the Earth's surface. Yet even when these events are precisely defined certainties, such as the physical location of physical objects, we can only realistically hope to make imperfect measurements of their spatial attributes. Fundamental reasons for this include the effects of seismic motions, continental drift, and the wobbling of the Earth's axis. Moreover, the Earth is not a perfect sphere, and a range of different mathematical approximations are made of it for different purposes (see Chapter 4).

There are many reasons why measurement of position on the Earth's surface is inherently uncertain.

Measurement error is also sometimes deliberate. Until May 2000 NAVSTAR GPS signals were only available to civilian applications in degraded form (this policy was known as *selective availability*), in order to protect the military interests of the USA and its allies (see Box 10.1). More generally, it is common for a controlled amount of uncertainty to be added to published results from social survey data, particularly for small areas, in order to maintain respondent confidentiality. For example, Canada uses a process of random rounding, by randomly forcing the rightmost digit in any aggregate count to 0 or 5. It is also the case that much GIS-based analysis continues to be carried out using legacy data from the pre-digital era. Accuracy is often lost when legacy data are digitized from paper documents, while the measures themselves were often captured using previous, less sophisticated, technologies.

It follows that there will inevitably be discrepancies between recorded measurements and the truth, and thus measurements will only be *accurate* to a limited extent. Statistical methods for dealing with such discrepancies, or *errors*, are discussed in Chapter 15, along with the relationship between accuracy and *precision*, a term used to describe both the repeatability of a measurement and the level of detail used in its reporting (see Section 15.1.2).

6.3.3 Measurement error

Uncertainty, in the straightforward sense of measurement error, is also generated by the procedures of digital data capture. In the early days of GIS, most vector data were captured using digitizing tables or tablets. The first stage of digitizing is to attach a paper map to the table or

tablet and to register its corners as control points using a cursor (see Box 10.2). Having thus established the coordinate frame, in the case of polygon data collection, the cursor is then used manually to trace a succession of boundary points, arranged together in line segments. Topological information is added in a tag file, which records the zones lying to either side of each segment. Together these data enable polygons to be reconstructed, and attributes to be assigned to them.

Difficulties in making accurate measurements from maps were a major impediment to early GIS.

This time-consuming procedure presented a data capture bottleneck to the development of GIS from the 1970s to the early 1990s. Maps have always been selective and generalized abstractions of reality, and a range of uncertainties are present in paper map sources. Yet because digitizing is a tedious and hence error-prone practice, it presents a source of additional measurement errors – as in when the operator fails to position the cursor correctly, or fails to record line segments. Figure 6.8 presents a typology of human errors that are commonly introduced in the digitizing procedure. They are: over- and undershoots where line intersections are inexact (Figure 6.8(A)); invalid polygons which are topologically inconsistent because of omission of one or more lines or omission of tag data (Figure 6.8(B)); and sliver polygons, in which multiple digitizing of the interstices of adjacent polygons leads to the creation of an additional polygon (Figure 6.8(C)).

Most GIS and computer-assisted cartography packages include standard software functions, which can be used to restore integrity and clean (or rather obscure, depending upon your viewpoint!) obvious measurement errors. Such operations are best carried out immediately after digitizing, in order that omissions may be easily rectified. Data cleaning operations require sensitive setting of threshold values, or else damage can be done to real-world features, as Figure 6.9 shows. The processes of error detection and correction are difficult if not impossible to automate, and the methods of cartographic generalization used by expert cartographers have proven very difficult to formalize and replicate in digitizing. Although these issues have been raised here in relation to point-by-point digitizing, broadly similar problems hold for semi-automated modes of capture that attempt to follow lines automatically. We examine the technical remedies to these

(A)

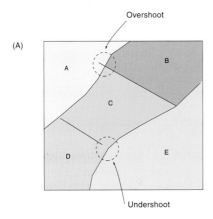

(B)

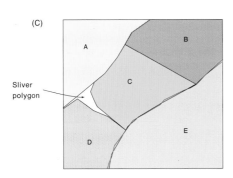

(C)

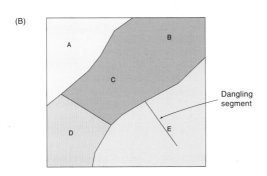

Figure 6.8 A typology of human errors in digitizing: (A) undershoots and overshoots; (B) invalid polygons; and (C) sliver polygons

problems in Section 10.3.2, but it is important to be aware that such remedies are rarely perfect solutions.

Many errors in digitizing can be remedied by appropriately designed software.

Further classes of problems arise when the products of digitizing adjacent map sheets are

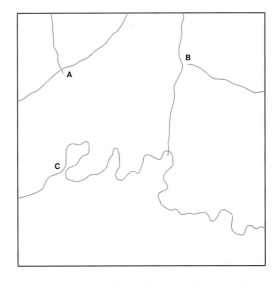

Figure 6.9 Error induced by data cleaning. If the tolerance level is set large enough to correct the errors at A and B, the loop at C will also (incorrectly) be closed (after Flowerdew 1991).

merged together. Stretching of paper base maps, coupled with errors in rectifying them on a digitizing table, give rise to the kinds of mismatches shown in Figure 6.10. *Rubber-sheeting* is the term used to describe methods for removing such errors on the assumption that strong spatial autocorrelation exists among errors. If errors tend to be spatially auto-correlated up to a distance of x, say, then rubber-

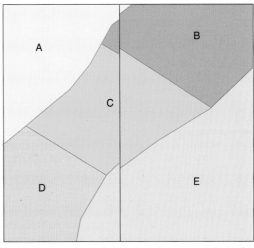

Figure 6.10 Mismatches of adjacent spatial data sources that require rubber-sheeting

sheeting will be successful at removing them, at least partially, provided control points can be found that are spaced less than x apart. For the same reason, the shapes of features that are less than x across will tend to have little distortion, while very large shapes may be badly distorted. The results of calculating areas (Section 13.3), or other geometric operations that rely only on relative position, will be accurate as long as the areas are small, but will grow rapidly with feature size. Thus it is important for the user of a GIS to know which operations depend on *relative* position, and over what distance; and where *absolute* position is important (of course the term *absolute* simply means relative to the Earth frame, defined by the Equator and the Greenwich Meridian, or relative over a very long distance: see Section 4.6). Analogous procedures and problems characterize the rectification of raster datasets – be they scanned images of paper maps or satellite measurements of the curved Earth surface.

GPS is the first system that allows accurate, direct, and inexpensive measurement of absolute position on the Earth's surface.

The advent of Global Positioning Systems (GPS) technology has revolutionized the capture of vector information (see Section 10.2.2.2). GPS has increased the accuracy of most data capture, has obviated the need for paper maps as a storage medium, and in certain important respects has burst the data capture bottleneck. Yet GPS data will never attain perfect accuracy. Thus GIS will always have to deal with uncertainty of position, with the distinctions between relative and absolute accuracy, and the complex implications of these factors for analysis.

6.3.4 Data integration and shared lineage

Goodchild and Longley (1999) use the term *concatenation* to describe the integration of two or more different data sources, such that the contents of each are accessible in the product. The polygon overlay operation that will be discussed in Section 13.4.3 is one simple form of concatenation. The term *conflation* is used to describe the range of functions that attempt to overcome differences between datasets, or to merge their contents (as with rubber-sheeting in Section 6.3.3 above). Conflation thus attempts to replace two or more versions of the same information with a single version that reflects the pooling, or weighted averaging, of the sources.

The individual items of information in a single geographic dataset often share lineage, in the sense that more than one item is affected by the same error. This happens, for example, when a map or photograph is registered poorly, since all of the data derived from it will have the same error. One indicator of shared lineage is the persistence of error – because all points derived from the same misregistration will be displaced by the same, or a similar, amount. Because neighboring points are more likely to share lineage than distant points, errors tend to exhibit strong positive spatial autocorrelation (see the text by Goodchild and Gopal 1989 for examples of this property, and for other issues associated with the quality of geographic databases).

Conflation combines the information from two data sources into a single source.

When two datasets that share no common lineage are concatenated (for example, they have not been subject to the same misregistration), then the relative positions of objects inherit the absolute positional errors of both, even over the shortest distances. While the shapes of objects in each dataset may be accurate, the relative locations of pairs of neighboring objects may be wildly inaccurate when drawn from different datasets. The anecdotal history of GIS is full of examples of datasets which were perfectly adequate for one application, but which failed completely when an application required that they be merged with some new dataset that had no common lineage. For example, merging GPS measurements of point positions with streets derived from the US Bureau of the Census TIGER files may lead to surprises where points appear on the wrong sides of streets. If the absolute positional accuracy of a dataset is 50m, as it is with parts of the TIGER database, points located less than 50 m from the nearest street will frequently appear to be misregistered.

Datasets with different lineages often reveal unsuspected errors when overlaid.

Figure 6.11 shows an example of the consequences of overlaying data with different lineages. In this case, two datasets of streets produced by different commercial vendors using their own processes fail to match in position by amounts of up to 100 m, and also fail to match the names of many streets, and even the existence of streets. In a similar vein, the positional accuracy of the georeferences in the UK Central Postcode Directory match only imprecisely with recent census geographies, such that concatenation of composite geodemographic indicators (requiring data referenced from each source) is

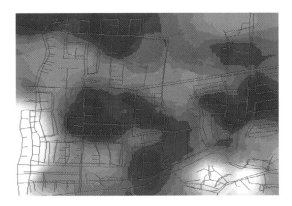

Figure 6.11 Overlay of two street databases for part of Goleta, California, USA. The red and green lines fail to match by as much as 100 m. Note also that in some cases streets in one data set fail to appear in the other, or have different connections. The background is dark where the fit is best, and white where it is poorest (it measures the average distance locally between matched intersections)

an error-prone task. Box 11.4 also details similar problems in the measurement of geodemographic characteristics.

The integrative functionality of GIS makes it an attractive possibility to generate multivariate indicators from diverse sources. Yet such data are likely to have been collected at a range of different scales, and for a range of areal units as diverse as census tracts, river catchments, land ownership parcels, travel-to-work areas, and market research surveys. Established procedures of statistical inference are used to reason from geographically representative samples to the populations from which they were drawn (Moser and Kalton 1985). Yet these procedures do not regulate the assignment of inferred values to (usually smaller) zones, or their apportionment to *ad hoc* regional categorizations. There is an emergent tension within the socio-economic realm, for there is a limit to the domains of inference that can be made from conventional, scientifically valid data sources which are frequently out-of-date, zonally coarse, and irrelevant to what is happening in modern societies. Yet the alternative of using new rich sources of marketing data is profoundly unscientific in its inferential procedures, for reasons illustrated in Box 6.2.

Box 6.2 *Living with uncertainty and lifestyles data*

Lifestyles is a broad term that has been used to describe data pertaining to the consumption of a wide range of goods and services by identifiable individuals and households. Lifestyles data originate from a diverse range of sources, such as guarantee card returns, questionnaires attached to nationally circulated prize draw entries, and market research surveys. They are usually georeferenced through the postcode system (e.g. in the UK to the unit postcode, equivalent to ZIP + 4 in the US, and which typically comprises 15 or so addresses in urban areas). At least one UK "data warehouse" estimates that it holds up-to-date information on 11 million UK households, and individual survey *waves* can gain responses from around 10 per cent of all resident households. Such data have evident use for direct marketing, because past consumption habits are key guides to future behavior. They also provide current small-area estimates of income and material standards of living (see Figure 6.12). In recent years, lifestyles approaches have gained some ground as tools for geomarketing at the expense of the use of census and composite *geodemographic* indicators (see the discussion of the MOSAIC geodemographic system in Box 2.1). This is because the census data may be out of date and expensive to use (because of royalty structures, as in the UK), and because the census contains too few variables that are of much interest to marketeers and policy analysts.

However, the lineage of geodemographics data is clear, while that of lifestyles data is very uncertain. Geodemographics is based on tried and trusted techniques and derives from a dataset (the census) which has been designed and implemented using the most rigorous research design principles. It is by no means straightforward to reconcile lifestyles data with conventional sources: sampling theory tells us that reweighting of largely self-selecting samples on the basis of sub-group response rates is foolhardy. Yet survey research practice also tells us that quantitative indicators should be direct, relevant, timely, and transparent. Science, ethical issues of confidentiality (Section 17.3.1), and public funding of relevant and up-to-date surveys are all considerations in creating the data infrastructure to GIS-based spatial analysis.

A middle path between census and lifestyle approaches lies in the use of lifestyle descriptors as a wrapper to add depth to the labels assigned to different geodemographic groups (see Figure 6.13). Yet the lineage of some of the variables used to create such categories may be uncertain, and the best scientific practice for dealing with scale and aggregation issues may not always be adhered to.

**Box 6.2
Continued**

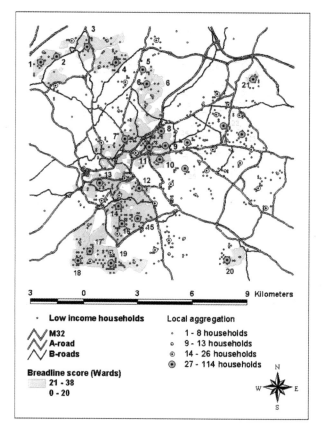

Figure 6.12 Composite local area deprivation ("Breadline" index) scores for central Bristol, UK (Courtesy: Richard Harris)

Figure 6.13 "Pen portraits" are used to illustrate the characteristics of geodemographic types, such as "Affluent Achievers". However, the lineage of some of the variables used to create such classifications is often uncertain

6.3.5 Ambiguity revisited

GIS has fundamentally upgraded our abilities to generalize about spatial distributions. Yet our abilities to do so may be constrained by the different taxonomies that are used by data-collecting organizations within our overall study area. A study of wetland classification in the USA found no fewer than six agencies engaged in mapping the same phenomena over the same geographic areas, and each with their own definitions of wetland types. If wetland maps are to be used in regulating the use of land, as they are in many areas, then uncertainty in mapping clearly exposes regulatory agencies to potentially damaging and costly lawsuits. How might soils data classified according to the UK national classification be assimilated within a pan-European soils map, which uses a classification honed to the full range and diversity of soils found across the European Continent rather than just those characteristic of an offshore island (see Fisher 1999)? How might different national geodemographic classifications be combined into a form suitable for a pan-European marketing exercise? These are all variants of the question:

● How may mismatches between the categories of different classification schema be reconciled?

Differences in definitions are a major impediment to integration of geographic data over wide areas.

Like the process of pinning down the different nomenclatures developed in different cultural settings, the process of reconciling the semantics of different classification schema is an inherently *ambiguous* procedure. Ambiguity arises in data concatenation when we are unsure regarding the *meta-category* to which a particular class should be assigned (Fisher 1999).

6.4 U3: Further Uncertainty in the Analysis of Geographic Phenomena

6.4.1 Spatial analysis and uncertainty

In Chapter 1 we identified one remit of GIS as the resolution of scientific or decision-making problems through spatial analysis, which we defined in Section 1.7 as "the process by which we turn raw spatial data into useful spatial information". Good science needs secure foundations, yet Sections 6.2 and 6.3 have shown the conception and measurement of many geographic phenomena to be inherently uncertain. How can

the outcome of spatial analysis be meaningful if such uncertainty obscures the precise nature of spatial variation?

Uncertainties in data lead to uncertainties in the results of analysis.

Once again, there are no easy answers to this question, although we can begin by re-examining the effects of aggregating true elements of analysis into artificial geographic individuals (as when people are aggregated by census tracts, or disease incidences are aggregated by county). Like much research into the problems of geographic analysis, our approach here will be to proceed by *illustration*. The conception and measurement of geographic individuals may distort the outcome of spatial analysis by masking or accentuating apparent variation across space, and constraining the set of ways in which problems may be framed. In this section we will illustrate that although we can only rarely tackle the *source* of distortion, we can nevertheless quantify the way in which it *operates* and the magnitude of its likely impacts. Moreover, although aggregation operates as a constraint upon spatial analysis, GIS allows us to model within-zone spatial distributions in order to ameliorate its worst effects. Taken together, GIS provides a medium for sensitivity analysis and modeling of the effects of scale and aggregation. This is known as *internal validation* of the effects of scale and spatial partitioning.

The abilities of GIS to conflate and concatenate diverse data sources also provide a means of *external validation* of the effects of zonal averaging. As the range and detail of digital georeferenced data sources magnify, so the range of ancillary sources that can be used to sustain a consistent base to analysis increases. Different internal and external validating measures are provided through the statistical and geocomputational paradigms. In Chapter 12 we will also consider the ways in which GIS provides a medium for visualizing models of spatial distributions and patterns of homogeneity and heterogeneity.

6.4.2 Aggregation and analysis

We have already seen that a fundamental difference between geography and other scientific disciplines is that the definition of its objects of study is rarely unambiguous and, in practice, rarely precedes our attempts to measure their characteristics. In socio-economic GIS applications, these objects of study (geographic individuals) are usually aggregations, since the

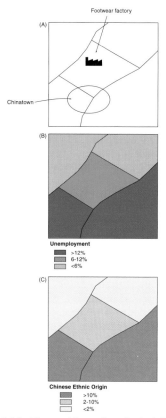

Figure 6.14 The problem of ecological fallacy. Before it closed down, the Anytown footwear factory drew its labor from blue-collar residential areas to the south and west of it. Its closure led to high local unemployment, but not amongst the residents of Chinatown, who remain employed in service industries. Yet comparison of choropleth maps B and C suggests a spurious relationship between Chinese ethnicity and unemployment

spaces that human individuals occupy are geographically unique, and confidentiality restrictions usually dictate that uniquely attributable information must be anonymized in some way. Even in natural environment applications, the nature of sampling in the data collection process (Section 5.4) often makes it expedient to collect data pertaining to aggregations of one kind or another. Thus in socio-economic and environmental applications alike, the measurement of geographic individuals is unlikely to be determined with the endpoint of particular spatial analysis applications in mind. As a consequence, we cannot be certain in ascribing even dominant characteristics *of* areas to true individuals or point locations *in* those areas. This source of uncertainty is known as the *ecological fallacy*,

and has long bedevilled the analysis of spatial distributions (the opposite of ecological fallacy is atomistic fallacy, in which the individual is considered in isolation from his or her environment). This is illustrated in Figure 6.14.

> Inappropriate inference from aggregate data about the characteristics of individuals is termed the ecological fallacy.

We have also seen that the *scale* at which geographic individuals are conceived conditions our measures of association between the mosaic of zones represented within a GIS. Yet even when scale is fixed, there is a multitude of ways in which basic areal units of analysis can be *aggregated* into zones, and the requirement of spatial contiguity represents only a weak constraint upon the huge combinatorial range. This gives rise to the related *aggregation* or *zonation* problem, in which different combinations of a given number of geographic individuals into coarser-scale areal units can yield widely different results. In a classic short monograph, Stan Openshaw (1984) applied correlation and regression analysis to the attributes of a succession of zoning schemes (see Box 6.3). He demonstrated that the constellation of elemental zones within aggregated areal units could be used to manipulate the results of spatial analysis to a high degree of prespecified precision. Similar arguments and applications are developed by Openshaw and Alvanides (1999). These numerical experiments have some sinister counterparts in the real world, the most notorious example of which is the political gerrymander of 1812 (see Section 13.3.2 and Box 13.4). Chance or design might therefore conspire to create apparent spatial distributions which are unrepresentative of the scale and configuration of real-world geographic phenomena. The outcome of multivariate spatial analysis is also very sensitive to the particular zonal scheme that is used (Fotheringham and Wong 1991). Taken together, the effects of scale and aggregation are generally known as the *modifiable areal unit problem* (MAUP: Openshaw 1984; Openshaw and Alvanides 1999).

The ecological fallacy and the MAUP have long been recognized as problems in applied spatial analysis and, through the concept of spatial autocorrelation (Section 5.2), they are also understood to be related problems. Increased technical capacity for numerical processing and innovations in scientific visualization have refined the quantification and mapping of these measurement effects, and have also focused interest on the effects of within-area spatial distributions upon analysis.

6.4.3 From MAUP to zone design

Reformulation of the MAUP into a *geocomputational* (Box 6.3) approach to zone design forms the focus of an extended discussion by Openshaw and Alvanides (1999). Briefly, they assess the prospects for developing and implementing a geocomputational pattern-seeking paradigm within GIS, in order to appraise and interpret the effects of zone measurement upon spatial analysis. They anticipate that the proliferation of digital spatial databases, coupled with the ever-wider use of GIS, will allow users to create their own customized zonal representations of spatial distributions, and that this will make the MAUP still more central to the zonal measurement task. Their approach to calculating zonal averages is avowedly inductive, and uses the sheer computational power of GIS to identify the full range and diversity of outcomes that zone design can generate through repeated scaling and aggregation experiments.

The Modifiable Areal Unit Problem can be investigated through simulation of large numbers of alternative zoning schemes.

In some senses this is merely playing with the problem. The conception and measurement of elemental zones, the geographic individuals, may be *ad hoc*, but it is rarely wholly random either. In measuring the distribution of zonally averaged outcomes there is thus no tenable analogy with the established procedures of statistical inference and its concepts of precision and error. Classic statistical inference allows us to draw inference from a random (or similar) sample to the population from which it was drawn. Yet the size, shape, and configuration of zones is not *a priori* known or prespecified, and so there can be no analogous concept of simple random zoning.

Zoning seems similar to sampling, but its effects are very different.

The way forward is to match the recently developed ability to customize zoning schemes in GIS with clearer thinking in each application about the amount of heterogeneity within zones. In this deductive view, espoused by Stan Openshaw, MAUP will disappear if GIS analysts understand the particular areal units that they wish to study. But what does "understanding" entail? If this kind of approach is not possible, the approach adopted in the reference companion by Openshaw and Alvanides provides a fall-back position of using zoning systems to act as spatial pattern detectors. From this latter interpretation, repeated measurements of spatial patterns might

be used to identify zoning systems which support or refute an hypothesis of interest.

There is a sense in which sensitivity analysis of the effects of zones upon the calculation of spatial measures implies recognition of the importance, not just of the objective spatial structure of space, but also its particular qualitative character. There is also a practical recognition here that the areal objects of study are ever-changing, and our perceptions of what constitutes their appropriate definition will change. And finally, within the socio-economic realm, the act of defining zones can also be self-validating if the allocation of individuals affects the interventions they receive (be they a mail-shot about a shopping opportunity or aid under an areal policy intervention). Spatial discrimination affects spatial behavior, hence also changing the geographic individuals themselves.

6.5 Consolidation

Uncertainty is certainly much more than error. Just as the amount of available digital data and our abilities to process it have developed, so our understanding of the quality of digital depictions of reality has broadened. It is one of the supreme ironies of contemporary GIS that as we accrue more and better data and have more computational power at our disposal, so we seem to become more uncertain about the quality of our digital representations. The advent of high-power computing has confirmed the suspicion that there will never be any general analytical solution to the modifiable areal unit problem. Richness of representation and computational power only make us more aware of the range and variety of established uncertainties, and challenge us to integrate new ones. The only way beyond this impasse is to advance hypotheses about the structure of data, in a spirit of humility rather than conviction. But this implies greater *a priori* understanding about the structure in spatial as well as attribute data. There are some general rules to guide us here and statistical measures of spatial autocorrelation can provide further structural clues (Section 5.6). The developing range of context-sensitive spatial analysis methods provides a bridge between such general statistics and methods of specifying place or local (natural) environment (Fotheringham *et al* 2000). Geocomputation helps too, by allowing us to gauge the sensitivity of outputs to inputs, but, unaided, is unlikely to provide any unequivocal best solution. The fathoming of uncertainty requires a combination of cumulative devel-

Box 6.3 Stan Openshaw, spatial analyst and father of geocomputation

Stan Openshaw (Figure 6.15) trained as a geographer but began his career as a planner in Newcastle-upon-Tyne, UK. He subsequently returned to academic geography, and over the last 25 years has changed much of the landscape of spatial analysis as we see it today. Although a quantitative geographer by training, he became frustrated that geographers were using statistical and mathematical tools that were at the same time too complex and too restrictive. While embracing GIS as a useful and relevant technology for the capture and management of mappable information, at the same time he found it lacking in theory and spatial analysis capabilities.

His research has consistently advocated the intensive use of high power computing in geographical problem-solving. One of his early classic books (Openshaw 1984) introduced and popularized the *Modifiable Areal Unit Problem*, discussed in Sections 6.4.2 and 6.4.3. His subsequent research developed the *Geographical Analysis Machine* (GAM: 1987), which was innovative in the way that it searched datasets for localized event clusters – at a time when most spatial statistical methods still concentrated on global measures of pattern. Subsequent work sought to develop map-based associations between clusters and other variables (the 1997 Geographical Explanations Machine, GEM) and advanced techniques of data mining using so-called genetic algorithms (the 1998 Map Explorer, MAPEX).

In the late 1990s, Stan and his colleagues at Leeds University, UK, coined the term "geocomputation", to describe the application of computationally intensive approaches to the problems of physical and human geography in particular and the geosciences in general. In some important respects the term geocomputation is synonymous with geographic information science (Section 1.6), although it has often put greater emphasis upon the use of high-performance computers.

Although his work has very much been at the "coal face" of technical and methodological development, Stan is a geographer in the widest sense. His various books and research papers consider the geographical correlates of cancer, flood forecasting, modeling the effect of nuclear attack, crime forecasting, and the siting and safety of nuclear power plants. He is one of the most concerned and passionate geographers of his generation. Stan suffered a severe stroke in 1999.

Figure 6.15 Stan Openshaw, spatial analyst

---

opment of *a priori* knowledge (we should expect scientific research to be cumulative in its findings) and inductive generalization in the fluid, eclectic data-handling environment that is contemporary GIS.

Questions for Further Study

1. Summarize the relative merits of geodemographic indicators of consumer behavior, and lifestyles shopping survey data that are based on unconventional data sources.
2. "Insofar as the statements of geometry speak about reality, they are not certain, and insofar as they are certain, they do not speak about reality" (Einstein A 1921, *Geometrie und Erfahrung*, 3). Discuss!
3. You are a senior retail analyst for Safemart, which is contemplating expansion from its home US state to three others in the Union. Assess the relative merits of your own company's store loyalty card data (which you can assume are similar to those collected by any retail chain with which you are familiar) and of data from the latest available census in planning this strategic initiative. Pay particular attention to issues of survey content, the representativeness of population characteristics, and problems of scale and aggregation.

Suggest ways in which the two data sources might complement one another in an integrated analysis.

4. Using aggregate data for Iowa counties, Openshaw (1984) found a strong positive correlation between the proportion of people over 65 and the proportion who were registered voters for the Republican party. What if anything does this tell us about the tendency for older people to register as Republicans?

5. Are there any instances where geographic data create perfect representations of real phenomena?

Online Resources

NCGIA Core Curricula (www.ncgia.ucsb.sb.edu/pubs/core.html):
Core Curriculum in GIScience, Section 2.10 (Handling Uncertainty, edited by Gary Hunter), 2.10.1 (Managing Uncertainty in GIS, Gary Hunter), 2.10.3 (Detecting and Evaluating Errors by Graphical Methods, Kate Beard), 2.10.4 (Data Quality Measurement and Assessment, Howard Veregin);
Core Curriculum in GIS, 1990, Units 45 (Accuracy of Spatial Databases) and 46 (Managing Error)
ESRI Virtual Campus course (campus.esri.com):
See Online Resources to Chapter 15 for relevant technical material

Reference Links

Maguire D J, Goodchild M F, and Rhind D W (eds) 1991 *Geographical Information Systems: Principles and applications.* Harlow, UK: Longman (Text available online from 'Links

to Big Book 1' at www.wiley.com/gis and www.wiley.co.uk/gis).
Chapter 11, Language issues for GIS (Frank A U, Mark D M)
Chapter 24, Spatial data integration (Flowerdew R)
Longley P A, Goodchild M F, Maguire D J, and Rhind D W (eds) 1999 *Geographical Information Systems: Principles, techniques, management and applications.* New York: John Wiley.
Chapter 7, Spatial representation: a cognitive view (Mark D M)
Chapter 13, Models of uncertainty in spatial data (Fisher P F)
Chapter 18, Applying geocomputation to the analysis of spatial distributions (Openshaw S, Alvanides S)
Chapter 35, Multi-criteria evaluation and GIS (Eastman J R)
Chapter 40, The future of GIS and spatial analysis (Goodchild M F, Longley P A)

References

Burrough P A and Frank A U (eds) 1996 *Geographic Objects with Indeterminate Boundaries.* London: Taylor and Francis.
Fotheringham A S, Brunsdon C, and Charlton M 2000 *Quantitative Geography: Perspectives on spatial data analysis.* London: Sage.
Fotheringham A S and Wong D W S 1991 The modifiable areal unit problem in multivariate statistical analysis. *Environment and Planning A* **23**: 1025–1044.
Goodchild M F and Gopal S (eds) 1989 *Accuracy of Spatial Data.* London: Taylor and Francis.
Moser C and Kalton G 1985 *Survey Methods in Social Investigation.* Aldershot: Gower
Openshaw S 1984 *The Modifiable Areal Unit Problem.* Concepts and Techniques in Modern Geography 38. Norwich, UK: GeoBooks.
Yule G U and Kendall M G (1950) *An Introduction to Statistics.* New York: Hafner.

GENERALIZATION, ABSTRACTION, AND METADATA

7

The infinite complexity of the Earth's surface creates a major problem for any digital representation, because it implies a perfect representation necessarily will be infinitely large. But despite recent progress in developing massive and cheap storage devices, capacity is very quickly exhausted by even modestly accurate representations. It follows that any practical GIS must make use of generalization, approximation, and other techniques to achieve representations of useful accuracy. This chapter is about those methods. In the extreme, generalization attempts to summarize the entire contents of a dataset, and this is the basis of cataloging, and other methods designed to make it possible to search large archives for suitable data. So somewhat unconventionally, the chapter ends with a section on dataset description, or what is now known as metadata.

Learning objectives

On finishing this chapter you will know:

● Why generalization and abstraction are needed;

● The basic principles of cartographic generalization;

● Ways of describing degrees of generalization and levels of detail;

● The concept of metadata, and its formalization into standards;

● How metadata support search for datasets over the Internet.

7.1 Introduction

In Chapter 3 we saw how thinking about geographic information as a collection of atomic facts of the form $\langle x, t, z \rangle$ – in other words, links between a place, a time (not always, because many geographic facts are stated as if they were permanently true), and a property – led to an immediate problem, because the potential number of such atomic facts is infinite. If seen in enough detail, the Earth's surface is unimaginably complex, and its effective description impossible. So instead, humans have devised numerous ways of simplifying their view of the world. In the previous chapter, we saw how instead of making statements about each and every point, we describe entire areas, attributing uniform characteristics to them, even when areas are not strictly uniform; we identify features on the ground and describe their characteristics, again assuming them to be uniform; or we limit our descriptions to what exists at a finite number of sample points, hoping that these samples will be adequately representative of the whole, as we saw in Chapter 5.

> A geographic database cannot contain a perfect description – instead, its contents must be carefully selected to fit within the limited capacity of computer storage devices.

Consider the statement "Iowa is flat" (with the implication that it is permanently flat, since there is no reference to time). Instead of describing the terrain at every point in the state, this statement captures a vast amount of such information in a single statement, about the aggregate region of Iowa. Formally, an infinite number of atoms of information, or *tuples*, of the form $\langle x, \text{"flat"} \rangle$, have been replaced by one, $\langle \text{"Iowa"}, \text{"flat"} \rangle$, along with a definition of Iowa in terms of the points that lie within it. Given a definition of Iowa, in the form of a defined polygonal boundary, it is possible to transform the single tuple $\langle \text{"Iowa"}, \text{"flat"} \rangle$ into an infinite number of tuples of the atomic form $\langle x, \text{"flat"} \rangle$ where x is a point inside the polygon representing Iowa. This is data compression on a massive scale, from an infinite quantity to a single tuple. Note also that the definition of Iowa is based in law, uniformly agreed between everyone, and obtainable from many sources. In other words, a definition of Iowa does not need to be sent along with the fact, because the receiver already has that information. By agreeing on such definitions, we humans create the potential for such massive data compression and efficiency in communication.

> Efficient compression is made possible because many definitions are already shared, and do not need to be communicated.

Chapter 6 discussed the downside of this process. Since Iowa is not in fact flat, an accurate description would produce an infinite number of tuples $\langle x, \text{slope} \rangle$ where slope is almost never exactly 0. So the statement $\langle \text{"Iowa"}, \text{"flat"} \rangle$ is a distortion or approximation to the truth, and the compression process that produces it must involve loss of information – it is "lossy", in the language of data compression. Geographic description is full of examples like this – the formulation of statements that are not exactly true, but achieve massive levels of data compression. Also although Iowa is legally defined, there is likely to be variation among the available definitions of its polygon, which provides an additional source of uncertainty.

This chapter is about the process of compression, and its relationship to the level of detail in a representation of the Earth's surface. Cartographers have long practiced *generalization*, or the process of representing the Earth's surface on maps at given levels of detail. The process of compression is often described as *aggregation*, particularly if it occurs through the lumping of data items. *Abstraction* is a more general term, referring to any process that attempts to capture the essence of a dataset in a smaller volume. The chapter reviews many aspects of these processes, using examples from GIS applications.

The later sections of the chapter deal with *metadata*, or the formalized description of data. Consider a conversation between two GIS users that goes along the following lines:

User A "I'm looking for data on Iowa, do you have any?"
User B "I have some, what exactly are you looking for?"
User A "Data on Iowa's topography"
User B "I can give you a 30 m DEM, will that help?"
User A "I think so, can you send it to me as an e-mail attachment?"

Conversations like this take time, especially if A has to ask many Bs to find the needed data. The Internet has made it possible to work much more efficiently, by searching collections of data on major servers instead of contacting potential suppliers individually by phone. Each dataset in such collections is accompanied by metadata, with entries that describe such useful properties as the area covered, level of detail, accuracy, and availability. Today there exist widely used standards for metadata that allow users to conduct searches with speed and efficiency.

Metadata are formal descriptions of data.

Discussions of metadata are not normally linked to generalization and abstraction, and our decision to deal with them all in a single chapter is unconventional. But metadata are essentially an abstraction of a dataset, and in that sense represent an extremely high level of compression. Of course they do much more, by describing properties like the dataset's origins that are not evident from its contents. But it makes sense to us to discuss metadata alongside other forms of abstraction, and we hope it makes sense to you too.

7.2 Generalization Basics

Each of the basic data models used in GIS (see Chapter 9) is a generalization, because it allows geographic variation to be represented in a compressed form. By thinking of variation in terms of discrete objects, we are able to compress representation into a finite collection, with associated attributes that necessarily apply to each entire object. By thinking of variation in terms of a field, we compress representation by using one of six basic approaches (Figure 7.1):

A. Capturing the value of the field variable at each of a set of irregularly spaced sample points (for example, variation in surface temperature captured at weather stations).
B. Capturing the value of the variable at each of a grid of regularly spaced sample points (for example, elevations at 30 m spacing in a DEM).
C. Capturing a single value of the variable for a regularly shaped cell (for example, values of reflected radiation in a remotely sensed scene).
D. Capturing a single value of the variable over an irregularly shaped area (for example, vegetation cover class or the name of a parcel's owner).
E. Capturing the linear variation of the field variable over an irregularly shaped triangle (for example, elevation captured in a triangulated irregular network or TIN, Section 9.2.3.4).
F. Capturing the isolines of a surface, as digitized lines (for example, digitized contour lines representing surface elevation).

GIS utilize six ways of representing a field.

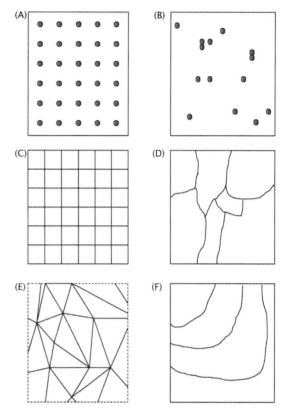

Figure 7.1 The six approximate representations of a field used in GIS. (A) Regularly spaced sample points. (B) Irregularly spaced sample points. (C) Rectangular cells. (D) Irregularly shaped polygons. (E) Irregular network of triangles, with linear variation over each triangle (the triangulated irregular network or TIN model; the bounding box is shown dashed in this case because the unshown portions of complete triangles extend outside it). (F) Polylines representing contours (see the discussion of isopleth maps in Box 5.4)

Each of these options achieves a high level of compression, but also results in information loss. For example, methods (C) and (D) capture a single value for each cell or polygon. In the case of field variables defined on interval or ratio scales this single value might be the mean over the area, or perhaps the minimum or maximum, but in all cases the replacement of true values at every point in the area by a single value results in information loss. In the case of variables measured on nominal scales the single value might be the mode, or the commonest value observed – and in the case of ordinal scales it might be the median, or the value such that half of the area has higher values, and half has lower values. Again, information loss is almost inevitable. Only in rare cases, such as the representation of the elevation of a lake, or the representation of ownership of land parcels, is the field variable essentially constant over the area, allowing us to avoid information loss.

From this perspective some degree of generalization is almost inevitable in all geographic data. But cartographers often take a somewhat different approach, for which this observation is not necessarily true. Suppose we are tasked to prepare a map at a specific scale, say 1:25,000, using the standards laid down by a national mapping agency, such as the Institut Géographique National (IGN) of France. Every scale used by IGN has its associated rules of representation. For example, at a scale of 1:25,000 the rules lay down that individual buildings will be shown only in certain circumstances, and similar rules apply to the 1:24,000 series of the US Geological Survey. These rules are known by various names, including *terrain nominal* in the case of IGN, which translates

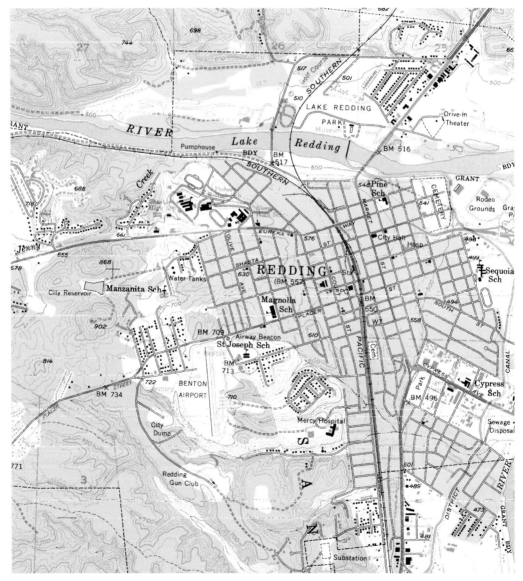

Figure 7.2 An extract from a US Geological Survey 1:24,000 topographic map. In one view maps like this can be perfectly accurate if they adhere exactly to the rules of the agency's specification. For example, the specification may call for features to be moved from their true positions on the map in the interests of clarity (Source: USGS)

roughly but not very helpfully to "nominal ground", and is perhaps better translated as "specification". From this perspective a map that represents the world by following the rules of a specification precisely can be perfectly accurate *with respect to the specification*, even though it is not a perfect representation of the full detail on the ground.

A map's specification defines how real features on the ground are selected for inclusion on the map.

Consider the representation of vegetation cover using the rules of a specification. For example, the rules might state that at a scale of 1:100,000, a vegetation cover map should not show areas of vegetation that cover less than 1 hectare. But smaller areas of vegetation almost certainly exist, so deleting them inevitably results in information loss. But under the principle discussed above, a map that adheres to this rule must be accurate, *even though it differs substantively from the truth as observed on the ground*.

7.3 Methods of Generalization

McMaster and Shea (1992) identify many different forms of generalization, and similar lists have appeared elsewhere in the cartographic literature. Generalization rules may involve:

- *Simplification*, for example by weeding out points in the outline of a polygon to create a simpler shape.
- *Smoothing*, or the replacement of sharp and complex forms by smoother ones.
- *Collapse*, or the replacement of an area object by a combination of point and line objects.
- *Aggregation*, or the replacement of a large number of distinct symbolized objects by a smaller number of new symbols.
- *Amalgamation*, or the replacement of several area objects by a single area object.
- *Merging*, or the replacement of several line objects by a smaller number of line objects.
- *Refinement*, or the replacement of a complex pattern of objects by a selection that preserves the pattern's general form.
- *Exaggeration*, or the relative enlargement of an object to preserve its characteristics when these would be lost if the object were shown to scale.
- *Enhancement*, through the alteration of the physical sizes and shapes of symbols.

- *Displacement*, or the moving of objects from their true positions to preserve their visibility and distinctiveness.

The differences between these types of rules are much easier to understand visually, and Figure 7.3 reproduces McMaster's and Shea's original example drawings.

In addition, they describe two forms of generalization, as distinct from geometric forms of generalization. *Classification* generalization reclassifies the attributes of objects into a smaller number of classes, while *symbolization* generalization changes the assignment of symbols to objects. For example, it might replace an elaborate symbol including the words "Mixed Forest" with a color identifying that class. Figure 7.4 shows an example of how reclassification of attributes and merging of small areas can both produce similar effects on a map of classified land cover.

They also discuss two forms of generalization, depending on the stage at which generalization occurs. *Database* generalization occurs during the creation of the database, when rules are applied to simplify and compress the contents of a database. *Cartographic* generalization occurs at the point when the data are displayed, on a computer screen or by printing or plotting a map. The criteria can be quite different in the two cases. Cartographers often devise generalization rules to simplify the map, so it is not too confusing to the user, by avoiding excessively dense detail, or *clutter*. But this argument may not apply to a database, which may be designed for many purposes that do not involve display, including statistical analysis (Chapters 13 and 14). Database generalization is likely to be permanent, whereas cartographic generalization may occur "on the fly", when and as needed for the purposes of display. This issue is explored further in Chapter 12.

Generalization can occur temporarily, for purposes of display, or can be applied permanently to a database.

7.3.1 Weeding

One of the commonest forms of generalization in GIS is the process known as weeding, or the simplification of the representation of a line. Suppose a line is represented as a polyline. It may turn out that the representation is excessively detailed, especially if the line is to be shown on a map at a coarse scale. Because costs can be reduced by storing less detail, or by spending

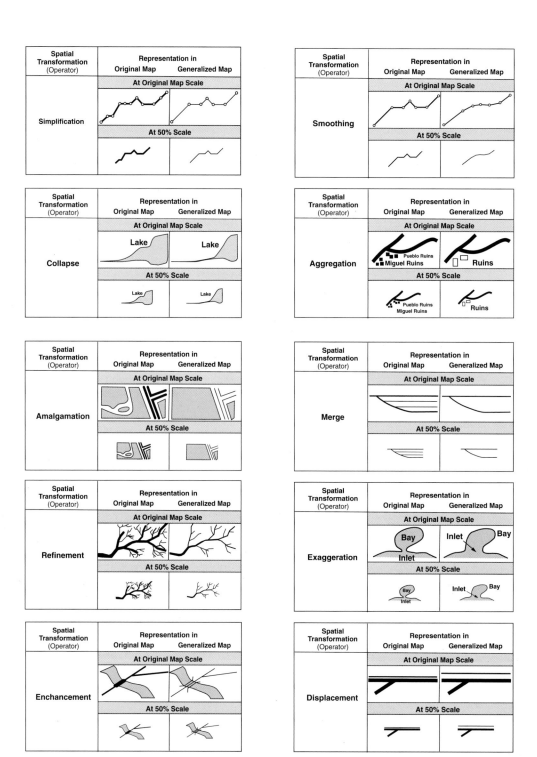

Figure 7.3 Illustrations from McMaster and Shea (1992) of their ten forms of generalization (see text). The original feature is shown at full scale, and below it at half scale. Each generalization technique resolves a specific problem of display at half scale, and results in the acceptable version shown in the lower right of each diagram (Source: Association of American Geographers)

less time displaying or plotting the line, it is often desirable to reduce the complexity of the line by weeding out unnecessary points. The process is an instance of McMaster and Shea's simplification (Figure 7.3, top left). Standard methods exist in GIS for doing this, and the commonest by far is the method known as the Douglas-Poiker algorithm after its inventors, David Douglas and Tom Poiker (Douglas and Peucker 1973 – Tom changed the spelling of his name after this article was published). The operation of the Douglas-Poiker weeding algorithm is shown in Figure 7.6.

> Weeding is the process of simplifying a line or area by reducing the number of points in its representation.

Note that the algorithm relies entirely on the assumption that the line is represented as a polyline, in other words as a series of straight line segments. GIS increasingly support other representations, including arcs of circles, arcs of ellipses, and Bézier curves, but there is little consensus to date on appropriate methods for weeding or generalizing them, or on methods of analysis that can be applied to them.

7.3.2 Merging

Another very common form of generalization consists of reducing the number of primitive objects by aggregating or merging them. Figure 7.4 shows a detailed map of land cover in an area. The smallest area is less than a hectare in size. An obvious way to generalize this map would be to establish some threshold size larger than this smallest area, say 1 hectare, and to remove all areas smaller than this threshold by merging them with their neighbors. A common term for this threshold area is the *minimum mapping unit*, or MMU, and many such maps are drawn to some known MMU.

Merging requires that rules be established to determine which neighboring area to choose. It makes most sense if the chosen neighbor is the one with the greatest similarity. For example, a small forest stand of pure Douglas Fir would be better merged with a neighbor classified as Mixed Conifer, and it would make much less sense to merge it with a neighbor classified as Oak Savannah. It helps greatly if these rules are made explicit, because knowledge of them is essential if the resulting generalized map is to be checked against ground truth. In this example, suppose a ground check identifies an area of pure Douglas Fir, but the database indicates that the area is classified as Mixed Conifer. According to the generalization rules this area is not misclassified, because the class found in the database is consistent with the specified rules for the generalized map.

> The minimum mapping unit is the smallest area allowed to appear in the database: all smaller areas have been merged with larger neighbors.

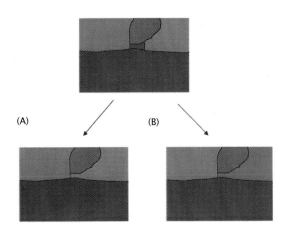

(A)   (B)

Figure 7.4 The effects of reclassification and aggregation of areas are often similar. (A) areas smaller than 1 hectare have been merged with their most similar neighboring areas. (B) the light blue and brown classes have been amalgamated, into a single class denoted by pink, and neighboring areas that are now of the same class have been merged

Figure 7.5 The Douglas-Poiker algorithm is designed to simplify complex objects like this shoreline, by reducing the number of points in its polyline representation

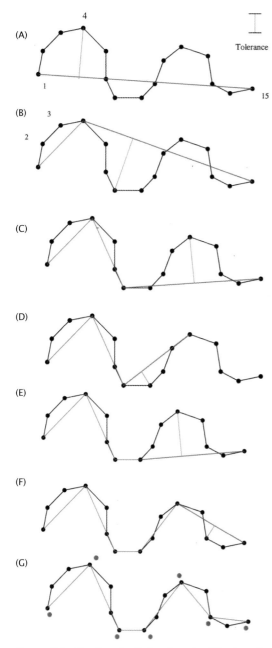

Tolerance

Figure 7.6 The Douglas-Poiker line simplification algorithm in action. The original polyline has 15 points. In (A) Points 1 and 15 are connected (red), and the furthest distance of any point from this connection is identified (blue). This distance to Point 4 exceeds the user-defined tolerance. In (B) Points 1 and 4 are connected (green). Points 2 and 3 are within the tolerance of this line. Points 4 and 15 are connected, and the process is repeated. In the final step 7 points remain (identified with green disks), including 1 and 15. No points are beyond the user-defined tolerance distance from the line

7.4 Measuring the Degree of Generalization

It is often important to be able to measure the degree of generalization present in a geographic dataset. Such measures allow us to describe generalization very simply, using standard methods that are understood by others; to compare datasets with different degrees of generalization; and to decide whether datasets meet the requirements of specific applications. This section describes the various measures that are in widespread use.

7.4.1 Representative fraction

The representative fraction, also often known as the scale, is defined for a paper map as the ratio between distance on the map and the corresponding distance on the ground (see Sections 3.7 and 5.3). Section 3.7 described the representative fraction as a measure fundamentally associated with analog modeling, or models that are scaled-down versions of real phenomena. In that sense the representative fraction is akin to the size relationship between a toy car and a real one, or between a model ship and a real one. Representative fractions associated with standard map series, such as 1:24,000 in the USA, have become standard bases for description of maps, and map users have become accustomed to the link between representative fraction and the types of features shown on maps.

> The representative fraction is the preferred measure of the degree of detail in a map's representation.

But as we began to suggest in Section 3.7, the trend towards digital representation of geographic information has created a problem, because digital models are fundamentally different from analog ones. There is no distance in a digital database that can be compared to distance on the ground. If databases are created by digitizing maps, then it makes sense to describe the degree of generalization present in the digital version by quoting the representative fraction of the map. But what if a database never went through a paper stage, but was created entirely digitally? And what if a database is manipulated, for example by automated aggregation of features, so that its degree of generalization changes?

Goodchild and Proctor (1997) discuss the many conventions that have evolved in the past three decades for assigning a representative fraction to digital data. The conventions are complex, and can be very confusing. For example, one convention assigns a representative fraction based on positional accuracy, on the grounds that national map accuracy standards often prescribe the positional accuracy required for a map at a certain representative fraction. Suppose the features in a database are located with a positional accuracy of 6 m. According to the US National Map Accuracy Standards, this positional accuracy is required of maps having a representative fraction of approximately 1:12,000. Then by this convention we say that the database is at 1:12,000.

Although the representative fraction is not well-defined for digital data, numerous conventions allow it to be estimated.

Such practices are instances of how an older technology – the paper map – can impose its legacy on a new one, the digital database. They allow us to work with familiar measures, but only by the use of a complex system of conventions that is difficult for the novice to learn and understand. A far better approach would be to find a measure of degree of generalization that is valid both for the old and the new.

7.4.2 Minimum mapping unit

The MMU described earlier applies to a specific set of data models, those that represent either discrete objects or fields as collections of primitive areas. Typically these areas are in turn represented as polygons. In the case of discrete objects, increasing the MMU results in the deletion of any objects with area less than the MMU. In the case of fields, it results in the merger of areas with the most appropriate neighbors, thus preserving the essential property of fields represented as mosaics of areas, that every point in the plane lies in exactly one area.

The minimum mapping unit (MMU) is the preferred measure for the level of detail in a map of classes of land use, soils, or vegetation.

Although the MMU is widely used and cited as a measure of the degree of generalization of maps and databases, it suffers from significant problems. First, it may be impossible to determine the MMU for a map or dataset unless it is specifically cited, because the smallest area actually present may be significantly larger than the MMU that was used during the creation or generalization process. Although an MMU of 1

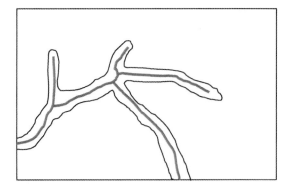

Figure 7.7 The area of a polygon can often be a misleading indicator of its relative importance. In this illustration, a river totaling 50 km in length is surrounded by a narrow riparian zone. The total area of the zone might amount to 10 sq km, though its width might average only 100 m on either side of the river

hectare may have been used to create a map, for example, there may simply not be any areas of 1 hectare, and instead the smallest actual area may be much larger. Second, MMU is expressed as an area, which can be misleading for areas that are very distended. For example, maps of vegetation often show what are known as *riparian* zones, or areas of distinct vegetation immediately adjacent to streams and rivers. Such zones are often very long and thin. The riparian zone on both sides of 50 km of river might average only 100 m in width on both sides of the river, but would have a total area on the order of 10 sq km (see Figure 7.7). Suppose we applied an MMU of 1 sq km to this dataset. Circular areas of radius 564 m would be deleted, but the riparian zone, which is only 200 m across, would not. In such cases area seems a crude and sometimes inappropriate basis for generalization.

7.4.3 Spatial resolution

The spatial resolution of a dataset is defined as the minimum distance over which change is recorded. It has dimensions of length (it is measured in units of length), unlike MMU, which has dimensions of area (it is measured in units of area). A simple way of connecting the two is to take the square root of MMU, but the result would be misleading in cases like the riparian zone example cited above, where it would be better to think of the width of the riparian zone as the measure of spatial resolution, a number that is much less than the square root of riparian zone area (200 m compared to 3.2 km in the example).

Spatial resolution is clear in the case of raster datasets, since observations are uniformly spaced. For square cells the spatial resolution is simply the length of each cell side, but for rectangular cells spatial resolution is coarser in the direction of the longer cell side and this is the measure that should be quoted, not the length of the shorter cell side (see Figure 7.8).

Resampling is a common practice for raster datasets (Figure 7.9). Its primary use is to bring one raster dataset into congruence with another, so that the two can be compared cell by cell. For example, a raster remotely-sensed image with a cell size of 60 m might be resampled to make it fit a DEM with 30 m spacing, in effect quadrupling the number of cells. This raises an important issue concerning spatial resolution: has the spatial resolution improved as a result of the resampling? Suppose the rules used to resample were as follows (very simplistic rules compared with the methods in common use – see for example the rules used to create the global GTOPO30 DEM, summarized in Box 7.1):

● The new 30 m cells are centered as shown in Figure 7.10.

● Where the new 30 m cell center falls on an old 60 m cell the 60 m cell's value is used.

● Where the new 30 m cell falls on a cell boundary, its value is the mean of the two old cell values.

The spatial resolution of rectangular cells is defined by the length of the longer cell side.

It is possible that these rules exactly match the real world, that is, if a 30 m cell had been used in the first place, the values obtained by measuring point elevations on the ground would have been exactly the values produced by these rules. In that case, the true spatial resolution of the 30 m data is 30 m. On the other hand it is much more likely

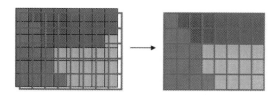

Figure 7.9 Example of resampling. The original cells outlined in black have been resampled to the cells outlined in red. New attributes of each cell have been assigned using the largest area rule

that the new values miss the true values, in some cases by wide margins. In that case, it makes no sense to say that the spatial resolution is now 30 m – in effect, it is still 60 m. Because resampling is widely practiced, it is important to know when it has been used, and to have some idea of the accuracy of the rules with respect to the real world.

Unfortunately the definition of spatial resolution is much more problematic for vector datasets (see also Section 5.8 and Box 5.5). For example, consider a digitized section of a coastline. The true coastline is much more wiggly, and the digitized polyline version is a generalization. One way to define spatial resolution would be as the minimum distance between digitized points, or *vertices* of the polyline. Since no change is recorded between such points this definition seems to fit the overall definition of spatial resolution given earlier. But this definition can be misleading if the line being digitized is truly straight, as in the case of a street network,

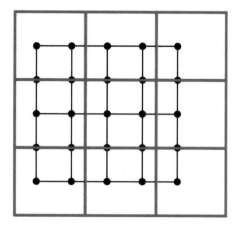

Figure 7.10 Example of resampling an existing DEM to obtain a new DEM with shorter spacing between sample points. The black dots are the new DEM sample points, and the existing DEM provided mean elevations for each red square (see text). The apparent improvement in spatial resolution as a result of resampling may not be justified

(A)   (B)

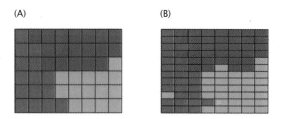

Figure 7.8 The spatial resolution of the square pixels (A) is defined by the length of each pixel's side. For the rectangular pixels (B) the side length that defines spatial resolution is the longer, or horizontal side, not the shorter, or vertical side

Box 7.1 *An example of rules used to generalize DEMs*

Preparation of the GTOPO30 global 1 km resolution DEM required use of several different methods of generalization and resampling, depending on the source. When the source was of higher spatial resolution, the method used was designed to capture and emphasize breaklines, including ridges and stream channels. In this example the input data points are spaced 3 arc seconds (3 seconds of latitude and longitude, or roughly 100 m) apart, and each output data value is obtained from a block of 100 input data values. Instead of averaging the 100 data points in each block, or selecting the central data point, the block is analyzed to determine whether it is dominated by a ridge. If it is, the selected value is the mean elevation of the ridge. Similarly if the block is dominated by a valley the valley's mean elevation is selected. The three images in Figure 7.11 show the valleys detected in an input dataset, and the resulting generalized output.

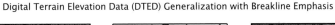

Digital Terrain Elevation Data (DTED) Generalization with Breakline Emphasis

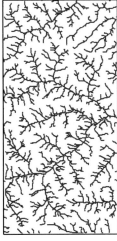

| Full resolution DTED (3 arc second grid spacing) | Extracted stream channels used as breaklines in the generalization process | Generalized DTED (30 arc second grid spacing) |

Figure 7.11 The process used to create the GTOPO30 DEM by generalizing from a spatial resolution of about 30 m to a spatial resolution of about 1 km (Source: USGS)

because now the minimum distance between digitized vertices is a property of the street network, not of its representation. In this definition, lines that have long straight segments appear to be more generalized, which is not necessarily the case.

This problem, of confusing properties of the phenomenon with properties of the representation, is not limited to lines, as the MMU example showed earlier. Consider Figure 7.1(B) and suppose it represents the locations of a network of rain gauges. An obvious measure of spatial resolution here is the distance between rain gauges, since spacing determines the accuracy with which the sample points capture the variation across the entire precipitation field. But which measure should we use? One candidate is the shortest distance between any two gauges, but this misrepresents the spatial resolution in areas where gauges are spaced further apart. Perhaps the only rational approach is to think of spatial resolution as varying spatially, with the greatest detail in areas where the sampling is densest. But it is also important to ask why the gauges were located in this particular pattern in the first place. Presumably the pattern is less dense in some areas because there was prior knowledge that the precipitation field varied less in these areas, and gauges were concentrated in areas where the field varied more.

> The effective spatial resolution of many vector datasets varies from one location to another.

In summary, spatial resolution is an attractive property, in the sense that it can be defined as a measure of degree of generalization for both maps and databases, and thus avoids the legacy issues associated with the representative fraction. But it is readily defined only for raster data, and for vector data is best thought of as a property that varies spatially, rather than as a single measure for an entire dataset. Figure 7.12 illustrates a common instance of varying spatial resolution. In this case, the representation of a city is achieved by using much smaller areas in the core, and much larger areas in the periphery. This reflects variation in population density, and is used by census agencies to create zones for reporting statistics that have roughly equal populations. But it can also reflect mobility, since car-based travel is generally easier in the periphery than in the core.

Because travel is easier in the periphery the concept of neighborhood is also larger there, as car users find it easy to overcome larger distances to visit friends, shop, or go to school.

7.5 Metadata

Strictly defined, metadata are data about data. We need information about data for many purposes, and metadata try to satisfy them all. First, we need metadata to automate the process illustrated in Section 7.1 in the example of a dialog between two users, in other words, to enable the process of search and discovery over distributed archives. In that sense metadata are similar to a library's catalog, which organizes the library's contents by author, title, and subject, and makes it easy for a user to find a book. But metadata are potentially much more powerful, because a computer is much more versatile than the traditional catalog in its potential for re-sorting items by a large number of properties, going well beyond author, title, and subject. Second, we need metadata to determine whether a dataset, once discovered, will satisfy the user's requirements – in other words, to assess the fitness of a dataset

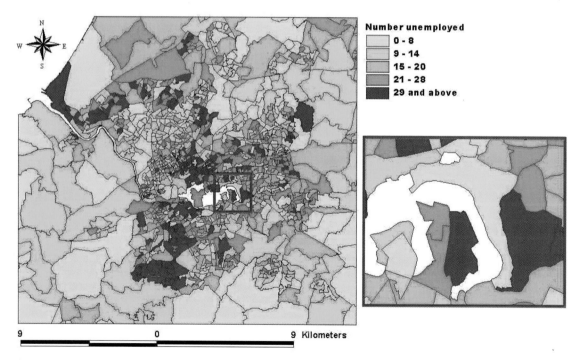

Figure 7.12 Varying spatial resolution in the zones used to describe Bristol, UK. Smaller zones are found in the center and in other areas of high density, and larger zones are in the periphery (Courtesy: Richard Harris)

for a given use. Does it have sufficient spatial resolution, and acceptable quality? Such metadata may include comments provided by others who tried to use the data, or contact information for such previous users (users often comment that the most useful item of metadata is the phone number of the person who last tried to use the data). Third, metadata must provide the information needed to handle the dataset effectively. This may include technical specifications of format, or the names of software packages that are compatible with the data, along with information about the dataset's location, and its volume. Finally, metadata may provide useful information on the dataset's contents. In the case of remotely – sensed images, this may include the percentage of cloud obscuring the scene, or whether the scene contains particularly useful instances of specific phenomena, such as hurricanes.

Metadata must serve many different purposes, from description of contents to instructions for handling.

Metadata generalize and abstract the contents of datasets, and in that sense are extreme instances of the generalization processes identified earlier, which is why we have included a discussion of metadata in this chapter. By definition, then, metadata should be smaller in volume than the data they describe. This is paradoxical, because complete description of a dataset can generate a greater volume of information than the actual contents. Metadata are also expensive to generate, because they represent a level of under-standing of the data that is difficult to assemble, and requires a high level of professional expertise. Libraries have compiled much useful information on the time required to catalog a book. Generation of metadata for geographic datasets can easily take much longer, particularly if it has to deal with technical issues such as the precise geographic coverage of the dataset, its projection and datum details (Chapter 4), and other properties that may not be easily accessible. Thus the cost of metadata generation, and the incentives that motivate people to provide metadata, are important issues.

7.5.1 The FGDC standard

For metadata to be useful it is essential that they follow widely accepted standards. If two users are to be able to share data, they must both understand the rules used to create metadata, so that the custodian of the data can first create the description, and so that the potential user can understand it. The most widely used standard for metadata is the US Federal Geographic Data Committee's Content Standards for Digital Geospatial Metadata, or CSDGM, first published in 1993 and now the basis for many other standards worldwide. Box 7.2 lists some of its major features. As a *content standard* CSDGM describes the items that should be in a metadata archive, but does not prescribe exactly how they should be formatted or structured. This allows developers to implement the standard in ways that suit their own software environments, but guarantees that one implemen-

---

Box 7.2 Major features of the US Federal Geographic Data Committee's Content Standards for Digital Geospatial Metadata

1. Identification Information – basic information about the dataset.
2. Data Quality Information – a general assessment of the quality of the dataset.
3. Spatial Data Organization Information – the mechanism used to represent spatial information in the dataset.
4. Spatial Reference Information – the description of the reference frame for, and the means to encode, coordinates in the dataset.
5. Entity and Attribute Information – details about the information content of the dataset, including the entity types, their attributes, and the domains from which attribute values may be assigned.
6. Distribution Information – information about the distributor of and options for obtaining the dataset.
7. Metadata Reference Information – information on the currentness of the metadata information, and the responsible party.
8. Citation Information – the recommended reference to be used for the dataset.
9. Time Period Information – information about the date and time of an event.
10. Contact Information – identity of, and means to communicate with, person(s) and organization(s) associated with the dataset.

tation will be understandable to another – in other words, that the implementations will be *interoperable*. For example, ESRI's ArcGIS 8 provides several formats for metadata, including one using the XML standard and the other using ESRI's own metadata format.

7.5.2 Other standards

CSDGM was devised as a system for describing geographic datasets, and most of its elements make sense only for data that are accurately georeferenced and represent the spatial variation of phenomena over the Earth's surface. As such, its designers did not attempt to place CSDGM within any wider framework. But in the past decade a number of more broadly based efforts have also been directed at the metadata problem, and at the extension of traditional library cataloging in ways that make sense in the evolving world of digital technology.

One of the best known of these is the Dublin Core (see Box 7.3), the outcome of an effort to find the minimum set of properties needed to support search and discovery for datasets in general, not only geographic datasets. Dublin Core treats both space and time as instances of a single property, coverage, and unlike CSDGM does not lay down how such specific properties as spatial resolution, projection, or datum should be described.

The principle of establishing a minimum set of properties is sharply distinct from the design of CSDGM, which was oriented more toward the capture of all knowable and potentially important properties of geographic datasets. Of direct relevance here is the problem of cost, and specifically the cost of capturing a full CSDGM metadata record. While many organizations have wanted to make their data more widely available, and have been driven to create metadata for their datasets, the cost of determining the full set of CSDGM elements is often highly discouraging. There is interest therefore in a concept of *light metadata*, a limited set of properties that is both comparatively cheap to capture, and still useful to support search and discovery. Dublin Core represents this approach, and thus sits at the opposite end of a spectrum from CSDGM. Every organization must somehow determine where its needs lie on this spectrum, which ranges from light and cheap to heavy and expensive.

Light, or stripped-down metadata, provide a short but useful description of a dataset that is cheaper to create.

Box 7.3   *The 15 basic elements of the Dublin Core metadata standard*

1. TITLE. The name given to the resource by the CREATOR or PUBLISHER.
2. AUTHOR or CREATOR. The person(s) or organization(s) primarily responsible for the intellectual content of the resource.
3. SUBJECT or KEYWORDS. The topic of the resource, or keywords, phrases, or classification descriptors that describe the subject or content of the resource.
4. DESCRIPTION. A textual description of the content of the resource, including abstracts in the case of document-like objects or content description in the case of visual resources.
5. PUBLISHER. The entity responsible for making the resource available in its present form, such as a publisher, a university department, or a corporate entity.
6. OTHER CONTRIBUTORS. Person(s) or organization(s) in addition to those specified in the CREATOR element who have made significant intellectual contributions to the resource, but whose contribution is secondary to the individuals or entities specified in the CREATOR element.
7. DATE. The date the resource was made available in its present form.
8. RESOURCE TYPE. The category of the resource, such as home page, novel, poem, working paper, technical report, essay, dictionary.
9. FORMAT. The data representation of the resource, such as text/html, ASCII, Postscript file, executable application, or JPEG image.
10. RESOURCE IDENTIFIER. String or number used uniquely to identify the resource.
11. SOURCE. The work, either print or electronic, from which this resource is delivered, if applicable.
12. LANGUAGE. Language(s) of the intellectual content of the resource.
13. RELATION. Relationship to other resources.
14. COVERAGE. The spatial locations and temporal durations characteristic of the resource.
15. RIGHTS MANAGEMENT. The content of this element is intended to be a link (a Universal Resource Locator (URL) or other suitable Universal Resource Identifier (URI) as appropriate) to a copyright notice, a rights-management statement, or perhaps a server that would provide such information in a dynamic way.

7.5.3 Geolibraries

The use of digital technology to support search and discovery opens up many options that were not available in the earlier world of library catalogs and book shelves. Books must be placed in a library on a permanent basis, and there is no possibility of reordering their sequence, but in a digital catalog it is possible to reorder the sequence of holdings in a collection almost instantaneously. So while a library's shelves are traditionally sorted by subject, it would be possible to re-sort them digitally by author name or title, or by any property in the metadata catalog. Similarly the traditional card catalog allowed only three properties to be sorted – author, title, and subject – and discouraged sorting by multiple subjects. But the digital catalog can support any number of subjects.

Of particular relevance to GIS users is the possibility of sorting a collection by the coverage properties: location and time. Both the spatial and temporal dimensions are continuous, so it is impossible to capture them in a single property analogous to author that can then be sorted numerically or alphabetically. But in a digital system this is not a serious problem, and it is straightforward to capture the coverage of a dataset, and to allow the user to search for datasets that cover an area or time of interest defined by the user. Moreover, the properties of location and time are not limited to geographic datasets, since there are many types of information that are associated with specific areas on the Earth's surface, or with specific time periods. Searching based on location or time would enable users to find information about any place on the Earth's surface, or any time period, and to find reports, photographs, or even pieces of music, as long as they possessed geographic and temporal *footprints*.

The term *geolibrary* has been coined to describe digital libraries that can be searched for information about any user-defined geographic location. A US National Research Council report (NRC 1999) describes the concept and its implementation, and many instances of geolibraries can be found on the Web (see Box 7.4). Digital Earth, discussed in Box 3.4, is a vision of a geolibrary in which the user interacts with a visual representation of the surface of the Earth, rather than with a digital representation of the traditional card catalog.

7.5.4 Collection-level metadata

Thus far we have seen how metadata can be useful in the description of individual datasets. There are several other forms of metadata of potential interest to GIS users, and this section describes the issues associated with one of them: metadata that describe entire collections of datasets, rather than individual datasets.

Many different collections or archives of geographic datasets exist on the Internet, and Box 7.4 identified a few examples. The US National Geospatial Data Clearinghouse is structured as a mechanism that allows custodians to make their data visible to users who access the clearinghouse, and to search across multiple servers using a simple protocol. The U.S. Geological Survey supports several major collections of geographic data, notably at the EROS Data Center in Sioux Falls, South Dakota (EDC; www.cr.usgs.gov). Major commercial collections of data can be found at the sites of Space Imaging (www.spaceimaging.com), Terraserver (www.terraserver.com), and GlobeXplorer (www.globexplorer.com), and at many other locations. Clearinghouses are also being sponsored by many other countries and by several US states.

But this abundance of readily accessible data presents a problem: how does the user know which collection to search for a given dataset? The term *collection-level metadata* (CLM) defines the information needed to make an intelligent choice, based on general knowledge of each collection's contents. For example, the EROS Data Center is a good site to search for images and maps of the USA. Terraserver is a good source for imagery of any part of the world, and for digital orthophotography. In looking for maps of a specific country, it makes sense to try to find a site sponsored by that country's national mapping agency, and similarly information about a state may exist in a clearinghouse sponsored by that state.

Unfortunately there has been little consistent effort to develop standards for CLM, unlike its individual dataset counterpart, or *object-level metadata* (OLM). Knowledge of CLM is still largely unorganized, and tends to accumulate slowly in the minds of geographic information specialists (or SAPs, see Section 1.4). Knowing where to look is still largely a matter of personal knowledge and luck.

Various possible solutions to this problem can be identified. In the future, we may find the means to develop a new generation of Web search engines that are able to discover geographic datasets automatically, and catalog their OLM. Then a user needing information on a particular area could simply access the search engine's results, and be provided with a list of appropriate collections. That capability is well beyond the reach of today's

Box 7.4 Some prototype geolibraries

The Alexandria Digital Library (alexandria.ucsb.edu) provides access to over 2 million maps and images using a simple method of search based on user interaction with a world map, placenames, or coordinates. Many of the maps and images are stored in the Map and Imagery Laboratory at the University of California, Santa Barbara, but the Alexandria catalog also identifies materials in other collections. Figure 7.13 shows a screen shot of the library's WWW interface.

(A)

(B)

Figure 7.13 Example of a geolibrary: the Alexandria Digital Library (alexandria.ucsb.edu). (A) shows the interface to the library in a standard Netscape browser. The user has selected an area of Southern California to search, and the system has returned the first 50 items in the library whose titles include the word ''spotview'' (images from the SPOT satellite) and whose footprints match that area. (B) shows one of the images, of the lower Colorado River (the contrast between Mexican and US field patterns is clear in the lower left)

Box 7.4   *Continued*

The Federal Geographic Data Committee's National Geospatial Data Clearinghouse (www.fgdc.gov) is a collection of over 100 spatial data servers that offer digital geographic data primarily for use in GIS, image processing systems, and other modeling software. These data collections can be searched through a single interface based on their metadata. Several portals to the clearinghouse have been established.

The Terraserver site (www.terraserver.com) provides access to a large store of imagery and digital orthophotography. Users can select an area of interest, query for the availability of imagery, and download relevant information. Figure 7.14 shows a screen shot of the site.

The Geography Network (www.geographynetwork.com) is a site sponsored by ESRI to serve the needs of users who wish to share geographic datasets. Users can register datasets for use by others, and also search the archive for useful data. The site also serves the needs of owners and users of geographic software, by similarly allowing users to contribute or use transformation algorithms, and to provide other kinds of useful services related to geographic information. A screen shot appears in Figure 1.19.

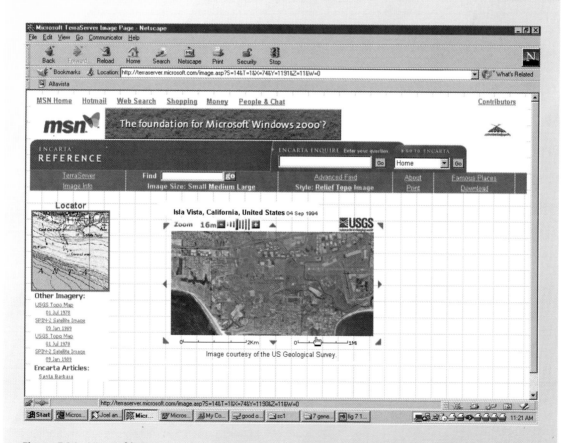

Figure 7.14   Microsoft's Terraserver, another example of a geolibrary. In this example a search for data for an area of Goleta, California, USA has returned an orthophotograph produced by the US Geological Survey (Source: USGS)

generation of search engines, although several format standards, including GeoTIFF, already add the tags to geographic datasets that clever search engines could recognize. Alternatively, some scheme might be developed that authorizes a uniform system of servers, such that each server's contents are defined precisely, using geographic, thematic, or other criteria. In such a system all data about locations within the server's nominal coverage area, such as a state or a range of latitude and longitude, would be found on that server and no other. Such a system would need to be hierarchical, with very detailed data in collections with small coverage areas, and coarser data in collections with larger coverage areas. Finally, it may be possible in the future to provide a single portal to all geographic data – a single massive collection that provides links to the custodians of datasets, rather than storing the datasets themselves [the term Mother of All Databases (MOADB) seems particularly apt]. But at this time the CLM problem remains a major impediment to effective sharing of geographic data.

cluttering detail – that has found new focus and motivation in the era of GIS. The other is a brand-new issue that is no older than the Web – the problem of searching vast resources of digital geographic data, assessing the fitness of discovered data for specific uses, and retrieving the data in useful form. It is good to remind ourselves occasionally that the vast resources now available for discovery hardly existed at all before 1994, when the Web began its stunning impact on human activity.

Generalization is a longstanding issue, and it is important to recognize that very little progress has been made on its resilient problems as a result of GIS. Today, 30 years into the active use of GIS, we still have only very limited methods for automatic generalization that emulate the skills of the experienced or well-trained cartographer, despite extensive research. It is not likely that this situation will change, because generalization is essentially *very hard* (Box 7.5). Instead, GIS users are likely to continue to find clever ways around the problem, through better ways of using the power of computer display for interactive zooming, so that GIS databases continue to provide better and more accurate representations of the world around us.

7.6 Summary

This chapter has looked at two related topics. One is a longstanding problem in cartography – the effective generalization of geographic information to isolate its broad patterns and suppress

Questions for Further Study

1. This chapter made use of McMaster and Shea's classification of generalization methods. What

---

other classifications exist in the literature, and how do they compare? Is the McMaster and Shea classification complete, and if not, what forms of generalization does it omit?

2. Using a readily available map, such as a road map or a single plate from an atlas, go through the exercise of building a description using the FGDC standard. How long did the process take, and how might it be shortened? What percentage of the items in the standard were you able to complete?

3. One interpretation of spatial resolution is that it defines the size of the smallest objects recorded in the database. Find examples of familiar objects that are just visible at each of the following spatial resolutions: 1 m, 10 m, 100 m, 1 km.

4. Find an example of a topographic map sheet (e.g. a sheet from the USGS 1:24,000 National Topographic Series), and note its representative fraction. What definitions of spatial resolution are meaningful for this sheet, and what values would they assign?

Online Resources

NCGIA Core Curricula (www.ncgia.ucsb.edu/pubs/core.html):
Core Curriculum in GIScience, Section 1.6.2 (Line generalization), 3.2.1. (Public Access to Geographic Information, Albert Yeung), and 3.2.3 (Digital Libraries, Albert Yeung)
Core Curriculum in GIS, 1990, Unit 48 (Line generalization)
ESRI Virtual Campus course (campus.esri.com)
Protecting Your Investment in Data with Metadata, by George Shirey
Section 7.5.4 and Box 7.4 contain references to several prototype geolibraries
Metadata standards:
FGDC Content Standards for Digital Geospatial Metadata: www.fgdc.gov/metadata.contstan.html
OpenGIS Consortium metadata specification: www.opengis.org/techno/specs.htm

Sites related to the Dublin Core metadata standard: dublincore.org www.xml.com, www.purl.org/dc

Reference Links

Maguire D J, Goodchild M F, and Rhind D W (eds) 1991 *Geographical Information Systems: Principles and applications.* Harlow, UK: Longman (Text available online from 'Links to Big Book 1' at www.wiley.com/gis and www.wiley.co.uk/gis).
Chapter 30, Generalization of spatial databases (Müller J-C)
Longley P A, Goodchild M F, Maguire D J, and Rhind D W (eds) 1999 *Geographical Information Systems: Principles, techniques, management and applications.* New York: John Wiley.
Chapter 10, Generalizing spatial data and dealing with multiple representations (Weibel R, Dutton G)
Chapter 49, Metadata and data catalogues (Guptill S)
Chapter 64, Applying GIS in libraries (Adler P, Larsgaard M)

References

Douglas D H and Peucker T K 1973 Algorithms for the reduction of points required to represent a digitized line or its caricature. *Canadian Cartographer* 10: 112-122.
Goodchild M F and Proctor J 1997 Scale in a digital geographic world. *Geographical and Environmental Modelling* 1(1): 5-24.
McMaster R B and Shea K S 1992 *Generalization in Digital Cartography.* Washington, DC: Association of American Geographers.
National Research Council 1999 *Distributed Geolibraries: Spatial information resources.* Washington, DC: National Academy Press.

GIS SOFTWARE

8

GIS software is the geoprocessing engine of GIS. It is made up of integrated collections of computer programs that encapsulate geographic processing functions. The three key parts of GIS software are the user interface, the tools, and the data manager. All three parts may be located on a single computer or they may be spread over multiple machines in a department or enterprise. GIS software implementations can be based on either the desktop or Internet model. There are six main types of GIS software: professional, desktop, hand-held, component, viewer, and Internet. The main GIS vendors offer products in most, if not all, of these categories. The major GIS vendors are: Autodesk, ESRI, GE Smallworld, Intergraph, and MapInfo.

Learning objectives

After reading this chapter you will be able to:

● Understand the architecture of GIS software systems
  – Project, departmental, and enterprise
  – Three-tier architecture
  – GUI, tools, and data access parts;

● Describe the process of GIS customization;

● Describe the types of commercial software
  – Internet
  – Viewer
  – Component
  – Hand-held
  – Desktop
  – Professional;

● Outline the software vendor product family offerings.

8.1 Introduction

In Chapter 1 the four technical parts of a geographic information system were defined as the network, the hardware, the software, and the database, which together functioned with reference to people and the institutional structures within which people work (Section 1.5). This chapter is concerned with GIS software, the geoprocessing engine of a complete, working GIS. The functionality or capabilities of GIS software will be discussed later in this book (especially in Chapters 9, 13, and 14), and the focus here is on the different ways in which these capabilities are realized in the products of the various GIS software vendors.

This chapter takes a fairly narrow view of GIS software, concentrating on systems with a range of generic capabilities to collect, store, manage, query, analyze, and present geographic information. It excludes atlases, simple mapping systems, route finding software, image processing systems, and spatial extensions to database management systems (DBMS). For a wider review of GIS software that includes many of these see Elshaw, Thrall, and Thrall (1999) and the popular GIS magazines.

Earlier chapters, especially Chapter 3, introduced several fundamental computer concepts, including digital representations, data, and information. The instructions that are used to manipulate digital data in computers are called programs. Integrated collections of programs are referred to as software packages or systems (or just software for short). Modern software packages are built from software modules or components that are self-contained, reusable building blocks for bigger systems.

GIS software builds on the foundation of a computer operating system – the instruction set that controls all the activities of a computer. GIS software vendors – the companies that design, develop, and sell GIS software – rely on the basic operating system capabilities such as security, file management, peripheral drives (controllers), printing, and display management. GIS software builds on these capabilities to provide a controlled environment for geographic information management, analysis, and interpretation. GIS software adds support for geographic data types (objects) and processing capabilities (functions) to facilitate geographic information science. The unified architecture and consistent approach to representing and working with geographic information in a GIS software package aim to provide users with a standardized approach to GIS.

GIS software packages aim to provide a unified approach to working with geographic information.

8.2 The Evolution of GIS Software

In the early days, GIS software consisted simply of collections of computer routines that a skilled programmer could use to build an operational GIS. During this period each and every GIS was unique in terms of its capabilities, and significant levels of resource were required to create a working system.

As software engineering techniques advanced and the GIS market grew in the 1970s and 1980s, demand increased for higher-level applications with a standard user interface. In the late 1970s and early 1980s the standard means of communicating with a GIS was to type in command lines. User interaction with a GIS entailed typing instructions to, for example, draw a map, query the attributes of a forest stand object, or summarize the length of highways in a project area. A GIS software package was essentially a toolbox of geoprocessing operators or commands that could be applied to a dataset to create another dataset.

Dataset + GIS operator = New Dataset

To make the software easier to use and more generic, there were two key developments in the late 1980s. First, command line interfaces were supplemented and eventually replaced by graphical user interfaces (GUIs). Second, a customization capability was added to allow specific-purpose applications to be created from the generic toolboxes. Together these stimulated enormous interest in GIS, and led to much wider adoption and expansion into new areas. In particular, the ability to create custom application solutions allowed developers to build focused applications for end users in specific market areas. This led to the creation of GIS applications specifically tailored to the needs of the major markets (e.g. local government, utilities, transportation, and forestry). New terms were developed to distinguish these sub-types of GIS software: planning information systems, automated mapping/facility management (AM/FM) systems, land information systems, and more recently, location-based services systems. The GIS software of today still embodies these same principles of an easy-to-use menu-driven interface, and a customization capability.

Application-specific GIS often have domain-specific names such as planning information systems, automated mapping/facility management systems, and land information systems.

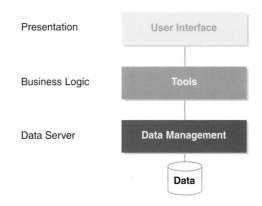

Figure 8.1   Classical three-tier architecture of a GIS software system

8.3   Architecture of GIS Software

8.3.1   Project, departmental, and enterprise GIS

GIS is usually first introduced into organizations in the context of a single, fixed-term project. The technical components (network, hardware, software, and database) of an operational GIS are assembled for the duration of the project, which may be from several months to a few years. Data are collected specifically for the project and typically little thought is given to reuse of software, data, and human knowledge. In larger organizations multiple projects may run one after another or even in parallel. The "one off" nature of the projects, coupled with an absence of organizational vision, often leads to duplication, as each project develops using different hardware, software, data, people, and procedures. Sharing data and experience is usually a low priority.

As interest in GIS grows, to save money and encourage sharing and resource reuse, several projects in the same department may be amalgamated. This often leads to the creation of common standards, development of a focused GIS team, and procurement of a new or replacement GIS. Yet it is also quite common for different departments to have different GIS standards.

As GIS becomes more pervasive, organizations learn more about it and begin to become dependent on it. This leads to the realization that GIS is a useful way to structure many of the organization's assets, processes, and workflows. Through a process of natural growth, and possibly further major procurement (e.g. purchase of upgraded software), GIS gradually becomes accepted as an important enterprise-wide information system. At this point GIS standards are accepted across multiple departments. Resources to support and manage the GIS are often centrally funded and managed.

A fourth type of societal implementation has additionally been identified in which hundreds or thousands of users become engaged in GIS and connected by a network. Today there are only a few examples of societal implementations with perhaps the best being the State of Qatar in the Middle East where more than 16 government departments have joined together to create a comprehensive and nationwide GIS with thousands of users.

8.3.2   Three-tier architecture

From an information systems perspective there are three key parts to a GIS software package: the user interface, the tools, and the data management system (Figure 8.1). The user interacts with the graphical user interface (GUI), an integrated collection of menus, tool bars, and other controls. The GUI provides access to the GIS tools. The tool set defines the capabilities or functions that the GIS software has for processing data. The data are stored in files or databases organized by data management software. In standard information system terminology this is a *three-tier architecture* with the three tiers being called: presentation, business logic, and data server.

GIS software systems deal with user interfaces, geoprocessing tools, and data management.

In the simplest, single-user project GIS these three parts are all installed on a single piece of hardware and users are usually unaware of their existence (Figure 8.2). In larger and more advanced multi-user departmental GIS, the tiers can be installed on multiple machines to improve performance and flexibility. In this type of configuration, the users in a single department (for example, the planning or public works department in a typical local government organization) still interact with a GUI on their desktop computer, but the data management software and data may be located on another machine connected over a network. This type of computing architecture is usually

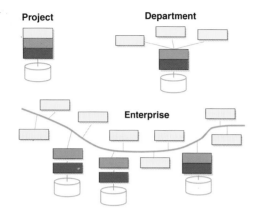

Figure 8.2 Three types of GIS implementation configuration (see Figure 8.1 for key)

referred to as *client–server*, because clients request data or processing services from servers that perform work to satisfy client requests. Clients and servers can communicate over local area, wide area, or Internet networks.

> In a client-server GIS, clients request data or processing services from servers that perform work to satisfy client requests.

Large mature organizations that have been using GIS for several years often split the software tiers over three machines and have multiple clients and multiple servers. This type of implementation is common in enterprise GIS. Large, enterprise GIS may involve more than 10 servers, and hundreds or even thousands of clients that are widely dispersed geographically.

Although organizations often standardize on either a project, departmental, or enterprise system, it is also common for large organizations to have project, department, and enterprise configurations all operating in parallel or as subparts of a full-scale system.

8.3.3   Software data models and customization

In addition to the three-tier model, two further topics are relevant to an understanding of software architecture: data models and customization.

GIS data models will be discussed in detail later in Chapter 9, and so the discussion here will be brief. From a software perspective, a data model defines how the real world is represented in a GIS and the organization and mode of operation of software tools. For example, it describes how

the different tools are grouped together, how they can be used, and how they interact with data. This is largely transparent to end users whose interaction with a GIS is via the user interface, but it becomes very important when customizing or extending software.

Customization is the process of modifying GIS software to create a specific-purpose application. It can be as simple as deleting unwanted controls (for example menu choices or buttons) from a GUI, or as sophisticated as adding a major new extension to core software for network analysis, high-quality cartography, forestry management, etc.

To facilitate customization, GIS software must provide an overview of its data model and expose capabilities to use, modify, and supplement existing functions. In the late 1980s when customization capabilities were first added to GIS software, each vendor had to provide a proprietary customization system simply because no standard customization systems existed. Nowadays, with the widespread adoption of Microsoft's Visual Basic and Javasoft's Java programming languages for customization of commercial systems, there is increasing adoption of them in GIS.

Both Visual Basic and Java are examples of integrated development environments (IDEs). The term IDE refers to the fact that the packages combine several software development tools. Many of the so-called visual programming languages support the development of form-based GUIs containing forms, dialogs, buttons, and other controls. Program code can be entered and attached to the GUI elements using the integrated code editor. An interactive debugger will help identify syntactic problems in the code, for example, misspelled commands and missing instructions. Finally, there are tools to support profiling programs. These show where resources are being consumed and how programs can be speeded up or improved in other ways.

> Contemporary GIS typically use an industry-standard programming language like Microsoft Visual Basic or Javasoft Java for customization.

Typically, a GIS vendor will license the right to include one of these IDEs within their GIS software package. A particularly popular choice for desktop GIS is Microsoft's Visual Basic. Figure 8.3 shows a screenshot of the customization environment within ESRI's ArcInfo/ArcView 8 GIS. To support customization using open, industry-standard IDEs, a GIS vendor must expose details of the package's object model and functionality. This can be done by creating and documenting a set of application programming interfaces (APIs).

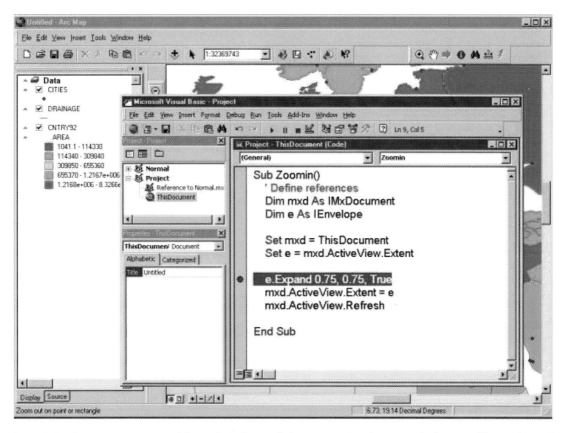

Figure 8.3 The customization capabilities of ESRI's ArcInfo/ArcView 8. ESRI chose to embed Microsoft's Visual Basic for Applications as the scripting and GUI integrated development environment. The window at the front is the Visual Basic Integrated Development Environment (IDE). The window at the back is ArcMap, the main map-centric application of ArcInfo and ArcView (see also Box 8.2) (Source: ESRI)

These are interfaces that allow GIS functionality to be called by the programming tools in an IDE. In recent years, second-generation interfaces have been developed for accessing software functionality in the form of independent building blocks called components. For further details about GIS data models and how they are documented see Chapter 9, especially Section 9.2.

In recent years, three technology standards have emerged for defining and reusing software functionality. For building interactive desktop applications, Microsoft's Component Object Model (COM) and its .Net successor have developed into the *de facto* standard. On the Internet, and possibly as a future serious rival to COM on the desktop, Javasoft's Java Beans is very viable standard technology. At a coarser grain appropriate for application-level integration rather than working with fine-grain functions, the Object Management Group (OMG) – a cross-industry alliance – has specified the Common Object Request Broker Architecture (CORBA). Components are important to software developers because they allow the creation of reusable, self-contained, software building blocks. They allow many programmers to work together to develop a large software system incrementally. The standard, open (published) format of components means that they can easily be assembled into larger systems. In addition, because the functionality within components is exposed, developers can reuse them in many different ways, even supplementing or replacing functions if they wish. Users also benefit from this approach because GIS software can evolve incrementally and support multiple third party extensions. In the case of GIS products this includes tools for charting, reporting, and table management.

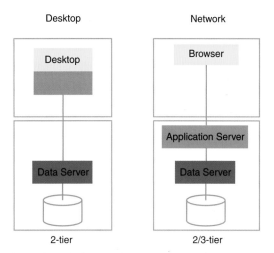

Desktop   Network

2-tier   2/3-tier

Figure 8.4 Desktop and Internet GIS paradigms

8.3.4 GIS on the desktop and on the Internet

The discussion in Section 8.3.3 essentially summarizes how GIS is implemented on the desktop. In such systems, the PC (personal computer) is the main hardware platform and Microsoft Windows the operating system (Figure 8.4 and Table 8.1). In the desktop paradigm clients tend to be functionally rich and substantial in size, and are often referred to as *thick* clients. Using the Windows standard facilitates inter-operability (interaction) with other desktop applications, such as word processors, spread-sheets, and databases. As noted earlier, most sophisticated and mature GIS sites have adopted client–server implementations approach by adding either a thin or thick server application running on the Windows or Unix operating system. The terms *thin* and *thick* are less widely used in the context of servers, but they mean essentially the same as when applied to clients. Thin servers perform relatively simple tasks, such as serving data from files or databases, whereas thick servers also offer more extensive analytical capabilities such as geocoding, routing, mapping,

and spatial analysis. In desktop GIS implementations, Local Area Networks (LANs) and Wide Area Networks (WANs) tend to be used for client–server communication. It is natural for developers to select Microsoft's Component Object Model (COM) technology to build the underlying components making up these systems given the preponderance of the Windows operating system, although other component standards could also be used. The Windows-based client–server system architecture is a good platform for hosting interactive, high performance GIS applications. Examples of applications well suited to this platform include those involving geographic data editing, map production, 2D and 3D visualization, spatial analysis, and modeling. It is currently the most practical platform for general-purpose systems because of its wide availability, good performance for a given price, and common usage in business, education, and government.

> GIS users are standardizing their systems on the desktop and Internet implementation models.

In the last few years there has been increasing interest in harnessing the power of the Internet for GIS. Although desktop GIS have been and continue to be very successful in their own right, users are constantly looking for lower costs of ownership and improved access to geographic information. Internet GIS allows previously inaccessible information resources to be made more widely available. The Internet GIS model intrigues many organizations because it is based on centralized software and data management, which can dramatically reduce initial implementation, and ongoing support and maintenance costs. The continued rise in Internet GIS will not signal the end of desktop GIS, indeed quite the reverse, since it is likely to stimulate the demand for content and professional GIS skills in geographic database automation and administration, and application development.

In contrast to desktop GIS, Internet GIS can use the cross-platform Web Browser to host the viewer

Table 8.1 Comparison of Desktop and Internet (Network) GIS

Feature	Desktop	Network
Client size	Thick	Thin
Client platform	Windows	Browser
Server size	Thin/thick	Thick
Server platform	Windows/Unix	Windows/Unix
Component standard	COM	Java
Network	LAN/WAN	Internet

user-interface. Currently, clients are typically very thin, often with simple display and query capabilities, although it is to be expected that they will become more functionally rich in the future. Server-side functionality may be encapsulated on a single server, although in medium and large systems it is more common to have two servers, one containing the business logic (application server), the other the data manager (data server). The server applications typically contain all the business logic and are comparatively thick. The server applications may run on a Windows or UNIX platform. Internet applications are frequently built using the Java programming language, which makes them good for exploiting the capabilities of the Internet.

An interesting new development is the integration of the desktop and Internet models. Many GIS vendors are now offering the ability to stream geographic data over Internet connections. Additionally, Internet application servers are now able to provide geographic services to a range of clients from thin browsers to traditional thick desktop and professional GIS software. This approach is also at the heart of Microsoft's so-called .Net strategy which is concerned with developing the next generation of Windows by combining the best of desktop and Internet technologies.

8.4 Building GIS Software Systems

GIS software systems are built and released as GIS software products by GIS-vendor software development and product teams. The key parts of a GIS software architecture – user interface, tools, data manager, data model, and customization environment – were outlined in the previous section.

The key GIS software components deal with user applications, tools, and data access.

GIS software vendors start with a formal design for the software system and then build each part or component separately before assembling the whole system. Typically, development will be iterative with creation of an initial prototype framework containing a small number of partially functioning parts, followed by increasing refinement of the system. An extended discussion of the process of software development and customization can be found in Maguire (1999). Core GIS software systems are usually written in a modern programming language like Visual C++, with Visual Basic or Java sometimes used for

operations that do not involve significant amounts of computer processing like the GUI.

As standards for software development become more widely adopted, so the prospect of reusing software components becomes a reality. A key choice that then faces all software developers or customizers is whether to *design* a software system by buying in components, or to *build* it more or less from scratch. There are advantages to both options: building components gives greater control over system capabilities and enables specific-purpose optimization; buying components can save valuable time and money. Examples of components which have been purchased and licensed for use in GIS software systems include: Seagate Crystal Reports in ESRI ArcView GIS and MapInfo Professional; and Microsoft Visual Basic for Applications in Autodesk World and ESRI ArcInfo/ArcView.

A key GIS implementation issue is whether to buy a system or to build one. Increasingly, users prefer to buy complete systems.

A modern GIS software system comprises an integrated suite of software components. Figure 8.5 shows the main groups of components. At the highest level there are three types of components: end user applications – a group of menu-driven user interfaces that implement the main system tasks (locate data, make maps, geocode addresses, etc.); geographic tools – the main engines that constitute the main functions or capabilities of the software; and data access components – the relatively low level components that store and manage access to geographic data.

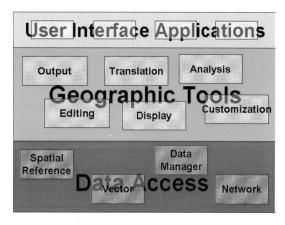

Figure 8.5 The main components of a GIS software system

Box 8.1 Jack Dangermond, Founder and President, ESRI

ESRI was founded in 1969 by Jack and Laura Dangermond as a privately held consulting group. The business began with $1100 from their personal savings and operated out of an historical home located in Redlands, California, the city where Jack grew up.

During the 1970s, ESRI focused on the development of fundamental ideas of GIS and their application in real-world projects such as developing a plan for rebuilding the City of Baltimore, Maryland, and helping Mobil Oil select a site for the new town of Reston, Virginia.

According to Dangermond:

"The more projects we completed, the more we learned how GIS methods could contribute to integrating information in new ways. It was during these years that we first observed the potential of GIS technology. We saw how GIS could influence the way decisions were made. We learned that using geography as a framework for integration could benefit the way people approach problem-solving in land management and fundamental scientific research and education. We knew GIS could make a difference."

The early years were a struggle because most of the work was being done by Jack and Laura Dangermond and a small staff. They made the decision early not to take out loans, use venture capital, or go public to raise money.

In its second decade, ESRI moved from a company primarily concerned with carrying out projects to a company building software tools and products. Dangermond realized that to leverage the methods and technologies developed for project work during the 1970s, the company needed to develop commercial software products that others could use and rely on for doing their own projects. The company hired several software engineers, including Scott Morehouse as Chief Software Architect, who came from the Harvard Lab, and created ESRI's first software product ArcInfo.

As the company shifted from being a consulting and project-based company to a software company, ESRI staff had to learn new skills: software product development, manufacturing and shipping, education and customer service, technical support, team management, and sales and marketing. The company developed a unique corporate structure that emphasizes a team approach. Corporate philosophy continued to focus on developing software and creating useful applications for companies and government organizations. For many years ESRI was a single-product company, marketing and supporting ArcInfo and a wide variety of related GIS services, but in the late 1980s and 1990s it embarked on a new strategy to diversify its interests by creating a family of GIS products.

In addition to maintaining relationships with major hardware companies, DBMS vendors, and systems integrators, ESRI has developed formal ties with more than 1000 application developers and value-added resellers who create new applications or combine ESRI technology with their data, hardware, and consulting services. Today, ESRI employs more than 2600 full-time staff, more than 1300 of whom are based at the Redlands corporate headquarters. In 2000 ESRI worldwide revenues exceeded $400 million. ESRI is still wholly owned by Jack and Laura Dangermond: there are no plans for the company to go public or change ownership.

Figure 8.6 Jack Dangermond, Founder and President of ESRI (courtesy ESRI)

At a more detailed level the tools can be sub-divided into tools for editing, translation, display, analysis, output, and customization. More information on these topics is provided in Chapter 13.

The main data access components are: vector, raster, spatial reference, and data manager. The vector and raster subsystems are responsible for the creation and low-level manipulation of these three object types (see Chapter 3 for an overview of raster and vector data structures). It is here that algorithms for point-in-polygon, polygon adjacency, proximity analysis, and other spatial operators are implemented. The spatial reference subsystem performs coordinate transformation and map projection. Last, but not least, the data manager is responsible for storing geographic objects in files or databases (see Chapter 11). The data manager also provides sophisticated services for multi-user access to geographic databases.

Some GIS software systems additionally include a database management system. For example, ESRI ArcInfo and Autodesk World include Microsoft Access, while Smallworld GIS has its own proprietary data management technology. All GIS software packages also provide interfaces to popular file formats and DBMS (DB2, Informix, Oracle, SQL Server, etc. – see Chapter 11).

8.5 Types of GIS Software Systems

Over 100 software systems claim to have mapping and GIS capabilities. Reviews of GIS software packages can be found in Elshaw Thrall and Thrall (1999), and periodic updates appear in the various GIS magazines (see Box 1.3). It is convenient to classify the main GIS software packages into

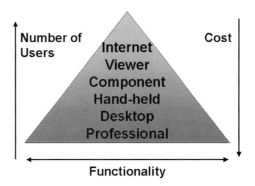

Figure 8.7 Classification of GIS software packages

six groups, based on their functionality and type: professional, desktop, hand-held, component, viewer, and Internet (Figure 8.7 and Table 8.2). For completeness this section also concludes with some comments about three other types of GIS software systems: CAD-based; raster-based, and GIS application servers.

8.5.1 Professional GIS

The term *professional* relates to the full-featured nature of this class of software. The distinctive features of professional GIS include data collection and editing, database administration, advanced geoprocessing and analysis, and other specialist tools. Professional GIS offer a superset of the capabilities of the systems in other classes. Examples of professional GIS include: ESRI ArcInfo (see Boxes 8.1 and 8.2) and Smallworld GIS. The people who use these systems are typically

Table 8.2 Product families of the main GIS software vendors.

	Autodesk	ESRI	Intergraph	MapInfo	GE Smallworld
Internet	MapGuide	ArcIMS	GeoMedia Web Map, GeoMedia Web Enterprise	MapXtreme, MapXSite	Smallworld Internet Application Server
Viewer	AutoCAD LT	ArcExplorer	GeoMedia Viewer	ProViewer	Custom
Component	Featured in several products	MapObjects	Part of GeoMedia	MapX, MapJ	Part of Smallworld GIS
Hand-held	OnSite	ArcPad	In development	MapXtend	Scout
Desktop	World	ArcView	GeoMedia	MapInfo Professional	Spatial Intelligence
Professional	AutoCAD/World	ArcInfo	GeoMedia Pro	MapInfo Professional	Smallworld GIS
Database server	Vision	ArcSDE	Uses Oracle Spatial	SpatialWare	Part of Smallworld GIS
CAD	AutoCAD MAP	ArcCAD	Featured in several products	Featured in several products	Part of Smallworld GIS

Box 8.2 Professional GIS: ESRI ArcInfo

ArcInfo is ESRI's full-featured professional GIS software product (Figure 8.8). It supports the full range of GIS functions from data collection and import, to editing, restructuring, and transformation, to display, query, and analysis. It is also the platform for a suite of analytic extensions for 3D analysis, network routing, geostatistical, and spatial (raster) analysis, among others (see Figure 8.8).

ArcInfo was originally released in 1981 on minicomputers. The early releases offered very limited functionality by today's standards and the software was basically a collection of subroutines that a programmer could use to build a working GIS software application. A major breakthrough came in 1987 when ArcInfo 4 was released with AML (Arc Macro Language), a scripting language that allowed ArcInfo to be easily customized. This release also saw the introduction of a port to Unix workstations (the software was adapted to function on this new platform) and the ability to work with data in external databases like Oracle, Informix, and Sybase. In 1991, with the release of ArcInfo 6, ESRI again re-engineered ArcInfo to take better advantage of Unix and the X-Windows windowing standard. The next major milestone was the development of a menu-driven user interface in 1993 called ArcTools. This made the software considerably easier to use and also defined a standard for how developers could write ArcInfo-based applications. ArcInfo was ported to Windows NT at the 7.1 release in 1996. About this time ESRI also took the decision to re-engineer ArcInfo from first principles. This vision was realized in the form of ArcInfo 8, released in 1999.

ArcInfo 8 is quite unlike earlier versions of the software because it was designed from the outset as a collection of re-usable, self-contained software components, based on Microsoft's COM standard. ESRI used these components to create an integrated suite of menu-driven, end-user applications: ArcMap – a map-centric application supporting integrated editing and viewing; ArcCatalog – a data-centric application for browsing and managing geographic data in files and databases; and ArcToolbox – a tool-oriented application for performing geoprocessing tasks such as proximity analysis, map overlay, and data conversion. ArcInfo 8 is customizable using either the in-built Microsoft Visual Basic for Applications (VBA) or any other COM-compliant programming language. The software is also notable because of the ability to store and manage all data (geographic and attribute) in standard commercial off-the-shelf DBMS (e.g. DB2, Informix, SQL Server, and Oracle). Other GIS software vendors had adopted this approach as early as 1986, but were initially plagued by poor performance.

Another interesting aspect of ArcInfo 8 is that for compatibility reasons ESRI included a fully working version of the existing ArcInfo technology and applications. This has allowed ESRI users to migrate their existing databases and applications to the new version in their own time.

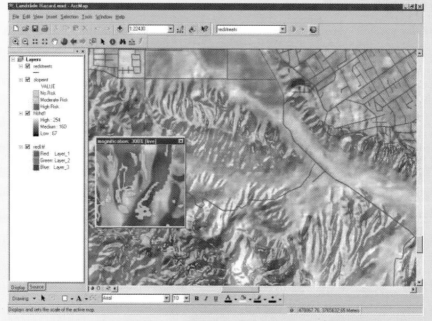

Figure 8.8 Screenshot of ArcInfo 8. A landslide hazard map of southern California derived from combined raster image and shaded surface relief, with overlain vector street centerlines (Source: ESRI)

technically literate and think of themselves as GIS professionals (career GIS staff) with degrees and, in many cases, advanced degrees in GIS or related disciplines. Prices for professional GIS are typically in the range $8000 – $20 000 per user (these and other prices mentioned later typically have discounts for multiple purchases).

Professional GIS are high-end, fully functional systems.

8.5.2 Desktop GIS

In the last few years of the 20th century desktop GIS (and so-called desktop mapping systems) grew to become the most widely used category of GIS software package. With the focus on data *use*, rather than data *creation*, and excellent tools for making maps, reports, and charts, they represent most people's experience of GIS today. The successful systems have all adopted the Microsoft

standards for interoperability and user interface style. Well-known examples of desktop GIS include Autodesk World, ESRI ArcView, Intergraph GeoMedia (Figure 8.9), Clark Lab's Idrisi, and MapInfo Professional (Box 8.3). Users often see a Desktop GIS as simply a tool to enable them to do their full-time job faster, more easily, or more cheaply. Desktop GIS users work in planning, engineering, teaching, the army, the airforce, marketing, and other professions. Desktop GIS software prices typically range from $1000 – $2000.

Desktop GIS are the mainstream workhorses of GIS today.

8.5.3 Hand-held GIS

As hardware design and miniaturization have improved dramatically over the past few years, so it has become possible to develop GIS software for

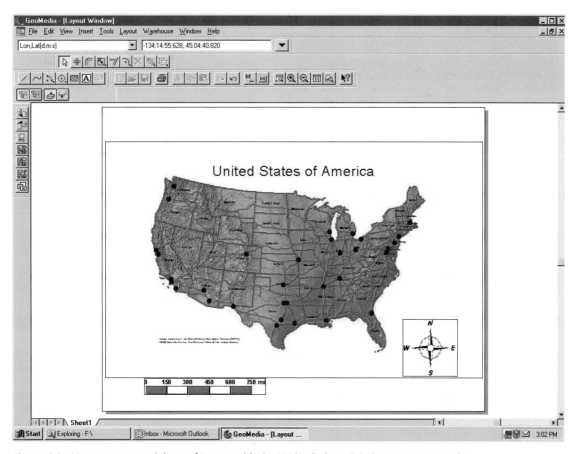

Figure 8.9 Map creation capabilities of Intergraph's GeoMedia desktop GIS (Courtesy: Intergraph)

mobile and personal use on hand-held systems. With capabilities similar to the desktop systems of just a few years ago, these palm and pocket devices can support many display, query, and simple analytical applications, even on displays of 320 × 240 pixels (quarter VGA, a 640 × 480 pixel screen resolution standard). An interesting characteristic of these systems at the present time is that all programs and data are held in memory because of the lack of a hard disk. This provides fast access, but because of the cost of memory compared to disk systems, designers have had to develop compact data storage structures.

Hand-held GIS are lightweight systems designed for mobile and field use.

Hand-held GIS are now available from many vendors, and include Autodesk OnSite, ESRI ArcPad, and Smallworld Scout. Many of these systems are designed to work with Internet products (see below). Costs are typically around $500. Figure 8.11(A) shows a list of work orders that has been downloaded to an employee's mobile device from a corporate database. Updates in the form of a redline sketch or notes can be captured in the field along with other tabular data and posted to the central database.

8.5.4 Component GIS

With the advent of component-based software development (see Section 8.4), a number of GIS

Box 8.3 Desktop GIS: MapInfo Professional

MapInfo Professional is essentially a desktop mapping and GIS package, although it overlaps considerably with ArcInfo and other full-featured professional systems. It too possesses a wide range of tools for creating and maintaining databases, but it is in the area of data display, query, and use that MapInfo Professional and other Desktop Mapping systems excel.

MapInfo Professional owes its origins and rise to fame to the advent of Windows and the personal computer (PC). The initial idea behind MapInfo Professional was that as GIS expanded and more data became available, users would be interested in the 20% of professional GIS functionality that would satisfy 80% of their needs. Furthermore, many of the new users coming to GIS did not have the technical expertise or interest necessary to learn the technical details of programming, algorithms, and storage strategies. Instead their interest was in "getting the job done". Right from the outset MapInfo Professional was menu-driven and had a task-based interface that made it easy to make maps, to geocode addresses, and to query and visualize data. These capabilities, together with the "look and feel" of the Windows standard made MapInfo Professional a compelling solution for many application areas and markets (see Figure 8.10).

Over the past 10 years MapInfo Professional has developed rapidly, but the developers have remained true to their original vision and design principles. The introduction of a customization capability, MapBasic, third-party charting and reporting, and many additional tools and capabilities, have further ensured that MapInfo remains at the forefront of desktop GIS.

Another key innovation from MapInfo was the realization that many GIS users, particularly those in business GIS (banking, insurance, retail, and associated areas: see Section 2.3.5), wanted to buy pre-packaged datasets instead of building a database themselves. Initially, this was only possible in the USA where datasets were available at a reasonable price, but subsequently it has become possible in Europe and other parts of the world, albeit on a more restricted scale.

MapInfo Professional 5 is a self-contained application with capabilities for editing, accessing, analyzing, visualizing, and presenting geographic information. The easy ability to integrate data from multiple sources makes MapInfo Professional a simple-to-use mapping and GIS package, which allows more informed decisions to be made quickly. A key feature of MapInfo is its compatibility with Microsoft Windows applications. This includes adherence to Windows standards for the user interface, communication between applications ("Drag and Drop" and Object Linking and Embedding (OLE)), and data access.

Like other vendors, in recent years MapInfo has added a number of functional extensions to the core software for network analysis, surface modeling, and database management. MapInfo has also built some specific-purpose applications on top of the core platform.

In conclusion, MapInfo's introduction defined a new class of desktop GIS. Though others have followed with packages that have comparable functionality (see Table 8.2), MapInfo remains a leading product in what is now the biggest revenue-generating sector of the GIS software market.

Box 8.3   Continued

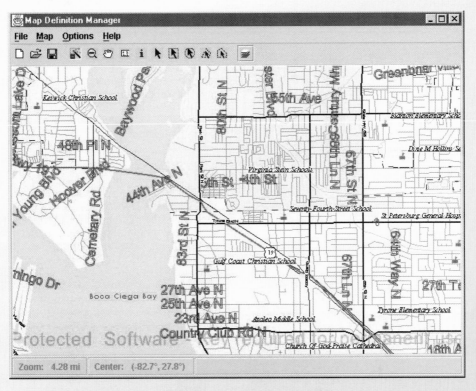

Figure 8.10   Screenshot of MapInfo Professional (Courtesy: MapInfo)

(A)

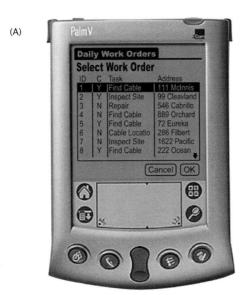
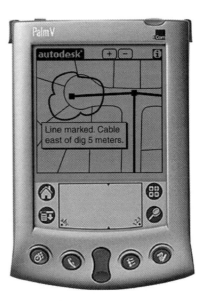

Figure 8.11   Examples of hand-held GIS: (A) Autodesk OnSite (Courtesy: Autodesk) (Cntd.)

(B)

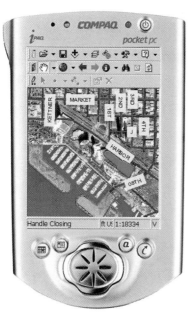

Figure 8.11 (cntd.) Examples of hand-held GIS: (B) ESRI ArcPad (Courtesy: ESRI)

vendors have released collections of GIS software components. These are really tool kits of GIS functions (components) that a reasonably knowledgeable programmer can use to build a full GIS software system. They are of interest to developers because they can use these components to create highly customized and optimized applications that can either stand alone or can be embedded within other software systems. Typically, component GIS packages offer strong display and query capabilities, but only limited cartography and analysis tools.

Component GIS are used by developers to create focused applications.

Examples of component GIS include Blue Marble Geographics GeoObjects, ESRI MapObjects, and MapInfo MapX. The typical cost is $1000–2000 for the developer kit and $100 per deployed application. The people who use deployed applications may not even realize that they are using a GIS, because it is embedded in other applications (e.g. customer care systems, routing systems, or interactive atlases).

8.5.5 GIS viewer

In the late 1990s a number of vendors released free GIS viewers that are able to display and query popular file formats. These include ESRI's

introduced ArcExplorer, Intergraph's GeoMedia Viewer and MapInfo's ProViewer. Today the GIS viewer represents a significant product category. The rationale behind these products is that they help to establish market share, and specific vendor terminology and data formats as *de facto* standards. GIS users often work with viewers on a casual basis, often in conjunction with more sophisticated GIS software products. GIS viewers have limited functional capabilities restricted to display, query, and simple mapping. They do not support editing, sophisticated analysis, modelling, or customization.

8.5.6 Internet GIS

The GIS products with the highest potential user base and lowest cost per user are those in the Internet GIS category. Stimulated by the widespread availability of the Internet and market demand for greater access to geographic information, GIS software vendors have been quick to release products that harness the power of the Internet. Most vendors have chosen to exploit the unique characteristics of the WWW by developing GIS technology that integrates with Web browsers and servers, and uses the hypertext transmission protocol (http) for communication.

Internet GIS have the highest number of users, although Internet users typically focus on simple display and query tasks.

Initially, Internet GIS applications focused on display and query application – making simple things simple and cost-effective – with more advanced applications becoming available as user awareness and technology expanded. Today, it is possible to perform standard GIS operations like making maps (Microsoft Expedia has online interactive maps; www.expediamaps.com), routing (MapQuest offers pathfinding with directions (www.mapquest.com: see Section 1.4.3), publishing census data (US Census Bureau, AmericanFactFinder has online census data and maps for the whole US; www.census.gov), and suitability analysis (the US National Association of Realtors has a site for locating homes for purchase based on user-supplied criteria; www.realtor.com: see Section 2.1 especially Figure 2.6). A recent trend has been the development of Internet-based online data networks such as the Geography Network (www.GeographyNetwork.com: see Box 7.4 and Figure 1.19). Over time it is to be expected that the capabilities of systems will grow significantly and there are those that foresee the Internet becoming the dominant GIS delivery mechanism.

Examples of Internet GIS products include: Autodesk MapGuide (Box 8.4), ESRI ArcIMS, Intergraph GeoMedia Web Map, and MapInfo MapXtreme. The cost of Internet GIS varies from around $5000 to $25 000, for small to medium-sized systems, to well beyond for large multi-function, multi-user systems.

8.5.7 Other types of GIS software

Raster-based GIS, as the name suggests, focus primarily on raster data and raster analysis. Chapters 3 and 9 provide a discussion of raster and other data models, while Chapters 13 and 14 present an overview and examples of specific capabilities. Just as many vector-based systems have raster analysis extensions (for example, ESRI ArcInfo and ArcView have Spatial Analyst, and MapInfo Professional has Vertical Mapper), in recent years raster systems have added vector capabilities (for example ERDAS Imagine and Clarke Labs Idrisi now have vector capabilities

built in). The distinction between raster-based and other categories is increasingly blurred as a consequence. The users of raster-based GIS are primarily interested in working with imagery and undertaking spatial analysis. The prices for raster-based GIS range from $500 to $10 000.

CAD-based GIS are systems that started life as computer-aided design (CAD) packages and then had GIS capabilities added. Typically, a CAD system is supplemented with database, spatial analysis, and cartography capabilities. Not surprisingly, these systems appeal mainly to users whose primary focus is in typical CAD fields such as architecture, engineering, and construction, but who also want to use geographic information and geographic analysis in their projects. The best-known examples of CAD-based GIS are: Autodesk Map and ESRI ArcCAD, which extend Autodesk AutoCAD; and Bentley GeoGraphics and Intergraph Modular GIS Environment (MGE), which extend Bentley Microstation. CAD-based GIS typically cost $1000–3000, excluding the core CAD system.

Box 8.4 Internet GIS: Autodesk MapGuide

In the late 1990s, a time when desktop GIS had become dominant, a small Canadian company called Argus released an Internet GIS product called MapGuide (Figure 8.12). Subsequently purchased by Autodesk, MapGuide marked the start of another chapter in the history of GIS software. Just as desktop GIS evolved to complement professional GIS, so Internet GIS are able to integrate well with both these types of systems. The emphasis in MapGuide and other Internet GIS products is very much on map display and use. Indeed, these systems have few if any data editing capabilities.

MapGuide is an important innovation for the many users who have spent considerable amounts of time and money creating valuable databases, and who want to make them available to other users inside or outside their organizations. Autodesk MapGuide allows users to leverage their existing GIS investment by publishing dynamic, intelligent maps at the point at which they are most valuable – in the field, at the job site, or on the desks of colleagues, clients, and the public.

There are three key components to MapGuide: the viewer – a relatively easy-to-use Web application with a browser style interface; the author – a menu-driven authoring environment used to create and publish a site for client access; and the server – the administrative software that monitors site usage, and manages requests from multiple clients and to external databases. MapGuide works directly with Internet browsers and servers, and uses the http protocol for communication. It makes good use of standard Internet tools like HTML (hypertext markup language) and JScript (Java Script) for building client applications, and ColdFusion (an Internet site generation and management tool from Allaire Corp.) for managing data and queries on the server.

Typical features of MapGuide sites include the display of raster and vector maps, map navigation (pan and zoom), geographic and attribute queries, simple buffering, report generation, and printing. Like other advanced Internet GIS, MapGuide also has some limited tools for redlining (drawing on maps) and editing geographic objects.

To date, MapGuide has been used most widely in existing mature GIS sites that want to publish their data internally or externally, and in new sites that want a way to publish dynamic maps quickly to a widely dispersed collection of users (for example, maps showing election results).

In conclusion, MapGuide and the other Internet GIS products are growing in importance. Their cost-effective nature and focus on ease of use will help to disseminate geographic information even more widely and will introduce many new users to the field of GIS.

Box 8.4 *Continued*

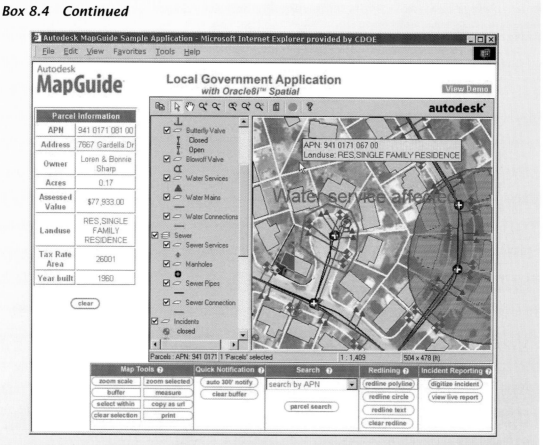

Figure 8.12 Screenshot of Autodesk MapGuide (Courtesy: Autodesk)

Many enterprise-wide GIS include GIS data and application servers. Their purpose is to manage multiple users accessing continuous geographic databases, which are stored and managed in commercial-off-the-shelf (COTS) database management systems (DBMS). This technology offers centralized management of data, the ability to process data on a server (which delivers good performance for certain types of applications), and control over database editing and update (see Chapter 11 for further details). To assist in managing data in standard DBMS, some vendors – notably IBM, Informix, and Oracle – have developed technology to extend their DBMS servers so that they are able to store and process geographic information efficiently. A number of GIS vendors have also developed technology that fulfils this function. Examples of GIS application servers include: Autodesk Vision, ESRI ArcSDE,

and MapInfo SpatialWare. These systems typically cost $10 000–25 000 or more depending on the number of users.

> Raster-based, CAD-based, and GIS application servers are also important types of GIS software.

8.5.8 GIS product families

Recognizing that there are many different types of GIS user and that their requirements are very different, a recent trend has been towards the evolution of GIS software product families. These are integrated collections of GIS products. Table 8.2 summarizes the product families of the main vendors. Because not all the vendors' product strategies are identical and products evolve,

Box 8.5   Using GIS in Honduras hurricane disaster rescue and relief

In 1998 Hurricane Mitch strolled the length of Honduras at a leisurely two miles per hour, destroying everything in its broad path during a five-day rampage. The statistics are grim: more than 7000 dead, 8000 missing, and more than 1 400 000 persons, or about 30% of the entire population, directly affected by the disaster. Honduras was virtually destroyed. After a night of frenzied rescue efforts, the dawn light of October 31, 1998 revealed a scene of unimaginable devastation to the residents of Tegucigalpa, the Honduran capital. Some reports indicated that the city looked as if it had been struck by an atomic bomb.

One of the first acts of President Flores in the hours immediately following the disaster was to appoint a National Emergency Committee. From this evolved the country's National Center for Disaster Information (NCDI). Reconstruction activities were facilitated by high-quality data from the Government of Honduras, US federal agencies, and private-sector partners including ESRI, Silicon Graphics, Microsoft, and EarthSat (Figure 8.13). GIS was used quickly to coordinate rescue and relief services so that resources could be allocated to those in immediate need. It was also used to minimize the duplication of efforts as international groups rushed forward with assistance, foodstuffs, and medical supplies. Additionally, it was also employed to brief many groups on the progress of the relief effort. GIS is quite commonly used for avoiding and dealing with the consequence of natural and human disasters. After disasters, when confusion reigns, a central information system can greatly assist planning and operational management. GIS can also help to identify at-risk groups, plan evacuation routes, and simulate the likely locations of disasters.

The Honduras disaster relief effort was based on a combination of professional, desktop, and Internet GIS software systems. The professional system was used for database creation and management, and high-end map production and analysis. The desktop systems were used in the laboratory and field for briefings and focused projects. The Internet GIS was used to provide up-to-date data and information to many users, managers, and members of the public.

Imagine if the country had been equipped with GIS prior to the hurricane. In parts of the USA and Far East, hurricane, typhoon, and cyclone warning systems now transmit real-time updates about the status of storm events using Internet GIS. Embedded component GIS are used for high-performance, special-purpose systems such as evacuation routing. Hand-held GIS are employed to collect information about damage assessments in the field following storms, and GIS viewers provide free data and educational materials to schools and members of the public to bring the importance of these life-threatening events to people's attention.

(A)

(B)

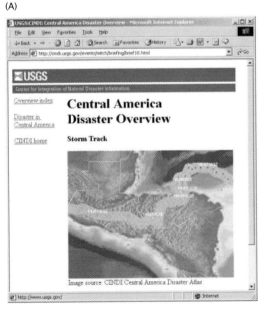

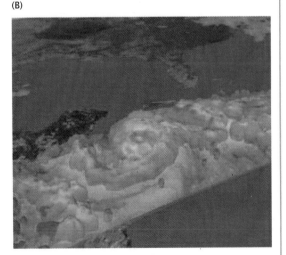

Figure 8.13 Data were provided for the Hurricane Mitch relief effort in the form of a Digital Atlas CD produced by USGS and others, and a Web site (see USGS Center for Integration of Natural Disaster Information site – www.cindi.usgs.gov/): (A) Storm track; (B) Rainfall map – notice that rainfall (red) is highest close to the eye of the hurricane (Source: NASA)

occasionally rapidly, some latitude should be used when interpreting this table. Nevertheless, it can be seen that there is increasing similarity in the way vendor products are evolving, at least at the macro level. In part this is a reflection of the increasing maturity of GIS, but also the standardization on underlying base technology (e.g. Windows and Internet servers).

The main GIS software vendors have created integrated product families to meet the needs of a diverse user community.

Box 8.5 shows how different types of GIS software can be used together in an application context. The example application in this case is managing a hurricane disaster relief effort in Honduras.

8.6 GIS Software Usage

It is surprisingly difficult to obtain accurate figures for the size of the GIS market and the market share of specific vendor products. Daratech estimates that in 2000 the worldwide GIS market was $6.9 billion, of which $1.05 billion was core software. Estimates of the size of the GIS market by type of system, based on the authors' knowledge, are shown in Figure 8.14. In 2000 the total size of the GIS market, measured in terms of the numbers of GIS software users, is about 2 million spread over 1 million sites. If the number of Internet GIS users is also taken into consideration then the GIS user population rises to about 5 million in total. This excludes users of GIS products such as hard copy maps, charts, and reports.

Conclusion

GIS software is a fundamental sub-part of any operational GIS. Today there are many types of GIS software product to choose from and a number of ways to configure implementations. One of the exciting and at times unnerving characteristics of GIS software is the very rapid rate of development. This is a trend that seems set to continue as the software industry pushes ahead with significant R&D efforts. In the following chapters we shall explore in more detail the functionality of GIS software and how it can be applied in a real-world context.

Questions for Further Study

1. Go to the Websites of the main GIS software vendors and compare their product families. In what ways are they similar/dissimilar?
   Autodesk www.autodesk.com
   ESRI www.esri.com
   Intergraph www.intergraph.com
   MapInfo www.mapinfo.com
   Siemens www.sicad.com
   Smallworld www.smallworld.com
2. Describe the evolution of GIS software in relation to developments in hardware (see also Section 20.3.2).
3. Review the characteristics of three Web sites containing geographic information about your local area.
4. Compare the main types of functions in two popular GIS software packages (choose one from each of the main groups, e.g. professional and Internet).
5. Characterize the type of use that different users in a large organization would make of GIS and determine what type of GIS software would be most suitable for them.

Online Resources

NCGIA Core Curricula (www.ncgia.ucsb.edu/pubs/core. html):
   Core Curriculum in GIScience, Sections 2.1.1 (Fundamentals of Data Storage, Carol Jacobson), and 2.14.5 (WebGIS, Ken Foote and Anthony Kirvan)
   Core Curriculum in GIS, 1990, Units 3 (Introduction to Computers), 4 (The Raster GIS), 13 (The Vector or Object GIS), 18 (Modes of User/GIS Interaction), and 24 (The GIS Marketplace)

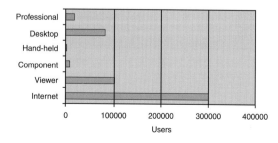

Figure 8.14   Estimated size of the different GIS software sectors (Scale: x10)

ESRI Virtual Campus courses on ESRI software (campus.esri.com):
 Migrating to ArcView 8, by ESRI
 Introduction to ArcInfo using ArcTools, by ESRI
 What's New in ArcInfo 8, by ESRI
 Introduction to ArcInfo using ArcMap, ArcCatalog, and ArcToolbox, by ESRI
 Programming with Avenue, by ESRI
 MapObjects Basics, by ESRI
 Introduction to Visual Basic for ESRI Software, by ESRI
 Introduction to ArcView GIS, by ESRI (also Einführung in ArcView GIS by ESRI and Josef Strobl, and Introdução ao ArcView GIS by ESRI and Mirna Cortopassi Lobo)
 Introduction to ArcView 3D Analyst, by ESRI
 Introduction to ArcView Tracking Analyst, by LittonTASC
Web sites of the main GIS vendors (see Questions for Further Study above)
General definitions of software concepts:
 www.whatis.com
Reviews of GIS software products and vendors
 www.geospatial-online.com
 GeoSpatial Solutions magazine site
 www.geoplace.com
 GeoWorld and sister magazine site
 www.giscafe.com GIS Café
 www.geocomm.com GeoCommunity
 www.tenlinks.com TenLinks

Reference Links

Maguire D J, Goodchild M F, and Rhind D W (eds), 1991 *Geographical Information Systems: Principles and applications*. Harlow, UK: Longman (Text available online from 'Links to Big Book 1' at www.wiley.com/gis and www.wiley.co.uk/gis).
 Chapter 3, The technological setting of GIS (Goodchild M F)
 Chapter 4, The commercial setting of GIS (Dangermond J)
Longley P A, Goodchild M F, Maguire D J, and Rhind D W (eds), 1999 *Geographical Information Systems: Principles, techniques, management and applications*. New York: John Wiley.
 Chapter 22, Geographical information systems in networked environments (Coleman D J)
 Chapter 23, Desktop GIS software (Elshaw Thrall S, Thrall G I)
 Chapter 25, GIS customization (Maguire D J)

Reference

Daratech 2000 *Geographic Information Systems: Markets and opportunities*. Cambridge, Massachusetts: Daratech

GEOGRAPHIC DATA MODELING

9

This chapter discusses the technical issues involved in modeling the real world in a GIS. It describes the process of data modeling and the various data models that have been used in GIS. A data model is a set of constructs for describing and representing parts of the real world in a computer system. Data models are vitally important to GIS because they control the way that data are stored and have a major impact on the type of analytical operations that can be performed. Early GIS were based on CAD, simple graphical, and image data models. In the 1980s and 1990s the hybrid geo-relational model came to dominate GIS. In the last few years major software systems have been developed based on more advanced and standardized geographic object data models that include elements of all earlier models.

Learning objectives

After reading this chapter you will be able to:

● Define what geographic data models are and discuss their importance in GIS;

● Understanding how to undertake GIS data modeling;

● Outline the main geographic data models used in GIS;

● Understand topology and why it is useful for data validation and analysis;

● Read data model notation;

● Describe how to model the world and create a useful geographic database.

9.1 Introduction

This chapter builds on the material on geographic representation presented in Chapter 3. By way of introduction it should be noted that the terms *representation* and *model* describe essentially the same thing. Representation is more widely used to describe the conceptual and scientific issues, whereas model is used in practical and database circles. In this chapter the term "model" will be used, given the practical approach. This chapter focuses on how geographic reality is modeled (abstracted or simplified) in a GIS, with particular emphasis on choosing one particular style of data model over another.

9.1.1 What is a data model?

The heart of any GIS is the data model, which is a set of constructs for representing objects and processes in the digital environment of the computer (Figure 9.1). People (GIS users) interact with operational GIS in order to perform tasks like making maps, querying databases, and performing site suitability analyses. Because the types of analyses that can be undertaken are strongly influenced by the way the real world is modeled, decisions about the type of model to be adopted are vital to the success of a GIS project.

> A data model is a set of constructs for describing and representing selected aspects of the real world in a computer.

As described in Chapter 3, geographic reality is infinitely complex, but computers are finite. Therefore, difficult choices have to be made

about what and how things are modeled using GIS. Because different types of people use GIS for different purposes, and the type of phenomena people study have different characteristics, there is no single type of GIS data model that is best for all circumstances.

9.1.2 Levels of data model abstraction

When representing the real world in a computer, it is helpful to think in terms of the four different levels of abstraction (levels of generalization or simplification) that are shown in Figure 9.2. First, *reality* is made up of real-world phenomena (buildings, streets, wells, lakes, people, etc.), and includes all aspects that may or may not be perceived by individuals, or deemed relevant to a particular application. Second, the *conceptual model* is a human-oriented, often partially structured, model of selected objects and processes that are thought relevant to a particular problem domain. Third, the *logical model* is an implementation-oriented representation of reality that is often expressed in the form of diagrams and lists. Fourth, the *physical model* portrays the actual application in a GIS, and often comprises tables stored as files or databases (see Chapter 11). Use of the term physical here is actually misleading because the models are not physical, they only exist digitally in computers.

In data modeling, users and system developers participate in a process that successively engages with each of these levels. The first phase of modeling begins with definition of the main types of objects to be represented in the GIS and concludes with a conceptual description of the main types of objects and relationships between

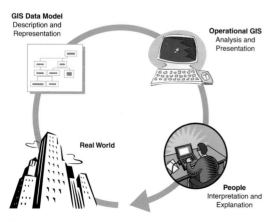

Figure 9.1 The role of a data model in GIS

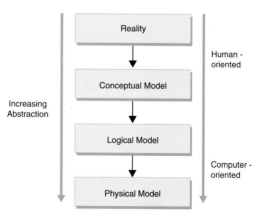

Figure 9.2 Levels of abstraction relevant to GIS data models

them. Once this phase is complete, further work will lead to the creation of diagrams and lists describing the names of objects, their behavior, and the type of interaction between objects. This type of logical data model is very valuable for defining what a GIS will do and the type of domain over which it will extend. Logical models are implementation independent, and can be created in any GIS with appropriate capabilities. The final data modeling phase involves creating a model showing how the objects under study can be digitally implemented in a GIS. Physical models describe the exact files or database tables used to store data, the relationships between object types, and the precise operations that can be performed. For more details about the practical steps involved in data modeling see Sections 9.3 and 9.4.

A data model provides system developers and users with a common understanding and reference point. For developers a data model is the means to represent an application domain in terms that may be translated into a design and implementation of a system. For users, it provides a description of the structure of the system, independent of specific items of data or details of the particular application (Worboys 1995).

The discussion of geographic representation in Chapter 3 introduced discrete objects and fields, the two fundamental conceptual models for representing real-world objects geographically. In the same chapter the raster and vector logical models were also introduced. Figure 9.3 shows two representations of raster and vector objects in a GIS. Notice the difference in the objects

represented. Major roads, in false color green, and areas cleared of vegetation in false color red, are more clearly visible in the vector representation, whereas smaller roads and built-up areas can best be seen on the scanned raster aerial photograph. The next sections in this chapter focus on the logical and physical representation of raster, vector, and related models in GIS software systems.

9.2 GIS Data Models

In the past half-century many GIS data models have been developed and deployed in GIS software systems. The key types of geographic data models and their main areas of application are listed in Table 9.1. All are based in some way on the conceptual discrete object/field and logical vector/raster geographic data models.

All GIS software systems include a core data model that is built on one or more of the GIS data models described below. As discussed earlier, the system data model is the means to represent geographic aspects of the real world and defines the type of geographic operations that can be performed. It is the responsibility of the GIS implementation team to populate this generic model with information about a particular problem (e.g. utility outage management, military mapping, or natural resource planning). Some GIS software packages come with a fixed data model, while others have models that can be easily extended. Those that can easily be extended are

(A)   (B)

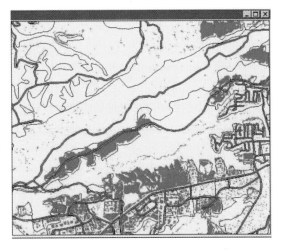

Figure 9.3 Representations of the same area (San Diego, California): (A) panchromatic SPOT raster satellite image collected in 1990 at 10 m resolution (Courtesy: ERDAS); (B) vector objects digitized from the image (Courtesy: ERDAS)

Table 9.1 Geographic data models used in GIS

Data Model	Application
Computer-aided design (CAD)	Engineering design
Graphical (non-topologic)	Simple mapping
Image	Image processing and simple grid analysis
Raster/grid	Spatial analysis and modeling especially in environmental and natural resources applications
Vector/geo-relational topologic	Many operations on vector geometric features in cartography, socio-economic and resource analysis, and modeling
Network	Network analysis in transportation, hydrology, and automated mapping/facilities management (AM/FM)
Triangulated irregular network (TIN)	Surface/terrain analysis and modeling
Object	Many operations on all types of entities in all types of applications

better able to model the richness of geographic domains, and in general are the easiest to use and are the most productive systems.

When modeling the real world for representation inside a GIS it is convenient to group entities of the same geometric type together (for example, all point entities such as lights, garbage cans, dumpsters, etc. might be stored together). A collection of entities of the same geometric type (dimensionality) is referred to as a class or layer. It should also be noted that the term layer is quite widely used in GIS as a general term for a distinct dataset. It derived from the process of entering different types of data into a GIS from paper maps, which was undertaken one plate at a time (all entities of the same type were represented in the same color and, using printing technology, were reproduced together on film or printing plates). Grouping entities of the same geographic type together makes the storage of geographic databases more efficient (for further discussion of this see Section 11.3). It also makes it much easier to implement rules for validating edit operations (e.g. addition of a new building or census administrative area) and for building relationships between entities. All of the data models discussed below use layers in some way to handle geographic entities.

A layer is a collection of geographic entities of the same geometric type (e.g. points, lines, or polygons). Grouped layers may combine layers of different geometric types.

9.2.1 CAD, graphical, and image GIS data models

The earliest GIS were based on very simple models derived from work in the fields of CAD (computer-

aided design), computer cartography, and image analysis. In a CAD system real-world entities are represented symbolically as simple point, line, and polygon vectors. This basic CAD data model never became widely popular in GIS because of three severe problems for most applications at geographic scales. First, because CAD models typically use local drawing coordinates instead of real-world coordinates for representing objects they are of little use for map-centric applications. Second, because individual objects do not have unique identifiers it is difficult to tag them with attributes. As the following discussion shows this is a key requirement for GIS applications. Third, because CAD data models are focused on graphical representation of objects they do not store details of any relationships between objects (e.g. topology), the type of information essential in many spatial analytical operations.

A second type of simple GIS geometry model was derived from work in the field of computer cartography. The main requirement for this field in the 1960s was the automated reproduction of paper maps and the creation of simple thematic maps. Techniques were developed to digitize maps and store them in a computer for subsequent plotting and printing. All paper map entities were stored as points, lines, and polygons, with annotation used for placenames. Like CAD systems there was no requirement to tag objects with attributes or to work with object relationships.

At about the same time that CAD and computer cartography systems were being developed, a third type of data model emerged in the field of image processing. Because the main data source for geographic image processing is scanned aerial photographs and digital satellite images it was natural that these systems would use rasters or grids to represent the patterning of real-world

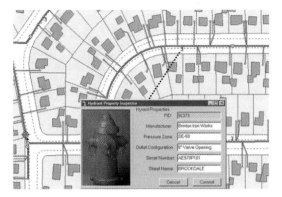

Figure 9.4 An image of a hydrant used as an object attribute in a water facility system

objects on the Earth surface. The image data model is also well suited to working with pictures of real-world objects, such as photographs of water valves and scanned building floor plans that are held as attributes of geographically referenced entities in a database (Figure 9.4).

In spite of their many limitations GIS still exist based on these simple data models. This is partly for historical reasons – the GIS may have been purchased before newer, more advanced models became available – but also because of lack of knowledge about the newer approaches described below.

9.2.2 Raster data model

The raster data model uses an array of cells, or pixels, to represent real-world objects (Figure 3.7). The cells can hold any attribute values based on one of several encoding schemes including categories, and integer and floating point numbers (see Box 3.3 for details). In the simplest case a binary representation is used (for example presence or absence of vegetation), but in more advanced cases floating point values are preferred (for example, height of terrain above sea level in meters). In some systems multiple attributes can be stored for each cell in a type of value attribute table where each column is an attribute and each row either a pixel, or a pixel class (Figure 9.5).

Raster data are usually stored as an array of grid values, with metadata about the array held in a file *header*. Typical metadata include the geographic coordinate of the upper-left corner of

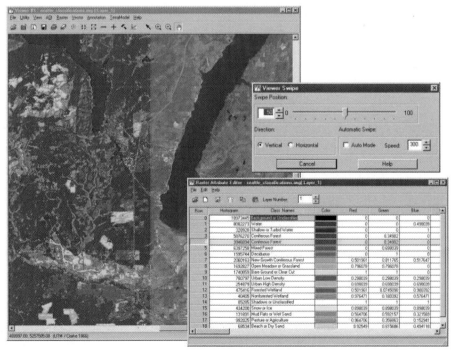

Figure 9.5 Raster data of the Olympic Peninsula, Washington State, USA, with associated value attribute table. Bands 4, 3, 2 from Landsat 5 satellite with land cover classification overlain (ERDAS Imagine: screenshot courtesy of ERDAS Inc.; data courtesy of EROS Data Center)

Box 9.1   *Raster compression techniques*

Although the raster data model has many uses in GIS, one of the main operational problems associated with it is the sheer amount of raw data that must be stored. To improve storage efficiency many types of raster compression technique have been developed such as run-length encoding, block encoding, wavelet compression, and quadtrees (see Section 11.7.2.2 for another use of quadtrees as a means to index geographic data). Table 9.2 presents a comparison of file sizes and compression rates for three compression techniques based on the image in Figure 9.6. It can be seen that even the comparatively simple run-length encoding technique (Figure 3.11) compresses the file size by a factor of 5. The more sophisticated wavelet compression technique results in a compression rate of almost 40, reducing the file from 90.5 to 2.3 MB.

Figure 9.6   Shaded digital elevation model of North America used for comparison of image compression techniques in Table 9.2. Original image is 8726 by 10619 pixels, 8 bits per pixel. The inset shows part of the image at a zoom factor of 1000 for the San Francisco Bay area (Source: ESRI)

Table 9.2   Comparison of file sizes and compression rates for raster compression techniques (using image shown in Figure 9.6)

Compression technique	File size (MB)	Compression rate
Uncompressed original	90.5	
Run-length	17.7	5.1
Wavelet	2.3	39.3

Box 9.1 *Continued*

Run-length encoding

Run-length encoding is perhaps the simplest compression method, and for this reason it has already been used to introduce the concept of raster data compression (see Section 3.6.2 and Figure 3.11). It involves encoding row cells that have the same value, with a pair of values indicating the number of cells with the same value, and the actual value.

Block encoding

Block encoding is a little like a two-dimensional version of run-length encoding in which the array is defined as a series of square blocks of the largest size possible. Recursively, the array is divided using blocks of smaller and smaller size. It is sometimes described as a *quadtree* data structure.

Wavelet

Wavelet compression techniques invoke principles similar to those discussed in our treatment of fractals, as discussed in Section 5.8. They remove information by recursively examining patterns in datasets at different scales. A useful by-product of this for geographic applications is that wavelet-compressed raster layers can be quickly viewed at different scales with appropriate amounts of detail. MrSID (Multiresolution Seamless Image Database) from LizardTech is an example of a wavelet compression technique that is widely used in geographic applications, especially for compressing aerial photographs. Similar wavelet compression algorithms are available from other public and private sources.

Run-length and block encoding both result in lossless compression of raster layers, that is, a layer can be compressed and decompressed without loss of detail. In contrast, the MrSID wavelet compression technique is *lossy* since information is irrevocably discarded during compression. Although MrSID compression results in very high compression ratios, because information is lost its use is limited to applications that do not need to use the raw digital numbers for processing or analysis. It is not appropriate for compressing DEMs for example, but many organizations use it to compress scanned maps and aerial photographs when access to the original data is not necessary.

the grid, the cell size, and the number of row and column elements. The array itself is usually stored as a compressed file or as a record in a database management system (see Section 11.3). Techniques are described in Box 9.1 (see also Section 3.6.2 for the general principles involved).

Data encoded using the raster data model are particularly useful as a *backdrop* map display because they look like conventional maps and can communicate a lot of information quickly. They are also widely used for analytical applications such as disease dispersion modeling, surface water flow analysis, and store location modeling.

9.2.3 Vector data model

The raster data model discussed above is most commonly associated with the field conceptual data model. The vector data model on the other hand is closely linked with the discrete object view. Both of these conceptual perspectives were introduced in Section 3.5. To date, the vector data model has been very widely implemented in GIS. This is because of the precise nature of its representation method, its storage efficiency, the quality of its cartographic output, and the availability of functional tools for operations like map projection, overlay, and analysis.

In the vector data model each object in the real world is first classified into a geometric type: point, line, or polygon (Figure 9.7). Points (e.g. wells, soil pits, and retail stores) are recoded as single coordinate pairs, lines (e.g. roads, streams, and geologic faults) as a series of ordered coordinate pairs, and polygons (e.g. census tracts, soil areas, and oil license areas) as one or more line segments that close to form an area. The coordinates that define the geometry of each object may have 2, 3, or 4 dimensions: 2 (*x*, *y*: row and column, or latitude and longitude), 3 (*x*, *y*, *z*: the addition of a height value), or 4 (*x*, *y*, *z*, *m*: the addition of another value to represent time or some other property — perhaps the offset of road signs from a road centerline).

For completeness it should also be said that in some data models linear features can be represented not only as a series of ordered coordinates, but also as curves defined by a mathematical function (e.g. a spline or Bézier curve). These are

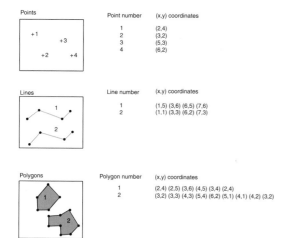

Points

Point number	(x,y) coordinates
1	(2,4)
2	(3,2)
3	(5,3)
4	(6,2)

Lines

Line number	(x,y) coordinates
1	(1,5) (3,6) (6,5) (7,6)
2	(1,1) (3,3) (6,2) (7,3)

Polygons

Polygon number	(x,y) coordinates
1	(2,4) (2,5) (3,6) (4,5) (3,4) (2,4)
2	(3,2) (3,3) (4,3) (5,4) (6,2) (5,1) (4,1) (4,2) (3,2)

Figure 9.7 Representation of point, line, and polygon objects using the vector data model

particularly useful for repre-senting artificial entities like road curbs and some buildings.

9.2.3.1 Simple features

Geographic entities encoded using the vector data model are often called features and this will be the convention adopted here. Features of the same geometric type are stored in a geographic database as a feature class, or when speaking about the physical (database) representation the term *feature table* is preferred. Here each feature occupies a row and each property of the feature occupies a column. GIS commonly deal with two types of feature: simple and topologic. The structure of simple feature line and polygon datasets is sometimes called spaghetti because, like a plate of cooked *spaghetti*, lines and polygons (strands of spaghetti) can overlap and there are no relationships between any of the objects.

> Features are vector objects of type point, line, or polygon.

Simple feature datasets are useful in GIS applications because they are easy to create and store, and because they can be retrieved and rendered on screen very quickly. On the other hand the lack of any connectivity relationships between the features means that operations like shortest-path network analysis and polygon adjacency cannot be performed without additional time-consuming calculations. Simple feature polygon layers are also inefficient for modeling phenomena conceptualized as fields

because the boundary between adjacent features (for example, two potato fields) must be digitized and stored twice. They are also inflexible, as it is not easy to dissolve common boundaries when amalgamating zones. Furthermore, because simple feature polygons can overlap this limits their use in certain applications. For example, in land ownership applications land parcels cannot overlap, for otherwise two people could own the same piece of land (land ownership is conceptualized as a field that necessarily has one owner at every point)!

9.2.3.2 Topologic features

Topologic features are essentially simple features structured using topologic rules. Topology is the science and mathematics of relationships. In GIS topology is used to validate the geometry of vector datasets (e.g. check that polygons close and that all lines in a network are joined together), and for certain types of operations (e.g. network tracing and tests of polygon adjacency). Topologic structuring of vector layers introduces some interesting and very useful properties, especially for line (also called 1-cell, arc, edge, and link) and polygon (also called 2-cell, area, and face) data. Topological structuring of line layers forces all line ends that are within a user-defined distance to be snapped together so that they are given exactly the same coordinate value (see also Section 6.3.3). A node is placed wherever the ends of lines meet. Following on from the earlier analogy this type of data model is sometimes referred to as spaghetti with meatballs (the nodes being the meatballs on the spaghetti lines).

> Topology is the science and mathematics of relationships used to validate the geometry of vector entities, and for operations such as network tracing and tests of polygon adjacency.

In a topologically structured polygon data layer each polygon is defined as a collection of lines, that in turn are made up of an ordered list of coordinates (vertices). The list of lines that make up a polygon is stored with the geometry data. Figure 9.8 shows an example of a polygon dataset comprising six polygons (including the "outside world": polygon 1). A number in a circle identifies a polygon. The lines that make up the polygons are shown in the polygon-line list. For example, polygon 2 can be assembled from lines 4, 6, 7, 10, and 8. In this particular implementation example the 0 before the 8 is used to indicate that line 8 actually defines an "island" inside polygon

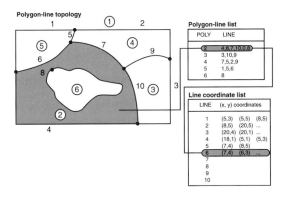

Figure 9.8 A topologically structured polygon data layer. The polygons are made up of the lines shown in the polygon-line list. The lines are made up of the coordinates shown in the line coordinate list (Source: after ESRI 1997)

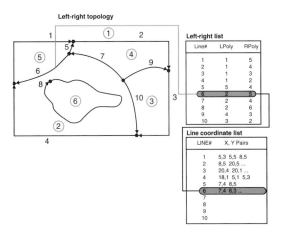

Figure 9.9 The contiguity of a topologically structured polygon data layer. For each line the left and right polygon is stored with the geometry data (Source: after ESRI 1997)

2. The list of coordinates for each line is also shown in Figure 9.8. For example, line 5 begins with coordinates 7,4 and 6,3 — other coordinates have been omitted for brevity. A line may appear in the polygon-line list more than once (for example, line 6 is used in the definition of both polygons 2 and 5), but the actual coordinates for each line are only stored once in the line-coordinate list. Storing common boundaries between adjacent polygons avoids the potential problems of gaps (slivers) or overlaps between adjacent polygons. It has the added bonus that there are fewer coordinates in a topologically structured polygon feature layer compared with a simple feature layer representation of the same entities. The downside, however, is that drawing a polygon requires that multiple lines must be retrieved from the database and the boundary assembled, a process that can be time consuming when repeated for each polygon in a large dataset.

Moving on from the one-dimensional case of lines to the two-dimensional case of polygons requires introduction of the concept of planar enforcement. In simple terms planar enforcement means that all the space on a map must be filled and that any point must fall in one polygon alone, that is, polygons must not overlap. Planar enforcement implies that the phenomenon is conceptualized as a field.

The contiguity (adjacency) relationship between polygons is also defined during the process of topologic structuring. This information is usually stored as a list of the polygons on the left- and right-hand side of each line, in the direction defined by the list of coordinates (Figure 9.9). In Figure 9.9 polygon 2 is on the left of line 6 and polygon 5 is on the right. Thus we can deduce

from a simple look-up operation that polygons 2 and 5 are adjacent.

Software systems based on the vector topologic feature data model, also called the geo-relational model, have become popular over the years. The term *geo-relational* derives from the way the topologic feature model is implemented. In this model, the geometry and associated topologic information is stored in regular computer files, whereas the associated attribute information is held in relational database management system (RDBMS) tables. The GIS software maintains the intimate linkage between the geometry, topology, and attribute information. This hybrid data management solution was developed to take advantage of RDBMS to store and manipulate attribute information. Geometry and topology were not placed in RDBMS because, until relatively recently, RDBMS were unable to store and retrieve geographic data efficiently. Figure 9.10 is an

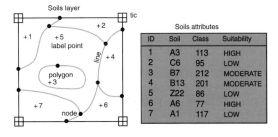

Figure 9.10 An example of a geo-relational polygon dataset. Each of the polygons is linked to a row in an RDBMS table. The table has multiple attributes, one in each column (Source: after ESRI 1997)

example of a geo-relational or coverage polygon dataset showing file-based geometry and topology information linked to attributes in an RDBMS table. The ID (identifier) of the polygon, the label point, is linked (related or joined) to the ID column in the attribute table (see also Chapter 11). Thus in this soils dataset polygon 3 is soil B7, of class 212, and its suitability is moderate.

The topologic feature geographic data model has been extensively used in GIS applications over the last 20 years, especially in government and natural resources applications based on polygon representations (see Sections 2.3.2 and 2.3.4). Typical government applications include: cadastral management, tax assessment, parcel management, zoning, planning, and building control. In the areas of natural resources and environment key applications include site suitability analysis, integrated land use modeling, license mapping, natural resource management, and conservation.

The tax appraisal case study discussed in Section 2.3.2.2 is an example of a GIS based on the topologic feature data model. Because the developers of this system wanted to avoid overlaps and gaps in tax parcels (polygons), to ensure that all parcel boundaries closed (were validated), and that they were stored in an efficient way, they chose this model. This is in spite of the fact that there is an overhead in creating and maintaining parcel topology, as well as a slight degradation in draw and query performance for large databases.

9.2.3.3 Network data model

The network data model is really a special type of topologic feature model. It is discussed here separately because it raises several new issues and has been widely applied in GIS studies.

Networks can be used to model the flow of goods and services. There are two primary types

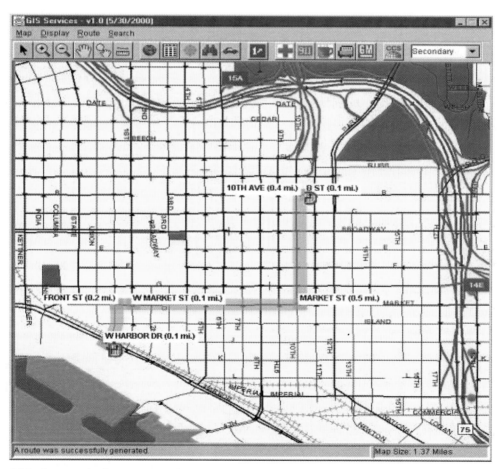

Figure 9.11   An example of a street network

of networks: radial and looped. In radial or tree networks flow always has an upstream and downstream direction. Stream and storm drainage systems are examples of radial networks. In looped networks self-intersections are common occurrences. Water distribution networks are looped by design to ensure that service interruptions affect the fewest customers.

In GIS software systems networks are modeled as points (for example, street intersections, fuses, switches, water valves, and the confluence of stream reaches: usually referred to as nodes in topologic models), and lines (for example, streets, transmission lines, pipes, and stream reaches). Network topologic relationships define how lines connect with each other at nodes. For the purpose of network analysis it is also useful to define rules about how flows can move through a network. For example, in a sewer network, flow is directional from a customer (source) to a treatment plant (sink), but in a pressurized gas network flow can be in any direction. The rate of flow is modeled as impedances on the nodes and lines. Figure 9.11 shows an example of a street network. The network comprises a collection of nodes (types of street intersection) and lines (types of street). The topologic relationship between the features is maintained in a connectivity table. By consulting information stored in the connectivity table it is possible to trace the flow of traffic through the network and to examine the impact of street closures. The impedance on the intersections and streets determines the speed at which traffic flows. Typically, the rate of flow is proportional to the street speed limit and number of lanes, and the timing of stoplights at intersections. Although this example relates to streets, the same basic principles also apply to, for example, electric, water, and railroad networks.

In geo-relational implementations of the topologic feature model, the geometry and topologic information is typically held in ordinary computer files and the attributes in a linked database. The GIS software tools are responsible for creating and maintaining the topologic information each time there is a change in the feature geometry.

There are many applications that utilize networks. Prominent examples include: calculating power load drops over an electricity network; routing emergency response vehicles over a street network; optimizing the route of mail deliveries over a street network; and tracing pollution upstream to a source over a stream network.

Network data models are also used to support another data model variant called *linear referencing* (Section 4.4). The basic principle of linear referencing is quite simple. Instead of recording the location of geographic entities as explicit *x*, *y*, *z* coordinates, they are stored as distances along a network (called a route system) from a point of origin. This is a very efficient way of storing information such as road pavement (surface) wear characteristics (e.g. the location of pot holes and degraded asphalt), geological seismic data (e.g. shockwave measurements at sensors along seismic lines), and pipeline corrosion data. An interesting aspect of this is that a two-dimensional network is reduced to a one-dimensional linear route list. The location of each entity (often called an event) is simply a distance along the route from the origin. Offsets are also often stored to indicate the distance from a network centerline. For example when recording the surface characteristics of a multi-carriageway road several readings may be taken for each carriageway at the same linear distance along the route. The offset value will allow the data to be related to the correct carriageway. Dynamic segmentation is a special type of linear referencing. The term derives from the fact that event data values are held separately from the actual network route in database tables (still as linear distances and offsets) and then dynamically added to the route (segmented) each time the user queries the database. This approach is especially useful in situations in which the event data change frequently and need to be stored in a database due to access from other applications (e.g. traffic volumes or rate of pipe corrosion). For further discussion of linear referencing see Section 4.4.

9.2.3.4 TIN data model

The geographic data models discussed so far have concentrated on one- and two-dimensional data. There are several ways to model three-dimensional data, such as terrain models, sales cost surfaces, and geologic strata. Strictly speaking, surfaces are said to be 2.5D structures. A true 3D structure will contain multiple *z* values at the same *x*, *y* location and thus is able to support volumetric calculations like cut and fill (a term derived from civil engineering applications that describes cutting earth from high areas and placing it in low areas to construct a flat surface, as is required in railroad construction). For further discussion of 2.5D see Box 3.5. Both grids and triangulated irregular networks (TINs) are used to create and represent surfaces in GIS. A regular grid surface is really a type of raster dataset as discussed earlier in Section 9.2.2. Each grid cell stores the height of the surface at a given location. The TIN structure, as the name suggests, represents a surface as contiguous non-overlapping triangular elements (Figure 9.12). A TIN is created from a set of mass points, that is, points with x, y, and z coordinate values. A key advantage of the TIN structure is that the density of sampled points, and therefore the size of triangles, can be adjusted to reflect the relief of the surface being modeled, with more points sampled in areas of variable relief (see Section 5.4). TIN surfaces are frequently created by performing what is called a *Delaunay triangulation* of the points to create a series of triangular areas (also called faces) that touch their neighbors at each edge.

A TIN is a topologic data structure that manages information about the nodes that comprise each triangle and the neighbors of each triangle. Figure 9.13 shows the topology of a simple TIN. As with other topologic data structures, information about a TIN may be conveniently stored in a file or database table.

TINs offer many advantages for surface analysis. First, they incorporate the original sample points, providing a useful check on the accuracy of the model. Second, the variable density of triangles means that a TIN is an efficient way of storing surface representations. Third, the data structure makes it easy to calculate elevation, slope, aspect, and line-of-sight between points. The combination of these factors has led to the widespread use of the TIN data structure in applications such as volumetric calculations for roadway design, drainage studies for land development, and visualization of urban forms. Figure 9.14 shows two example applications of TINs. Figure 9.14(A) is a shaded landslide risk TIN of Pisa district, Italy with building objects draped on top to give a sense of landscape. Figure 9.14(B) is a TIN of the Yangtse River, China greatly exaggerated in the z dimension. It shows how TINs draped with images can provide photo-realistic views of landscapes.

Like all 2.5D and 3D models TINs are only as good as the input sample data. They are especially susceptible to extreme high and low values because there is no smoothing of original data. Given that many DEMs are provided as grid data structures, the usage of TINs is not as high as it might be.

9.2.4 Object data model

All the geographic data models described so far are geometry-centric, that is they model the world as collections of points, lines, and polygons. Any operations to be performed on the geometry (and, in some cases, associated topology) are created as separate procedures (programs or scripts).

(A)

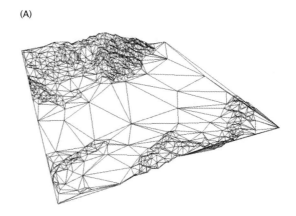

(B)

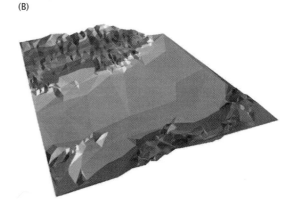

(C)

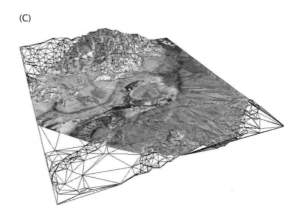

Figure 9.12 TIN surface of Death Valley, California: (A) "wireframe" showing all triangles; (B) shaded by elevation; (C) draped with satellite image

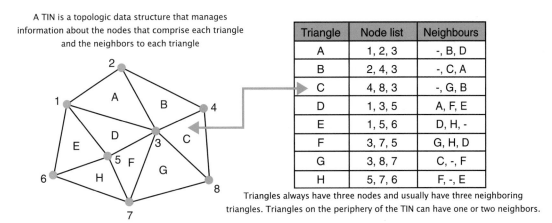

A TIN is a topologic data structure that manages information about the nodes that comprise each triangle and the neighbors to each triangle

Triangle	Node list	Neighbours
A	1, 2, 3	-, B, D
B	2, 4, 3	-, C, A
C	4, 8, 3	-, G, B
D	1, 3, 5	A, F, E
E	1, 5, 6	D, H, -
F	3, 7, 5	G, H, D
G	3, 8, 7	C, -, F
H	5, 7, 6	F, -, E

Triangles always have three nodes and usually have three neighboring triangles. Triangles on the periphery of the TIN can have one or two neighbors.

Figure 9.13 The topology of a TIN (Source: after Zeiler 1999)

Unfortunately, this approach can present several limitations for modeling geographic systems. All but the simplest of geographic systems contain many entities with large numbers of properties, complex relationships, and sophisticated behavior. Modeling such entities as simple point, line, and polygon geometry types is overly simplistic and does not easily support the sophisticated characteristics required for modern analysis. Additionally, separating the state of an entity (attributes or properties defining what it *is*) from the behavior of an entity (methods defining what it *does*) makes software and database development tedious, time-consuming, and error prone. To try to address these problems geographic object data models were developed. These allow the full richness of geographic systems to be modeled in an integrated way in a GIS.

The central focus of an object data model is the collection of geographic objects and the relationships between the objects (see Box 9.3). Each geographic object is an integrated package of geometry, properties, and methods. As such it is a software representation of the extent and function of the *geographic individuals* described in Chapter 5. In the object data model geometry is treated like any other attribute of the object and not as its primary characteristic. Geographic objects of the same type are grouped together as object classes, with individual objects in the class referred to as instances. In modern GIS software systems each object class is stored in the form of a database table, with each row an object and each property a column. The methods that apply are attached to the object instances when they are created in memory for use in the application.

An object is the basic atomic unit in an object data model and comprises all the properties that define the state of an object, together with the methods that define an object's behavior.

All geographic objects have some type of relationship to other objects in the same object class and, possibly, to objects in other object classes. Some of these relationships are inherent in the class definition (for example, a topologic polygon dataset will not have overlapping polygons) while other interclass relationships are user-definable. Three types of relationships are commonly used in geographic object data models: topologic, geographic, and general.

A class is a template for creating objects.

Topologic relationships are generally built into the class definition. For example, modeling real-world entities as a network class will cause network topology to be built for the nodes and lines participating in the network. Similarly, real-world entities modeled as topologic polygon classes will be structured using the node–line model described in Section 9.2.3.2.

Geographic relationships between object classes are based on geographic operators (such as overlap, adjacency, inside, and touching) that determine the interaction between objects. In a model of an agricultural system, for example, it might be useful to ensure that all farm buildings are within a farm boundary using a test for geographic containment.

Geographic relationships are useful to define other types of relationship between objects. In a

(A)

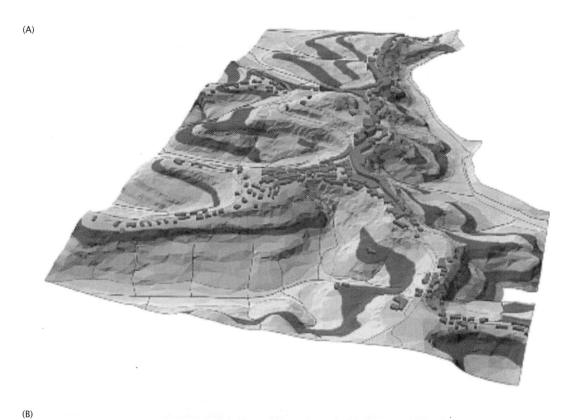

(B)

Figure 9.14 Examples of applications that use the TIN data model: (A) Landslide risk map for Pisa, Italy (Courtesy: Earth Science Department, University of Siena, Italy); (B) Yangtse River, China (Courtesy: Human Settlements Research Center, Tsinghua University, China)

Box 9.2   Object-oriented concepts in GIS

An object is a self-contained package of information describing the characteristics and capabilities of an entity under study. In a geographic object data model the real world is modeled as a collection of objects and the relationships between the objects. Each entity in the real world to be included in the GIS is an object. A collection of objects of the same type is called a *class*. In actual fact classes are a more central concept than objects from the implementation point of view. A class can be thought of as a template for objects. When creating an object data model the data model designer specifies classes and the relationships between classes. Only when the data model is used to create a database are objects (instances or examples of classes) actually created.

Examples of objects include oil wells, soil bodies, stream catchments, and aircraft flight paths. In the case of an oil well's class, each oil well object might include properties defining its state – annual production, owner name, date of construction, and type of geometry used for representation at a given scale (perhaps a point on a small scale map and a polygon on a large-scale one). The well class could have connectivity relationships with a pipeline class that represents the pipeline used to transfer oil to a refinery. There could also be a relationship defining the fact that each well must be located on a drilling platform. Finally, each well object might also have methods or behavior defining what it can do. Example behavior might include how to draw itself on a computer screen, how to create and delete itself, and editing rules about how the well snaps to pipelines.

There are three key facets of object data models that make them especially good for modeling geographic systems: encapsulation, inheritance, and polymorphism. *Encapsulation* describes the fact that each object packages together a description of its state and behavior. The state of an object can be thought of as its properties or attributes (e.g. for a forest object it could be the dominant tree type, average tree age, and soil pH). The behavior is the methods or operations that can be performed on an object (for a forest object these could be create, delete, draw, query, split, and merge). For example, when splitting a forest polygon into two parts, perhaps following a part sale, it is useful to get the GIS to automatically calculate the areas of the two new parts. Combining the state and behavior of an object together in a single package is a natural way to think of geographic entities, and a useful way to support the re-use of objects.

Inheritance is the ability to re-use some or all of the characteristics of one object in another object. For example, in a gas facility system a new type of gas valve could easily be created by overwriting or adding a few properties or methods to a similar existing type of valve. Inheritance provides an efficient way to create models of geographic systems by reusing objects and also a mechanism to extend models easily. New object classes can be built to reuse parts of one or more existing object classes and add some new unique properties and methods. The example described in Section 9.3 below shows how inheritance and other object characteristics can be used in practice.

Polymorphism describes the process whereby each object has its own specific implementation for operations like draw, create, and delete. One example of the benefit of polymorphism is that a geographic database can have a generic object creation component that issues requests to be processed in a specific way by each type of object class (e.g. gas pipes, valves, and service lines). This mechanism will work for a new object class because the new class is responsible for implementing the object creation method.

tax assessment system, for example, it is advantageous to define a relationship between land parcels (polygons) and ownership data that is stored in an associated DBMS table. Similarly, an electric distribution system relating light poles (points) to text strings (called annotation) allows depiction of pole height and material of construction on a map display. This type of information is very valuable for creating work orders (requests for change) that alter the facilities. Establishing relationships between objects in this way is useful because if one object

is moved then the other will move as well, or if one is deleted then the other is also deleted. This makes maintaining databases much easier and safer.

In addition to supporting relationships between objects (strictly speaking, between object classes), object data models also allow several types of rules to be defined. Rules are a valuable means of maintaining database integrity during editing tasks because they enforce validation constraints. The most popular types of rules used in object data models are attribute, connectivity, relationship, and geographic.

Attribute rules are used to define the possible attribute values that can be entered for any object. Both range and coded value attribute rules are widely employed. A range attribute rule defines the range of valid values that can be entered. Examples of range rules include: highway traffic speed must be in the range 25–70 miles (40–120 km) per hour; forest compartment average tree height must be in the range 0–50 meters. Coded attribute rules are used for categorical data types. Examples include: land use must be of type commercial, residential, park, or other; or pipe material must be of type steel, copper, lead, or concrete.

(A)

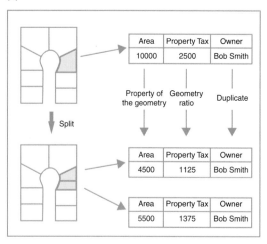

(B)

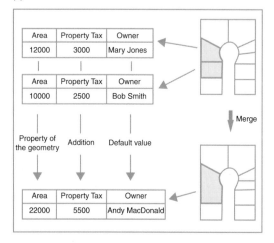

Figure 9.15 Example of split and merge rules for parcel objects: (A) split; (B) merge (Source: after MacDonald 1999)

Connectivity rules are based on the specification of valid combinations of features, derived from the geometry, topology, and attribute properties. For example, in an electric distribution system a 28.8 kV conductor can only connect to a 14.4 kV conductor via a transformer. Similarly, in a gas distribution system it should not be possible to add pipes with free ends (that is, with no fitting or cap).

Geographic rules define what happens to the properties of objects when an editor splits or merges them (Figure 9.15). In the case of a land parcel split following the sale of part of the parcel it is useful to define rules to determine the impact on properties like area, land use code, and owner. In this example, the original parcel area value should be divided in proportion to the size of the two new parcels, the land use code should be transferred to both parcels, and the owner name should remain for one parcel, but a new one should be added for the part that was sold off. In the case of a merge of two adjacent water pipes, decisions need to be made about what happens to attributes like material, length, and corrosion rate. In this example, the two pipe materials should be the same, the lengths should be summed, and the new corrosion rate determined by a weighted average of both pipes.

9.3 Example of a Water Facility Object Data Model

The goal of this section is to describe an example of a geographic object model. It will discuss how many of the concepts introduced earlier in this chapter are used in practice. The example selected is that of an urban water facility model. The types of issues raised in this example apply to all geographic object models, although of course the actual objects, object classes, and relationships under consideration will differ. The role of data modeling, as discussed in Section 9.1, is to represent the key aspects of the real world inside the digital computer for management, analysis, and display purposes.

Figure 9.16 is a diagram of part of a water distribution system, a type of pressurized network controlled by several devices. A pump is responsible for moving water through pipes (mains and laterals) connected together by fittings. Meters measure the rate of water consumption at houses. Valves and hydrants control the flow of water.

The purpose of the object model is to support asset management, mapping, and network analysis applications. Based on this it is useful to classify the objects into two types: the landbase

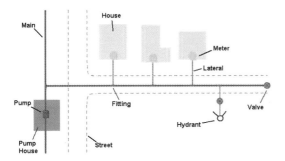

Figure 9.16 Water distribution system water facility object types and geographic relationships

and the water facilities. Landbase is a general term for objects like houses and streets that provide geographic context, but are not used in network analysis. The landbase object types are Pump-House, House, and Street. The water facilities object types are: Main, Lateral (a smaller type of WaterLine), Fitting (water line connectors), Meter, Valve, and Hydrant. All of these object types need to be modeled as a network in order to support network analysis operations like network isolation traces and flow prediction. A network isolation trace is used to find all parts of a network that are unconnected (isolated). Using the topologic connectivity of the network and information about whether pipes and fittings support water flow it is possible to determine connectivity. Flow prediction is used to estimate the flow of water through the network based on network connectivity and data about water availability and consumption. Figure 9.17 shows all the main object types and the implicit geographic relationships to be incorporated into the model. The arrows indicate the direction of flow in the network. When digitizing this network using a GIS editor it will be useful to specify topologic connectivity and attribute rules to control how objects can be connected (see Section 9.2.3.3 above). Before this network can be used for analysis it will also be necessary to

add flow impedances to each link (for example, pipe diameter).

Having identified the main object types, the next step is to decide how objects relate to each other and the most efficient way to implement them. Figure 9.18 shows one possible object model that uses the Unified Modeling Language (UML) to show objects and the relationships between them. Some additional color-coding has been added to help interpret the model. UML models are a type of tree structure where each box is an object class and the lines define how one class reuses (inherits) part of the class above it in the tree.

Object class names in an italic font are abstract classes; those with regular font names are used to create (instantiate) actual object instances. Abstract classes exist for efficiency reasons. It is sometimes useful to have a class that implements some capabilities once, so that several other classes can then be reused. For example, Main and Lateral are both types of *Line*, as is Street. Because Main and Lateral share several things in common – such as ConstructionMaterial, Diameter, and InstallDate properties, and connectivity and draw behavior – it is efficient to implement these in a separate abstract class, called *WaterLine*. The triangles indicate that one class is a type of another class. For example, PumpHouse and House are types of *Building*, and Street and *WaterLine* are types of *Line*. The diamonds indicate composition. For example, a network is composed of a collection of *Line* and *Node* objects. In the water facility object model, object classes without any geometry are colored pink. The Equipment and OperationsRecord object classes have their location determined by being associated with other objects (e.g. valves and mains). The Equipment and OperationsRecord classes are useful places to store properties common to many facilities, such as EquipmentID, InstallDate, ModelNumber, and SerialNumber.

Once this logical geographic object model has been created it can be used to generate a physical data model. One way to do this is to create the model using a computer-aided software engineering (CASE) tool. A CASE tool is a software application that has graphical tools to draw and specify a logical model (Figure 9.19). A further advantage of a CASE tool is that physical models can be generated directly from the logical models, including all the database tables and much of the supporting code for implementing behavior (Figure 9.20). Once a database structure (schema) has been created, it can be populated with objects and the intended applications put into operation.

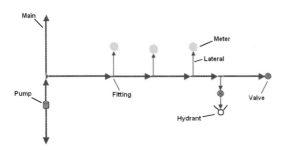

Figure 9.17 Water distribution system network

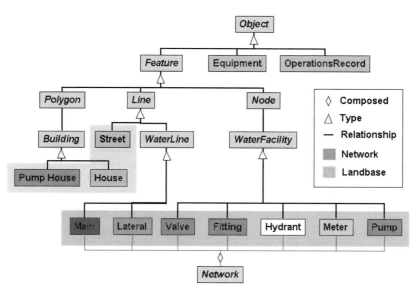

Figure 9.18 A water facility object model

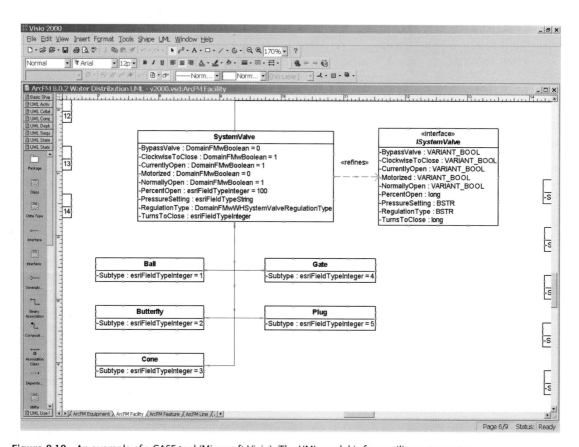

Figure 9.19 An example of a CASE tool (Microsoft Visio). The UML model is for a utility water system

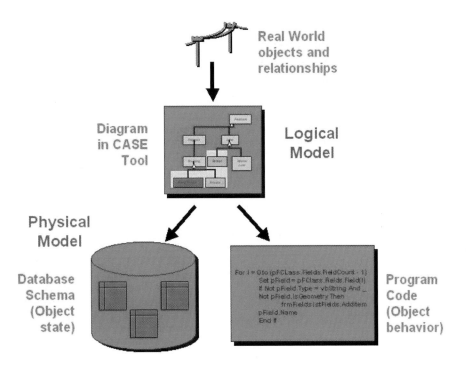

Figure 9.20   Use of a CASE tool in geographic data modeling (Source: ESRI)

9.4   Geographic Data Modeling in Practice

Geographic information science is only as good as the geographic database on which it is based and a geographic database is only as good as the geographic data model from which it is derived. Geographic data modeling begins with a clear definition of the project goals and progresses through an understanding of user requirements, a definition of the objects and relationships, formulation of a logical model, and then creation of a physical model. These steps are a prelude to database creation and, finally, database use.

No step in data modeling is more important than understanding the purpose of the data modeling exercise. This understanding can be gained by collecting user requirements from the main users. Initially, user requirements will be vague and ill defined, but over time they will become clearer. Project goals and user requirements should be precisely specified in a list or narrative.

Formulation of a logical model necessitates identification of the objects and relationships to be modeled. Both the attributes and behavior of objects are required for an object model. A useful graphic tool for creating logical data models is a CASE tool and a useful language for specifying

models is UML. It is not essential that all objects and relationships be identified at the first attempt, because logical models can be refined over time. The key objects and relationships for the water distribution system object model are shown in Figure 9.18 above.

Once an implementation-independent logical model has been created, this model can be turned into a system–dependent physical model. A physical model will result in an empty database schema — a collection of database tables and the relationships between them. Sometimes, for performance optimization reasons or because of changing requirements, it is necessary to alter the physical data model. Even at this relatively late stage in the process, flexibility is still necessary.

It is important to realize that there is no such thing as the *correct* geographic data model. Every problem can be represented with many possible data models. Each data model is designed with a specific purpose in mind and is sub-optimal for other purposes. A classic dilemma is whether to define a general-purpose data model that has wide applicability, but that can, potentially, be complex and inefficient, or to focus on a narrower highly optimized model. A small prototype can often help resolve some of these issues.

Geographic data modeling is both an art and a science. It requires a scientific understanding of the key geographic characteristics of real-world systems, including the state and behavior of objects, and the relationships between them. Geographic data models are of critical importance because they have a controlling influence over the type of data that can be represented and the operations that can be performed. As we have seen, object models are the best type of data model for representing rich object types and relationships in facility systems, whereas simple feature models are sufficient for very elementary applications such as a map of the body. In a similar vein raster models are good for field-based data such as soils, vegetation, pollution, and population counts. There is further discussion on geographic data modeling in the next chapter.

Questions for Further Study

1. Figure 9.21 is an oblique aerial photograph of part of the city of Kfar-Saba, Israel. Take 10 minutes to list all the object classes (including their attributes and behavior) and the relationships between the classes that you can see in this picture that would be appropriate for a city information system study.
2. Why are there so many GIS data models?
3. Why is it useful to include the conceptual, logical, and physical levels in geographic data modeling?
4. Describe five key differences between the topologic vector and raster geographic data models. It may be useful to consult Figure 9.3 and Chapter 3.
5. Review the terms encapsulation, inheritance, and polymorphism and explain with geographic examples why they make object data models superior for representing geographic systems.

Online Resources

NCGIA Core Curricula (www.ncgia.ucsb.edu/pubs/core.html

Core Curriculum in GIScience, Sections 2.3.1 (Information Organization and Data Structure, Albert Yeung), 2.3.2 (Non-Spatial Database Models, Thomas Meyer), 2.4.1 (Rasters, Michael Goodchild), 2.4.2 (TINs), 2.4.3 (Quadtrees and Scan Orders, Michael Goodchild), 2.6 (Representing Networks, Benjamin Zhan), and 2.9.1 (Transportation Networks, Val Noronha)

Core Curriculum in GIS, 1990, Units 4 (The Raster GIS), 11 (Spatial Objects and Database Models), 12 (Relationships among Spatial Objects), 13 (The Vector or Object GIS), 21 (The Raster/Vector Database Debate), 30 (Storage of Complex Objects), 31 (Efficient Storage of Lines – Chain Codes), 35 (Raster Storage), 36 (Hierarchical Data Structures), 37 (Quadtree Algorithms and Spatial Indexes), 38 (Digital Elevation Models), 39 (The TIN Model),

Figure 9.21 Oblique aerial view of Kfar-Saba, Israel (Courtesy: ESRI)

42 (Temporal and Three-Dimensional Representations), 43 (Database Concepts I), and 44 (Database Concepts II)

ESRI Virtual Campus course on ESRI software (campus.esri.com):

Introduction to ArcView GIS, by ESRI (especially the "Getting Data into ArcView" lesson of the Module "Basics of ArcView": see also Einführung in ArcView GIS by ESRI and Josef Strobl, and Introdução ao ArcView GIS by ESRI and Mirna Cortopassi Lobo)

Understanding Geographic Data by David DiBiase

General:

www.infogoal.com/dmc/dmcdmd.htm General data modeling resources

www.utexas.edu/cc/database/datamodeling/ Introduction to data modeling

www.fairview-industries.com/intro.htm Example of cadastral data modeling

www.rational.com/university/index.jsp Rational University – e-University from a commercial vendor

www.innovativegis.com/education/primer/concepts.html Introduction to GIS data models

Reference Links

Maguire D J, Goodchild M F, and Rhind D W (eds) 1991 *Geographical Information Systems: Principles and applications.* Harlow, UK: Longman (Text available online from 'Links to Big Book 1' at www.wiley.com/gis and www.wiley.co.uk/gis).

Chapter 16, High-level spatial data structures for GIS (Egenhofer M J, Herring J)

Chapter 19, Digital terrain modeling (Weibel R, Heller M)

Longley P A, Goodchild M F, Maguire D J, and Rhind D W (eds) 1999 *Geographical Information Systems: Principles, techniques, management and applications.* New York: John Wiley.

Chapter 25, GIS customization (Maguire, D J)

Chapter 26, Relational databases and beyond (Warboys M F)

Chapter 29, Principles of spatial database design and analysis (Bédard, Y)

Chapter 36: Spatial tessellations (Boots, B)

References

ESRI 1997 *Understanding GIS: the ArcInfo Method.* Redlands, California: ESRI Press.

ESRI 2000 *Inside ArcFM Water.* Redlands, California: ESRI Press.

Heywood I, Cornelius S, and Carver S 1998 *An Introduction to Geographical Information Systems.* Harlow: Longman.

MacDonald A 1999 *Building a Geodatabase.* Redlands, California: ESRI Press.

Worboys M F 1995 *GIS: A computing perspective.* London: Taylor and Francis.

Zeiler M 1999 *Modeling Our World: The ESRI guide to geodatabase design.* Redlands, California: ESRI Press.

Speed

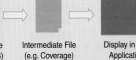

Price

Translation

Intermediate File
(e.g. Coverage)

Display in GIS
Application

Direct Read

Display in GIS
Application

GIS DATA COLLECTION

10

Data collection is one of the most time-consuming and expensive, yet important of GIS tasks. There are many diverse sources of geographic data and many methods available to enter them into a GIS. The two main methods of data collection are data capture and data transfer. It is useful to distinguish between primary (direct measurement) and secondary (indirect derivation from other sources) data capture, for both raster and vector data types. Data transfer involves importing digital data from other sources. There are many practical issues associated with planning and executing an effective GIS data collection plan. This chapter reviews the main methods of GIS data capture and transfer, and introduces key practical management issues.

Learning objectives

After reading this chapter you will be able to:

● Describe data collection workflows;

● Understand the primary data capture techniques in remote sensing and surveying, and understanding the role of GPS:

● Understand the secondary data capture techniques of scanning, manual digitizing, vectorization, photogrammetry and COGO

● Understand the important subjects of data transfer
  – Sources of digital geographic data
  – Geographic data formats;

● Analyse practical issues associated with managing data capture projects.

10.1 Introduction

GIS can contain a wide variety of geographic data types originating from many diverse sources. From the perspective of creating geographic databases, it is convenient to classify raster and vector geographic data as primary and secondary (Table 10.1). Primary data sources are those collected specifically for use in GIS. Typical examples of primary GIS sources include raster SPOT and IKONOS Earth satellite images, and vector building survey measurements captured using a total survey station. Secondary sources are those that were originally captured for another purpose and need to be converted into a form suitable for use in a GIS project. Typical secondary sources include raster scanned color aerial photographs of urban areas, and USGS and IGN paper maps that can be scanned and vectorized.

> Primary geographic data sources are captured specifically for use in GIS by direct measurement. Secondary sources are those reused from earlier studies.

Geographic data may be obtained in either digital or analog format. Analog data must always be digitized before being added to a geographic database. Depending on the format and characteristics of the digital data, considerable reformatting and restructuring may be required prior to import.

This chapter describes the data sources, techniques, and workflows involved in GIS data collection. The processes of data collection are also variously referred to as data capture, data automation, data conversion, data transfer, data translation, and digitizing. Although there are subtle differences between these terms, they essentially describe the same thing, that is, adding geographic data to a database. Data capture refers to direct entry; data transfer is the importing of existing digital data across the Internet, WANs, or LANs; or using CD ROMs, zip disks, or diskettes. This chapter focuses on the techniques of data collection. Of equal, perhaps more, importance to a real-world GIS

implementation are project management, cost, legal, and organization issues. These are covered briefly in Section 10.6 of this chapter as a prelude to more detailed treatment in Chapters 16 through 19.

> Data capture costs can account for up to 85% of the cost of a GIS.

In the early days of GIS, when geographic data were very scarce, data collection was the main project task and it typically consumed the majority of the available resources. Data collection still remains a time consuming, tedious, and expensive process. Usually it accounts for 15–50% of the total cost of a GIS project (Table 10.2). Data capture costs can in fact be much more significant because in many organizations (especially those that are public) staff costs are often assumed to be fixed and are not used in budget accounting. Furthermore, as the majority of data capture effort and cost tends to fall at the start of projects, they receive greater scrutiny from senior managers. If staff costs are excluded from a GIS budget then in cash expenditure terms data collection can be as much as 60–85% of costs.

After an organization has completed basic data collection, their emphasis moves on to data maintenance. Over the multi-year lifetime of a GIS project, data maintenance often turns out to be a far more complex and expensive activity than initial data collection. This is because of the high volume of update transactions in many systems (for example, changes in land parcel ownership, maintenance work orders on a highway transport network, or logging military operational activities) and the need to manage multi-user access to operational databases. For more information about data maintenance, see Chapter 11.

10.1.1 Data collection workflow

Data collection projects involve a series of sequential stages (Figure 10.1). The workflow commences with planning, followed by preparation, digitizing (here taken to mean a range of techniques such as table digitizing,

Table 10.1 Classification of geographic data for data collection purposes

	Raster	Vector
Primary	• Digital remote-sensing images • Digital aerial photographs	• GPS measurements • Survey measurements
Secondary	• Scanned maps or photographs • Digital elevation models from maps	• Topographic maps • Toponymy (placename) databases

Table 10.2 Breakdown of costs (in $1000s) for two typical client–server GIS implementations. Hardware costs include desktop clients and servers only (i.e. not network infrastructure). Data costs assume the purchase of a landbase and digitizing assets such as pipes and fittings (water utility), conductors and devices (electrical utility), or land and property parcels (local government). Staff costs assume that all core GIS staff will be full-time, but that users will be part-time

	10 seats		100 seats	
	$	%	$	%
Hardware	30	3.4	250	8.6
Software	25	2.8	200	6.9
Data	400	44.7	450	15.5
Staff	440	49.1	2000	69.0
Total	895	100	2900	100

survey entry, scanning, and photogrammetry) or transfer, editing and improvement and, finally, evaluation. Planning is obviously important to any project and data collection is no exception. It includes establishing user requirements, garnering resources (staff, hardware, and software) and developing a project plan. Preparation is especially important in data collection projects. It involves many tasks such as obtaining data, redrafting poor-quality map sources, editing scanned map images, and removing noise (unwanted data such as speckles on a scanned map image). Digitizing and transfer are the stages where the majority of the effort will be expended. It is naïve to think that data collection is really just digitizing, when in fact it involves very much more. Editing and

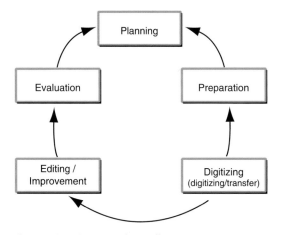

Figure 10.1 Stages in data collection projects

improvement follows digitizing/transfer. This covers many techniques designed to validate data, as well as correcting errors and improving quality (Section 6.3.3 and Figure 6.8). Evaluation, as the name suggests, is the process of identifying project successes and failures. Since all large data projects involve multiple stages, this workflow is iterative with earlier phases (especially a first, pilot, phase) helping to improve subsequent parts of the overall project.

10.2 Primary Geographic Data Capture

Primary geographic capture involves the direct measurement of objects. Both raster and vector data capture methods are available.

10.2.1 Raster data capture

Much the most popular form of primary raster data capture is remote sensing. Broadly speaking, remote sensing is a technique used to derive information about the physical, chemical, and biological properties of objects without direct physical contact. Information is derived from measurements of the amount of electromagnetic radiation reflected, emitted, or scattered from objects. A variety of sensors, operating throughout the electromagnetic spectrum from visible to microwave wavelengths, are commonly employed to obtain measurements (see Section 3.6.1 and Lillesand and Kiefer 1999). Passive sensors are reliant on reflected solar radiation or emitted terrestrial radiation; active sensors (such as synthetic aperture radar) generate their own source of electromagnetic radiation. The platforms on which these instruments are mounted are similarly diverse. Although Earth-orbiting satellites and fixed-wing aircraft are by far the most common, helicopters, balloons, masts, and booms are also employed (Figure 10.2). As used here, the term remote sensing subsumes the fields of satellite remote sensing and aerial photography.

Remote sensing is the measurement of physical, chemical, and biological properties of objects without direct contact.

From the GIS perspective, resolution is the key physical characteristic of remote sensing systems. There are three aspects to resolution: spatial, spectral, and temporal. All sensors need to trade off spatial, spectral, and temporal properties because of storage, processing, and bandwidth considerations. For further discussion

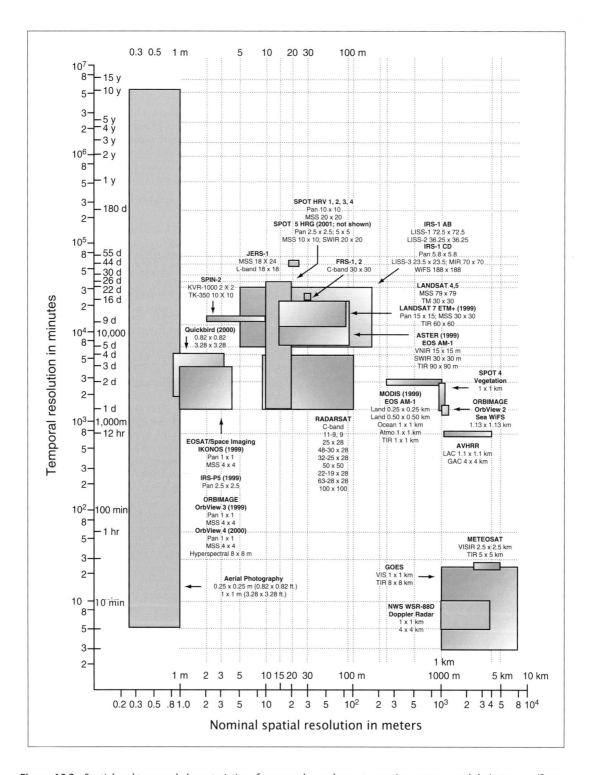

Figure 10.2 Spatial and temporal characteristics of commonly used remote sensing systems and their sensors (Source: after Jensen and Cowen 1999)

of the important topic of resolution see also Sections 3.4, 3.6.1, 4.1, 6.4.2 and 7.1.

Three key aspects of resolution are: spatial, spectral, and temporal.

Spatial resolution refers to the size of object that can be resolved and the most usual measure is the pixel size. Satellite remote sensing systems typically provide data with pixel sizes in the range 1 m – 1 km. The cameras used for capturing aerial photographs usually range from 0.1 m – 5 m. Image (scene) sizes vary quite widely between sensors – typical ranges include 1000 by 1000 to 3000 by 3000 pixels. The total coverage of remote sensing images is usually in the range 10 by 10 – 200 by 200 km.

Spectral resolution refers to the parts of the electromagnetic spectrum that are measured. Since different objects emit and reflect different types and amounts of radiation, selecting which part of the electromagnetic spectrum to measure is critical for each application area. Figure 10.3 shows the spectral signatures of water, green vegetation, and dry soil. Remote sensing systems may capture data in one part of the spectrum (referred to as a single band) or simultaneously from several parts (multi-band or multi-spectral). See also Figure 3.8. The radiation values are usually normalized and resampled to give a range of integers from 0–255 for each band, for each pixel, in each image.

Temporal resolution, or repeat cycle, describes the frequency with which images are collected for the same area. There are essentially two types of commercial remote sensing satellite: Earth orbiting and geostationary. Earth orbiting satellites collect information about different parts of the Earth surface at regular intervals. To maximize utility, orbits are typically polar, at a fixed altitude and speed, and are Sun synchronous. The French SPOT (Système Probatoire d'Observation de la Terre) satellite, for example, passes virtually over the poles at an altitude of 832 km sensing the same location on the Earth surface during daylight every 26 days. The SPOT platform carries two sensors: a single band panchromatic sensor measuring radiation in the visible part of the electromagnetic spectrum at a resolution of 10 by 10 m: and a multi-spectral sensor measuring green, red, and reflected infrared radiation at a resolution of 20 by 20 m. Each SPOT scene covers an area of 60 by 60 km.

Much of the discussion so far has focused on commercial satellite remote sensing systems. Of equal importance, especially in medium to large-scale GIS projects, is aerial photography. Although the data products resulting from remote sensing satellites and aerial photography systems are technically very similar (i.e. they are both images) there are some significant differences in the way data are captured and can be interpreted. The most notable difference is that aerial photographs are normally collected using analog optical cameras (although digital cameras are becoming more widely used) and then later rasterized, usually by scanning a film negative. The quality of the optics of the camera and the mechanics of the scanning process both affect the spatial and spectral characteristics of the resulting images. Most aerial photographs are collected on an *ad hoc* basis using cameras mounted in airplanes flying at low altitudes (3000–10 000 m) and are either panchromatic

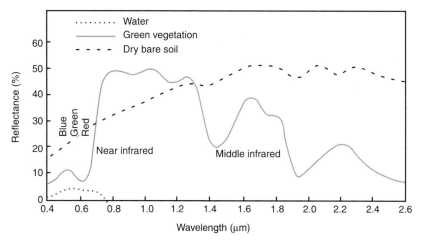

Figure 10.3 Typical reflectance signatures for water, green vegetation, and dry soil (Source: after Jones 1997)

(black and white) or color, although multi-spectral cameras/sensors operating in the non-visible parts of the electromagnetic spectrum are also used. Aerial photographs are very suitable for detailed surveying and mapping projects.

An important feature of satellite and aerial photography systems is that they can provide stereo imagery from overlapping pairs of images. These images are used to create a 3D analog or digital model from which 3D coordinates, contours, and digital elevation models can be created (see Section 10.3.2.3 below).

Satellite and aerial photograph data offer a number of advantages for GIS projects. The consistency of the data and the availability of systematic global coverage make satellite data especially useful for large area projects (for example, mapping landforms and geology at the river catchment area level) and for mapping inaccessible areas. The regular repeat cycles of commercial systems and the fact that they record radiation in many parts of the spectrum makes such data especially suitable for assessing the condition of vegetation (for example, the moisture stress of wheat crops). Aerial photographs in particular are very useful for detailed surveying and mapping of, for example, urban areas and archaeological sites, especially those applications requiring 3D data.

On the other hand, the spatial resolution of commercial satellites is too coarse for many large area projects and the data collection capability of many sensors is restricted by cloud cover. The data volumes from both satellites and aerial cameras can be very large and create storage and processing problems for all but the most modern systems. The cost of data can also be prohibitive for a single project or organization.

10.2.2 Vector data capture

Primary vector data capture is a major source of geographic data. The two main branches of vector data capture are ground surveying and GPS.

10.2.2.1 Surveying

Ground surveying is based on the principle that the 3D location of any point can be determined by measuring angles and distances from other known points. Surveys begin from a benchmark point. If the coordinate system of this point is known, all subsequent points can be collected in this coordinate system. If it is unknown then the survey will use a local or relative coordinate system (see Section 4.7).

Ground surveying uses measurements to determine the locations of objects.

Since all survey points are obtained from survey measurements their locations are always relative to other points. Any measurement errors need to be apportioned between multiple points in a survey. For example, when surveying a field boundary, if the last and first points are not identical in survey terms (within the tolerance employed in the survey) then errors need to be apportioned between all points that define the boundary (see Section 6.3.4). As new measurements are obtained these may change the locations of points. For this reason it is necessary to store both the measurements and the points inside a GIS database until the survey is complete.

Traditionally, surveyors used equipment like transits and theodolites to measure angles, and tapes and chains to measure distances. Today these have been replaced by electro-optical devices called total stations that can measure both angles and distances to an accuracy of 1 mm (Figure 10.4). Total stations automatically log data and the most sophisticated can create vector point, line, and polygon objects in the field, thus providing direct validation.

Figure 10.4 A tripod-mounted Leica TPC1100 Total Station (Courtesy: Leica)

The basic principles of surveying have changed very little in the past 100 years. Two people are usually required to perform a survey, one to operate the total station and the other to hold a reflective prism that is placed at the object being measured. On some remote-controlled systems a single person can control both the total station and the prism.

Ground survey is a very time-consuming and expensive activity, but it is still the best way to obtain highly accurate point locational data. Surveying is typically used for capturing buildings, land and property boundaries, manholes and other objects that need to be located accurately. It is also used to obtain reference marks for other data capture methods. For example, large-scale aerial photographs and satellite images are frequently georeferenced using points obtained from ground survey.

10.2.2.2 GPS

The Global Position System (GPS) is a collection of 27 NAVSTAR satellites orbiting the Earth at a height of 12,500 miles, five monitoring stations, and individual receivers (Steede-Terry 2000). The GPS was originally funded by the US Department of Defense, and for many years military users had access to only the most accurately data.

Fortunately this *selective availability* was removed in May 2000 so that now civilian and military users can fix the x, y, z location of objects relatively easily to an accuracy of better than 10 m with standard equipment.

> The GPS is a network of satellites, monitoring stations, and inexpensive receivers used for primary GIS data capture.

In many respects GPS has revolutionized primary data capture, especially since the development of Differential GPS (Box 10.1), the removal of selective availability, and the creation of low-cost, low-power receivers. Today units costing less than $100 can easily provide locational data at better than 10 m accuracy. One of the drawbacks of GPS, however, is that it is necessary to have three or more satellites in unobstructed view in order to collect measurements. This can especially be a problem in forests and urban areas with tall buildings. GPS is very useful for recording ground control points for other data capture projects, for locating objects that move (for example, combine harvesters, tanks, cars, and shipping containers), and for direct capture of the locations of many types of objects such as utility assets, buildings, geological deposits, and stream sample points.

Box 10.1   *Principles of GPS*

GPS works according to a simple principle – the length of time it takes a signal to travel from a satellite to a receiver on the ground (Figure 10.5). The GPS satellites constantly transmit a coded radio signal that indicates their exact position in space and time. The receiver measures how long it takes the signal to travel from the satellites. By measuring the distance from three or more satellites, the location of the receiver can be obtained by triangulation. If a signal can be obtained from a fourth satellite, then the elevation of the receiver can also be determined.

Although standard GPS receivers (Figure 10.6) can provide locations at accuracies of 5–10 m, it is important to understand that there are several possible sources of error inherent in these locations. Some of the errors are random in nature, while others are systematic and can therefore be corrected. Errors arise from signal degradation due to atmospheric effects, minor variations in the location of the satellites, inaccuracies in the timing clocks, errors in receivers, and variations in the reflection of signals from local objects.

A number of techniques are available to improve the accuracy of GPS measurements. Many GPS receivers perform averaging of measurements to improve apparent accuracy. Others snap measurements to map features. So, for example, in-car navigation systems snap the location of the vehicle to a road centerline.

The accuracy of measurements can also be improved by using Differential GPS. This technique uses two receivers. One is fixed and the other is used to collect measurements. If the location of the fixed (base) receiver is known accurately, comparing the exact location with the location reported by GPS will provide an estimate of error. This error can be used to correct measurements obtained from the roving receiver provided that it is within about 300 km. In some countries, the differential correction information is broadcast freely over airwaves and can be received using a standard radio receiver. Differential GPS can improve accuracy to allow locations to be determined to better than 1 m.

Box 10.1
Continued

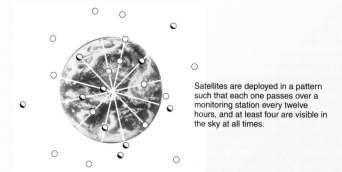

Satellites are deployed in a pattern such that each one passes over a monitoring station every twelve hours, and at least four are visible in the sky at all times.

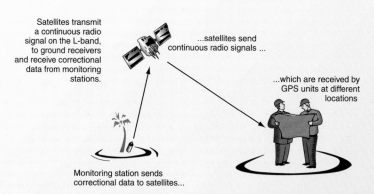

Satellites transmit a continuous radio signal on the L-band, to ground receivers and receive correctional data from monitoring stations.

...satellites send continuous radio signals ...

...which are received by GPS units at different locations

Monitoring station sends correctional data to satellites...

Figure 10.5 The components involved in GPS (Source: after Steede-Terry 2000)

(A)

(B)

Figure 10.6 GPS data collection: (A) using GPS in the field – a hand-held PC is used to log data; (B) Leica GPS GS50 receiver (Courtesy: Leica)

Strictly speaking, the term GPS refers only to the US Department of Defense System. GLONASS is the Russian version of GPS offering similar coverage and accuracy; Galileo is the European Union's proposed equivalent.

10.3 Secondary Geographic Data Capture

Geographic data capture from secondary sources is the process of creating raster and vector files and databases from maps and other hardcopy documents. Scanning is used to capture raster data. Table digitizing, heads-up digitizing, stereo-photogrammetry, and COGO data entry are used for vector data.

10.3.1 Raster data capture using scanners

A scanner is a device that converts hardcopy analog media into digital images by scanning successive lines across a map or document and recording the amount of light reflected from a local data source (Figure 10.7). The differences in reflected light are normally scaled into bi-level black and white (1 bit per pixel), or multiple gray levels (8, 16, or 32 bits). Color scanners output

Figure 10.7 A large-format roll-feed image scanner (Courtesy: GTCO Calcomp)

data into 8-bit red, green, and blue color bands. The spatial resolution of scanners varies widely from as little as 100 dpi (4 dots per mm) to 1800 dpi (72 dots per mm) and beyond. Most GIS scanning is in the range 400–1000 dpi (16–40 dots per mm). Depending on the type of scanner and the resolution required, it can take from 30 seconds to 30 minutes or more to scan a map. Litton (1998) provides an excellent introduction to scanning for GIS.

> Scanned maps and documents are used extensively in GIS as background maps and data stores.

There are three reasons to scan hardcopy media for use in GIS:

● Documents, such as building plans, CAD drawings, property deeds, and equipment photographs are scanned to reduced wear and tear, improve access, provide integrated database storage, and to index them geographically (e.g. building plans can be attached to building objects in geographic space).

● Film and paper maps, aerial photographs, and images are scanned and georeferenced so that they provide geographic context for other data (typically vector layers). This type of unintelligent image or background *geographic wallpaper* is very popular in systems that manage equipment and land and property assets (Figure 10.8).

● Maps, aerial photographs, and images are also scanned prior to vectorization (see below).

An 8 bit (256 gray level) 400 dpi (16 dots per mm) scanner is a good choice for scanning maps for use as a background GIS reference layer. For a color aerial photograph that is to be used for subsequent photo-interpretation and analysis, a color (8 bit for each of three bands) 1000 dpi (40 dots per mm) scanner is more appropriate. The quality of data output from a scanner is determined by the nature of the original source material, the quality of the scanning device, and the type of preparation prior to scanning (e.g. redrafting key features or removing unwanted marks will improve output quality).

10.3.2 Vector data capture

Secondary vector data capture involves digitizing vector objects from maps and other geographic data sources. The most popular methods are manual digitizing, heads-up digitizing and vectorization, photogrammetry, and COGO data entry.

Figure 10.8 An example of raster background data (black and white aerial photography) underneath vector data (land parcels)

(A)

(B)

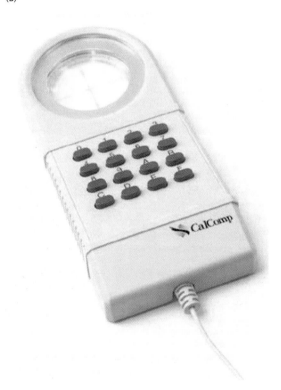

Figure 10.9 Digitizing equipment: (A) Digitizing table (B) cursor (both courtesy: GTCO Calcomp)

10.3.2.1 Manual digitizing

Manually operated digitizers are much the simplest, cheapest, and most commonly used means of capturing vector objects from hardcopy maps. Digitizers come in several designs, sizes, and shapes. They operate on the principle that it is possible to detect the location of a cursor or puck passed over a table inlaid with a fine mesh of wires. Accuracies typically range from 0.003 inch (0.075 mm) to 0.010 inch (0.25 mm). Small digitizing tablets up to 12 by 24 inches (30 by 60 cm) are used for small tasks, but bigger (typically 50 by 32 inches (120 by 80 cm)) freestanding table digitizers are preferred for larger tasks (Figure 10.9). Both types of digitizer usually have cursors with cross hairs mounted in glass and buttons to control capture.

> Manual digitizing is still the simplest, easiest, and cheapest method of capturing vector data from existing maps.

Vertices defining point, line, and polygon objects are captured using manual or stream digitizing methods. Manual digitizing involves placing the center point of the cursor cross hairs at the location for each object vertex and then clicking a button on the cursor to record the location of the vertex. Stream mode digitizing partially automates this process by instructing the digitizer control

(A)

(B)

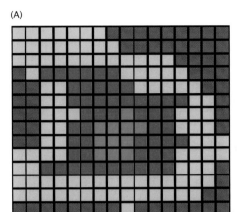
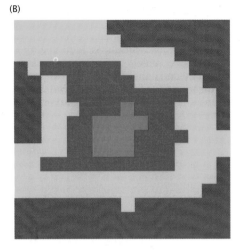

Figure 10.10 Batch vectorization of a scanned map: (A) original raster file; (B) vectorized polygons. Adjacent raster cells with the same attribute values are aggregated. Class boundaries are then created at the intersection between adjacent classes in the form of vector lines

software automatically to collect vertices every time a distance or time threshold is crossed (e.g. every 0.02 inch (0.5 mm) or 0.25 second). Stream-mode digitizing is a much faster method, but it typically produces larger files with many redundant coordinates.

10.3.2.2 Heads-up digitizing and vectorization

One of the main reasons for scanning maps (see 10.2.1 above) is as a prelude to vectorization. The simplest way to create vectors from raster layers is to digitize vector objects manually straight off a computer screen using a mouse or digitizing cursor. This method is called heads-up digitizing because the map is vertical and can be viewed without bending the head down. It is widely used for selective capture of, for example, land parcels, buildings, and utility assets.

> Vectorization is the process of converting raster data into vector data. The reverse is called rasterization.

A faster and more consistent approach is to use software to perform automated vectorization in either batch or interactive mode. Batch vectorization takes an entire raster file and converts it to vector objects in a single operation. Vector objects are created using software algorithms that build simple (spaghetti) line strings from the original pixel values (Figure 10.10). Depending on the size of the raster file, it

typically takes 1–100 minutes to complete vectorization.

Unfortunately, batch vectorization software is far from perfect and post-vectorization editing is required to clean up errors. To avoid large amounts of vector editing, prior to vectorization it is useful to undertake a little raster editing of the original raster file to remove unwanted noise that may affect the vectorization process. For example, text that overlaps lines should be deleted and dashed lines are best converted to solid lines. Following vectorization, topological relationships are usually created for the vector objects. This process may also highlight some previously unnoticed errors that require additional editing.

Batch vectorization is best suited to simple bi-level maps of, for example, contours, streams, and highways. For more complicated maps and where selective vectorization is required (for example, digitizing electric conductors and devices or water mains and fittings off topographic maps), interactive vectorization (also called semi-automatic vectorization, line following, or tracing) is preferred. In interactive vectorization, software is used to automate digitizing. The operator snaps the cursor to a pixel, indicates a direction for line following, and the software then automatically digitizes lines. Typically, many parameters can be tuned to control the density of points (level of generalization), the size of gaps (blank pixels in a line) that will be jumped, and whether to pause at junctions for operator intervention or always to trace in a specific direction (most systems require that all polygons are ordered either clockwise or counterclockwise). Although quite labor intensive, interactive vectorization generally results in much greater productivity than manual or heads-up digitizing. It also produces high-quality data, as software is able to represent lines more accurately and consistently than can humans. It is for these reasons that specialized data capture groups much prefer vectorization to manual digitizing.

10.3.2.3 Photogrammetry

Photogrammetry is the science and technology of making measurements from pictures, aerial photographs, and images (see Box 10.3 for an unusual example). Although in the strict sense it includes 2D measurements taken from single aerial photographs, today in GIS it is almost exclusively concerned with capturing 2.5D and 3D measurements from models derived from stereo-pairs of photographs and images. In the case of aerial photographs, it is usual to have 60% overlap along each flight line and 30% overlap between flight lines. Similar layouts are used by remote sensing satellites. The amount of overlap defines the area for which a 3D model can be created.

Photogrammetry is used to capture measurements from photographs and other image sources.

To obtain true georeferenced coordinates from a model it is necessary to georeference photographs using control points (the procedure is essentially analogous to that described for manual digitizing in Box 10.2 below). Control points can be defined by ground survey or nowadays more usually with GPS (see Sections 10.2.2.1 and 10.2.2.2 above for discussion of these techniques).

Measurements are captured from overlapping pairs of photographs using stereoplotters. These build a model and allow 3D measurements to be captured, edited, stored, and plotted. Stereoplotters have undergone three major generations of development: analog (optical), analytic, and digital. Mechanical analog devices are seldom used today, whereas analytical (combined mechanical and digital) and digital (entirely computer-based) are much more common. It is likely that digital (softcopy) photogrammetry will eventually replace mechanical devices entirely.

The options for extracting vector objects from 3D models are directly analogous to those available for manual digitizing as described above: namely batch, interactive, and manual (Sections 10.3.2.1 and 10.3.2.2). The obvious difference, however, is that there is a requirement for capturing z (elevation) values. In the case of manual and interactive methods, this requires a 3D cursor.

Photogrammetric techniques are particularly suitable for highly accurate capture of contours,

Box 10.2 *Manual digitizing*

Manual digitizing involves five basic steps.

1. The map document is attached to the center of the digitizing table using sticky tape.
2. Because a digitizing table uses a local rectilinear coordinate system, the map and the digitizer must be registered so that vector data can be captured in real-world coordinates. This is achieved by digitizing a series of four or more well-distributed control points (also called reference points or tic marks) and then entering their real-world values. The digitizer control software (usually the GIS) will calculate a transformation matrix and then automatically apply this to any future coordinates that are captured.
3. Before proceeding with data capture it is useful to spend some time examining a map to determine rules about which features are to be captured at what level of generalization. This type of information is often defined in a data capture project specification.
4. Data capture involves recording the shape of vector objects using manual or stream mode digitizing as described in Section 10.3.2.1. A common rule for vector GIS is to press button 2 on the digitizing cursor to start a line, button 1 for each intermediate vertex, and button 2 to finish a line. There are other similar rules to control how points and polygons are captured.
5. Finally, after all objects have been captured it is necessary to check for any errors. Easy ways to do this include using software to identify geometric errors (such as polygons that do not close or lines that do not intersect – see Figure 6.9), and producing a test plot that can be overlaid on the original document.

Box 10.3 *Preserving medieval art with GIS in Bucharest, Romania*

GIS is widely used in conservation, but a slightly unusual application is the conservation of medieval art in Bucharest, Romania. In an effort to expedite conservation, as well as to reduce costs and improve the quality and consistency of restorations, the University of Fine Arts in Bucharest has been using digital survey data collection methods and spatial referencing technologies to create a database for mural conservation.

To restore a mural, conservators have to know in detail how original wall paintings were created and the nature of past interventions. In addition, they must determine, as accurately as possible, the current condition of the mural – including all types of deterioration, such as exfoliated regions, salted areas, and humidity damage – as well as establish a correlation with other information, such as chemical analyses, humidity measurements, and operational tests. All these data must be spatially referenced to the mural to ensure not only that the best conservation approach is being implemented, but also that current interventions will facilitate any need for future restoration.

A variety of data collection techniques was used in Bucharest for this purpose including photography, surveying, and attribute recording. A GIS was developed by the University to integrate and manage all this information (Figure 10.11), including geometric correction and superimposition of information from cameras and other sensors. Desktop GIS software was used to document critical areas of murals, such as deteriorated regions, fissure lines, and injection points, and to create composite reference layers for each unique deterioration phenomenon. Laboratory analyses and microclimate measurements were also spatially referenced to the mural maps.

Although at first sight this is an unusual use of GIS, because of its scale and antiquity, it is actually a good example of the versatility and wide applicability of GIS. From a small mural to the whole universe, GIS can be applied wherever space can be used as a framework for investigation and analysis. Critical to the success of this and other GIS projects is the accurate collection of spatial and attribute information about the objects of interest.

Figure 10.11 An example of part of a mural stored in a medieval art GIS in Bucharest, Romania (Courtesy: ESRI)

digital elevation models, and almost any type of object that can be identified on an aerial photograph or image. One type of popular specialist photogrammetry product is the orthophotograph. Orthophotographs result from using a DEM to correct distortions in an aerial photograph derived from varying land elevation. They have become popular because of their relatively low cost of creation (when compared with topographic maps) and ease of interpretation as base maps. They can also be used as accurate data sources for heads-up digitizing (see Section 10.3.2.2 above).

In summary, photogrammetry is a very cost-effective data capture technique that is sometimes the only practical method of obtaining detailed topographic data about an area of interest. Unfortunately, the complexity and high cost of equipment have restricted its use to large-scale primary data capture projects and specialist data capture organizations.

10.3.2.4 COGO data entry

COGO, a contraction of the term *coordinate geometry*, is a methodology for capturing and representing geographic data. COGO uses survey-style bearings and distances to define each part of an object in much the same way as described in Section 10.2.2.1 above. The COGO system is widely used in North America to represent land records and property parcels (also called lots). Coordinates can be obtained from COGO measurements by geometric transformation (i.e. bearings and distances are converted into x, y coordinates). Although COGO data obtained as part of a primary data capture activity are used in some projects, it is more often the case that secondary measurements are captured from hard copy maps and documents. Source data may be in the form of legal descriptions, records of survey, tract (housing estate) maps, or similar documents.

COGO stands for coordinate geometry. It is a vector data structure and method of data entry.

COGO data are very precise measurements and are often regarded as the only legally acceptable definition of land parcels. Measurements are usually very detailed and data capture is often time consuming. Furthermore, commonly occurring discrepancies in the data must be manually resolved by highly qualified individuals.

10.4 Obtaining Data from External Sources (Data Transfer)

One major decision that needs to be faced at the start of a GIS project is whether to build or buy a database. All the preceding discussion has been concerned with techniques for building databases from primary and secondary sources. This section focuses on how to import or transfer data captured by others. Some of these data are freely available, but many of them are sold as a commodity from a variety of outlets including, increasingly, Internet sites.

There are many sources and types of geographic data. Space does not permit a comprehensive review of geographic data sources here, but a small selection of key sources is listed in Table 10.3. In any case, the characteristics and availability of datasets are constantly changing so those seeking an up-to-date list should consult one of the good online sources described below.

The best way to find geographic data is to search the Internet using one of the specialist geographic search engines such as the US NSDI Clearinghouse or the Geography Network.

The best way to find geographic data is to search the Internet. Several types of resources are available to assist searching. These include specialist geographic data catalogs and stores, as well as specific geographic data vendors (some Web sites are shown in Table 10.4). Particularly good sites are the Data Store (www.data-store.co.uk) and the AGI (Association for Geographic Information) Resource List (www.geo.ed.ac.uk/home/giswww.html). These sites provide access to information about the characteristics and availability of geographic data. Some also have facilities to purchase and download data directly. Probably the most useful resource for locating geographic data is the network of Web sites that make up the US NSDI Clearinghouse system (see Section 7.5.1 and Box 7.4 for an overview of the Clearinghouse network: www.fgdc.gov). Clearinghouse nodes provide access to metadata about available geographic data; they do not provide direct access to actual data.

One of the good things about data standards is that there are many to choose from!

An interesting new trend initiated by the *Geography Network* project (see Figure 1.19) is the idea of providing data online in ready-to-use GIS formats. The Geography Network is a global

Table 10.3 Examples of some digital data sources that can be imported into a GIS

Type	Source	Details
Basemaps		
Geodetic framework	Many National Mapping Organizations (NMOs)	Definition of framework, map projections, and geodetic transformations
General topographic map data	NMOs, NIMA, and other military agencies	Many types of data at fine to medium scales
Elevation	NMOs, NIMA, EOSAT, SPOT Image, NASA	DEMs, contours at local, regional, and global levels
Transportation	National governments, GDT, NavTech	Highway/street centerline databases at national levels
Hydrology	NMOs and government agencies	National hydrological databases are available for many countries
Toponymy	NMOs and other government agencies	Gazetteers of placenames at global and national levels
Satellite images	Commercial and military providers, e.g. Landsat, SPOT, and IRS	See Figure 10.2 for further details
Aerial photographs	Many private and public agencies	Scales vary widely, typically from 1:500 – 1:20,000
Environmental		
Wetlands	National agencies, e.g. US National Wetlands Inventory	Government wetlands inventory
Toxic release sites	National Environmental Protection Agencies, e.g. US EPA	Details of thousands of toxic sites
World eco-regions	World Wide Fund for Nature (WWF)	Habitat types, threatened areas, biological distinctiveness
Flood zones	Federal Emergency Management Agency (FEMA)	National flood risk areas
Socio-economic		
Population census	National governments	Typically every 10 years with annual estimates
Lifestyle classifications	Private agencies (e.g. CACI and Experian)	Derived from population censuses and other socio-economic data
Geodemographics	Private agencies (e.g. Claritas and NDS)	Many types of data at many scales and prices
Land and property ownership	National governments	Street, property, and cadastral data
Administrative areas	National governments	Obtained from maps at scales of 1:5000–1:750,000

collection of data users and providers connected by the Internet. Information about available data sources can be found by consulting the Geography Network Web site (www.GeographyNetwork.com). Once a useful data source has been located, the actual data can be streamed directly into a browser or desktop GIS. Much of the content on the Geography Network is accessible without charge, but additional commercial content is also provided and maintained by its owners. This information is accessible in the same way as free content, but every time a map is viewed, an online service utilized (for example a retail site suitability or flood risk mapping application), or a dataset downloaded, a charge is recorded by the Geography Network e-commerce system. The Geography Network management organization is responsible for maintaining the e-commerce system and for billing users and paying providers.

A critical requirement for providing online geographic data indexing, searching, access, and

Table 10.4 Selected Web sites containing information about geographic data sources

Source	URL	Description
AGI GIS Resource List	www.geo.ed.ac.uk/home/giswww.html	Indexed list of several hundred sites
The Data Store	www.data-store.co.uk/	UK, European, and worldwide data catalog
National Spatial Data Infrastructure	www.fgdc.gov	Worldwide list of data sources indexed by metadata
GISLinx.com	www.gislinx.com/Data/	More than 70 links to GIS data sources
EROS Data Center	edcwww.cr.usgs.gov/	US government data archive
Geography Network	www.GeographyNetwork.com	Global online data and map services
National Geographic Society	www.nationalgeographic.com	Worldwide maps
GEOWorld Data Directory	www.geoplace.com	List of GIS data companies
The Data Depot	www.gisdatadepot.com	Free geographic data depot

download is good quality metadata. This is discussed in detail in Section 7.5.

10.4.1 Geographic data formats

One of the biggest problems with data obtained from external sources is that they can be encoded in many different formats. There are so many different geographic data formats because no single format is appropriate for all tasks and applications. It is not possible to design a format that supports, for example, both fast rendering in police command and control systems, and sophisticated topological analysis in natural resource information systems: the two are mutually incompatible. The many different formats have evolved in response to diverse user requirements.

Given the high cost of creating databases many people have asked for tools to move data between systems and to re-use data through open application programming interfaces (APIs). In the former case, the approach has been to develop software that is able to translate data (Figure 10.12), either by a direct read into memory, or via an intermediate file format. In the latter case, software developers have created open interfaces to allow access to data.

Many GIS software systems are now able to read directly AutoCAD DWG and DXF (see Table 10.5 for an explanation of acronyms), Microstation DGN, and Shapefile, VPF, and many image formats. Unfortunately, direct read support can only easily be provided for relatively simple product-oriented formats. Complex formats, such as SDTS and UK NTF, were designed for exchange purposes and require more advanced processing before they

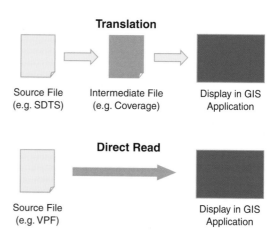

Figure 10.12 Comparison of data access by translation and direct read

can be viewed (e.g. multi-pass read and feature assembly from several parts).

Data can be transferred between systems by direct read into memory or via an intermediate file format.

More than 25 organizations are involved in the standardization of various aspects of geographic data and geoprocessing (for a list see, for example, www.diffuse.org). Several of these are country and domain specific. At the global level, ISO (the International Standards Organization) is responsible for coordinating efforts through the work of technical committees TC 211 and 287. In Europe, CEN (Commission Européan Normalisation) is engaged in geographic standardization. At the national level, there are many complementary bodies. One other standards-forming organization of particular note is OGC (Open GIS Consortium: www.opengis.org), a group of vendors, academics, and users interested in the interoperability of geographic systems. To date there have been promising OGC-coordinated efforts to standardize on simple feature access (simple geometric object types), metadata catalogs, and Web access. Table 10.5 lists some popular vector and raster geographic data formats.

The most efficient way to translate data between systems is via a common intermediate file format.

Having obtained a potentially useful source of geographic information the next task is to import it into a GIS database. If the data are already in the native format of the target GIS software system, or the software has a direct read capability for the format in question, then this is a relatively straight-forward task. If the data are not compatible with the target GIS software then the alternatives are to ask the data supplier to convert the data to a compatible format, or to use a third-party translation software system, such as the Feature Manipulation Engine from Safe Software (www.safe.com lists over 60 supported geo-

graphic data formats) to convert the data. Geographic data translation software must address both syntactic and semantic translation issues. Syntactic translation involves converting specific digital symbols (letters and numbers) between systems. Semantic translation is concerned with converting the meaning inherent in geographic information. While the former is relatively simple to encode and decode, the latter is much more difficult and has seldom met with much success to date.

Although the task of translating geographic information between systems was described earlier as relatively straightforward, those that have tried this in practice will realize that things on the ground are seldom quite so simple. Any number of things can (and do!) go wrong. These range from corrupted media, to incomplete data files, wrong versions of translators, and different interpretations of a format specification, to basic user error.

There are two basic strategies used for data translation: direct and via a neutral format. For small systems that involve the translation of a small number of formats, the first is the simplest. Directly translating data between the internal structures of two systems requires two new translators (A to B, B to A). Adding two further

Table 10.5 Popular geographic data formats

Vector	Raster (Image)
Automated Mapping System (AMS)	Arc Digitized Raster Graphics (ADRG)
ESRI Coverage	Band Interleaved by Line (BIL)
Computer Graphics Metafile (CGM)	Band Interleaved by Pixel (BIP)
Digital Feature Analysis Data (DFAD)	Band SeQuential (BSQ)
Encapsulated Postscript (EPS)	Windows Bitmap (BMP)
Microstation drawing file format (DGN)	Device-Independent Bitmap (DIB)
Dual Independent Map Encoding (DIME)	Compressed Arc Digitized Raster Graphics (CADRG)
Digital Line Graph (DLG)	Controlled Image Base (CIB)
AutoCAD Drawing Exchange Format (DXF)	Digital Terrain Elevation Data (DTED)
AutoCAD Drawing (DWG)	ERMapper
MapBase file (ETAK)	Graphics Interchange Format (GIF)
ESRI Geodatabase	ERDAS IMAGINE (IMG)
Land Use and Land Cover Data (GIRAS)	ERDAS 7.5 (GIS)
Interactive Graphic Design Software (IGDS)	ESRI GRID file (GRID)
Initial Graphics Exchange Standard (IGES)	JPEG File Interchange Format (JFIF)
Map Information Assembly Display System (MIADS)	Multi-resolution Seamless Image Database (MrSID)
MOSS Export File (MOSS)	Tag Image File Format (TIFF; GeoTIFF tags are supported)
TIGER/Line file: Topologically Integrated Geographic Encoding and Referencing (TIGER)	Portable Network Graphics (PNG)
Spatial Data Transfer Standard/Topological Vector Profile (SDTS/TVP)	
ESRI ArcView GIS (Shapefile)	
Vector Product Format (VPF)	
UK National Transfer Format (NTF)	

systems will require 12 translators to share data between all systems (A to B, A to C, A to D, B to A, B to C, B to D, C to A, C to B, C to D, D to A, D to B, and D to C). A more efficient way of solving this problem is to use the concept of a data *switchyard* and a common intermediate file format. Systems now need only to translate to and from the common format. The four systems will now need only eight translators instead of 12 (A to Neutral, B to Neutral, C to Neutral, D to Neutral, Neutral to A, Neutral to B, Neutral to C, Neutral to D). The more systems there are the more efficient this becomes. This is one of the key principles underlying the need for common file interchange formats.

10.5 Capturing Attribute Data

All geographic objects have attributes of one type or another. Although attributes can be collected at the same time as vector geometry, it is usually more cost-effective to capture attributes separately. In part, this is because attribute data capture is a relatively simple task that can be undertaken by lower-cost clerical staff. It is also because attributes can be entered by direct data loggers, manual keyboard entry, optical character recognition (OCR) or, increasingly, voice recognition, which do not require expensive hardware and software systems. Much the most common method is direct keyboard data entry into a spreadsheet or database. For some projects, a custom data entry form with in-built validation is preferred. On small projects single entry is used, but for larger, more complex projects data are entered twice and then compared as a validation check.

An essential requirement for separate data entry is a common identifier (also called a key) that can be used to relate object geometry and attributes together following data capture (see Figure 11.2 for a diagrammatic explanation of relating geometry and attributes).

Metadata are a special type of non-geometric data that are increasingly being collected. Some metadata are derived automatically by the GIS software system (for example, length and area, extent of data layer, and count of features), but some must be explicitly collected (for example, owner name, quality estimate, and original source). Explicitly collected metadata can be entered in the same way as other attributes, as described above. For further information about metadata see Section 7.5.

10.6 Managing a Data Capture Project

The subject of managing a GIS project is given extensive treatment later in this book in Chapters 16–19. The management of data capture projects is discussed briefly here both because of its critical importance and because there are several unique issues. That said, most of the general principles for any GIS project apply to data capture: the need for a clearly articulated plan, adequate resources, appropriate funding, and sufficient time.

In any data capture project there is a fundamental tradeoff between quality, speed, and price. Capturing high quality data quickly is possible, but it is very expensive. If price is a key consideration then lower quality data can be captured over a longer period (Figure 10.13).

GIS data capture projects can be carried out intensively or over a longer period.

A key decision facing managers of data capture projects is whether to pursue a strategy of incremental capture or "Blitzkrieg" – that is, to capture all data as rapidly as possible. Incremental data capture involves breaking the data capture project into small manageable subprojects. This allows data capture to be undertaken with lower annual resource and funding levels (although total project resource requirements may be larger). It is a good approach for inexperienced organizations that are embarking on their first data capture project because they can learn and adapt as the project proceeds. On the other hand, these longer-term projects run the risk of employee turnover and

Figure 10.13 Relationship between quality, speed, and price in data capture (Source: after Hohl 1998).

burnout, as well as changing data, technology, and organizational priorities.

Whichever approach is preferred, a pilot project carried out on part of the study area and a selection of the data types can prove to be invaluable. A pilot project can identify problems in workflow, database design, personnel, and equipment. A pilot database can also be used to test equipment and to develop procedures for quality assurance. Many projects require a test database for hardware and software acceptance tests and to facilitate software customization. It is essential that project managers are prepared to discard all the data obtained during a pilot data capture project, so that the main phase can proceed unconstrained.

A further important decision is whether data capture is to use in-house or external resources. Three factors influencing this decision are: cost/ schedule, quality, and long-term ramifications (Struck and Dilks 1998). Specialist external data capture agencies can often perform work faster, cheaper, with higher quality than in-house staff, but because of the need for real cash to pay external agencies this may not be possible. In the short term, project costs, quality, and time are the main considerations, but over time dependency on external groups may become a problem.

Questions for Further Study

1. What are the most important data collection techniques for capturing different forest objects at scales of 1:1000, 1:10 000 and 1:100 000?
2. What are the advantages of batch vectorization over manual digitizing?
3. What quality assurance steps would you build into a data capture project?
4. Why are there so many data formats? Which ones are most suitable for selling vector data?

Online Resources

NCGIA Core Curricula (www.ncgia.ucsb.edu/pubs/core.html):
Core Curriculum in GIScience, Sections 2.9.2 (Natural Resources Data, Peter Schut), and 2.9.2.1 (Soil Data for GIS, Peter Schut)
Core Curriculum in GIS, 1990, Units 7 (Data Input), 8 (Socio-Economic Data), 9 (Environmental and Natural Resource Data), and 66 (Database Creation)
ESRI Virtual Campus courses (campus.esri.com):
Getting Started with Census Data, by Christopher Williamson

Understanding Geographic Data, by David DiBiase (Modules consider ''Land Surveys and GPS Data'', ''Air Photos and Planimetric Data'', ''Elevation Data'', ''Satellite Image Data'', and ''Census Data'')
Protecting your Investment in Data with Metadata by George Shirey
See Table 10.4 for a list of major data sources; the geolibraries listed in Chapter 7; and the spatial data infrastructure sites listed in Chapter 19
List of example data vendors:
www.geoplace.com/bg/2000/0300/data_vendors. asp
Case study on Digitized data collection at Great Britain Ordnance Survey:
www.agi.org.uk/pag-es/case-stru/cac-os.htm
Selected National Mapping Agencies:
www.usgs.gov USGS
www.ordsvy.gov.uk Ordnance Survey
www.nima.mil NIMA
www.ign.fr IGN, France
www.auslig.gov.au (AUSLIG)

Reference Links

Maguire D J, Goodchild M F, and Rhind D W (eds) 1991 *Geographical Information Systems: Principles and applications*. Harlow, UK: Longman (Text available online from 'Links to Big Book 1' at www.wiley.com/gis and www.wiley.co.uk/gis).
Chapter 17, GIS data capture hardware and software (Jackson M, Woodsford P)
Chapter 34, Spatial data exchange and standardization (Guptill S)
Longley P A, Goodchild M F, Maguire D J, and Rhind D W (eds) 1999 *Geographical Information Systems: Principles, techniques, management and applications*. New York: John Wiley.
Chapter 32, Digital remotely-sensed data and their characteristics (Barnsley, M)
Chapter 33, Using GPS for GIS data capture (Lange A, Gilbert C)
Chapter 47, Characteristics and sources of framework data (Smith N S, Rhind D W)
Chapter 48, Characteristics, sources, and management of remotely-sensed data (Estes J E, Loveland T R)
Chapter 50, National and international data standards (Salgé F)

References

Hohl P (ed) 1999 *GIS Data Conversion: Strategies, techniques and management*. Santa Fe, New Mexico, OnWord Press,

Jensen J R and Cowen D C 1999 Remote sensing of urban/suburban infrastructure and socio-economic attributes. *Photogrammetric Engineering and Remote Sensing* **65** 611–622.

Jones C 1997 *Geographic Information Systems and Computer Cartography.* Reading, Mass.: Addison-Wesley Longman.

Kennedy M 1995 *The Global Positioning Systems and GIS: An Introduction.* Ann Arbor, Michigan: Ann Arbor Press.

Lillesand T M and Kiefer R W 1999 *Remote Sensing and Image Interpretation.* (fourth edition). New York: John Wiley.

Litton A 1998 Scanning. In Hohl P (ed.) *GIS Data Conversion: Strategies, techniques and management.* Mexico, Santa Fe, New Mexico: OnWord Press, 259–301.

Steede-Terry K 2000 *Integrating GIS and the Global Positioning System.* Redlands, California: ESRI Press.

Struck K and Dilks K M 1998 Project planning and management. In Hohl P (ed.) *GIS Data Conversion: Strategies, techniques and management.* Santa Fe, New Mexico: OnWord Press, 63–98.

CREATING AND MAINTAINING GEOGRAPHIC DATABASES

11

All large operational GIS are built on the foundation of a geographic database. The database is arguably the most important part of a GIS because of its costs of collection and maintenance, and because the database forms the basis of all analysis and decision-making. Today, virtually all large GIS implementations store data in a database management system (DBMS), a specialist piece of software designed to handle multi-user access to an integrated set of data. Databases need to be designed with great care, and structured and indexed to provide good query and transaction performance. A comprehensive security and transactional access model is necessary to ensure multiple users can access the database at the same time. On-going maintenance is also an essential, but very resource-intensive activity.

Learning objectives

After reading this chapter you will:

- Understand the role of database management systems (DBMS) in GIS;

- Recognize SQL language statements;

- Understand the key geographic database data types and functions;

- Be familiar with the stages in geographic database design;

- Understand the key techniques for structuring geographic information, specifically creating topology and indexing;

- Understand the issues associated with multi-user editing and versioning.

11.1 Introduction

A *database* can be thought of as an integrated set of data on a particular subject. Geographic (also frequently referred to as spatial) databases are simply databases containing geographic data for a particular area and subject. A geographic database is a critical part of an operational GIS. This is both because of the cost of creation and maintenance, and because of the impact of a geographic database on all analysis, modeling, and decision-making activities. The database approach to storing geographic data offers a number of advantages over traditional file-based datasets (Date 1995):

- Collecting all data at a single location reduces redundancy and duplication.
- Maintenance costs decrease because of better organization and decreased data duplication.
- Applications become data independent so that multiple applications can use the same data and they can evolve separately over time.
- User knowledge can be transferred between applications more easily because the database remains constant.
- Data sharing is facilitated and a corporate view of data can be provided to all managers and users.
- Security and standards for data and data access can be established and enforced.

A database is an integrated set of data on a particular subject.

In recent years geographic databases have become increasingly large and complex. This chapter describes how to create and maintain geographic databases, and the concepts, tools, and techniques that are available to manage geographic data in databases. Several other chapters provide additional information that is relevant to this discussion. In particular, the nature of geographic data and how to represent them in GIS were described in Chapters 3 and 5, and data modeling and data collection were discussed in Chapters 9 and 10. Later chapters introduce the tools and techniques that are available to query, model, and analyze geographic databases (Chapters 13 and 14). Finally, Chapters 16 through 19 discuss the important management issues associated with creating and maintaining geographic databases.

11.2 Database Management Systems

Small, simple databases that are used by a small number of people can be stored on computer disk in standard files. However, larger, more complex databases with many tens, hundreds, or thousands of users require specialist database management system (DBMS) software to ensure database integrity and longevity. A DBMS is a software application designed to organize the efficient and effective storage and access of data. To carry out this function DBMS provide a number of important capabilities. These are introduced briefly here and discussed further in this and other chapters.

- Data model. As discussed in Section 9.4, a data model is the mechanism used to represent real-world objects digitally in a computer system. All DBMS include standard general-purpose data models suitable for representing several types of object (e.g. integer and floating point numbers, dates, and text). In most cases they can be extended to support geographic object types.
- Data load. DBMS provide tools to load data into databases. Simple tools are available to load standard supported data types in well-structured formats. Other non-standard data formats can be loaded by writing custom software programs that convert the data into a structure that can be read by the standard loaders.
- Index. An index is a data structure used to speed up searching. All databases include tools to index standard database data types.
- Query language. One of the major advantages of DBMS is that they support a standard data query/manipulation language called SQL (Structured Query Language).
- Security. A key characteristic of DBMS is that they provide controlled access to data. This includes restricting user access to all or part of a database. For example, a casual GIS user might have read-only access to just part of a database, but a specialist user might have read and write (create, update, and delete) access to the entire database.
- Controlled update. Updates to databases are controlled through a transaction manager responsible for managing multi-user access and ensuring that updates affecting more than one part of the database are coordinated.
- Backup and recovery. It is important that the valuable data in a database are protected from system failure and incorrect (accidental or

deliberate) update. Software utilities are provided to back up all or a part of a database and to recover the database in the event of a problem.

● Database administration tools. The task of setting up the structure of a database (the schema), creating and maintaining indexes, tuning to improve performance, backing up and recovering, and allocating user access rights is performed by a database administrator (DBA). A specialized collection of tools and a user interface are provided for this purpose.

● Applications. Modern DBMS include standard, general-purpose tools for creating, using, and maintaining databases. These include applications for designing databases (CASE tools) and for building user interfaces for data access and presentations (forms and reports).

● Programmable API. Although most DBMS have good general-purpose applications for standard use, most large, specialist applications will require further customization using a commercial off-the-shelf programming language and a DBMS programmable API (application programming interface).

This list of DBMS capabilities is very attractive to GIS users and so, not surprisingly, virtually all large GIS databases are based on DBMS technology. Indeed, most GIS software vendors include DBMS software within their GIS software, or provide an interface that supports very close coupling to a DBMS. For further discussion of this see Chapter 8.

A DBMS is a software application designed to organize the efficient and effective storage and access of data. Today, virtually all large GIS use DBMS technology.

11.2.1 Types of DBMS

DBMS can be classified according to the way they store and manipulate data. Three main types of DBMS are used in GIS today: relational (RDBMS), object (ODBMS), and object-relational (ORDBMS).

A relational database comprises a set of tables, each a two-dimensional list (array) of records containing attributes about objects. This apparently simple structure has proven to be remarkably flexible and useful in a wide range of application areas, such that today over 95% of the data in DBMS is stored in RDBMS. Examples of RDBMS software include IBM DB2, Informix Dynamic Server, Microsoft Access, Microsoft SQL Server, and Oracle Universal Server. As RDBMS

and the underlying relational model are so important in GIS, these topics are discussed at length in Section 11.3 below.

Relational databases dominate GIS today, as they do in many other business areas.

Object databases were initially designed to address several of the weaknesses of RDBMS. These include the inability to store complete objects directly in the database (object state and behavior: see Section 9.1 for an introduction to objects and object technology). Because RDBMS were focused primarily on business applications such as banking, human resource management, and stock control and inventory, they were never designed to deal with rich data types, such as geographic objects, sound, and video. A further difficulty is the poor performance of RDBMS for many types of geographic query. These problems are compounded by the difficulty of extending RDBMS to support geographic data types and processing functions, which obviously limits their adoption for geographic applications. ODBMS can store objects persistently (semi-permanently on disk or other media) and provide object-oriented query tools. A number of commercial ODBMS have been developed including ObjectStore from eXcelon Corp and GemStone from GemStone Systems.

In spite of the technical elegance of ODBMS, they have not proven to be as commercially successful as some predicted. This is largely because of the massive installed base of RDBMS and the fact that RDBMS vendors have now added many of the important ODBMS capabilities to their standard RDBMS software systems to create hybrid object-relational DBMS (ORDBMS). An ORDBMS can be thought of as an RDBMS engine adapted to handle objects. That is, both the data describing what an object is (object attributes such as color, size, and age) and the behavior that determines what an object does (object methods or functions such as drawing instructions, query interfaces, and interpolation algorithms) are stored together as an integrated whole. The ideal geographic ORDBMS is one that has been extended to support geographic object types and functions in several ways (these topics are introduced here and discussed further later in this chapter):

● Query parser – the engine used to interpret SQL queries must be extended to deal with geographic types and functions.

● Query optimizer – the software query optimizer must be able to handle geographic queries efficiently. Consider a query to find all potential users of a new brand of premier wine that you need to market to wealthy households from your network of 100 retail stores:

Select all households within 3 km of a store that have an income greater than $100,000.

This could be carried out in two ways:

1. Select all households with an income greater than $100,000; from this selected set, select all households within 3 km of a store.
2. Select all households within 3 km of a store; from this selected set select all households with an income greater than $100,000.

Selecting households with an income greater than $100,000 is an attribute query that can be performed very quickly. Selecting households within 3 km of a store is a geometric query that takes much longer. Executing the attribute query first (the first option above) will result in fewer tests for store proximity and therefore the whole query will be completed much quicker.

● Query language – the query language must be able to handle geographic types and functions.

● Indexing services – standard unidimensional DBMS data indexes must be extended to support multidimensional geographic data types.

● Storage management – the large volume and variable nature of the size of geographic records (especially geometric and topological relationships) requires new, efficiently designed storage structures.

● Transaction services – standard DBMS are designed to handle short (sub-second) transactions and cannot deal with the long transactions common in many geographic records.

● Replication – services for replicating databases must be extended to deal with geographic types and problems of reconciling changes made by distributed users.

11.2.2 Geographic DBMS extensions

As of the beginning of 2001, three major commercial DBMS vendors have released spatial database extensions to their standard ORDBMS products: IBM DB2 Spatial Extender, Informix Spatial Datablade, and Oracle Spatial Option. Although there are differences in the technology, scope, and capabilities of these systems, they all provide basic capabilities to store, manage, and query geographic objects. It is important to

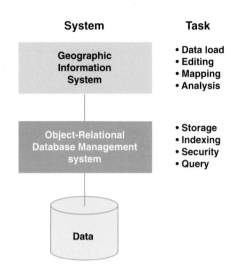

Figure 11.1 The roles of GIS and DBMS

realize, however, that none of these is a complete GIS software system in itself. The focus of these extensions is data storage, retrieval, and management, and they have no real capabilities for geographic editing, mapping, and analysis. Consequently, they must be used in conjunction with a GIS except in the case of the simplest query-focused applications. Figure 11.1 shows how GIS and DBMS software can work together and some of the tasks best carried out by each system.

> Several DBMS have now been extended so that they can support geographic data types and functions.

11.3 Storing Data in DBMS Tables

The lowest level of user interaction with a geographic database is usually the object class (also called a layer), which is an organized collection of data on a particular theme (e.g. all pipes in a water network, all soil polygons in a river basin, or all elevation values in a digital elevation model). Object classes are stored in database tables. A table is a two-dimensional row, column structure. Each object class is stored as a single database table in a database management system (DBMS). Table rows contain objects (instances of object classes) and the columns contain object properties or attributes as they are frequently called (Figure 11.2). The data stored at individual row, column intersections are usually referred to as values. Geographic database tables are distinguished from non-geographic tables by the presence of a geometry column (often called the shape column). To save space and improve performance, the actual coordinate values may be stored in a highly compressed binary form, in which case they cannot be read by the human eye in the same way as other data types (integer, float, date, etc.).

Relational databases comprise tables. Each geographic class is stored as a table.

Tables are joined together using common row/column values, or keys as they are known in the database world. Figure 11.2 shows parts of tables containing data about US states. The STATES table (Figure 11.2(A)) contains the geometry and some basic attributes, an important one being a unique STATE_FIPS (State FIPS [Federal Information Processing Standard] code) identifier. The POPULATION table (Figure 11.2(B)) was collected entirely independently, but also has a unique identifier column called STATE_FIPS. Using standard database tools the two tables can be joined together based on the common STATE_FIPS identifier column (the key) to create a third table, COMBINED STATES and POPULATION (Figure 11.2(C)). Following the join these can be treated as a single table for all GIS operations such as query, display, and analysis.

Database tables can be joined together to create new relations, or views of the database.

In a groundbreaking description of the relational model that underlies the vast majority of the world's geographic databases, Codd (1970) defined a series of rules for the efficient and effective design of database table structures. The heart of Codd's idea was that the best relational databases are made up of simple, stable tables that follow five principles:

1. Only one value is in each cell at the intersection of a row and column.
2. All values in a column are about the same subject.
3. Each row is unique.
4. There is no significance to the sequence of columns.
5. There is no significance to the sequence of rows.

Table 11.1 shows a land parcel tax assessor's database table that contradicts several of Codd's principles. Codd suggests a series of trans- formations called normal forms that successively improve the simplicity and stability, and reduce the redundancy of database tables by splitting them into sub-tables that are re-joined at query time. Unfortunately, joining large tables is

(A)

FID	Shape*	AREA	STATE_NAME	STATE_FIPS
41	Polygon	51715.656	Alabama	01
49	Polygon	576556.687	Alaska	02
35	Polygon	113711.523	Arizona	04
45	Polygon	52912.797	Arkansas	05
23	Polygon	157774.187	California	06
30	Polygon	104099.109	Colorado	08
17	Polygon	4976.434	Connecticut	09
27	Polygon	2054.506	Delaware	10
26	Polygon	66.063	District of Columbia	11
47	Polygon	55815.051	Florida	12
43	Polygon	58629.195	Georgia	13
48	Polygon	6381.435	Hawaii	15
7	Polygon	83340.594	Idaho	16
25	Polygon	56297.953	Illinois	17
20	Polygon	36399.516	Indiana	18
12	Polygon	56257.219	Iowa	19
32	Polygon	82195.437	Kansas	20
31	Polygon	40318.777	Kentucky	21
46	Polygon	45835.898	Louisiana	22
2	Polygon	32161.664	Maine	23
29	Polygon	9739.753	Maryland	24
13	Polygon	8172.482	Massachusetts	25
50	Polygon	57898.367	Michigan	26
9	Polygon	84517.469	Minnesota	27
42	Polygon	47618.723	Mississippi	28
34	Polygon	69831.625	Missouri	29
1	Polygon	147236.031	Montana	30

Record: 0 Show: All Selected Records (0 out of 51 Selected.)

(B)

STATE_FIPS	SUB_REGION	STATE_ABBR	POP1990	POP1996
53	Pacific	WA	4866692	5629613
30	Mtn	MT	799065	885762
23	N Eng	ME	1227928	1254465
38	W N Cen	ND	638800	633534
46	W N Cen	SD	696004	721374
56	Mtn	WY	453588	487142
55	E N Cen	WI	4891769	5144123
16	Mtn	ID	1006749	1201327
50	N Eng	VT	562758	587726
27	W N Cen	MN	4375099	4639933
41	Pacific	OR	2842321	3203820
33	N Eng	NH	1109252	1156932
19	W N Cen	IA	2776755	2831890
25	N Eng	MA	6016425	6066573
31	W N Cen	NE	1578385	1622272
36	Mid Atl	NY	17990455	18293435
42	Mid Atl	PA	11881643	12077607
09	N Eng	CT	3287116	3287604
44	N Eng	RI	1003464	993306
34	Mid Atl	NJ	7730188	7956917
18	E N Cen	IN	5544159	5801023
32	Mtn	NV	1201833	1532295
49	Mtn	UT	1722850	2000630
06	Pacific	CA	29760021	32218713
39	E N Cen	OH	10847115	11123416
17	E N Cen	IL	11430602	11731783
11	S Atl	DC	606900	550076

Record: 0 Show: All Selected Records (0 out of 51 Selec

Figure 11.2 GIS database tables for US States: (A) STATES table; (B) POPULATION table

(C)

Attributes of COMBINED STATES and POPULATION

FID	Shape*	AREA	STATE_NAME	STATE_FIPS	SUB_REGION	STATE_ABBR	POP1990	POP1996
0	Polygon	67286.875	Washington	53	Pacific	WA	4866692	5629613
1	Polygon	147236.031	Montana	30	Mtn	MT	799065	885762
2	Polygon	32161.664	Maine	23	N Eng	ME	1227928	1254465
3	Polygon	70810.156	North Dakota	38	W N Cen	ND	638800	633534
4	Polygon	77193.625	South Dakota	46	W N Cen	SD	696004	721374
5	Polygon	97799.492	Wyoming	56	Mtn	WY	453588	487142
6	Polygon	56088.066	Wisconsin	55	E N Cen	WI	4891769	5144123
7	Polygon	83340.594	Idaho	16	Mtn	ID	1006749	1201327
8	Polygon	9603.218	Vermont	50	N Eng	VT	562758	587726
9	Polygon	84517.469	Minnesota	27	W N Cen	MN	4375099	4639933
10	Polygon	97070.750	Oregon	41	Pacific	OR	2842321	3203820
11	Polygon	9259.514	New Hampshire	33	N Eng	NH	1109252	1156932
12	Polygon	56257.219	Iowa	19	W N Cen	IA	2776755	2831890
13	Polygon	8172.482	Massachusetts	25	N Eng	MA	6016425	6066573
14	Polygon	77328.336	Nebraska	31	W N Cen	NE	1578385	1622272
15	Polygon	48560.578	New York	36	Mid Atl	NY	17990455	18293435
16	Polygon	45359.238	Pennsylvania	42	Mid Atl	PA	11881643	12077607
17	Polygon	4976.434	Connecticut	09	N Eng	CT	3287116	3287604
18	Polygon	1044.850	Rhode Island	44	N Eng	RI	1003464	993306
19	Polygon	7507.302	New Jersey	34	Mid Atl	NJ	7730188	7956917
20	Polygon	36399.516	Indiana	18	E N Cen	IN	5544159	5801023
21	Polygon	110667.297	Nevada	32	Mtn	NV	1201833	1532295
22	Polygon	84870.187	Utah	49	Mtn	UT	1722850	2000630
23	Polygon	157774.187	California	06	Pacific	CA	29760021	32218713
24	Polygon	41192.863	Ohio	39	E N Cen	OH	10847115	11123416
25	Polygon	56297.953	Illinois	17	E N Cen	IL	11430602	11731783
26	Polygon	66.063	District of Columbia	11	S Atl	DC	606900	550076
27	Polygon	2054.506	Delaware	10	S Atl	DE	666168	724890

Record: 0 Show: All Selected Records (0 out of 51 Selected.) Options

Figure 11.2 (C) Joined table – COMBINED STATES and POPULATION

Table 11.1 Initial Assessor geographic database. OID (Object Identifier) is a system-defined unique identifier; Shape contains the object geometry; ParcelNumber is a user-defined identifier; AssessorName is the name of the Tax Assessor that performed the assessment; Zoning is the land use zoning code for each parcel; and Value is the assessed value for taxation purposes ($)

OID	Shape	ParcelNumber	AssessorName	Zoning	Value
001	Poly	673-100	John Smith Jane Doe	Commercial	1990 222000 1998 265225
002	Poly	673-101 673-101a 673-101b	Fred Bloggs	Residential	1984 98000
003	Poly	670-042	Jane Doe	Residential	1995 199500
004	Poly	567-013	F Bloggs	Commercial	1997 80000

Table 11.2 Assessor geographic database without repeating groups

OID	Shape	ParcelNumber	AssessorName	Zoning	Year	Value
001	Poly	673-100	John Smith	Commercial	1990	222000
001	Poly	673-100	Jane Doe	Commercial	1998	265225
002	Poly	673-101	Fred Bloggs	Residential	1984	98000
002	Poly	673-101a	Fred Bloggs	Residential	1984	98000
002	Poly	673-101b	Fred Bloggs	Residential	1984	98000
003	Poly	670-042	Jane Doe	Residential	1995	199500
004	Poly	567-013	Fred Bloggs	Commercial	1997	80000

Table 11.3 Assessor geographic database in partially normalized form

OID	Shape	ParcelNumber	AssessorName	Year
001	Poly	673-100	John Smith	1990
001	Poly	673-100	Jane Doe	1998
002	Poly	673-101	Fred Bloggs	1984
002	Poly	673-101a	Fred Bloggs	1984
002	Poly	673-101b	Fred Bloggs	1984
003	Poly	670-042	Jane Doe	1995
004	Poly	567-013	Fred Bloggs	1997

ParcelNumber	Zoning
673-100	Commercial
673-101	Residential
673-101a	Residential
673-101b	Residential
670-042	Residential
567-013	Commercial

Year	Value
1990	222000
1998	265225
1984	98000
1995	199500
1997	80000

computationally expensive and can result in complex database designs that are difficult to maintain. For this reason, non-normalized table designs are often used in GIS.

Removal of repeating groups to give a single value in each cell results in the table shown in Table 11.2. The potential confusion between Fred Bloggs and F Bloggs has also been resolved. Table 11.3 shows the same data in partially normalized form suitable for use in a GIS tax assessment application. The database now consists of three tables that can be joined together using common keys.

11.4 SQL

The standard database query language adopted by virtually all mainstream databases is SQL (Structured or Standard Query Language: ISO Standard ISO/IEC 9075). There are many good background books and system implementation manuals on SQL and so only brief details will be presented here. SQL may be used directly via an interactive command line interface; it may be compiled in a general-purpose programming language (e.g. C/C++, Java, or Visual Basic); or it may be embedded in a graphical user interface (GUI). There are two key types of SQL statements: DDL (data definition language) and DML (data manipulation language). The third major revision of SQL (SQL 3) defines spatial types and functions as part of a multi-media extension called SQL/MM.

SQL is the standard database query language. Recently, it has been extended with geographic capabilities.

In SQL, data definition language statements are used to create, alter, and delete relational database structures. The CREATE TABLE command is used to define a table, the attributes it will contain, and the primary key (the column used to identify records uniquely). For example the SQL statement to create a table to store data about Countries, with two columns (name and shape (geometry) is as follows:

```
CREATE TABLE Countries (
name         VARCHAR(200) NOT NULL
             PRIMARY KEY,
shape POLYGON NOT NULL
CONSTRAINT spatial reference
CHECK (SpatialReference(shape)=14)
)
CREATE INDEX countries_spatial_index
ON Countries (shape);
```

This SQL statement defines several table parameters. The name column is of type VARCHAR (variable character) and can store values up to 200 characters. Name cannot be NULL (that is, it must have a value) and it is defined as the PRIMARY KEY, which means that its entries must be unique. The shape column is of type POLYGON, and it cannot be NULL. It has an additional spatial reference constraint, meaning that a spatial reference is enforced for all shapes (Type 14 - this will vary by system, but could be Universal Transverse Mercator (UTM)).

Finally, a spatial index will be created called `countries_spatial_index` on the shape column in the countries table.

Data can be inserted into this table using the SQL INSERT command:

```
INSERT INTO Countries
 (Name, Shape)
 VALUES ('Kenya', Polygon('((x y,
 x y, x y, ., x y)),2))
```

Actual coordinates would need to be substituted for the *x*, *y* values. Several additions of this type would result in a table like this:

Name	Shape
Kenya	Polygon geometry
South Africa	Polygon geometry
Egypt	Polygon geometry

Data manipulation language statements are used to retrieve and manipulate data. Objects with a size greater than 10 000 can be retrieved from the countries table using a SELECT statement:

```
SELECT Countries.Name,
FROM Countries
WHERE Area(Countries.Shape) > 10000
```

Area is computed automatically from the shape field using a DBMS function and does not need to be stored.

11.5 Geographic Database Types and Functions

There have been several attempts to define the superset of geographic data types relevant for representing and processing geographic data in databases. Unfortunately space does not permit a review of them all. This discussion will focus on the practical aspects of this problem and will be based on the recently developed International Standards Organization (ISO) and the Open GIS Consortium (OGC) standards. For further background material on this subject see Chapter 3.

The GIS community working under the auspices of ISO and OGC has defined the core geographic types and functions to be used in a DBMS and accessed using the SQL query language (OGC 1999). The geometry types are shown in Figure 11.3. The Geometry class is the root class. It has an associated spatial reference (coordinate system and projection). The Point, Curve, Surface, and GeometryCollection classes are all subtypes of Geometry. For further explanation of how to interpret this object model diagram see the discussion in Section 9.2.

According to this standard, there are nine methods for testing spatial relations between these geometric objects. Each takes as input two geometries (collections of one or more geometric objects) and evaluates whether the relation is true or not. Two examples of possible relations are shown in Figure 11.4. The full set is:

● Equals – are the geometries the same?
● Disjoint – do the geometries share a common point?
● Intersects – do the geometries intersect?

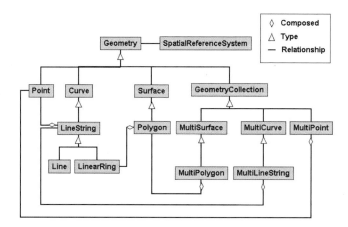

Figure 11.3 Geometry class hierarchy (Source: after OGC 1999, Open GIS Consortium, Inc.)

(A) Contains

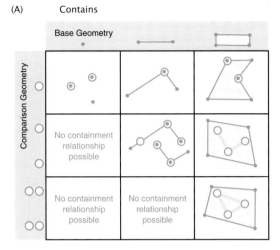

(B) Touches

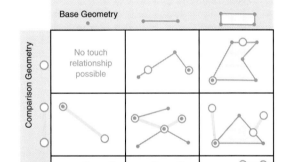

Figure 11.4 Examples of possible relations for two geographic database operators: (A) Contains; and (B) Touches operators (Source: after Zeiler 1999)

- Touches – do the geometries intersect at their boundaries?
- Crosses – do the geometries overlap (can be geometries of different dimensions, for example, lines and polygons)?
- Within – is one geometry within another?
- Contains – does one geometry completely contain another?
- Overlaps – do the geometries overlap (must be geometries of the same dimension)?
- Relate – are there intersections between the interior, boundary, or exterior of the geometries?

Seven methods support spatial analysis on these geometries. Four examples of these methods are shown in Figure 11.5.

- Distance – determines the shortest distance between any two points in two geometries (Section 13.3.1).
- Buffer – returns a geometry that represents all the points whose distance from the geometry is less than or equal to a user-defined distance (Section 13.4.1).
- ConvexHull – returns a geometry representing the convex hull of a geometry (a convex hull is the smallest polygon that can enclose another geometry without any concave areas).
- Intersection – returns a geometry that contains just the points common to both input geometries.
- Union – returns a geometry that contains all the points in both input geometries.
- Difference – returns a geometry containing the points that are different between the two geometries.
- SymDifference – returns a geometry containing the points that are in either of the input geometries, but not both.

11.6 Geographic Database Design

This section is concerned with technical aspects of logical and physical geographic database design. Chapter 9 provides an overview of these subjects, and Chapters 16 to 19 discuss the organizational, strategic, and business issues associated with designing and maintaining a database.

11.6.1 The database design process

Figure 9.2 in Chapter 9 shows three increasingly abstract stages in data modeling: conceptual, logical, and physical. The result of data modeling is a physical database design. This design will include specification of all data types and relationships, as well as the actual database configuration required to store them.

Database design involves three key stages: conceptual, logical, and physical.

Database design involves six practical steps (Figure 11.6):

(A)   Buffer

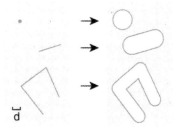

Given a geometry and a buffer distance, the buffer operator returns a polygon that covers all points whose distance from the geometry is less than or equal to the buffer distance.

(B)   Convex Hull

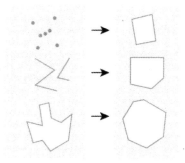

Given an input geometry, the convex hull operator returns a geometry that represents all points that are within all lines between all points in the input geometry.

A convex hull is the smallest polygon that wraps another geometry without any concave areas.

(C)   Intersection

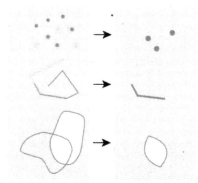

The intersect operator compares a base geometry (the object from which the operator is called) with another geometry of the same dimension and returns a geometry that contains the points that are in both the base geometry and the comparison geometry.

(D)   Difference

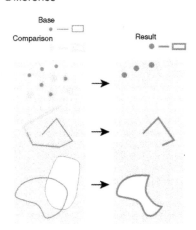

The difference operator returns a geometry that contains points that are in the base geometry and subtracts points that are in the comparison geometry.

Figure 11.5 Examples of spatial analysis methods on geometries: (A) Buffer; (B) Convex Hull; (C) Difference; (D) Intersection (Source: after Zeiler 1999)

Conceptual model

Model the user's view. This involves tasks such as identifying organizational functions (e.g. controlling forestry resources, finding vacant land for new building, and maintaining highways), determining the data required to support these functions, and organizing the data into groups to facilitate data management. This information can be represented in many ways – a report with accompanying tables is often used.

Define objects and their relationships. Once the functions of an organization have been defined, the object types (classes) and functions can be specified (Figure 11.6). The relationships between object types must also be described. This process usually benefits from the rigor of using object models and diagrams to describe a set of object classes and the relationships between them.

Select geographic representation. Choosing the types of geographic representation (discrete

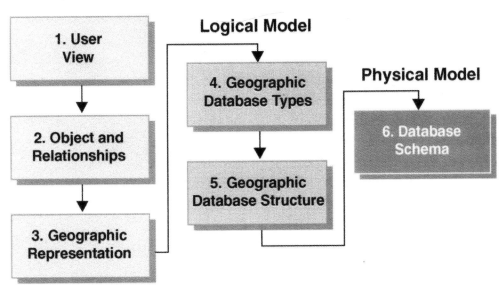

Figure 11.6 Stages in database design (Source: after Zeiler 1999)

Box 11.1 Using GIS for customer care at Telecom Italia Mobile

"Pronto" or "ready", the Italian salutation when answering the telephone, has particular significance in the battles currently being waged across Europe for mobile telephone market share. Telecom Italia Mobile S.pA. (TIM) is one of the largest mobile telephone operators in Europe, serving more than 10 million Italian customers with its Total Access and Communication System (TACS) and Global System for Mobile Communications (GSM) network. Headquartered in Rome, TIM operates 20 regional offices that are responsible for such strategic tasks as deployment, operations, and maintenance including seven service centers devoted to customer relations. Its aggressive technological development and marketing services allowed the company to add more than three million new users during the last 18 months alone.

One of the key operational systems devoted to customer care that serves this intricate wireless transmission web is TIM's Sistema Informativo Territoriale (SIT), a powerful GIS built on an integrated set of commercial off-the-shelf software products. The database for the system resides in an ORDBMS, accessed by an application server. The database design was optimized for fast server query and retrieval. SQL queries retrieve geographic data from the carefully designed geographic database tables.

The SIT database supports about 3000 operator call stations throughout its entire territory, providing real-time answers to customer questions related to GSM and TACS network coverage (Figure 11.7). Service is provided 24 hours a day, seven days a week. Approximately 100 000 calls are answered each day, 10% of which are questions regarding specific areas of coverage, loss of signal, or interference. In addition, SIT provides the location of public authorities, such as police stations, and the location of dealers selling TIM's range of equipment and services. A variety of tabular information is also available to the customer service operators including dealers' names, phone/FAX numbers, credit card information, business hours, and so forth. Those staffing the call centers need instant answers from SIT because customer response time is measured in seconds.

Box 11.1 Continued

(A)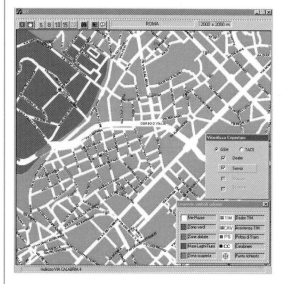

(B)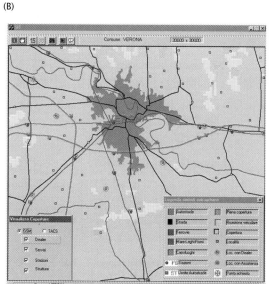

Figure 11.7 Mobile phone coverage maps produced with Telecom Italia's GIS, Sistema Informativo Territoriale: (A) Rome; and (B) Verona. Visualizza Coperture = View Coverages, Servizi = Services, Stazione = Stations, Strutture = Structures/Landmarks, Legenda simboli urbano = Urban Symbology Legend, Vie-Piazze = Streets/Squares, Zone Verdi = Green Zones/Areas, Zone Abitate = Populated Areas, Mare-Laghi-Fiumi = Ocean-Lakes-Rivers, Zona scoperta = Uncovered Areas, Visualizza Coperture = View Coverages, Servizi = Services , Stazione = Stations (Railroad), Strutture = Structures/Landmarks, Legenda simboli urbano = Rural Symbology Legend, Autostrade = Highways, Strade = Streets, Ferrovie = Railroads, Mare-Laghi-Fiumi = Ocean-Lakes-Rivers, Stazioni = Stations (Railroad), Uscite Autostrade = Highway Exits (Courtesy ESRI)

object – point, line, polygon – or field) will have profound implications for the way a database is used and so it is a critical database design task. It is, of course, possible to change between representation types, but this is computationally expensive and results in loss of information.

Logical model

Match to geographic database types. This involves matching the object types to be studied to specific data types supported by the GIS to be used to create and maintain the database. Because the data model of the GIS is usually independent of the actual storage mechanism (i.e. it could be implemented in Oracle, Microsoft Access, or a proprietary file system), this activity is defined as a logical modeling task.

Organize geographic database structure. This includes tasks such as defining topological

associations, specifying rules and relationships, and assigning coordinate systems.

Physical model

Define database schema. The final stage is definition of the actual physical database schema that will hold the database data values. This is usually created using the DBMS software's data definition language. The most popular of these is SQL with geographic extensions (see the Section 11.4), although some non-standard variants also exist in older GIS/DBMS.

11.7 Structuring Geographic Information

Once data have been captured in a geographic database according to a schema defined in a geographic data model, it is necessary to perform some structuring and organization in order to support efficient query, analysis, and mapping. There are two main structuring techniques relevant to geographic databases: topologic creation and indexing.

11.7.1 Topologic creation

The subject of topological creation was covered in Section 9.1.3, but it is revisited here briefly with a more practical treatment.

Topology can be created for vector data sets using either batch or interactive techniques. Batch topology builders are required to handle CAD, survey, simple feature, and other unstructured vector data imported from non-topological systems. Creating topology is usually an iterative process because it is seldom possible to resolve all data problems during the first pass and manual editing is required to make corrections. Some typical problems that need to be fixed are shown in Figures 6.8 and 6.9 and discussed in Section 6.3.2.

Interactive topology creation is performed dynamically at the time objects are added to a database. For example, when adding water pipes using interactive vectorization tools (see Chapter 9), before each object is committed to the database topological connectivity can be checked to see if the object is valid (that is, it conforms to some pre-established database object rules).

11.7.2 Indexing

Geographic databases tend to be very large and, because of this, geographic queries, such as finding all the customers (points) within a store trade area (polygon), can take a very long time (perhaps 10–100 seconds or more for a 10 million customer database). The point has already been made in Section 9.2.3.2 that topologic structuring can help speed up certain types of queries. A second way to speed up queries is to index a database. A database index is very similar to a book index; both are special representations of objects that speed up searching. The standard indexes in database management systems (DBMS) are one-dimensional and are very poor at indexing geographic objects.

A database index is a special representation of information about objects that improves searching.

Many types of geographic indexing techniques have been developed. Some of these are experimental and have been highly optimized for specific types of geographic data. Research shows that even a basic spatial index will yield very significant improvements in spatial data access and that further refinements often yield only marginal improvements at the costs of simplicity, and speed of generation and update. Three main methods of general practical importance have emerged in GIS: grid indexes, quadtrees, and R-trees.

11.7.2.1 Grid index

A grid index can be thought of as a regular mesh placed over a layer of geographic objects (Figure 11.8). The grid location(s) of each object is recorded in a list (the index). A query to locate an object searches the indexed list first to find the object and then retrieves the object geometry or attributes for further analysis (e.g. tests for overlap, adjacency, or containment with other objects on the same or another layer). These two tests are often referred to as primary and secondary filters. Secondary filtering, which involves geometric processing, is much more computationally expensive. The performance of an index is clearly related to the relationship between grid and object size, and object density. If the grid size is too large relative to the size of object, too many objects will be retrieved by the primary filter, and therefore a lot of expensive secondary processing will be needed. If the grid size is too small any large objects will be spread across many grid cells, which is inefficient for

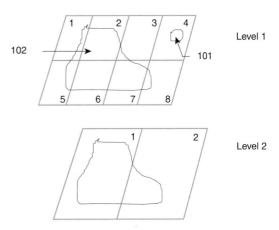

Figure 11.8 A multi-level grid geographic database index

draw queries (queries made to the database for the purpose of displaying objects on a screen).

A grid index is easy to create, can deal with a wide range of object types, and offers good performance.

For data layers that have a highly variable object density (for example, administrative areas tend to be smaller and more numerous in urban areas in order to equalize population counts) a uniform grid size can result in either too many or too few objects being retrieved for optimal performance. A simple solution to this problem is to create a hierarchy of grids at different sizes. Three grid levels are normally sufficient for good performance.

Figure 11.8 shows a layer that has two grid levels. In this particular index, an object is promoted to the next grid level if it is in more than four cells. The polygon object 101 is located in grid cell 4 on level 1. A record is added to the database geographic index table because the object lies within one of the grid cells. Object 102 on the other hand spans seven cells and is therefore promoted to level 2 where it fits within two cells. Object 102 is indexed at level 2 and two records are added to the index table.

Grid indexes are one of the simplest and most robust indexing methods. They are fast to create and update, and can handle widely ranging types and densities of data. For this reason they have been quite widely used in commercial GIS software systems.

11.7.2.2 Quadtree indexes

Quadtree is a generic name for several kinds of index that are built by recursive division of space into quadrants (van Oosterom 1999, Samet 1990). In many respects quadtrees are a special type of grid index. The difference here is that in quadtrees space is recursively split into four quadrants based on data density. Quadtrees can be conceptualized and represented in computer data structures in several different ways, as will be shown below. The term tree is used because the tree metaphor illustrates how objects are indexed (see Figure 11.9). Point and region quadtrees will be discussed here as exemplars because of their general applicability.

Quadtrees are data structures used for both indexing and compressing geographic database layers.

In a point quadtree, space is divided successively into four rectangles based on the location of the points (Figure 11.9). The root of the tree cor-

responds to the region as a whole. The rectangular region is divided into four usually irregular parts based on the x, y coordinate of the first point. Successive points subdivide each new sub-region into quadrants until all the points are indexed.

Region quadtrees are used to store lines, polygons, and rasters. First, each object is rasterized and enclosed by a square (Figure 11.10). The square is then progressively subdivided into blocks of four squares of equal size until the squares are completely inside (gray leaf) or outside (white leaf) the object or the maximum depth of the quadtree is reached.

An interesting variant of the region quadtree is the linear quadtree. Here the quadcodes belonging to each object are stored in a linear sequence (in

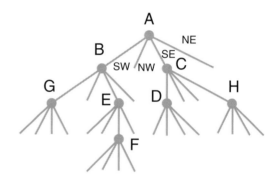

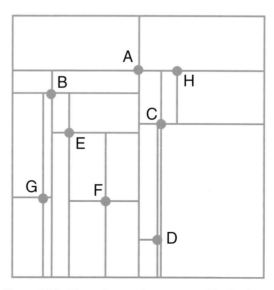

Figure 11.9 The point quadtree geographic database index (Source: after van Oosterom 1999)

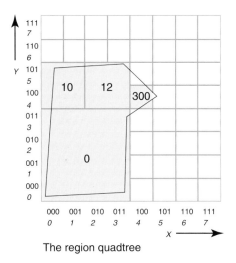

The region quadtree

Figure 11.10 The region quadtree geographic database index (Source: after van Oosterom 1999)

Figure 11.10: 0, 10, 12 and 300). Linear quadtrees are especially useful because of their ease of implementation and the speed at which data can be retrieved.

Quadtrees have found favor in GIS software systems because of their applicability to many types of data (both raster and vector), their ease of implementation and their good performance.

11.7.2.3 R-tree indexes

R-trees group objects using a rectangular approximation of their location called a minimum bounding rectangle (MBR) (see Box 11.2). Groups of point, line, or polygon objects are indexed based on their MBR. Objects are added to the index by choosing the MBR that would require the least

Branching factor M = 4

Figure 11.11 The R-tree geographic database index (Source: after van Oosterom 1999)

expansion to accommodate each new object. If the object causes the MBR to be expanded beyond some preset parameter then the MBR is split into two new MBR. This may also cause the parent MBR to be become too large, resulting in this also being split. The R-tree shown in Figure 11.11 has two levels. The lowest level contains three leaf nodes; the highest has one node with pointers to the MBR of the leaf nodes.

> R-trees are popular methods of indexing geographic data because of their flexibility and excellent performance.

R-trees are suitable for a range of geographic object types and can be relatively quickly created and updated, although they are more complex than the other methods discussed earlier. The spatial datablade extension to the Informix ORDBMS uses R-tree indexes.

11.8 Editing and Data Maintenance

Editing is the process of making changes to a geographic database by adding new objects or changing existing objects as part of data load or update and maintenance operations. A general-purpose geographic database will require many tools for geometry and attribute editing, database maintenance, creating and updating indexes and topology, importing and exporting data, and georeferencing objects.

> Modern GIS offer a wide range of tools for creating and editing vector and raster data. Simple and advanced COGO tools offer many ways to manipulate geographic objects.

11.8.1 Editing tools

Contemporary GIS come equipped with an extensive array of tools for creating and editing geographic object geometries and attributes. This section only introduces the main types of tools available, as space does not permit a comprehensive review. Object coordinates can be added to a geographic database using many methods (digitizing tables, on screen heads-up digitizing, survey, etc. – see Section 10.3.2), but the end result is always *x*, *y* coordinates with optional *z* and *m* values.

Box 11.2 Minimum bounding rectangle

Minimum bounding rectangles (MBR) are very useful structures widely implemented in GIS. An MBR essentially defines the smallest box whose sides are parallel to the axes of the coordinate system that encloses a set of one or more geographic objects. An MBR is defined by the two coordinates at the bottom left (minimum x, minimum y) and the top right (maximum x, maximum y) as is shown in Figure 11.12.

MBR can be used to generalize a set of data by replacing the geometry of the objects in the box with two coordinates defining the box. A second use is for fast searching. For example, all the polygon objects in a database layer that are within a given study area polygon can be found by performing a polygon on polygon contains test (see Figure 11.12) with each object and the study area. If the polygons have complex boundaries (as is normally the case in GIS), this can be a very time-consuming task. A quicker approach is to split the task into two parts. First screen out all the objects that are definitely in and definitely out by comparing their MBR. Because very few coordinate comparisons are required this is very fast. Then use the full geometry outline of the remaining objects to determine containment.

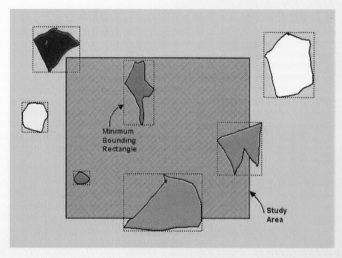

Figure 11.12 Polygon in polygon test using MBR. MBR can be used to determine objects definitely in study area (green), definitely out (yellow), or possibly in (blue). Objects possibly in can then be further analyzed using exact geometry. Note the dark blue object that is actually completely outside, although the MBR suggests it is partially within the study area

11.8.1.1 Simple editing tools

Simple editing tools for bulk digitizing, and basic additions and updates were described in Chapter 10: GIS Data Collection. Using these simple tools, point, line, or polygon geometries can easily be digitized for objects. Simple editing tools can be contrasted with more sophisticated COGO editing tools that create new object coordinates using existing coordinates.

11.8.1.2 COGO editing tools

Coordinate geometry (COGO) tools are used to construct new vertices that add or change objects. Construction tools are defined by the

type of geometric object they create. The tools described here are also described by Zeiler (1999) and are shown in Figure 11.13.

Points

These methods create a single point by specifying angles, distances, and relationships to existing geometries.

Construct Along: given a curve and a distance or ratio, a point is constructed along that curve.

Construct Angle Bisector: given a from-point, through-point, and to-point, this constructor bisects the angle subtended by the three

Construct Along

Construct Parallel

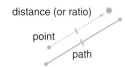

Construct Angle Bisector

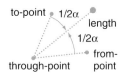

Line Construct Angle Bisector

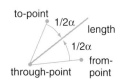

Construct Angle Intersection

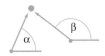

Construct Arc Distance

Construct Deflection

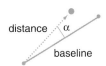

Construct Fillet

Construct Deflection Intersection

Construct Three Points

Construct Offset

Figure 11.13 Example COGO construction tools used to edit geographic databases (Source: after Zeiler 1999)

points and places a point along the bisector at the length specified.

Construct Angle Intersection: given two points and two angles, this constructor places a point at the intersection of the rays defined by the two points and angles.

Construct Deflection: given a line that serves as a baseline, a deflection angle, and a distance, this constructor places a point at the distance along a ray at the deflection angle.

Construct Deflection Intersection: given a line that serves as a baseline, a deflection angle measured from the start point of the baseline, and a deflection angle measured from the endpoint of the baseline, this constructor places a point at the intersection of the defined rays.

Construct Offset: given a path, a distance or ratio along the curve, and an offset distance, this constructor places a point at the offset.

Construct Parallel: given a path with straight lines, a reference point, and a distance, a new point is placed along a parallel curve.

Lines
Line Construct Angle Bisector: given a point, through-point, and to-point and length, bisect that angle and construct a line of that length.

Circular Arcs
A circular arc is a curve defined by a mathematical function (i.e. end point, origin, and curve radius), in contrast to a line that is defined explicitly by a series of coordinates.

Construct Arc Distance: given a center point, start point, and arc distance, this constructor builds a circular arc in a counterclockwise direction.

Construct Fillet: given two segments and a radius length, this constructor builds a circular arc tangent to the two segments.

Construct Three Points: given a start point, middle point, and end point, this constructor builds a circular arc that uniquely fits those points.

11.9 Multi-user Editing of Continuous Databases

For many years, one of the most challenging problems in GIS was how to allow multiple users

to edit the same continuous geographic database at the same time. It is relatively easy to provide multiple users with concurrent read and query access, but more difficult to avoid conflicts and potential database corruption when multiple users want write (update) access.

11.9.1 Transactions

A group of edits to a database, such as the addition of three new land parcels and changes to the attributes of a sewer line, is referred to as a transaction. In order to protect the integrity of databases, transactions are atomic, that is, transactions are either completely committed to the database or they are rolled back (not committed at all). Many of the world's GIS and non-GIS databases are multi-user and transactional, that is, they have multiple users performing edit/ update operations at the same time. For most types of database, transactions take a very short time (sub-second). For example, in the case of a banking system a transfer from a savings account to a checking account takes perhaps 0.01 second. It is important that the transaction is coordinated between the accounts and that it is atomic, otherwise one account might be debited and the other not credited. Multi-user access to banking and similar systems is handled simply by locking (preventing access to) the database records during the course of the transaction. Any attempt to write to the same record is simply postponed until the record lock is removed after the transaction is complete. Because banking transactions, like many others, take only a very short amount of time, users never even notice if a transaction is deferred.

> A transaction is a group of changes that are made to a database as a coherent group. All changes that form part of a transaction are either committed or the database is rolled back.

Although some geographic transactions have a short duration (short transactions), many last much longer. Consider, for example, the amount of time necessary to capture all the land parcels in a city subdivision. This might take a few hours for an efficient operator, for a small subdivision, but an inexperienced operator working on a large subdivision might take days or weeks. This may cause two problems. First, locking the whole or even part of a database for this length of time for a long transaction is unacceptable in many types of application, especially those involving frequent maintenance changes (e.g. utilities and land administration). Secondly, if a system failure occurs during the transaction, work may be lost unless there is a procedure for storing updates in the database.

11.9.2 Versioning

Short transactions use what is called a pessimistic locking concurrency strategy. That is, it is assumed that conflicts will occur in a multi-user database (concurrent users) and that the only way to avoid database corruption is to lock out all other user updates. The term pessimistic is used because this is a very conservative strategy assuming that update conflicts will occur. An alternative to pessimistic locking is optimistic versioning, which is based on the assumption that conflicts are very unlikely to occur, but if they do then software can be used to resolve them.

The two strategies for providing multi-user access to geographic databases are pessimistic locking and optimistic versioning.

Versioning sets out to solve the long transaction and optimistic versioning concurrency problem described above. It also addresses a second key requirement peculiar to geographic databases – the need to support alternative representations of the same objects in a database. In many applications, it is a requirement to allow designers to create and maintain multiple object designs. For example, when designing a new subdivision, the water department manager may ask two designers to lay out alternative designs for water and sewer systems. The two designers would work concurrently to add objects to the same database layers, snapping to the same objects. At some point, they may wish to compare designs and perhaps create a third design based on parts of their two designs. While this design process is taking place, operational maintenance editors could be changing the same objects they are working with. For example, updated surveys of roads and electric outside plant (conductors, devices, and infrastructure) following construction and maintenance may affect the water designs. Figure 11.14 compares linear short transactions and branching long transactions.

Within a versioned database, the different database versions are logical copies of their parents, i.e. only the additions, modifications, and deletions are stored in the database (in a version change table). The process of creating two versions based on the same parent version is called branching. In Figure 11.14(B), Version 4 is a branch from Version 2. Conversely, the process of

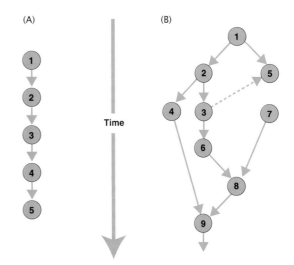

Figure 11.14 Database transactions: (A) Linear short transactions; (B) Branching version tree

combining two versions into one version is called merging; Figure 11.14(B) also illustrates the merging of Versions 6 and 7 into Version 8. A version can be updated at any time with any changes made in another version. Version reconciliation can be seen between Version 3 and 5. Since the edits contained within Version 5 were reconciled with 3, only the edits in Versions 6 and 7 are considered when merging to create Version 8.

There are no restrictions or locks placed on the operations performed on each version in the database. With optimistic versioning, it is possible for conflicting edits to be made within two separate versions, although normal working practice will ensure that the vast majority of edits made will not result in any conflicts (Figure 11.15).

In the event that conflicts are detected, the database management software will handle them either automatically or interactively. If interactive conflict resolution is chosen, the user is directed to each feature that is in conflict and must decide how to reconcile the conflict. The GUI will provide information about the conflict and display the objects in their various states. For example, if the geometry of an object has been edited in the two versions, the user can display the geometry as it was in any of its previous states.

An example will help to illustrate how versioning works in practice. In this scenario, a geographic database of a water network similar to that shown in Figure 9.16, called *Main Plant* (Figure 11.16), is edited by four users for different reasons. Users 1 and 2 are updating the *Main Plant* database with recent survey results

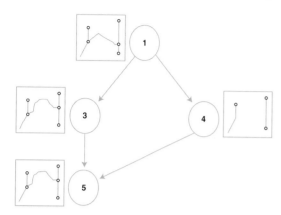

Figure 11.15 Version reconciliation. For Version 5 the user chooses via the GUI the geometry edit made in Version 3 instead of 1 or 4

(*Update 1* & *Update 2*), User 3 is performing some initial design prototype work (*Proposal 1*), and User 4 is following the progress of a construction project (*Plan A*).

Main Plant has a version (1) called *As Built* (this term is used in utility applications to denote the actual state of objects in the real world). This

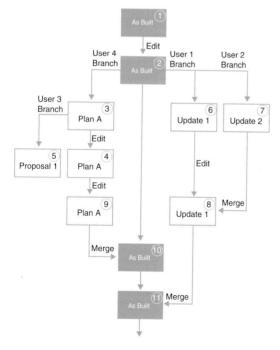

Figure 11.16 A version tree for the *Main Plant* geographic database. Version numbers are in circles

version is marked as the default version of the database. Edits made to this version (1) are saved as a second version (2). User 4 branches the database, basing the new version (3), *Plan A*, on the *As Built* version (2). As work progresses, edits are made to *Plan A* (3) and in so doing two further states of *Plan A* are created (Versions 4 and 9). Meanwhile User 3 creates a version (5) *Proposal 1*, which is based on User 4's *Plan A*, (3). After initial design work, no further action is taken on *Proposal 1* (5), but for a historical record the version is left undeleted. Users 1 and 2 create versions of *Main Plant*, in order for their edits to be made (6 and 7). When Users 1 and 2 are completed with their edits, User 1 reconciles the edits in versions *Update 1* (6) and *Update 2* (7), so that version *Update 1* (8) contains all the edits. On completion of the work by User 4 the changes made in *Plan A* (9) are merged into the *As Built* version (10). Finally, the updates held in version *Update 1* (8) are merged into the *As Built* version (11). Other users continue to read the *Main Plant* database while these edits are being performed, unaware of the other versions of *Main Plant* until they are made public, or are merged into the *As Built* version.

Although the *As Built* version is not directly edited by end users, conceptually each user is editing his or her own version of the *As Built* version. Since the versions of the *As Built* version are only logical copies, the system can be scaled to many more users without the need for vast increases in disk storage capacity. The changes to the database are actually performed with a GIS map editor using tools like those described earlier in Section 11.5.

11.10 Conclusion

Database management systems are now a vital part of all modern operational GIS. They bring with them standardized approaches for storing and, more importantly, accessing and manipulating geographic data using the SQL query language. GIS provide the necessary tools to load, edit, query, analyze, and display geographic data. DBMS require a database administrator (DBA) to control database structure and security, and to tune the database to achieve maximum performance.

Questions for Further Study

1. What are the advantages and disadvantages of storing geographic data in a DBMS?

2. Describe five database data types and five functions for manipulating geographic data in a database.
3. Why are there multiple methods of indexing geographic databases?
4. Locate a geographic database and draw a diagram showing the tables and the relationships between them. Which are the primary keys, and which keys are used to join tables?
5. Is SQL a good language for querying geographic databases?

Online Resources

NCGIA Core Curricula (www.ncgia.ucsb.edu/pubs/core.html):
Core Curriculum in GIScience, Sections 3.1 (Making it Work, Hugh Calkins and others)
Core Curriculum in GIS, 1990, Units 60 (System Planning Overview), 66 (Database Creation), 67 (Implementation Issues), and 68 (Implementation Strategies for Large Organizations)
Standards for geographic information:
www.opengis.org OpenGIS Consortium
www.statkart.no/isotc211/welcome.html ISO TC 211
www.safe.com Safe Software – data conversion
General definitions of database and other technology:
www.whatis.com

Reference Links

Maguire D J, Goodchild M F, and Rhind D W (eds) 1991 *Geographical Information Systems:*

Principles and applications. Harlow, UK: Longman (Text available online from 'Links to Big Book 1' at www.wiley.com/gis and www.wiley.co.uk/gis).
Chapter 18, Database management systems (Healey R G)
Longley P A, Goodchild M F, Maguire D J, and Rhind D W (eds) 1999 *Geographical Information Systems: Principles, techniques, management and applications.* New York: John Wiley.
Chapter 26, Relational databases and beyond (Worboys M F)
Chapter 27, Spatial access methods (van Oosterom P)
Chapter 29, Principles of spatial database analysis and design (Bédard Y)

References

Codd E 1970 A relational model for large shared data banks. *Communications of the ACM* **13**: 377–387.
Date C J 1995 *Introduction to Database Systems.* (6th edn) Reading, Mass.: Addison-Wesley.
ESRI 1997 *Understanding GIS: the ArcInfo method..* Redlands, California: ESRI Press.
OGC 1999 OpenGIS simple features specification for SQL, Revision 1.1. Available at www.opengis.org
Samet H 1990 *The Design and Analysis of Spatial Data Structures.* Reading, Mass.: Addison-Wesley.
Zeiler M 1999 *Modeling our World: The ESRI guide to geodatabase design.* Redlands, California: ESRI Press.

VISUALIZATION AND USER INTERACTION

12

This chapter reviews the ways that users interact with GIS in order to produce usable information about spatial distributions. The paper map is the most established and conventional means of displaying data, yet GIS provides a far richer and more flexible medium for portraying attribute distributions and transforming spatial objects. Standard cartographic conventions and graphic symbology are discussed, as are the range of transformations that are used in choropleth mapping and in cartograms. The visual interface of GIS allows users to interpret, validate, and explore univariate and multivariate data. The chapter also reviews the specialized types of mapping that are appropriate for particular applications areas, and the development of virtual reality systems.

Learning objectives

At the end of this chapter you will understand:

● The rudiments of effective data display;

● Some of the main ways in which mapping can mislead;

● The ways in which good user interfaces can help to resolve spatial queries;

● How displays are customized to the requirements of particular applications.

12.1 Introduction: Uses, Users, Messages, and Media

"Roll up that map; it will not be wanted these ten years." British Prime Minister William Pitt the Younger made this remark after hearing of the defeat of British forces at the Battle of Austerlitz in 1805, where it became clear that his country's military campaign in Continental Europe had been thwarted for the foreseeable future. The quote illustrates the crucial historic role of mapping as a tool of *decision support* in warfare, in a world in which nation states were far more insular than they are today. It also identifies two other defining characteristics of the use of geographic information in 19th Century society. First, a principal, straightforward purpose of terrestrial mapping was to further national interests by infantry warfare. Second, the time frame over which changes in geographic activity patterns unfolded was, by today's standards, incredibly slow – Pitt envisaged that no British citizen would revisit this territory for a quarter of a (then) average lifetime!

Today's human society is immeasurably more complicated, while scarcely a month goes by without some advance in our understanding of the intricate detail of the natural world. Mapping remains an important medium for charting and anticipating change, and, as we accelerate through the 21st Century, the functions that mapping fulfils are becoming ever more diverse and demanding (Figure 12.1). It is certainly inconceivable that any senior decision-maker today (much less a senior politician of any country) could function without mapping.

Maps are important decision support tools.

Effective decision support requires that the message of the map is readily interpretable in the mind of the decision-maker. Yet sometimes the principal function of a map is not simply to marshal and transmit known information about the world, but rather to create or reinforce a particular message.

The historic map shown in Figure 12.2, and the case study to which it relates (Box 12.1), was not a military application, but was nonetheless used as a tool of warfare.

Historically, the origins of many national mapping organizations can be traced to the need for mapping for the campaigns of infantry warfare, for colonial administration, and for defense. Today such organizations fulfil a far wider range of needs of many more types of users (see Section 17.1.2). Although the military remains a heavy user of mapping, such territorial changes as arise out of today's conflicts reflect a more subtle interplay of economic, political, and historical considerations – although the threat or actual deployment of force remains of course a pivotal consideration. Today, GIS-based terrestrial mapping supports a range of purposes – such as the deployment of humanitarian aid (Figure 12.3(A)), and the partitioning of territory through negotiation rather than force. The time frame over which events unfold is also very much more rapid: it is inconceivable to think of politicians, managers, and officials being able to neglect geography for months, weeks, or even days, never mind years. And, unlike Haushofer's map (Figure 12.2) the same mapped data can be used to support a range of different worldviews, as in negotiation of

With The Wrong Map, This Could Be The Road To Nowhere.

Figure 12.1 GIS-based mapping is a necessity for navigating through an increasingly complex world of artificial structures, as illustrated in this excerpt from a trade advertisement (Courtesy: Convergent/Schlumberge)

Box 12.1   GIS and geopolitics

In the early 20th century the Oxford, UK, geographer Halford Mackinder proposed a theory that different national groupings organized around their own historic settlement *heartlands*. He worked at a time when the development of railroads and roads for troop movement was opening up previously inaccessible areas of western Russia, and he envisioned this new territory as a new heartland at the center of an increasingly integrated world. Indeed, in geopolitical terms, he saw western Russia as the center of the new joined up world, and believed that whoever occupied it would come to dominate the emerging *world island*. The German geographer Friedrich Ratzel subsequently developed the concept of the *heartland* into that of *Lebensraum*, the essence of which was that every different cultural group needed to develop its own geographic living space. In the years immediately prior to World War Two (1939–45) these related ideas of geographic heartland and Lebensraum were developed by General Haushofer, Professor of Geography at the University of Munich. The Lebensraum of the German national grouping (the German Volk), he argued, had been overly constrained and dissected by the Treaty of Versailles at the end of World War One. Haushofer crudely mapped the spatial distribution of "German-speaking people" (Figure 12.2) to substantiate this thesis, with the implication that that there was a need to extend the German cultural realm deep into Russia. This argument was subsequently used by the Nazi Party to justify German incursions into Czechoslovakia, the annexation of Austria, and the invasion of Poland that ultimately triggered World War Two. The extensive use of mapping in setting out this geopolitical thesis (termed Geopolitik) provides early and powerful evidence of the ways that maps can be used as propaganda. Haushofer's maps portray information in a particular way in order to garner support for a particular worldview and course of decision-making. They illustrate the general point that the map creator is very much the arbiter and architect of what the map portrays. In Haushofer's case, the message of the map ultimately contributed to a pseudo-scientific justification of an ideology that history would rather forget – yet is periodically rekindled in various guises by extremist groups.

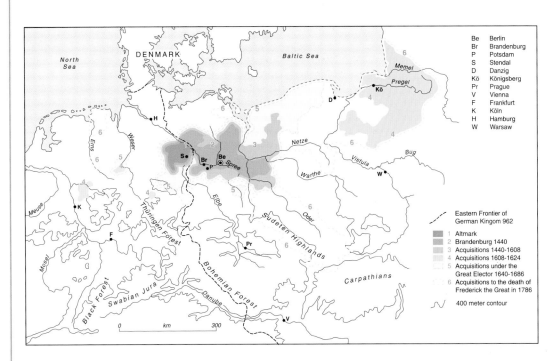

Figure 12.2   The geopolitical center of Germany (Source: Parker G 1988 *Geopolitics: Past, present, future.* London: Pinter)

(A)

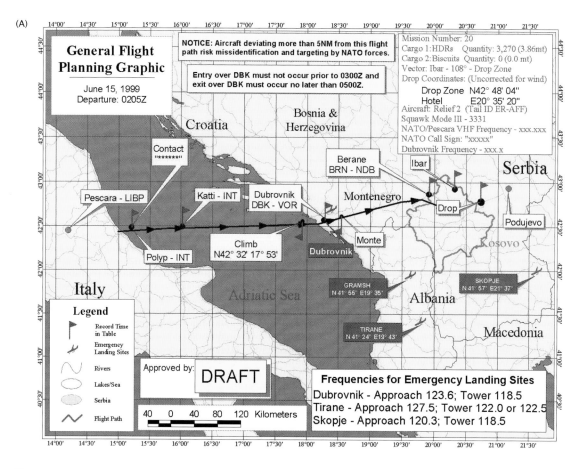

(B)

Figure 12.3 Today's military uses of mapping include (A) general flight planning and detailed drop locations in relief efforts, as in the 1999 Bosnian conflict. (B) Processing and analysis is performed using GIS, although paper maps are still the most convenient medium for some field operations (Courtesy ESRI)

the partitioning of Bosnia in 1999 (see Figure 1.1). Yet if today's conflicts are to be resolved in more protracted time scales, mapping must be comprehensive, up-to-date, clearly intelligible, and safe to use (see Chapters 17 and 19).

The maps used by Pitt and Haushofer were, of course, printed on paper. As we saw in Section 3.7, these particular forms of representations have been important in the past, and remain in wide use because of their transportability, their ease of use, and the straightforward application of printing press technology that they entail. They are also amenable to conveying straightforward messages and supporting decision making in narrowly-defined applications (Figure 12.3(B)). Yet the increasing detail of our understanding of the natural environment and the accelerating complexity of society means that the messages that mapping can convey are increasingly sophisticated and, for the reasons set out in Chapter 6, uncertain. Greater democracy and accountability, coupled with the increased spatial reasoning abilities that better education brings, mean that more people than ever feel motivated and able to contribute to all kinds of spatial policy. This makes the decision support role immeasurably more challenging, varied, and demanding of visual media. Today's mapping must be capable of communicating the widest range of messages and emulating the widest range of "what if" scenarios.

Figure 12.4 Representations can today be brought into direct contact with the real world using wearable computers

GIS makes mapping easy: but (GIS or conventional) mapping can mislead.

The visual medium of a given application must also be open to the widest community of users. Technology has led to the development of an enormous range of devices to bring visualization to the greatest range of users in the widest spectrum of decision environments. In-vehicle displays, palm top devices and wearable computers are all important in this regard (Figure 12.4). Most important of all, the innovation of the Internet makes "societal representations" of space a real possibility for the first time (see Harder (1998) for an overview and interpretation).

12.2 Visualization Conventions and Techniques

12.2.1 Maps and media

The paper map entails a particular set of conventions and constraints in depicting the world, which are less binding when GIS is used for mapping.

- The paper map is of fixed scale. *Generalization* procedures (Chapter 7) are invoked in order to maintain clarity during map creation. This detail is not recoverable, except by reference back to the data from which the map was compiled. The *zoom* facility of GIS can allow mapping to be viewed at a range of scales, and detail to be filtered out as appropriate at a given scale.

- The paper map is of fixed extent, and adjoining map sheets must be used if a single map sheet does not cover the entire area of interest. (An unwritten law of paper map usage is that the most important map features always lie at the intersection of four paper map sheets!) GIS, by contrast, can provide a seamless medium for visualizing space, and users are able to *pan* across wide swathes of territory.

- Most paper maps present a *static* view of the world. Although there are some very honorable exceptions (notably Minard's graphic of Napoleon's Russian campaign in 1812: Tufte 1983, 40-1), conventional paper maps and charts are not adept at portraying dynamics. GIS-based representations are able to achieve this through animation.

- The paper map is *flat* and hence limited in the number of perspectives that it can offer on three-dimensional data. *3D visualization* is much more effective within GIS.

Box 12.2 Visualizing continuous and discrete variation

Figure 12.5(A) presents an illustration of circumstances where attribute reclassi-fication may be desirable. It shows interval scale measures of income variation (modeled using sample data from a social survey) across the city of Bristol, UK, illustrated using continuous color variation. Imagine that you are required to use this map to assess the incidence of poverty at particular locations in the city. It is not easy for the human eye to detect the difference of the colour hue at specific points such as "A" and "B" on the map, and hence to identify the difference between incomes at these points with any degree of certainty. By contrast, in Figure 12.5(B) the continuous values have been categorized into five income classes. This enables a clearer comparison to be drawn between these two widely spaced locations. A more intelligible map still is presented in Figure 12.5(C) where incomes are classified simply as lying either above or below the deemed poverty line.

(A)

(B)

(C)

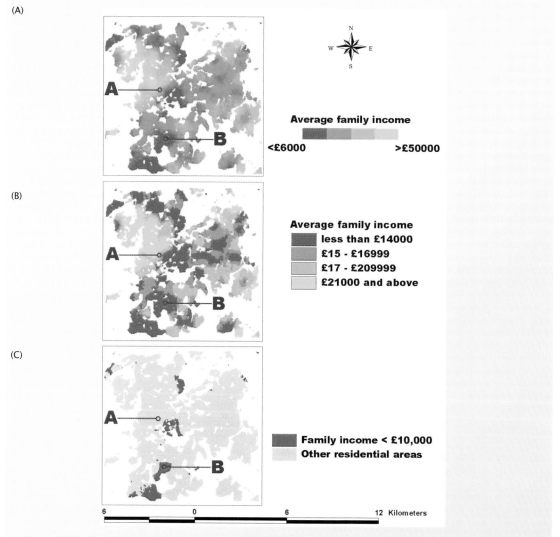

Figure 12.5 The geographical distribution of household incomes in Bristol, UK: (A) a continuous surface; (B) the spatial distribution of four income categories; and (C) areas with incomes beneath the poverty threshold (Courtesy Richard Harris)

- Paper maps provide a view of the world as essentially *complete*. GIS-based mapping allows the *supplementation* of base map material with further data.

- Paper maps provide a single, *map producer-centric*, view of the world. GIS users are able to create their own, user-centric, images in an *interactive* way.

In the remainder of this chapter, we will use the terms "visualization" and "mapping" interchangeably, although the latter sometimes has stronger connotations with paper mapping.

GIS is a flexible medium for the production of maps.

12.2.2 Measurement scales revisited

The medium of GIS is still quite new to mapping, yet many of the conventions that are used to represent spatial objects and their attributes are much more established. However, the objective of good mapping is not always a straightforward one. Whether using GIS or paper, mapping may entail reclassification or transformation of attribute measures, as illustrated by the example in Box 12.2.

Good mapping requires that spatial objects and their attributes can be readily interpreted in applications. In Chapter 3 attributes were measured on the nominal, ordinal, interval, ratio, or cyclic scales (Box 3.3), while in Chapter 5 spatial objects were classified into points, lines, areas, and surfaces (Box 5.2). Attribute measures that we think of as continuous can be discretized to levels of precision according to needs (Box 12.2). The representation of spatial objects is similarly imposed – cities might be captured as points, areas, mixtures of points, lines and areas (as in a street map), or 3D "walk-throughs", depending on the base scale of a representation and the importance of city objects to the application. Measurement scales and spatial object types are thus one set of conventions that are used to abstract reality.

The process of mapping attributes frequently entails further problems of *classification* because many spatial attributes are inherently uncertain (Chapter 6). For example, in order to create a map of occupational type, individuals' occupations will be classified first into socio-economic groups (e.g. "factory worker") and perhaps then into super groups, such as "blue collar". At every stage in the aggregation process we inevitably do injustice to many individuals who perform a mix of white- and blue-collar, intermediate and skilled

functions into a single group (what social class is a professional diver or frogman?). In practice, the validity and usefulness of an occupational classification will have become established over repeated applications, and the task of mapping is to convey thematic variation in as efficient a way as possible.

The classifications used in mapping are often subjective.

12.2.3 Attribute representation and transformation

Humans are good at interpreting visual data – much more so than interpreting numbers, for example – but conventions are still necessary to impart the map's message. Many of these conventions relate to the use of symbols (such as the use of highway shields to denote route number on many US medium- and fine-scale maps) and colors (blue for rivers, green for forested areas, etc.), and have developed over the past 200 years. Many originate in the use of paper maps in infantry warfare, and the 19th Century infantryman would still recognize their counterparts today. Mapping of different themes (such as vegetation cover, surface geology, and socio-economic characteristics of human populations) has a more recent history. Here too, however, mapping conventions have developed for particular applications classes.

Attribute mapping entails use of *graphic symbols*, which (in two dimensions) may be referenced by points (e.g. historic monuments, telecoms antennae), lines (e.g. roads), or areas (e.g. forests, urban areas). Basic point, line, and area symbols are modified in different ways in order to communicate different types of information. The ways in which these modifications take place adhere to cognitive principles and the accumulated experience of applications. The nature of these modifications was first explored by Bertin in 1967, and was extended to the typology illustrated in Figure 12.6 by MacEachren (1995). The *size* of point and line symbols is varied principally to distinguish between the values of ordinal and interval/ratio data using *graduated symbols*, such as the arrows shown in Figure 12.7. This figure also illustrates how *orientation* can be used to depict the inherent or associated properties of locations, such as ocean current direction, and how *value* and *saturation* of color represent variation in attribute data. *Hue* refers to the use of color, principally to discriminate between nominal categories, as in agricultural or urban land use maps (Figure 12.8). Different hues may be combined with different textures or shapes (see below) if there is a large number of categories, in order to avoid difficulties of interpretation. *Shape* of map symbols can be used either to communicate information about a spatial attribute (e.g. a viewpoint, or the start of a walking trail), or its spatial location (e.g. the location of a path or boundary of a particular type: Figure 12.8). *Arrangement*, *texture* and *focus* refer to within- and between-symbol properties that are used to signify pattern. A final graphic variable in the typologies of MacEachren and Bertin is *location* (not shown in Figure 12.6), which refers to the practice of offsetting the true coordinates of objects in order to improve map intelligibility, or changes in map projection. We discuss this in more detail in Section 12.3 below. Some of the common ways in which these graphic

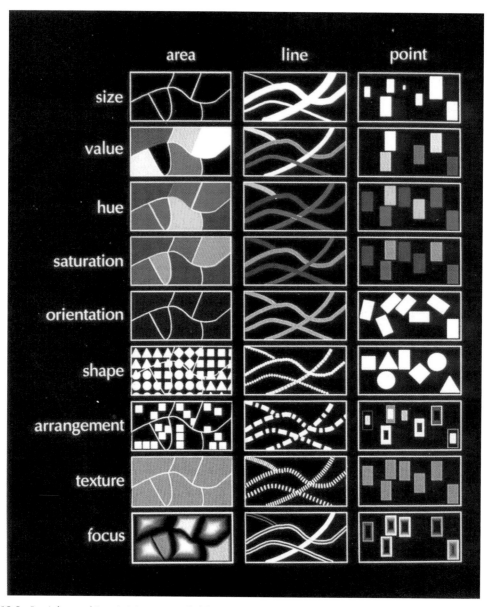

Figure 12.6 Bertin's graphic primitives, extended from seven to ten variables (the variable location is not depicted). (Source: MacEachren 1994)

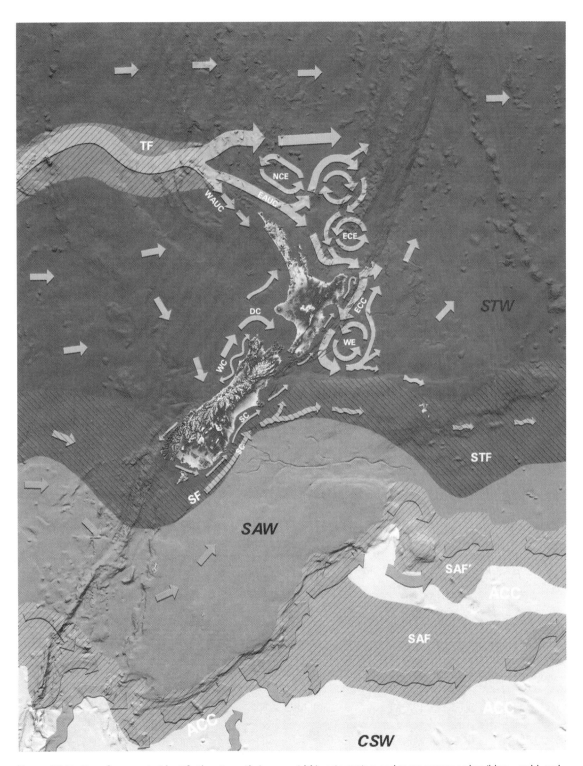

Figure 12.7 Use of arrows to identify the strength (arrow width), orientation and temperature value (blue = cold, red = warm) of ocean currents around New Zealand

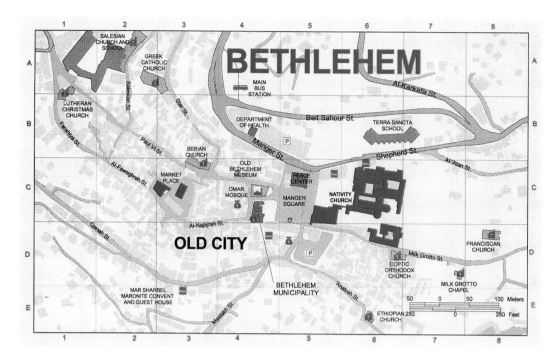

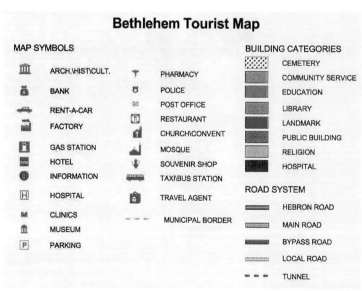

Figure 12.8 Use of hue to discriminate between urban land use categories, and of symbols and shape to communicate occasional and other attribute information (Source: ESRI)

variables are used to visualize spatial object types and attributes are shown in Table 12.1.

The selection of appropriate graphic variables to depict spatial locations and distributions presents one set of problems in mapping. A related task is how best to position them on the map, so as to optimize map interpretability. The representation of nominal data by graphic symbols and icons is apparently trivial, although in practice automating placement presents a

Table 12.1   Common methods of mapping spatial object types and attribute data

Attribute type	Nominal	Ordinal	Interval/Ratio
Spatial object type			
Point	Symbol map (each category a different class of symbol), and/or use of lettering	Hierarchy of symbols or lettering	Outline map with bar charts or proportional circles/symbols
Line	Network map (links between phenomena are present or absent)	Graduated thickness or graduated color lines	Flow map, with width of lines proportional to flows
Area	Unique category choropleth map	Graduated color or shading map	Continuous hue/shading choropleth map; dot density map; contoured surface
Volume/surface	Prism map	Prism map	Prism map; DEM, surface model with color intensity/shading of attributes

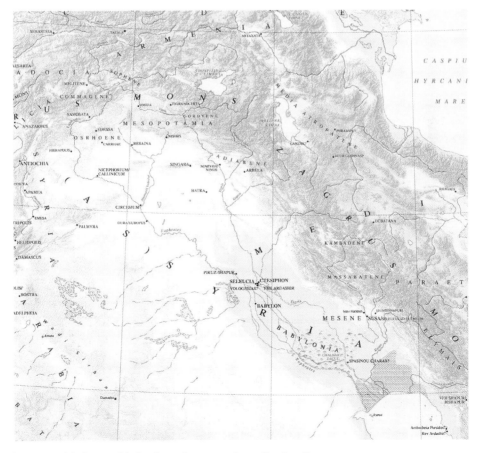

Figure 12.9   Feature labels are added to linear features using spline functions

range of challenging analytical problems. These include avoidance of overlap, alignment, and positioning. Most GIS include generic algorithms for positioning labels and symbols around geographic objects. Point labels are positioned to avoid overlap by creating a window (often invisible to the user) around text or symbols. Linear features, such as rivers, roads, and contours, are labeled using *splines* (Figure 12.9), or distinguished by use of color. Area labels are assigned to central points, using geometric algorithms similar to those used to calculate geometric *centroids* (Sections 13.3 and 14.2.1). These generic algorithms are frequently customized to accommodate common conventions and rules for particular classes of application – such as topographic, utility, and seismic maps, for example. Generic and customized algorithms also include color conventions for map symbolization and lettering.

Ordinal attribute data are assigned to point, line, and area objects in the same rule-based manner, with the ordinal property of the data accommodated through use of a hierarchy of graphic variables (symbol and lettering sizes, types, colors, intensities, etc.). Most users are unable to differentiate between more than seven ordinal categories, and this provides an upper limit on the extent of the hierarchy.

A wide range of conventions is used to visualize interval and ratio scale attribute data. Proportional circles and bar charts are often used to assign interval or ratio scale data to point locations. Figure 12.10 illustrates how a ratio scale measure, earthquake intensity on the Richter Scale, may be assigned to the epicenter of each occurrence. Variable line width (with increments that correspond to the precision of the interval measure) is a standard convention for representing continuous variation in flow diagrams (Figure 12.7). The range of descriptive summaries of quantitative data (scatterplots, histograms, and pie charts) is discussed in greater detail in Section 14.2.

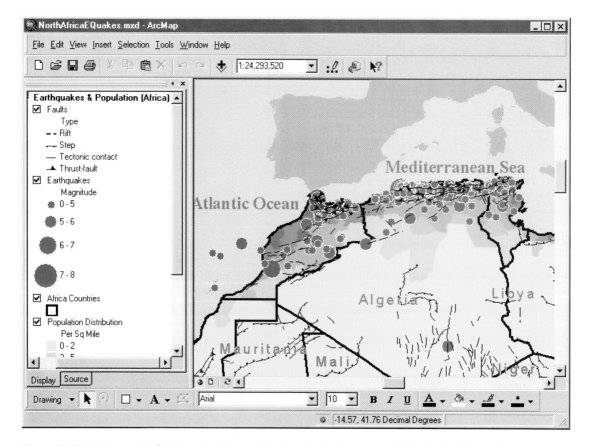

Figure 12.10 Assignment of ratio scale data to point locations. The centers of the circles identify the epicenters of earthquakes, and their sizes identify their intensity, as measured on the Richter Scale (Source: ESRI)

There is a variety of ways of ascribing continuous attribute data to areal entities that are pre-defined. In practice, none is unproblematic, however. The standard method of depicting areal data is in zones. However, as was discussed in Box 5.4, the choropleth map brings the dubious visual implication of within-zone uniformity of attribute value. Moreover, conventional choropleth mapping also allows any large (but possibly uninteresting) areas to visually dominate the map. A variant on the conventional choropleth map is the dot density map, which uses points as a more aesthetically pleasing means of representing the relative density of zonally averaged data – but not as a means of depicting the precise locations of point events. Proportional circles provide one way around this problem, since the magnitude of attribute counts is clearly scaled, and the circle can be centered on any convenient point within a zone. However, there is a tension between using circles that are of sufficient size to convey the variability in the data and the problems of overlapping circles on "busy" areas of the map (Figure 12.10). Circle positioning also entails the same kind of positioning problem as that of name and symbol placement, described above.

If the richness of the visualization is to be equivalent to that of the representation from which it was derived, the intensity of color or shading should directly mirror the intensity or magnitude of attributes, like the variability of incomes that was shown in Figure 12.5(A). The human eye is adept at discerning continuous variation in color and shading, and Waldo Tobler (Box 5.1) has advanced the view that continuous scales present the best means of representing geographic variation. There is no natural ordering implied by use of different colors, and the common convention is to represent continuous variation on the red–green–blue (RGB) spectrum. In a similar fashion, difference in the hue, lightness, and saturation (HLS) of shading is used in color maps to represent continuous variation. International standards on intensity and shading have been formalized.

Classification procedures are used in map production in order to ease user interpretation.

Four basic classification schemes have been developed to divide interval and ratio data into categories:

(a) Natural breaks, in which classes are defined according to apparently natural groupings of data values. The breaks may be imposed on the basis of break points that are known to be relevant to a particular application, such as fractions and multiples of mean income levels, or rainfall thresholds known to support different thresholds of vegetation ("arid", "semi-arid", "temperate", etc.). This is "top down" or deductive assignment of breaks. Inductive ("bottom up") classification of data values may be carried out by using GIS software to look for relatively large jumps in data values, as shown in Figure 12.11(A).

(b) Quantile breaks, in which each of a predetermined number of classes contains an equal number of observations. Quartile (four category) classifications are widely used in statistical analysis, while quintile (five category) classifications are well suited to the spatial display of linearly distributed data. Yet because the numeric size of each class is rigidly imposed, the result can be misleading. The placing of the boundaries may assign almost identical attributes to adjacent classes, or features with quite widely different values to the same class. The resulting visual distortion can be minimized by increasing the number of classes – assuming the user can assimilate the extra detail that this creates. See Figure 12.11(B).

(c) Equal interval breaks. These are best applied if the data ranges are familiar to the user of the map, such as temperature bands. See Figure 12.11(C).

(d) Standard deviation classifications show the distance of an observation from the mean. The GIS calculates the mean value and then generates class breaks in standard deviation measures above and below it. Use of a two-color ramp helps to emphasize values above and below the mean (shown in blue and red, respectively, in Figure 12.11(D)).

The choice of classification is very much the outcome of convenience and accumulated experience. The automation of mapping in GIS has made it possible to evaluate different classifications. Looking at distributions such as those in the figures allows us to see if the distribution is strongly skewed – which might justify using unequal class intervals in a particular application. (A study of poverty, for example, could quite happily class millionaires along with all those earning over $50,000, as they would be equally irrelevant to the study.)

The interactive environment of GIS also allows us to omit extreme observations if they are unhelpful to portraying the general characteristics of the spatial distribution, and to recalculate class intervals after "bad" data have been omitted (Sections 14.2 and 15.4).

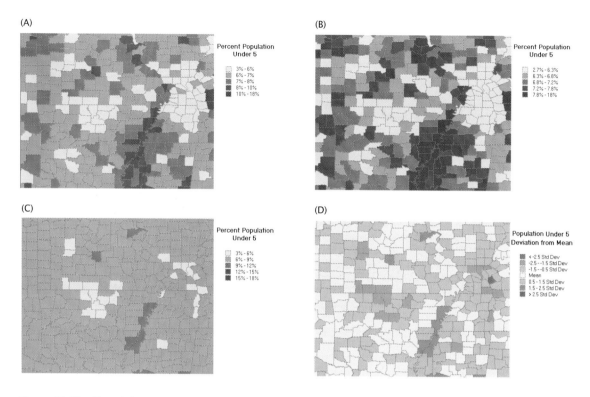

(A)

(B)

(C)

(D)

Figure 12.11 Class definition using (A) natural breaks; (B) quantile breaks; (C) equal interval breaks; and (D) standard deviation breaks (Source: ESRI)

12.2.4 Spatial object representation and transformation

General representational conventions have been adapted to make mapping fit for purpose – the width, color, and labeling of roads being to improve general navigation functions, for example. The scale at which a representation is visualized is also important – a major river might be represented as a dash on a fine-scale map, but as a patch on a coarser-scale representation. The degree of line and boundary generalization may be controlled by algorithms, which (like those for name placement) may be customized for particular classes of application. Similar algorithms are used to control the degree of terrain roughness in digital terrain modeling. Conversely, customized algorithms such as fractal techniques (Section 5.8) may be used to enhance the visual plausibility of the structure and character of boundaries and surfaces.

The usual state-of-affairs is that a mapped attribute takes on the proportions of the zone to which it pertains. Yet it is possible to manipulate the shape and form of mapped boundaries using GIS, and there are many circumstances in which

it may be appropriate to do so. Indeed, where a standard mapping projection obscures the message of the attribute distribution, map transformation becomes necessary. There is nothing untoward in doing so – remember that one of the messages of Chapter 4 was that most conventional mapping entails some transformation. The most obvious example of this was the task of representing extensive parts of the curved surface of the Earth on a flat screen or a sheet of paper. There is nothing sacrosanct about popular or conventional representations of the world, although the widely used transformations and projections do confer advantages in terms of wide user recognition, hence interpretability.

Such transformations are often termed *cartograms*. Cartograms lack planimetric correctness, and distort area or distance in the interests of some specific objective. The usual objective is to reveal patterns that might not be readily apparent from a conventional map or, more generally, to promote legibility. Thus the integrity of the spatial object, in terms of areal extent, location, contiguity, geometry, and topology, is made subservient to an emphasis

(A)

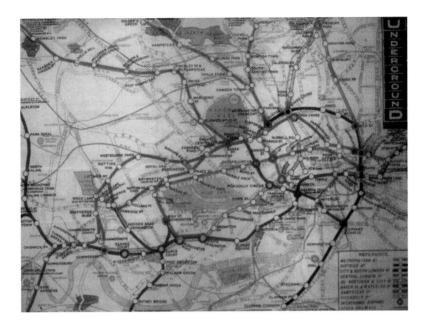

(B)

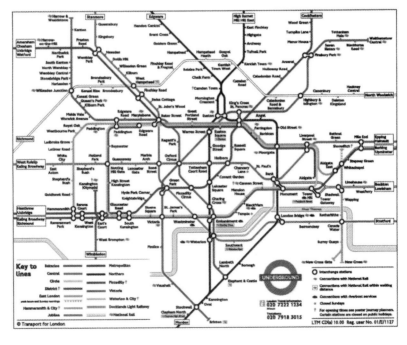

Figure 12.12 (A) A 1912 map of the London Underground; and (B) its post-1933 incarnation in cartogram form (Source: London Transport Museum)

upon attribute values or particular aspects of spatial relations. One of the best known cartograms (strictly speaking a *linear* cartogram) is the London Underground map. Figure 12.12(A) is an early map of the Underground, which shows

the locations of stations on the Circle Line in relation to the real-world pattern of streets. As the network developed, however, a draughtsman named Harry Beck was commissioned to devise a map to fulfil the specific purpose of helping travelers to navigate through the network. The London Underground map shown in Figure 12.12(B) is a descendant of Beck's 1933 map. It is a widely recognized representation of connectivity in London, and the conventions that it utilizes are well-suited to the attributes of spacing, configuration, scale, and linkage of the London Underground system. The attributes of public transit systems differ between cities, as do the cultural conventions of transit users, and thus it is unsurprising that cartograms pertaining to transit systems elsewhere in the world appear quite different. Similar diagrams are common in the utility industry (Section 9.3), where cartometric flow diagrams are useful in maintaining supply – although standard map representations are of more use for repair and maintenance functions.

Cartograms are map transformations that distort area or distance in the interests of some specific objective.

A central tenet of Chapter 3 was that all representations are abstractions of reality. Cartograms depict artificial realities, using particular exaggerations that are deliberately chosen. In Figure 12.13 each political constituency in the UK is represented using a hexagon of uniform size – which is not unreasonable, since political constituencies are of similar population size in reality. The effect of this is to make mapped area approximately proportional to population size. The result is that the city conurbations (the outlines of which are marked) and other densely populated urban areas swell in size, while the extent of sparsely populated rural constituencies (particularly the Highlands of Scotland and North and Mid Wales) diminish. Emphasis of the population size attribute as the base to map design eliminates the visual dominance of spatially extensive, but in attribute terms less significant, zones. However, the real world pattern of zone contiguity and topology is necessarily compromised in order to achieve this.

Throughout this book, one of the recurrent themes has been the value of GIS as a medium for data sharing, yet this is most easily achieved if coordinates constitute a common spatial reference framework. Cartometric transformations allow attribute values to exaggerate or diminish the representation of space rather than constraining it within conventional projections. There are many phenomena that we do not always conceptualize using the metrics and geometries that were discussed in Chapter 5. A range of cartometric transformations is possible to accommodate the way that humans think about

things, as with the measurement of distance as travel time (Mark 1999). Yet the geography of one attribute (e.g. adult population) is unlikely to map perfectly onto any other (e.g. juvenile population), and the user is likely to end up with a range of unfamiliar representations. However, just because one arbitrary representation is widely recognizable and is accepted does not mean that it is the most effective means of envisioning a particular spatial distribution. The visual transformations inherent in cartograms can thus provide improved means of envisioning spatial structure.

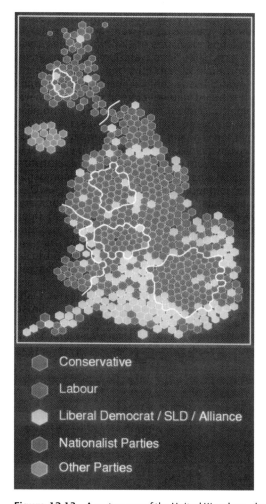

Figure 12.13 A cartogram of the United Kingdom, showing the borders between England, Scotland, and Wales and the extent of the major conurbations. Each political constituency is identified by a hexagon of equal size. The hexagons show which political party came second in the 1992 General Election (Source: Daniel Dorling)

12.3 Interacting with Representations through Visualization

12.3.1 Scientific visualization and representation

It is through mapping that the meaning of a spatial representation of the real world is communicated to users. Historically, the paper map was the only available interface between the map-maker and the user: it was impermeable, contained a fixed array of attributes, was of fixed and invariant scale, and rarely provided any quantitative or qualitative indications as to whether it was safe to use (Figure 12.14). These attributes severely limit the usefulness of the paper map in today's applications environment, which is of seemingly unfathomable complexity by comparison, and entails visualization of data which are richer, continuously updated, and scattered across the Internet.

For many, visual display has always been the essence of GIS, and indeed the origins of GIS as we understand it today partly lie in the development of computer-assisted cartography. More recently, faster processing hardware and more sophisticated computer graphics, including animation, has led to use of the term *visualization in scientific computing* (ViSC) to describe the use of new technology and media to convey the multi-faceted messages of today's mapping. ViSC provides an altogether more flexible, sophisticated, and interactive window on the world than the paper maps of the past. It is nevertheless guided by many of the same cognitive principles, that is, the ways that map users consciously think and reason about map data and perceive spatial phenomena through mapping.

The improved quality of visual images that new technology and media bring cannot compensate for the inherent limitations of a weak represen-tation, and should not be used to try to mask those aspects of representations that are in-herently uncertain (Chapters 5 and 6). Good ViSC minimizes the creation of further uncertainty at the eye–brain interface, which is where the user interprets a spatial representation. Moreover, ViSC has developed to the point where the original uncertainties in the creation of a representation

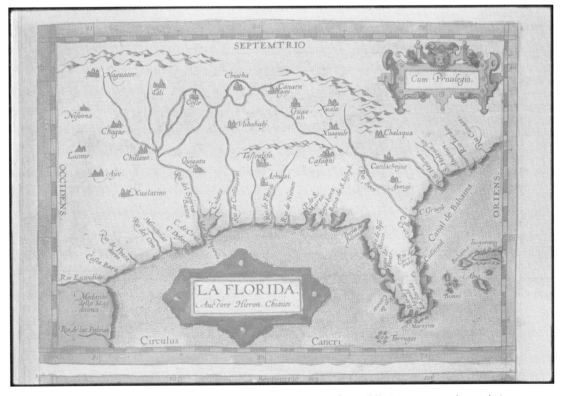

Figure 12.14 An historic paper map of Florida. Early mapping was often wildly inaccurate and speculative, yet even today's paper maps do not suggest that they may be fallible and contain errors

can be made apparent to the user, using a variety of media. ViSC now entails more than a one-way transfer of geographic informa-tion from machine to user – it not only supports the visual interpretation of spatial distributions, but also data quality assessment and data exploration. As such, visualization is very much part of the geographic information science mainstream.

In an ideal representation, we would like to locate every single relevant event or occurrence with perfect accuracy (Section 15.1.2.1), and to measure the full range of attributes of clearly definable *geographic individuals* (Section 6.2) at levels of detail commensurate with the require-ments of any given application. We would then need to devise a means of visualizing this ideal digital representation in a way that was clear and intelligible to the decision-maker. However, we have already shown in Chapters 5 and 6 that there are conceptual and practical reasons why a representation is inevitably a selective simplifica-tion of reality, and there are also good reasons why visualization is also inherently selective.

As a general scientific principle, we never discard data, unless we have grounds for believing that (for reasons of inaccuracy or imprecision) they make no net information contribution towards a representation. Even then, data should not be discarded just because we think they are "bad" – that is, they do not conveniently correspond to the general trend within the rest of a dataset. Yet "bad" data can obscure the principal message of a representation. Although the human eye is a sophisticated means of channeling infor-mation to the brain, it is not always realistic to seek to visualize in a single map all of the wealth of detail that is available through today's digital representations. Human cognition restricts the range and volume of data that can be readily assimilated from a single display.

Viewed in this context the principal purposes of visualization are threefold.

1. To communicate the message of the spatial and attribute data of a representation in an intel-ligible manner. If the range, complexity, and detail of data are too much for the decision maker to assimilate, good visualization should enable the user to see the wood for the trees.
2. To enable the user to understand the likely over-all quality of the representation.
3. To establish whether and to what extent the general message of the data is sensitive to inclusion or exclusion of particular data elements.

Here we will term these *interpretation*, *validation*, and *exploration*, respectively.

Scientific visualization allows users to interpret, validate, and explore their data in greater detail than was possible hitherto.

Interpretation, validation, and exploration may be considered in relation to the conceptual model of uncertainty that was presented in Figure 6.1. They encourage us to think of geographic analysis not as an end point, but rather as the start of an iterative process of feedbacks and *what if?* scenario testing. The eventual "best guess" reformulated model is used as a decision support tool, and the real world is changed as a consequence (see Chapters 16 and 17).

In many of these maps, the representation of attribute data and spatial object type can be thought of as modified by a "visualization filter" (U4 in Figure 12.15). Although this usually entails simplification of data, and hence suppression of the richness and detail of the representation, U4 is unlike the other filters originally presented in Figure 6.1, in that detail can be recovered through user interaction.

12.3.2 User interaction and decision support

Users need to ask generic spatial and temporal questions such as:

Where is . .?
What is at location . .?
What is the spatial relation between . . ?
What is similar to . .?
Where has . . occurred?
What has changed since . .?
Is there a general spatial pattern, and what are the anomalies?
(Egenhofer and Kuhn 1999).

The development of ViSC accompanied the in-novation of so-called *WIMP* interfaces – based upon windows, icons, menus, and pointers (Egenhofer and Kuhn 1999: Figure 12.16). The familiar actions of pointing, clicking, and dragging windows and icons are the most common ways of interrogating a geographic database and summarizing results in map and tabular form. These questions are answered through spatial *querying*, which is the process of retrieving data which have certain user-specified attributes. Today, research applications increas-ingly use multiple displays of maps, bar charts, and scatterplots, which enable a picture of the spatial and other properties of a representation to be built up, as shown in Section 13.2. Spatial query is thus integral to ViSC, and facilitates learning about a representation in a data-led way. Multiple displays make it possible to visualize the range of

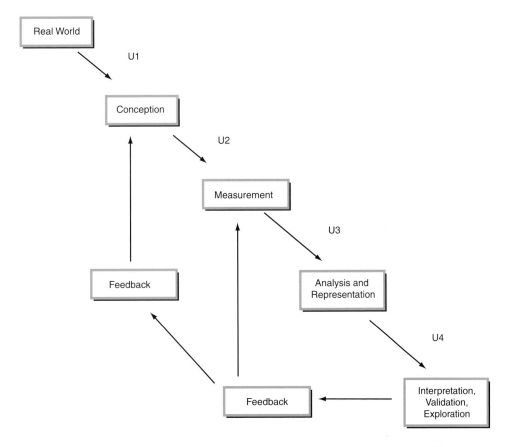

Figure 12.15 Filters U1–U4: conception, measurement, analysis and visualization. (See also Figure 6.1)

intrinsic spatial properties of data in any given application, and also to update them using ancillary sources (see Sections 12.4 and 12.5).

The WIMP interface allows spatial querying through pointing, clicking, and dragging windows and icons.

Spatial query functions are central to many Internet GIS applications. Such applications have tended to appear superficial and routine, in that spatial query is often an end in itself, rather than a precursor to more advanced geographic analysis. They are nevertheless more sophisticated than most conventional GIS operations in that the objects of spatial queries are often continuously updated (refreshed) in real time – as illustrated in the traffic application shown in Figures 2.15 and 12.20, and concerning weather conditions in Figure 12.17. Other common spatial queries are framed to identify the location of services, provide routing and direction information (see

Figure 1.16), facilitate rapid response in disaster management (see Section 2.3.4.2), and provide information about domestic property and neighborhoods to assist residential search (Figure 2.6).

The advent of the Internet has also led to the wider development of publicly available online information systems and has the potential to encourage greater public participation in the planning process. It has also been used to good effect in providing valuable electoral geography information (Box 12.3). Shiffer (1999) provides an overview of how combinations of GIS and multimedia can be used to facilitate the active involvement of many different groups in the discussion and management of urban change through planning, and may act as a bulwark against officialdom or big business.

New software is also facilitating interaction on grander scales. First, when users access virtual environments across networks, they are able to choose different perspectives and views of

POINTER   DROP DOWN MENU   ICONS

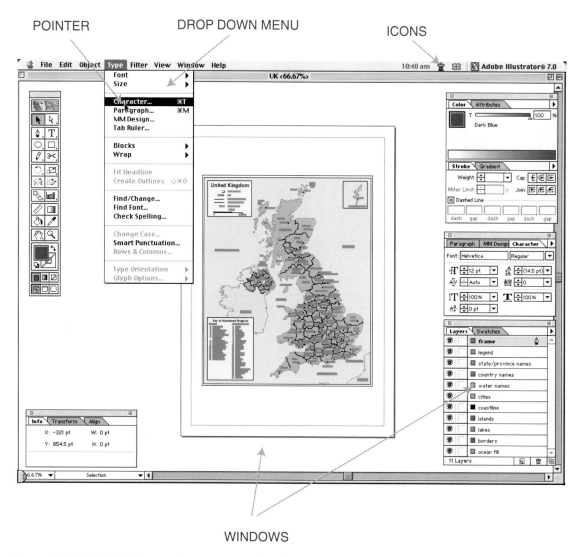

WINDOWS

Figure 12.16 The WIMP (windows, icons, menus, pointers) environment to computing

phenomena and to change these perspectives incrementally in real time in *fly-bys* (Figure 12.19). Second, such scenes can be decomposed into constituent wire frame objects (Figure 12.19(B)), which can then be rearranged by users (Figure 12.19(C)). Third, by extension, users can be represented graphically as *avatars*, that is, virtual representations of animate objects. Avatars can engage with others who are connected at different remote locations, and who are engaged in related tasks in modifying virtual worlds (e.g. see ''360 days in Activeworlds'' at www.casa.ucl.ac.uk/30days/). Networked virtual worlds are increasingly common applications of

ViSC, and they provide environments in which many new kinds of representation and modeling are able to take place. Fourth, the logical extension of interaction using avatars is linking such remote interaction with *virtual reality* (VR) systems. Semi-immersive and immersive VR is beginning to make the transition from the realm of science fiction to GIS application, with new methods of spatial object manipulation and exploration using head-mounted displays and sound. Fifth, miniaturization of hand-held, in-vehicle and wearable computer devices is ushering in a new era of field computing (Clarke 1998), which offers the prospect of reconciling

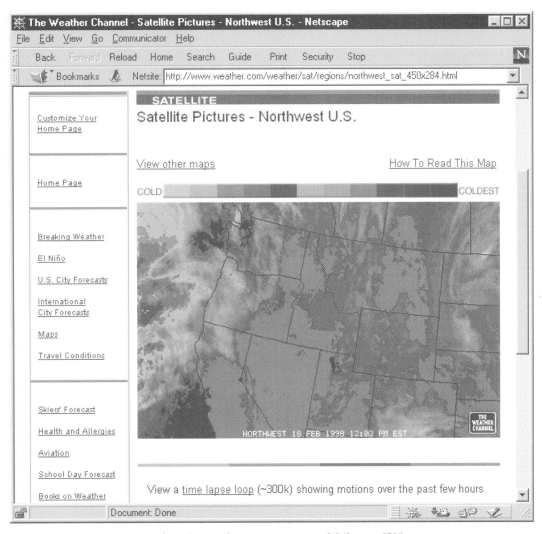

Figure 12.17 Dynamic updating of weather conditions using Internet GIS (Source: ESRI)

digital representations with reality in the field (Figure 12.4). The implications of moving real-time updates of spatial data into the field for applications such as disaster management are obvious (Section 2.3.4.2). Other more mundane functions, such as traffic management, can also be managed in real time (see Box 12.4). Each of these developments enables visualization to be centered around the user to a far greater extent than has ever been the case before.

12.4 Re-modeling Spatial Distributions

The cartogram offers a more radical means of transforming space, and hence restoring spatial balance, but at the expense of ditching the familiar spatial framework valued by most users. It seems that the user cannot have it both ways – there seems thus far to be a stark choice between being able to let attribute data speak for themselves, and being able to relate spatial objects to real locations on the Earth's surface. Yet there are also other ways in which GIS can be used to re-model spatial distributions, and hence assign spatial attributes to meaningful yet recognizable spatial objects.

One example of the way in which ancillary sources of information may be used to improve the model of a spatial distribution is known as *dasymetric mapping*. Here, the intersection of two datasets is used to obtain more precise

Box 12.3 Using the Internet to disseminate election results

Planning and executing a national population census is a massive task for any country, especially one the size of India. After the sudden dissolution of the House of the People in the Parliament in December 1997, an election was required within two and a half months. For the first time in its history, the Election Commission of India decided to make use of the Internet to provide comprehensive information on the electoral system, laws, rules, procedures, and results. An Elections India site was unveiled shortly before the election (www.eci.gov.in).

During the election counting data flowed over satellite and local networks to a central database from 1500 counting centers across the country. As the Commission's databases were updated following voting (Figure 12.18(A)), Web pages were created dynamically showing raw results and distribution maps and charts. The pages included details about party leads and results, at both the state and national level. Figure 12.18(B), for example, shows party voting in the 320 constituencies of the state of Madhya Pradesh. As new data arrived the maps were updated to show changing leads.

(A)

(B)

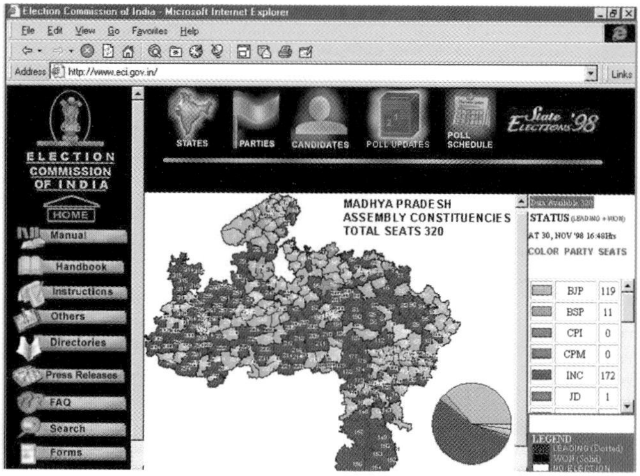

Figure 12.18 (A) Voting in and (B) electoral results for Madhya Pradesh State, India (Source: ESRI)

estimates of a spatial distribution. Figure 12.21(A) shows the census tract geography for which small area population totals are known, and Figure 12.21(B) shows the spatial distribution of built structures in an urban area (which might be obtained from a cadaster or very high resolution satellite imagery, for example). A reasonable assumption (in the absence of evidence of mixed land use or very different residential structures such as high rise apartments and widely spaced bungalows) is that all of the built structures house resident populations at uniform density. Figure 12.21(C) shows how this assumption, plus an overlay of the areal extent of built structures, allows population figures to be allocated to smaller areas than census tracts, and allows calculation of indicators such as of residential density. The potential usefulness of dasymetric

mapping is illustrated in Figure 12.22. Figures 12.22(A) and (B) show the location of an elongated census block (or enumeration district) in the city of Bristol, UK, which appears to have a high unemployment *rate*, but a low absolute *incidence* of unemployment. The resolution of this seeming paradox was explained in Section 6.3.3, but this would not help us much in deciding whether or not the elongated zone to the right of the river in the detail should be included in an inner city workfare program, for example. Use of GIS to overlay high resolution aerial photography (Figure 12.22(C)) reveals the tract to be largely empty of population apart from a small extension to a large housing estate. It would thus appear sensible to assign this zone the same policy status as the zone to its west.

(A)

(B)

(C)

Figure 12.19 (A) A three-dimensional representation of the Tottenham Court Road area of London; and (B) a wire frame model of the scene. The wire frame model may be manipulated to assess the visual impact of changes in the streetscape (C) (Courtesy Andy Smith)

Box 12.4 Use of Internet GIS to improve traffic flow in Southampton, England

Because international port cities function as gateways to distant destinations for holiday-makers, freight transporters, and a multitude of other travelers, they are often overwhelmed with unique vehicular traffic flow problems. Southampton, a historic English port city, has been involved in the development and application of part of a pan-European, GIS-based traffic monitoring and analysis system called the ROad MANagement SystEm for Europe (ROMANSE) for more than five years.

ROMANSE is based on desktop GIS software linked to an Urban Traffic Control (UTC) system. The GIS acts as the graphical front end to the UTC. The GIS software has been customized to create a simplified, specific-purpose display. A variety of current traffic information is overlaid on basemaps that originate from Great Britain's Ordnance Survey. The GIS creates on-the-fly status maps so that engineers at the Traffic and Travel Information Centre (TTIC) in Southampton can analyze existing road hazards and anticipate new ones. They can then issue appropriate advice and warning messages for immediate display on digital road signs in the area, or dispatch emergency services.

The information processed at the TTIC comes from a variety of sources, including closed-circuit television, roadside detectors, and satellite tracking systems, all of which are fed into the UTC. It is then processed and integrated by the GIS for instant analysis or display, providing the engineers with a good overview of what is happening on the road network. Information available to the engineers includes traffic flow per minute, traffic speeds, congestion (Figure 12.20(A)), and car park occupancy (Figure 12.20(B)).

Internet GIS has been used to provide current traffic information and maps on demand to Internet users visiting the ROMANSE Web site at www.romanse.org.uk. From this site members of the public can

Box 12.4 Continued

check the current occupancy in car parks before traveling. A city center map is displayed and by clicking on each car park it is possible to determine both the current occupancy status and what the occupancy is likely to be in the near future, based on historic records. Information for cyclists is also available at the Web site. Traffic information is also posted digitally on touch-screen displays at main transport interchanges, shopping centers, tourist information centers, and libraries. At bus stops, arrival times are electronically displayed and some stops feature an audio version of the information for visually impaired passengers.

This is one of the most advanced intelligent transport information systems in the world. A key part of the success was linking GIS software to a real-time traffic control systems for data collection, and to messaging systems and the Internet for data display.

(A)

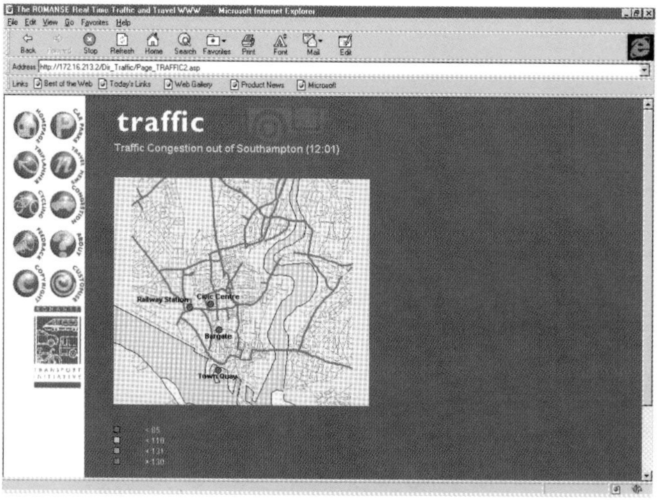

(B)

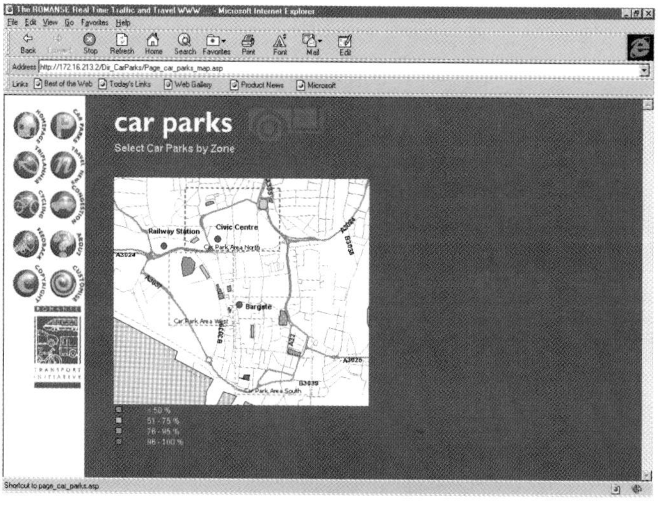

Figure 12.20 Monitoring (A) traffic flow and (B) car park occupancy in real time as part of the Web-based ROMANSE traffic management system (Crown copyright)

(A)

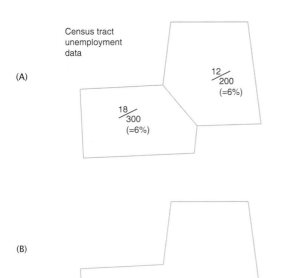

Census tract
unemployment
data

$\dfrac{12}{200}$
(=6%)

$\dfrac{18}{300}$
(=6%)

(B)

Built

Built

(C)

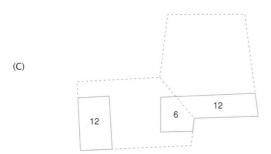

12

6

12

12

Figure 12.21 Modeling a spatial distribution in an urban area using dasymetric mapping: (A) incidence of unemployment amongst the population; (B) locations of (equal area) built structures; and (C) overlay of (A) and (B) to allocate unemployed and total persons proportional to built area.

Dasymetric mapping uses the intersection of two datasets (or layers in the same dataset) to obtain more precise estimates of a spatial distribution.

Dasymetric mapping, and related techniques developed in geographic research (Longley and Mesev 1997, Martin and Higgs 1996, Barr and Barnsley 1998) present a window on reality that looks more convincing than conventional choropleth mapping. However, it is important to remain aware that the visualization of reality is only as good as the assumptions that are used to create it. The information about population concentration used in Figure 12.21, for example, is subjectively defined and is likely to be error prone (some built forms may be offices and shops, for example), and this will inevitably feed through into inaccuracies in visualization. Interpolation functions cannot compensate for fundamental errors in representation. Other problems include inferring land use from land cover, the correct classification of (domestic versus non-domestic) built form, and so on.

12.5 Multivariate Mapping

The various symbolization conventions that were described in Section 12.2 may be used alongside choropleth map shading to present a multivariate picture of a spatial distribution. For example, in Figure 12.23 the size of the circle (representing degree of ecological stress) and the color of the choropleth background (ecology type) together establish the association between the two variables. However, the map assumes that one of the variables (ecology type) is uniform right across the zone, while the other is assigned to a more or less arbitrary point in each ecological unit. This is acceptable in a generalized small-scale map, where the intention is to identify a general pattern of association rather than a geographically specific inventory of incidences.

The more general case concerns the compilation of composite maps based upon a range of constituent indicators. Climate maps, for example, are compiled from direct measures such as amount and distribution of precipitation, diurnal temperature variation, humidity, and hours of sunshine, plus indirect measures such as vegetation coverage and type. There is unlikely ever to be a perfect correspondence between each of these components, and historically it has been the role of the cartographer to arbitrate between the different zonations. More generally, as we have seen, data may be averaged over zones which are devoid of any strong meaning, while the different components that make up a composite index may have been measured at a range of scales. Visualization can mask scale and aggregation problems where composite indicators are created at fine scales using components that were only intended for use at coarse scales. This was illustrated in the case of geodemographics in Box 6.2.

As we have already seen with geographic analysis (Section 6.4), there is a need to understand the nature (Chapter 5) and representational characteristics (Chapter 3) of what goes into

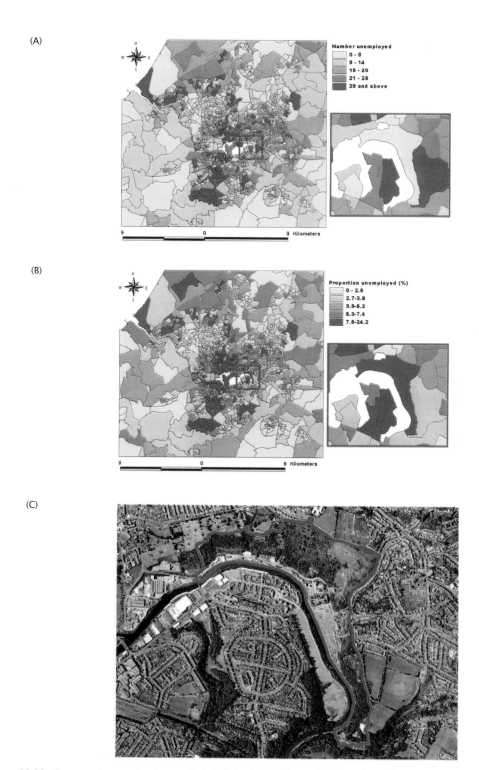

(A)

(B)

(C)

Figure 12.22 Dasymetric mapping in practice, in Bristol UK: (A) numbers employed by census enumeration district (block); (B) proportions unemployed using the same geography; and (C) orthorectified aerial photograph of part of the detailed area (Courtesy: Cities Revealed and Richard Harris)

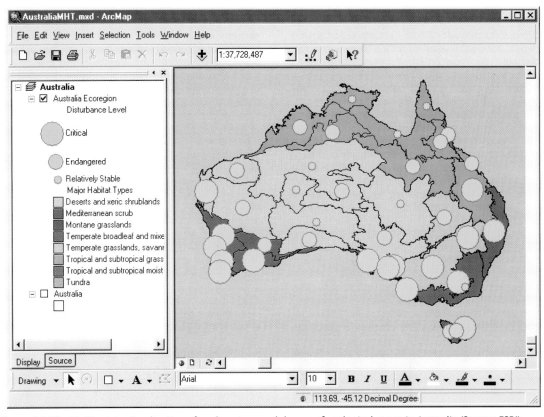

Figure 12.23 Multivariate visualization of ecology type and degree of ecological stress in Australia (Source: ESRI)

mapping if visualization is to provide robust and defensible aids to decision-making. Here there are often few hard and fast rules – today's ViSC often reuses and recycles different datasets, obtained over the Internet, that are rich in detail but may be un-systematic in collection and incompatible in terms of scale. This all underpins the importance of metadata to evaluate data in terms of scale, aggregation, and representativeness prior to visualization (see Chapter 7). More generally, the effects of combining data with different properties are creating interest in *fuzzy data* analysis, and the visualization of objects with indeterminate boundaries (Section 15.3).

12.6 Applications

The relative importance of representing space and attributes will vary within and between different applications, as will the ability to broker improved measures of spatial distributions through integration of ancillary sources. These tensions are not new. In a standard (typically scale 1:50,000) topographic map, for example, the width of a typical single carriageway road may be exaggerated by factors of five to ten. This is done in order to enhance the legibility of features that are central to general-purpose topographic mapping. In some instances, these prevailing conventions will have evolved over long periods of time, while in others the new-found capabilities of ViSC entail a distinct break from the past. As a general rule, where accuracy and precision of georeferencing are important, the standard conventions of topographic mapping will be applied. A range of cartometric transformations will, however, be appropriate in circumstances where attribute magnitude and attribute linkage are of greatest importance.

Utility applications use GIS that have come to be known as automated mapping and facilities management (AM/FM) systems. The prime objectives of such systems include the ability to manage edits on individual model entities (e.g. a single pipe, valve, or wire), known as feature-locking database management, and the ability for multiple users to edit or modify the same geographical data at the same time (see Section 11.9). A schematic user interface, which enables processing of multiple transactions, is used to provide a view of the way in which a system functions, and is used for operational activities such as identifying faults during power outage. Utilities applications also use large-scale mapping with scanned georeferenced background data for repair and maintenance functions. Other utilities applications use a hybrid of schematic and geographic visualization, known as *geoschematic* visualization, in which a synoptic view of the current state of the network is superimposed upon a background map with real-world coordinates.

Transportation applications use a procedure known as linear referencing (Section 4.4) to visualize point (such as street furniture), linear (such as parking restrictions and road surface quality measures), and continuous events (such as speed limits). In linear referencing two-dimensional geography is collapsed into one-dimensional linear space. Field measurements are collected as linear measures from the start of a route (a path through a network). The linear measures are usually dynamically added at display time, and segment the route into smaller sections (hence the term *dynamic segmentation*, which is sometimes used to describe this type of model). For more information on this topic see Sections 4.4 and 9.2.3.3.

We began this chapter with a discussion of the military uses of mapping, and some such applications also have special cartographic conventions – as in the operational overlay maps used to communicate battle plans. On these maps, friendly, enemy, and neutral forces are shown in blue, red, and green, respectively. The location, size, and capabilities of military units are depicted with special multivariate symbols that allow interpreters to understand their operational and tactical significance. Other features of significance, such as minefields, impassable vegetation, and direction of movement, also have special symbols. Animations of such maps can be used to show the progression of a battle, including future "what if?" scenarios.

There are particular conventions for mapping in utilities, transport, and military applications.

12.7 Consolidation

ViSC can make a powerful contribution to decision-

making and can be used to simulate changes to reality. Yet although visualization in GIS is governed by scientific principles, the limits of human cognition mean that it necessarily provides a further selective filter on the reality that it seeks to represent. Mapping is about seeing the detail as well as the big picture, yet the wealth of detail that is available in today's digital environment threatens to overwhelm the message of the map. The interactive nature of spatial query functions nevertheless means that the full richness and diversity of a representation may be retrieved at will through a range of ancillary displays. At its worst, the power of ViSC may be used to mask the inherent uncertainty in most spatial data: at its best, it allows the user to understand the full message of the data. The divergence between these best and worst cases has become wider with the advent of the Internet as a server of data from an ever-wider range of sources.

Ultimately good visualization and successful user interaction require the availability of data of quality commensurate with a given applications task. Previous chapters have identified the different ways in which the sharpness of a representation may be degraded by the selective availability of data about geographic objects of interest, and coarseness of attribute measurement scales. Techniques are available that can compensate for this, yet it is a role of visualization to represent the further uncertainties that this entails. GIS trivialize many of the problems of effective cartography. Yet the old adage that "seeing is believing" only holds if visualization and user interaction are properly attuned to the goal of conveying the message of the representation.

Questions for Further Study

1. Identify the criteria that you would have used in designing a military support system for use during the Bosnia Herzegovina hostilities. How might the system subsequently have been adapted for use in brokering the peace?
2. Carefully examine the cartograms of a major subway system (but not London, UK) with which you are familiar (see Figure 12.24). How do the conventions that underpin them differ from those used for the London Underground cartogram (Figure 12.12(B))? Examine maps of London and the city of your chosen system in a good atlas. To what extent do you think that differences between the cartograms reflect differences in the shapes and forms of the respective cities, and to what extent might

Figure 12.24   Navigating the Metro

they reflect differing attributes of the subway systems?

3. Assess the relative merits of the attempts to classify and measure urban structure in any two of the following from the References: Longley and Mesev 1997, Martin and Higgs 1996, and Barr and Barnsley 1998.

4. Figure 12.25 is a cartogram redrawn from a newspaper feature on the costs of air travel from London in 1992. Use current advertisements in the press and on the Internet to create a similar cartogram of travel costs, in local currency, from the nearest international air hub to your place of study.

Online Resources

NCGIA Core Curricula (www.ncgia.ucsb.edu/pubs/core.html):

Core Curriculum in GIScience, Sections 1.4.2 (Maps as Representations of the World, Judy Olson), 2.10.3 (Detecting and Evaluating Errors by Visual Means, Kate Beard), 2.11.1 (Cartographic Fundamentals), 2.14.2 (Exploratory Spatial Data Analysis, Robert Haining and Stephen Wise), 2.14.4 (Multimedia and Virtual Reality, George Taylor)

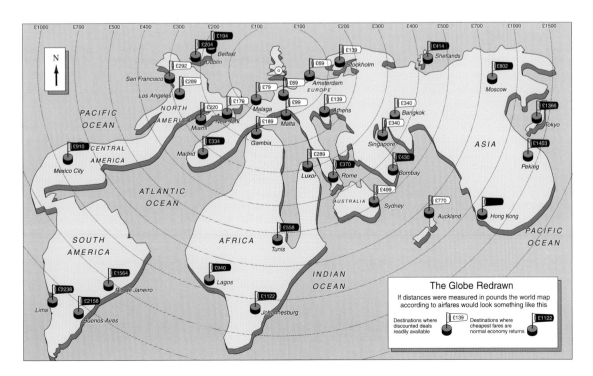

Figure 12.25   The globe redrawn in terms of travel costs from London

Core Curriculum in GIS, 1990, Units 2 (Maps and Map Analysis), 16 (Output), 17 (Graphic Output Design Issues), 18 (Modes of User/GIS Interaction), 49 (Visualization of Spatial Data), 50 (Color)

ESRI Virtual Campus courses (campus.esri.com):
Turning Data into Information, by Paul Longley, Michael Goodchild, David Maguire, and David Rhind (Module "Visualization and User Interaction")

Introduction to ArcView GIS, by ESRI (especially: lessons on "Introducing ArcView" and "Displaying Themes" in the Module "Basics of ArcView"; the Module "Working with Tables in ArcView"; and the lessons on "Creating Charts", "Creating Map Layouts", "Working with Frames", and "Adding the Finishing Touches to a Layout" in the Module "Presenting Information in ArcView". See also Einführung in ArcView GIS by ESRI and Josef Strobl, and Introdução ao ArcView GIS by ESRI and Mirna Cortopassi Lobo)

Introduction to ArcView 3D Analyst, by ESRI (especially the "Navigating in a 3D Scene" lesson in the Module "Basics of ArcView 3D Analyst", and lessons on "Setting 3D scene properties", "'Setting 3D Theme Properties", and "Symbolizing Grids and TINs" in the Module "Displaying Data in ArcView 3D Analyst")

Andy Smith's Web site at University College London's Centre for Advanced Spatial Analysis (www.casa.ucl.ac.uk/30days/) introduces a range of urban planning simulation activities

The Web site of Pennsylvania State University's GeoVISTA Center is a rich source of research on visualization of geographic information: www.geovista.psu.edu

Reference Links

Longley P A, Goodchild M F, Maguire D J, and Rhind D W (eds) 1999 *Geographical Information Systems: Principles, techniques,* *management and applications*. New York: John Wiley.
Chapter 7, Spatial representation: a cognitive view (Mark D)
Chapter 11, Visualising spatial distributions (Kraak, M-J)
Chapter 13, Models of uncertainty in spatial data (Fisher P F)
Chapter 28, Interacting with GIS (Egenhofer M J, Kuhn W)
Chapter 52, Managing public discourse: towards the augmentation of GIS with multi-media (Shiffer M).
Chapter 60, GIS in emergency management (Cova T)
Chapter 63, Military applications of GIS (Swann D).

References

Barr S L and Barnsley M J 1998 A syntactic pattern recognition paradigm for the derivation of second-order thematic information from remotely-sensed images. In Atkinson P and Tate N (eds) *Advances in Remote Sensing and GIS Analysis*. Chichester: John Wiley, 167–184.

Clarke K 1998 Visualizing different geofutures. In Longley P A, Brooks S M, McDonnell R, and Macmillan W D (eds) *Geocomputation: A primer*. Chichester: John Wiley, 119–137.

Harder C 1998 *Serving Maps on the Internet*. Redlands, California: ESRI Press

Longley P A and Mesev T V 1997 The use of diverse RS-GIS sources to measure and model urban morphology. *Geographical Systems* 4: 5–18

MacEachren A 1994 Time as a cartographic variable. In H M Hearnshaw and D J Unwin (eds) *Visualization in Geographic Information Systems*. Chichester: John Wiley, 115–130.

Martin D J and Higgs G 1996 Georeferencing people and places: a comparison of detailed datasets. In Parker D (ed) *Innovations in GIS 3*. London: Taylor and Francis, 37–47.

Tufte E R 1983 *The Visual Display of Quantitative Information*. Cheshire, Conn.: Graphics Press.

GEOGRAPHIC QUERY AND ANALYSIS: FROM DATA TO INFORMATION

13

The chapter begins with a review of definitions of spatial analysis. Methods are discussed under six broad headings, three in this chapter and three in the following one. Query methods allow users to interact with geographic databases using pointing devices and keyboards, and GIS have been designed to present data for this purpose in a number of standard views. The second area, measurement, includes algorithms for determining lengths, areas, shapes, slopes, and other properties of objects. Transformations allow new information to be created through simple geometric manipulation.

Learning objectives

After working through this chapter you will know:

- Definitions of spatial analysis, and tests to determine whether a method is spatial;

- The range of queries possible with a GIS, and the concept of reasoning with a GIS;

- Methods for measuring length, area, shape, and other properties, and their caveats;

- Transformations that manipulate objects to create new ones, or to determine geometric relationships between objects.

13.1 Introduction: What is Spatial Analysis?

Spatial analysis is in many ways the crux of GIS, because it includes all of the transformations, manipulations, and methods that can be applied to geographic data to add value to them, to support decisions, and to reveal patterns and anomalies that are not immediately obvious – in other words, spatial analysis is the process by which we turn raw data into useful information. If GIS is a method of communicating information about the Earth's surface from one person to another, then the transformations of spatial analysis are ways in which the sender tries to inform the receiver, by adding greater informative content and value, and by revealing things that the receiver might not otherwise see.

Some methods of spatial analysis were developed long before the advent of GIS, and carried out by hand, or by the use of measuring devices like the ruler. The term *analytical cartography* is sometimes used to refer to methods of analysis that can be applied to maps to make them more useful and informative, and spatial analysis using GIS is in many ways its logical successor.

Spatial analysis can reveal things that might otherwise be invisible – it can make what is implicit explicit.

In this and the next chapter we will look first at some definitions and basic concepts of spatial analysis. Following introductory material, the chapters include six sections, which look at spatial analysis grouped into six more-or-less distinct categories: queries and reasoning, measurements, transformations, descriptive summaries, optimization, and hypothesis testing. The six sections include a large amount of material, and so they are divided between this chapter and the next, with the simpler topics in this chapter and the more advanced ones in the next. Some of the methods discussed in these two chapters were introduced in Chapter 5 as ways of describing the fundamental nature of geographic data, so references will be made to that chapter as appropriate.

Spatial analysis is the crux of GIS, the means of adding value to geographic data, and of turning data into useful information.

Methods of spatial analysis can be very sophisticated, but they can also be very simple. A large body of methods of spatial analysis has been developed over the past century or so, and some methods are highly mathematical – so much so, that it might sometimes seem that mathematical complexity is an indicator of the importance of a technique. But the human eye and brain are also very sophisticated processors of geographic data, and excellent detectors of patterns and anomalies in maps and images. So the approach taken here is to regard spatial analysis as spread out along a continuum of sophistication, ranging from the simplest types that occur very quickly and intuitively when the eye and brain focus on a map, to the types that require complex software and sophisticated mathematical understanding. Spatial analysis is best seen as a *collaboration* between the computer and the human, in which both play vital roles, a theme that emerged in the discussion of the GIS application of gap analysis in Section 2.3.5.3.

Effective spatial analysis requires an intelligent user, not just a powerful computer.

There is an unfortunate tendency in the GIS community to regard the making of a map using a GIS as somehow less important than the performance of a mathematically sophisticated form of spatial analysis. According to this line of thought, *real* GIS involves number crunching, and users who just use GIS to make maps are not serious users. But every cartographer knows that the design of a map can be very sophisticated, and that maps are excellent ways of conveying geographic information and knowledge, by revealing patterns and processes to us. We agree, and believe that map making is potentially just as important as any other application of GIS.

Spatial analysis helps us in situations when our eyes might otherwise deceive us.

There are many possible ways of defining spatial analysis, but all in one way or another express the basic idea that information on locations is essential – that analysis carried out without knowledge of locations is not spatial analysis. One fairly formal statement of this idea is: "Spatial analysis is a set of methods whose results are not invariant under changes in the locations of the objects being analyzed". The double negative in this statement follows convention in mathematics, but for our purposes we can remove it: "Spatial analysis is a set of methods whose results change when the locations of the objects being analyzed change". On this test the calculation of an average income for a group of people is not spatial analysis, because it in no way depends on the locations of the people. But the calculation of the center of the

US population is spatial analysis, because the results depend on knowing where all US residents are located. GIS is an ideal platform for spatial analysis because its data structures accommodate the storage of object locations.

13.1.1 Examples

Spatial analysis can be used to further the aims of science, by revealing patterns that were not previously recognized, and that hint at undiscovered generalities and laws. Patterns in the occurrence of a disease may hint at the mechanisms that cause the disease, and some of the most famous examples of spatial analysis are of this nature, including the work of Dr John Snow in unraveling the causes of cholera (Box 13.1).

It is interesting to speculate on what would have happened today, if early epidemiologists like Snow had had access to a GIS. The rules governing research today would not have allowed Snow to remove the pump handle, except after lengthy review, because the removal constituted an

---

Box 13.1 Dr John Snow and the causes of cholera

In the 1850s cholera was very poorly understood, and massive outbreaks were a common occurrence in major industrial cities (today cholera remains a significant health hazard in many parts of the world, despite progress in understanding its causes and advances in treatment). An outbreak in London in 1854 in the Soho district was typical of the time, and the deaths it caused are mapped in Figure 13.1. The map was made by Dr John Snow (Figure 13.2), who had conceived the hypothesis that cholera was transmitted through the drinking of polluted water, rather than through the air, as was commonly believed. He noticed that the outbreak appeared to be centered on a public drinking water pump in Broad Street (Figure 13.3), and if his hypothesis was correct, the pattern shown on the map would reflect the locations of people who drank the pump's water. There appeared to be anomalies, in the sense that deaths had occurred in households that were located closer to other sources of water, but he was able to confirm that these households also drew their water from the Broad Street pump. Snow had the handle of the pump removed, and the outbreak subsided, providing direct causal evidence in favor of his hypothesis. The full story is much more complicated, of course; much more information is available at www.jsi.com

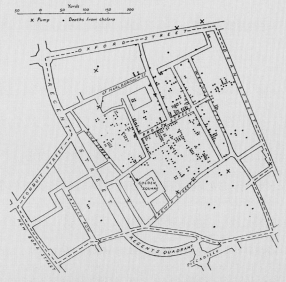

Figure 13.1 A redrafting of the map made by Dr John Snow in 1854, of the deaths that occurred in an outbreak of cholera in the Soho district of London. The existence of a public water pump in the center of the outbreak (the cross in Broad Street) convinced Snow that drinking water was the probable cause of the outbreak. Stronger evidence was obtained in support of this hypothesis when the water supply was cut off, and the outbreak subsided. (Source: Gilbert E W 1958 Pioneer maps of health and disease in England *Geographical Journal*, **124**: 172–183.)

Box 13.1 Continued

Figure 13.2 Dr John Snow (Source: John Snow, Inc. www.jsi.com)

Figure 13.3 A modern replica of the pump that led Snow to the inference that drinking water transmitted cholera, located in what is now Broadwick Street in Soho, London (Source: John Snow Inc. www.jsi.com)

Today, Snow is widely regarded as the father of modern epidemiology.

experiment on human subjects. To get approval, he would have had to have shown persuasive evidence in favor of his hypothesis, and it is doubtful that the map would have been sufficient, because several other hypotheses might have explained the pattern equally well. First, it is conceivable that the population of Soho was inherently at risk of cholera, perhaps by being comparatively elderly, or because of poor housing conditions. The map would have been more convincing if it had shown the *rate* of incidence, relative to the population at risk. For example, if cholera was highest among the elderly, the map could have shown the number of cases as a proportion of the population over 50 years. Second, it is still conceivable that the hypothesis of transmission through the air between carriers could have produced the same observed pattern, if the first carrier lived in the center of the outbreak. Snow could have eliminated this alternative if he had been able to produce a sequence of maps,

showing the locations of cases as the outbreak developed. Both of these options involve simple spatial analysis of the kind that is readily available today in GIS.

> GIS provides tools that are far more powerful than the map at suggesting causes of disease.

Today the causal mechanisms of diseases like cholera, which results in short, concentrated outbreaks, have long since been worked out. Much more problematic are the causal mechanisms of diseases that are rare, and not sharply concentrated in space and time. The work of Stan Openshaw of the University of Leeds, using one of his Geographical Analysis Machines, illustrates the kinds of applications that make good use of the power of GIS in this contemporary context. More about Openshaw's work can be found in Box 6.3.

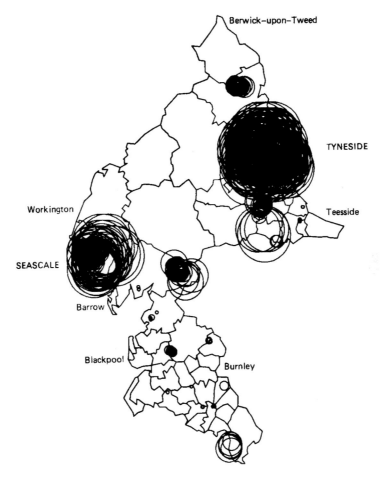

Figure 13.4 The map made by Openshaw and colleagues by applying their Geographical Analysis Machine to the incidence of childhood leukemia in northern England (1968–85). A very large number of circles of random sizes is randomly placed on the map, and a circle is drawn if the number of cases it encloses substantially exceeds the number expected in that area given the size of its population at risk (Source: Openshaw *et al* 1987)

Figure 13.4 shows an application of one of Openshaw's techniques to a comparatively rare but devastating disease whose causal mechanisms remain largely a mystery – childhood leukemia. The study area is northern England, from the Mersey to the Tyne. The analysis begins with two datasets: one of the locations of cases of the disease, and the other of the numbers of people at risk in standard census reporting zones. Openshaw's technique then generates a large number of circles, of random sizes, and places them (*throws* them) randomly over the map. The computer generates and places the circles, and then analyzes their contents, by dividing the number of cases found in the circle by the size of the population at risk. If the ratio is anomalously high, the circle is drawn. After a large

number of circles has been generated, and a small proportion have been drawn, a pattern emerges. Two large concentrations, or clusters of cases, are evident in the figure. The one on the left is located around Sellafield, the location of the British Nuclear Fuels processing plant and a site of various kinds of leaks of radioactive material. The other, in the upper right, is in the Tyneside region, and Openshaw and his colleagues discuss possible local causes.

Both of these examples are instances of *inductive* use of spatial analysis, to examine empirical evidence in the search for patterns that might support new theories or general principles, in this case with regard to disease causation. Other uses of spatial analysis are *deductive*, focusing on the testing of known theories or principles against data (see Sections 5.9 and 14.4). A third type of application is *normative*, using spatial analysis to develop or prescribe new or better designs, for the locations of new retail stores, or new roads, or new manufacturing plant. Examples of this type appear in Section 14.3.

13.1.2 Types of spatial analysis

The remaining sections of this chapter and the following chapter discuss methods of spatial analysis using six general headings:

Queries and reasoning are the most basic of analysis operations, in which the GIS is used to answer simple questions posed by the user. No changes occur in the database, and no new data are produced. The operations vary from simple and well-defined queries like "how many houses are found within 1 km of this point", to vaguer questions like "which is the closest city to Los Angeles going north", where the response may depend on the system's ability to understand what the user means by "going north" (see the extensive discussion of vagueness in Chapter 6). Queries of databases using SQL are discussed in Section 11.4.

Measurements are simple numerical values that describe aspects of geographic data. They include measurement of simple properties of objects, like length, area, or shape, and of the relationships between pairs of objects, like distance or direction.

Transformations are simple methods of spatial analysis that change datasets, combining them or comparing them to obtain new datasets, and eventually new insights. Transformations use simple geometric, arithmetic, or logical rules, and they include operations that convert raster data into vector data, or vice versa. They may also create fields from collections of objects, or detect collections of objects in fields.

Descriptive summaries attempt to capture the essence of a dataset in one or two numbers. They are the spatial equivalent of the descriptive statistics commonly used in statistical analysis, including the mean and standard deviation.

Optimization techniques are normative in nature, designed to select ideal locations for objects given certain well-defined criteria. They are widely used in market research, in the package delivery industry, and in a host of other applications.

Hypothesis testing focuses on the process of reasoning from the results of a limited sample to make generalizations about an entire population. It allows us, for example, to determine whether a pattern of points could have arisen by chance, based on the information from a sample. Hypothesis testing is the basis of inferential statistics and lies at the core of statistical analysis, but its use with spatial data is much more problematic.

These six are certainly not the only ways of classifying and organizing the numerous methods of spatial analysis into a simple scheme. Perhaps the most successful is the one developed by Dana Tomlin (1990) and known as *cartographic modeling*. It classifies all GIS transformations of rasters into four basic classes, and is used in several raster GISs as the basis for their analysis languages:

- *Local* operations examine rasters cell by cell, comparing the value in each cell in one layer with the values in the same cell in other layers.

- *Focal* operations compare the value in each cell with the values in its neighboring cells – most often eight neighbors.

- *Global* operations produce results that are true of the entire layer, such as its mean value.

- *Zonal* operations compute results for blocks of contiguous cells that share the same value, such as the calculation of shape for contiguous areas of the same land use, and attach their results to all of the cells in each contiguous block.

13.2 Queries and Reasoning

In the ideal GIS it should be possible for the user to interrogate the system about any aspect of its contents, and obtain an immediate answer. Interrogation might involve pointing at a map, or typing a question, or pulling down a menu and clicking on some buttons, or sending a formal SQL request to a database (Section 11.4). The visual aspects of query are discussed in Section 12.3. Today's user interfaces are very versatile, and have very nearly reached the point where it will be possible to interrogate the system by speaking to it – this would be extremely valuable in vehicles, where the use of more conventional ways of interrogating the system through keyboards or pointing devices can be too distracting for the driver.

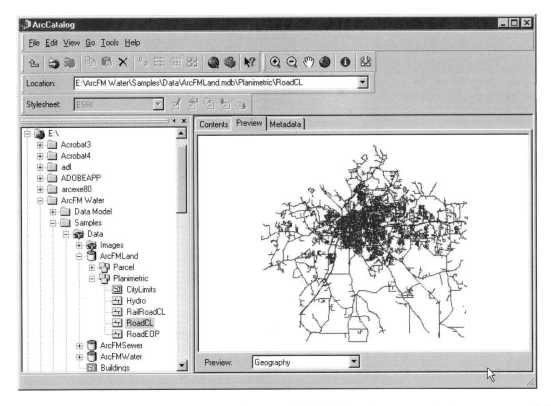

Figure 13.5 A catalog view of a GIS database (ESRI's ArcCatalog). The left window exposes the dataset structure of the database, and the right window in this view provides a geographic preview of the contents of a selected dataset (RoadCL). Other options for the right window include a view of the dataset's metadata, and its tabular structure (Source: ESRI)

When a GIS is used in a vehicle it is impossible to rely on the normal modes of interaction through keyboards or pointing devices.

The very simplest kinds of queries involve inter-actions between the user and the various *views* that a GIS is capable of presenting. A *catalog* view shows the contents of a database, in the form of storage devices (hard drives, Internet sites, floppies, CDs, or ZIP disks) with their associated folders, and the datasets contained in those folders. The catalog will likely be arranged in a hierarchy, and the user is able to expose or hide various branches of the hierarchy by clicking at appropriate points. Figure 13.5 shows a catalog view, in this case using ESRI's ArcCatalog software (a component of ArcGIS). Note how different types of datasets are symbolized using different icons, so the user can tell at a glance which files contain grids, polygons, points, etc.

In contemporary software environments, such as Microsoft's Windows or Windows NT, or the Macintosh or Unix environments, many kinds of interrogation are available through simple pointing and clicking. For example, in ArcCatalog simply pointing at a dataset icon and clicking the right mouse button exposes basic statistics on the dataset when the Properties option is selected. The metadata option exposes the metadata stored with the dataset, including its projection and datum details, the names of each of its attributes, and its date of creation.

Users query a GIS database by interacting with different views

The *map* view of a dataset shows its contents in visual form, and opens many more possibilities for querying. When the user points to any location on the screen the GIS should display the pointer's coordinates, using the units appropriate to the dataset's projection and coordinate system. For example, Figure 13.6 shows the most recent location of the pointer in the box below the map window in units of meters east and north, because the map projection in use is UTM. If the dataset is raster the system might display a cell's row and column number, or its coordinate system if the

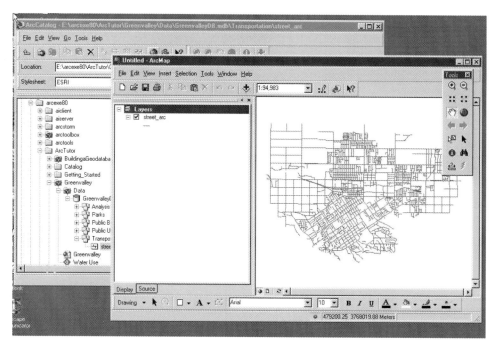

Figure 13.6 ESRI's ArcMap displays the most recent location of the pointer in a box below the map display, allowing the user to query location anywhere on the map. Location is shown in this case in meters east and north (using a Transverse Mercator projection), since this is the coordinate system of the data being queried (Source: ESRI)

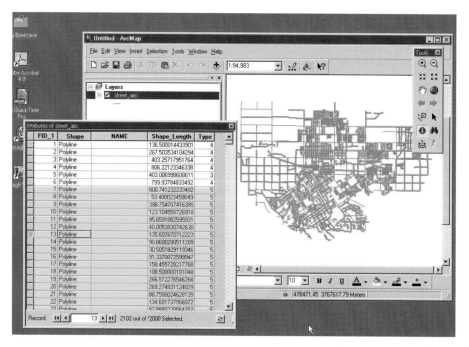

Figure 13.7 The objects shown in table view (left). In this instance all objects with Type equal to 5 have been selected, and the selected objects appear in blue in the map view (ESRI's ArcMap) (Source: ESRI)

raster is adequately georeferenced (tied to some Earth coordinate system). By pointing to an object the user should be able to display the values of its attributes, whether the object is a raster cell, a line representing a network link, or a polygon.

Finally, the *table* view of a dataset shows a rectangular array, with the objects organized as the rows and the attributes as the columns (Figure 13.7). This allows the user to see the attributes associated with objects at a glance, in a convenient form. There will usually be a table associated with each type of object – points, lines, areas, see Box 5.2 and Section 9.2.3 – in a vector database. Some systems support other views as well. In a *histogram* view, the values of a selected attribute are displayed in the form of a bar graph. In a *scatterplot* view, the values of two selected attributes are displayed plotted against each other. Scatterplots allow us to see whether relationships exist between attributes. For example, is there a tendency for the average income in a census tract to increase as the percentage of people with university education increases?

Today's GIS supports much more sophisticated forms of query than these. First, it is common for the various views to be linked, and the visual aspects of this are discussed in Section 12.3.2. Suppose both the map view and the table view are displayed on the screen simultaneously. Linkage allows the user to select objects in one view, perhaps by pointing and clicking, and to see the selected objects highlighted in both views, as in Figure 13.7. Linkage is often possible between other views, including the histogram and scatterplot views. For example, by linking a scatterplot with a map view, it is possible to select points in the scatterplot and see the corresponding objects highlighted on the map. This kind of linkage is very useful in examining *residuals* (Section 5.7), or cases that deviate substantially from the trend shown by a scatterplot (compare with the idealized scatterplots of

Figures 5.16 and 5.17). The term *exploratory spatial data analysis* is sometimes used to describe these forms of interrogation, which allow the user to explore data in interesting and potentially insightful ways.

> Exploratory spatial data analysis allows its users to gain insight by interacting with dynamically linked views.

Second, many methods are commonly available for interrogating the contents of tables, such as SQL (Section 11.4). Figure 13.7 shows the result of an SQL query on a simple table. The language becomes much more powerful when tables are linked, using common keys, as described in Section 11.3, and much more complex and sophisticated queries, involving multiple tables, are possible with the full language. More complex methods of table interrogation include the ability to average the values of an attribute across selected records, and to create new attributes through arithmetic operations on existing ones (e.g. create a new attribute equal to the ratio of two selected attributes).

> SQL is a standard language for querying tables and relational databases.

The term *reasoning* encompasses a collection of methods designed to respond to more complex forms of query and interrogation. Humans have sophisticated abilities to reason with spatial data, often learned in early childhood, and if computers could be designed to emulate these abilities then many useful applications would follow. One is in the area of navigation. Humans are very skilled at direction giving (though they are not always equally willing to seek directions), and computer emulation of these skills would be useful in the design of in-vehicle navigation systems. Figure 1.16 shows an example of this form of reasoning. Figure 1.16(A) is a map view of the best path from

Box 13.2 *Driving directions from 909 West Campus Lane to 1401 De La Vina St*

Go that way (giver of directions points) and turn right at the fence. When the road curves to the right head straight, under the big coral tree. Turn left at the first stop sign, pass the daycare center on the right, and turn right at the stop sign at the end. At the light head straight through, and follow Storke through two more lights (the second one is Hollister). At the third light turn right to take the ramp onto 101 South (it's actually heading east at this point). In about eight miles take the Mission Street exit, and turn left at the bottom of the ramp. Go through three lights, watch out for the sharp dip in the road at Bath, and turn right on De La Vina. 1401 is at the corner of Sola, one block after the light at Micheltorena.

Point A, the driver's current location, to Point B, the intended location. Figure 1.16(B) is a set of instructions generated by a standard GIS, through a simple route-finding analysis of the type described in Section 14.3.2. In Box 13.2 is a set of driving directions for the same route as they might be given by a human. The difference between the two sets of directions is obvious – the human's are given in familiar terms, and they use many more landmarks and hints designed to make the driver's task less error-prone and to allow the driver to recover from mistakes. They also use gestures such as pointing that cannot be easily represented in digital form.

One major difference between the two sets is in the use of *vague* terms. Computers are generally uncomfortable with vagueness (Section 6.2.2), preferring the precise terms used in Figure 1.16(B). But the world of human communication is inherently vague, and full of terms and phrases like "near", "north", "too far", or "watch out for" that defy precise definition. Very often the meaning of human terms depends on the context in which they are used. For example, Santa Barbara may be "near" Los Angeles in a conversation in London, but not in a conversation in Ventura (a city lying directly between them).

13.3 Measurements

Many types of interrogation ask for measurements – we might want to know the total area of a parcel of land, or the distance between two points, or the length of a stretch of road – and in principle all of these measurements are obtainable by simple calculations inside a GIS. Comparable measurements by hand from maps can be very tedious and error-prone. In fact it was the ability of the computer to make accurate evaluations of area quickly that led the Canadian government to fund the development of the world's first GIS, the Canada Geographic Information System, in the mid-1960s (see the brief history of GIS in Section 1.4.1), despite the primitive state and high costs of

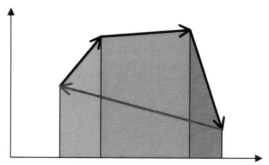

Figure 13.8 The algorithm for calculation of the area of a polygon given the coordinates of the polygon's vertices. The polygon consists of the three black arrows, plus the blue arrow forming the fourth side. Trapezia are dropped from each edge to the x axis and their areas are calculated as (difference in x) times average of y. The trapezia for the first three edges, shown in green, brown, and blue, are summed. When the fourth trapezium is formed from the blue arrow its area is negative because its start point has a larger x than its end point. When this area is subtracted from the total the result is the correct area of the polygon

computing at that time. Evaluation of area by hand is a messy and soul-destroying business. The *dot-counting* method uses transparent sheets on which randomly located dots have been printed – an area on the map is estimated by counting the number of dots falling within it. In the *planimeter* method a mechanical device is used to trace the area's boundary, and the required measure accumulates on a dial on the machine.

> Humans have never devised good manual tools for making measurements from maps, particularly measurements of area.

By comparison, measurement of the area of a digitally represented polygon is trivial and totally reliable. The common algorithm (Box 13.3) calculates and sums the areas of a series of trapezia, formed by dropping perpendiculars to the x axis as shown in Figure 13.8. By making a simple change to the algorithm it is also possible

Box 13.3 Definition of an algorithm

Algorithm: a procedure consisting of a set of unambiguous rules which specify a finite sequence of operations that provides the solution to a problem, or to a specific class of problems. Each step of an algorithm must be unambiguous and precisely defined, and the actions to be carried out must be rigorously specified for each case. An algorithm must always arrive at a problem solution after a finite and reasonable number of steps. An algorithm that satisfies these requirements can be programmed as software for a digital computer.

to use it to compute a polygon's centroid (Section 14.2.1).

13.3.1 Distance and length

A *metric* is a rule for the determination of distance between points in a space. Several kinds of metrics are used in GIS, depending on the application. The simplest is the rule for determining the shortest distance between two points in a flat plane, called the Pythagorean or straight-line metric. If the two points are defined by the coordinates (x_1, y_1) and (x_2, y_2), then the distance D between them is the length of the hypotenuse of a right-angled triangle (Figure 13.9), and Pythagoras's theorem tells us that the square of this length is equal to the sum of the squares of the lengths of the other two sides. So a simple formula results:

$$D = \sqrt{(x_2 - x_1)^2 + (y_2 - y_1)^2}$$

A metric is a rule for determining distance between points in space.

The Pythagorean metric gives a simple and straightforward solution for a plane, if the coordinates x and y are comparable, as they are in any coordinate system based on a projection, such as the UTM or State Plane, or UK National Grid (see Chapter 4). But the metric will not work for latitude and longitude, reflecting a common source of problems in GIS – the temptation to treat latitude and longitude as if they were equivalent to plane coordinates. This issue is discussed in detail in Section 4.7.1.

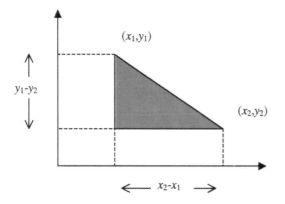

Figure 13.9 Pythagoras's Theorem and the straight-line distance between two points. The square of the length of the hypotenuse is equal to the sum of the squares of the lengths of the other two sides of the right-angled triangle

Distance between two points on a curved surface such as that of the Earth requires a more elaborate approach. The *shortest* distance between two points is the length of a taut string stretched between them, and if the surface is spherical that is the length of the arc of the great circle between them (the circle formed by slicing the sphere through the center and through the two points, see Section 4.6 and Figure 4.13). Given latitude and longitude for two points, the length of this arc is:

$$D = R \cos^{-1}[\sin\phi_1 \sin\phi_2 + \cos\phi_1 \cos\phi_2 \cos(\lambda_1 - \lambda_2)]$$

where R is the radius of the Earth (6378 km to the nearest km and assuming a spherical Earth). In some cases it may be necessary to use the ellipsoid model of the Earth, in which case the calculation of distance is more complex.

In many applications the simple rules – the Pythagorean and great circle equations – are not sufficiently accurate estimates of actual travel distance, and we are forced to resort to summing the actual lengths of travel routes. In GIS this normally means summing the lengths of links in a network representation, and many forms of GIS analysis use this approach. If a line is represented as a polyline, or a series of straight segments, then its length is simply the sum of the lengths of each segment, and each segment length can be calculated using the Pythagorean formula and the coordinates of its endpoints. But it is worth being aware of two problems with this simple approach.

First, a polyline is often only a rough version of the true object's geometry. A river, for example, never makes sudden changes of direction, and Figure 13.10 shows how smooth curves have to be approximated by the sharp corners of a polyline. Because there is a tendency for polylines to short-cut corners, *the length of a polyline tends to be shorter than the length of the object it represents*. There are some exceptions, of course – surveyed boundaries are often truly straight between corner points, and streets are often truly straight between intersections. But in general the lengths of linear objects estimated in a GIS, and this includes the lengths of the perimeters of areas represented as polygons, are often substantially shorter than their counterparts on the ground. Note that this is not similarly true of area estimates, because shortcutting corners tends to produce both underestimates and overestimates of area, and these tend to cancel out (Figure 13.10; see also Sections 5.7 and 5.8, and Box 5.5).

A GIS will almost always underestimate the true length of a geographic line.

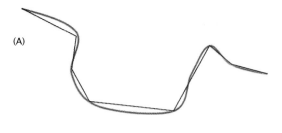

(A)

(B)

Figure 13.10 The polyline representations of smooth curves tend to be shorter in length (A). But estimates of area (B) tend not to show systematic bias because the effects of overshoots and undershoots tend to cancel out to some extent

Second, the length of a line in a two-dimensional GIS representation will always be the length of the line's planar projection, not its true length in three dimensions, and the difference can be substantial if the line is steep (Figure 13.11). In most jurisdictions the area of a parcel of land is the area of its horizontal projection, not its true surface area. A GIS that stores the third dimension for every point is able to calculate both versions of length and area, but not a GIS that stores only the two horizontal dimensions.

13.3.2 Shape

GIS are also used to calculate the *shapes* of objects, particularly area objects. In many countries the system of political representation is based on the concept of districts or constituencies, which are used to define who will vote for each place in the legislature. In the USA and the UK, and in many other countries that derived their system of representation from the UK, there is one place in the legislature for each district. It is expected that districts will be compact in shape, and the manipulation of a district's shape to achieve certain overt or covert objectives is termed Gerrymandering, after an early governor of Massachusetts, Elbridge Gerry (the shape of one of the state's districts was thought to resemble

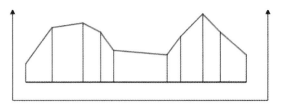

Figure 13.11 The length of a path as traveled on the Earth's surface (red line) may be substantially longer than the length of its horizontal projection as evaluated in a two-dimensional GIS

a salamander, with the implication that it had been manipulated to achieve a certain outcome in the voting; Gerry was a signatory both of the Declaration of Independence in 1776, and of the bill that created the offending districts in 1812). The construction of voting districts is an example of the principles of aggregation and zone design discussed in Sections 6.4.2 and 6.4.3.

Anomalous shape is the primary means of detecting gerrymanders of political districts.

Geometric shape was the aspect that alerted Gerry's political opponents to the manipulation of districts, and today shape is measured whenever GIS is used to aid in the drawing of political district boundaries, as must occur by law in the USA after every decennial census (see Box 13.4). An easy way to define shape is by comparing the perimeter length of an area to its area measure. Normally the square root of area is used, to ensure that the numerator and denominator are both measured in the same units. A common measure of shape or compactness is:

$$S = P/3.54\sqrt{A}$$

where *P* is the perimeter length and *A* is the area. The factor 3.54 (twice the square root of π) ensures that the most compact shape, a circle, returns a shape of 1.0, and the most distended and contorted shapes return much higher values.

13.3.3 Slope and aspect

The most versatile and useful representation of terrain in GIS is the *digital elevation model*, or DEM. This is a raster representation, in which each grid cell records the elevation of the Earth's surface, and reflects a view of terrain as a field of elevation values. The elevation recorded is often the elevation of the cell's central point, but sometimes it is the mean elevation of the cell, and other rules have been used to define the

Box 13.4 *Shape and the 12th Congressional District of North Carolina*

In 1992, following the release of population data from the 1990 Census, new boundaries were proposed for the voting districts of North Carolina, USA. For the first time race was used as an explicit criterion, and districts were drawn that as far as possible grouped minorities (notably African Americans) into districts in which they were in the majority. The intent was to avoid the historic tendency for minorities to be thinly spread in all districts, and thus to be unable to return their own representative to Congress. African Americans were in a majority in the new 12th District, but in order to achieve this the district had to be drawn in a highly contorted shape.

   The new district, and the criteria used in the redistricting, were appealed to the US Supreme Court. Writing for the 5–4 majority, and striking down the new districting scheme, Chief Justice William Rehnquist wrote that "A generalized assertion of past discrimination in a particular industry or region is not adequate because it provides no guidance for a legislative body to determine the precise scope of the injury it seeks to remedy. Accordingly, an effort to alleviate the effects of societal discrimination is not a compelling interest".

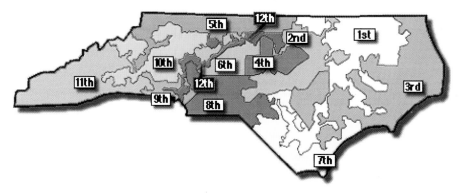

Figure 13.12 The boundaries of the 12th Congressional District of North Carolina drawn in 1992 show a very contorted shape, and were appealed to the US Supreme Court (Source: © Durham Herald Company, Inc., www.herald-sun.com)

cell's elevation (the rules used to define elevation in each cell of the US Geological Survey's GTOPO30 DEM, which covers the entire Earth's surface, vary depending on the source of data, see Box 7.1). Because of this variation, it is always advisable to read the available documentation to determine what exactly is meant by the recorded elevation in the cells of any DEM.

> The digital elevation model is the most useful representation of terrain in a GIS.

Knowing the exact elevation of a point above sea level is important for some applications, including prediction of the effects of global warming and rising sea levels on coastal cities, but for many applications the value of a DEM lies in its ability to produce derivative measures through trans-formation, specifically measures of slope and

aspect, both of which are also conceptualized as fields. Imagine taking a large sheet of plywood and laying it on the Earth's surface so that it touches at the point of interest. The magnitude of steepest tilt of the sheet defines the *slope* at that point, and the direction of steepest tilt defines the *aspect*.

   This sounds straightforward, but it is complicated by a number of issues. First, what if the plywood fails to sit firmly on the surface, but instead pivots, because the point of interest happens to be a peak, or a ridge? In mathematical terms, we say that the surface at this point *lacks a well-defined tangent*, or that the surface at this point is *not differentiable*, meaning that it fails to obey the normal rules of continuous mathematical functions and differential calculus. The surface of the Earth has numerous instances of sharp breaks of slope, rocky outcrops, cliffs, canyons, and deep

Box 13.5 Calculation of slope based on the elevations of a point and its eight neighbors

The equations used are as follows:

$$b = (z_3 + 2z_6 + z_9 - z_1 - 2z_4 - z_7)/8D$$

$$c = (z_1 + 2z_2 + z_3 - z_7 - 2z_8 - z_9)/8D$$

where b and c are tan(slope) in the x and y directions respectively, D is the grid point spacing, and z_i denotes elevation at the ith point, as shown below. These equations give the four diagonal neighbors of Point 5 only half the weight of the other four neighbors in determining slope at Point 5.

$$tan(slope) = \sqrt{b^2 + c^2}$$

where *slope* is the angle of slope in the steepest direction.

$$tan(aspect) = b/c$$

where *aspect* is the angle between the vertical and the direction of steepest slope, measured clockwise. Since *aspect* varies from 0 to 360, an additional test is necessary that adds 180 to *aspect* if c is positive, and 360 to aspect of c is negative and b is positive.

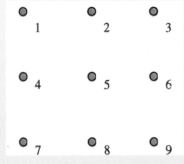

Figure 13.13 Calculation of the slope at Point 5 based on the elevation of it and its eight neighbors

gullies that defy this simple mathematical approach to slope, and this is one of the issues that led Benoît Mandelbrot to develop his theory of fractals, or mathematical functions that display behaviors of this nature. Mandelbrot argues in his books (Mandelbrot 1977, 1983) that many natural phenomena are fundamentally incompatible with traditional mathematics, and need a different approach. His concepts of fractals are discussed in detail in Section 5.8, and additional applications are described in Section 14.2.6.

A simple and satisfactory alternative is to take the view that slope must be measured at a particular resolution. To measure slope at a 30 m resolution, for example, we evaluate elevation at points 30 m apart and compute slope by comparing them. The value this gives is specific to the 30 m spacing, and a different spacing would have given a different result. In other words, *slope is a function of resolution*, and it makes no sense to talk about slope without at the same time talking

about a specific resolution or level of detail. This is convenient, because slope is easily computed in this way from a DEM with the appropriate resolution.

The spatial resolution used to calculate slope and aspect should always be specified.

Second, there are several alternative *measures* of slope, and it is important to know which one is used in a particular software package and application. Slope can be measured as an *angle*, varying from 0 to 90 degrees as the surface ranges from horizontal to vertical. But it can also be measured as a percentage or ratio, defined as *rise over run*, and unfortunately there are two different ways of defining run. Figure 13.14 shows the two options, depending on whether run means the horizontal distance covered between two points, or the diagonal distance (the *adjacent* or the *hypotenuse* of the right-angled

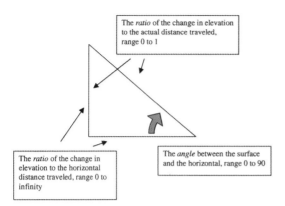

The *ratio* of the change in elevation to the actual distance traveled, range 0 to 1

The *ratio* of the change in elevation to the horizontal distance traveled, range 0 to infinity

The *angle* between the surface and the horizontal, range 0 to 90

Figure 13.14 Three alternative definitions of slope. To avoid ambiguity we use the angle, which varies between 0 and 90 degrees

triangle respectively). In the first case (opposite over adjacent) slope as a ratio is equal to the tangent of the angle of slope, and ranges from zero (horizontal) through 1 (45 degrees) to infinity (vertical). In the second case (opposite over hypotenuse) slope as a ratio is equal to the sine of the angle of slope, and ranges from zero (horizontal) through 0.707 (45 degrees) to 1 (vertical). To avoid confusion we will use the term slope only to refer to the measurement in degrees, and call the other options tan(slope) and sin(slope) respectively.

When a GIS calculates slope and aspect from a DEM, it does so by estimating slope at each of the data points of the DEM, by comparing the elevation at that point to the elevations of surrounding points. But the number of surrounding points used in the calculation varies, as do the weights given to each of the surrounding points in the calculation. Box 13.5 shows this idea in practice, using one of the commonest methods, which employs eight surrounding points and gives them different weights depending on how far away they are.

Slope and aspect are the basis for many interesting and useful forms of analysis. Slope is an input to many models of the soil erosion and runoff that result from heavy storms. Slope is also an important input to analyses that find the most suitable routes across terrain for power lines, highways, and military vehicles (see Section 14.3.3).

13.4 Transformations

In this section, we look at methods that transform GIS objects and databases into more useful

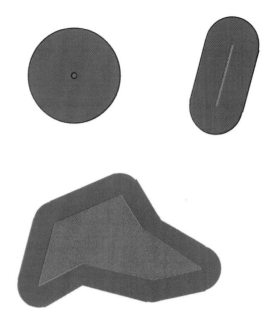

Figure 13.15 Buffers (dilations) of constant width drawn around a point, a polyline, and a polygon

products, using simple rules. These operations form the basis for many applications, because they are capable of revealing aspects that are not immediately visible or obvious.

13.4.1 Buffering

One of the most important transformations available to the GIS user is the *buffer* operation. Given any set of objects, which may include points, lines, or areas, a buffer operation builds a new object or objects by identifying all areas that are within a certain specified distance of the original objects. Figure 13.15 shows instances of a point, a line, and an area, and the results of buffering. Buffers have many uses, and they are among the most popular of GIS functions:

● The owner of a land parcel has applied for planning permission to rebuild – the owner could build a buffer around the parcel, in order to identify all homeowners who live within the legally mandated distance for notification of proposed redevelopments.

● A logging company wishes to clearcut an area, but is required to avoid cutting in areas within 100 m of streams – the company could build buffers 100 m wide around all streams to identify these protected riparian areas.

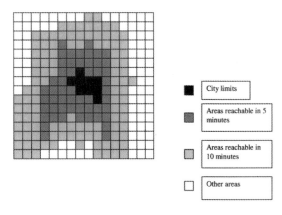

City limits

Areas reachable in 5 minutes

Areas reachable in 10 minutes

Other areas

Figure 13.16 A raster generalization of the buffer function, in which spreading occurs at rates controlled by a variable (travel speed; or friction, the inverse of travel speed) whose value is recorded in every raster cell

● A retailer is considering developing a new store on a site, of a type that is able to draw consumers from up to 4 km away from its stores – the retailer could build a buffer around the site to identify the number of consumers living within 4 km of the site, in order to estimate the new store's potential sales.

Buffering is possible in both raster and vector GIS – in the raster case, the result is the classification of cells according to whether they lie inside or outside the buffer, while the result in the vector case is a new set of objects (Figure 13.15). But there is an additional possibility in the raster case that makes buffering more useful in some situations. Figure 13.16 shows a city; average travel speeds vary in each cell of the raster outside the city. Rather than buffer according to distance from the city, we can ask a raster GIS to *spread* outwards from the city at rates determined by the travel speed values in each cell. Where travel speeds are high the spread will extend further, so we can compute how far it is possible to go from the city in a given period of time. This idea of spreading over a variable surface is easily implemented in raster representations, but impossible in vector representations. Another form of analysis that uses a raster surface to control rate of movement is discussed in Section 14.3.3.

Buffering is one of the most useful transformations in a GIS, and is possible in both raster and vector formats.

13.4.2 Point in polygon

In its simplest form, the point in polygon operation determines whether a given point lies inside or outside a given polygon. In more elaborate forms there may be many polygons, and many points, and the task is to assign points to polygons. If the polygons overlap, it is possible that a given point lies in one, many, or no polygons, depending on its location. Figure 13.17 illustrates the task. The operation is popular in GIS analysis because it is the basis for answering many simple queries:

● The points represent instances of a disease in a population, and the polygons represent reporting zones such as counties – the task is to determine how many instances of the disease occurred in each zone (in this case the zones should not overlap, and each point should fall into exactly one polygon).

● The points represent the locations of transmission line poles owned by a utility company, and the polygons are parcels of land – the task is to determine the owner of the land on which each pole lies, to verify that the company has the necessary easements and pays the necessary fees.

● The points represent the residential locations of voters, and the polygons represent voting districts – the task is to ensure that each voter receives the correct voting forms in the mail.

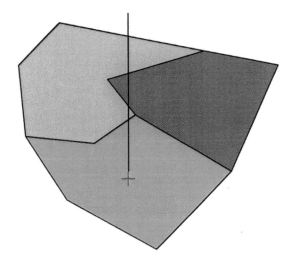

Figure 13.17 The point in polygon problem, shown in the field case (the point must by definition lie in exactly one polygon, or outside the project area). In only one instance (the pink polygon) is there an odd number of intersections between the polygon boundary and a line drawn vertically upward from the point

The point in polygon operation makes sense from both the discrete object and the field perspectives (see Section 3.5 for a discussion of these two perspectives). From a discrete object perspective both points and polygons are objects, and the task is simply to determine enclosure. From a field perspective, polygons representing a variable such as land ownership cannot overlap, since each polygon represents the land owned by one owner, and overlap would imply that a point is owned simultaneously by two owners. Similarly from a field perspective there can be no gaps between polygons. Consequently, the result of a point in polygon operation from a field perspective must assign each point to exactly one polygon.

> The point in polygon operation is used to determine whether a point lies inside or outside a polygon.

The standard algorithm for the point in polygon operation is shown in Figure 13.17. In essence, it consists of drawing a line vertically upwards from the point, and determining the number of intersections between the line and the polygon's boundary. If the number is odd the point is inside the polygon, and if it is even the point is outside. The algorithm must deal successfully with special cases, for example, if the point lies directly below a corner point of the polygon. Some algorithms extend the task to include a third option, when the point lies exactly on the boundary. But others ignore this, on the grounds that it is never possible to determine location with perfect accuracy, and so never possible to determine if an infinitely small point lies on an infinitely thin boundary line.

13.4.3 Polygon overlay

Polygon overlay is similar to point in polygon transformation in the sense that two sets of objects are involved, but in this case both are polygons. It exists in two forms, depending on whether a field or discrete object perspective is taken.

> The complexity of computing a polygon overlay was one of the greatest barriers to the development of vector GIS.

From the discrete object perspective, the task is to determine whether two area objects overlap, to determine the area of overlap, and to define the area formed by the overlap as one or more new area objects (the overlay of two polygons can produce a large number of distinct area objects,

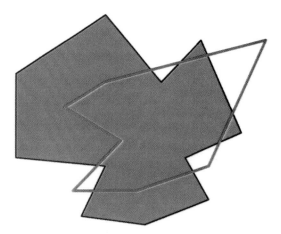

Figure 13.18 Polygon overlay, in the discrete object case. Here the overlay of two polygons produces nine distinct polygons. One has the properties of both polygons, four have the properties of the brown polygon but not the green polygon, and four are outside the brown polygon but inside the green polygon

see Figure 13.18). This operation is useful to determine answers to such queries as:

● How much of this proposed clearcut lies in this riparian zone?

● How much of this land parcel is affected by this easement?

● What proportion of the land area of the USA lies in areas managed by the Bureau of Land Management?

From the field perspective the task is somewhat different. Figure 13.19 shows two datasets, both

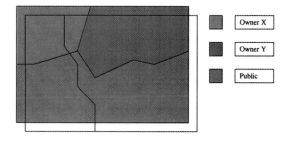

Owner X

Owner Y

Public

Figure 13.19 Polygon overlay in the field case. Here a dataset representing two types of land cover (A on the left, B on the right) is overlaid on a dataset representing three types of ownership. The result will be a single dataset in which every point is identified with one land cover type and one ownership type. It will have five polygons, since land cover A intersects with two ownership types, and land cover B intersects with three

representations of fields – one differentiates areas according to land ownership, and the other differentiates the same region according to land cover class. In the terminology of ESRI's ArcInfo, both datasets are instances of *area coverages*, or fields of nominal variables represented by non-overlapping polygons. The methods discussed earlier in this chapter could be used to inter-rogate either dataset separately, but there are numerous queries that require simultaneous access to both datasets, for example:

- What is the land cover class and who is the owner of the point indicated by the user?
- What is the total area of land owned by X and with land cover class A?
- Where are the areas that lie on publicly owned land and have land cover class B?

None of these queries can be answered by interrogating one of the datasets alone – the datasets must somehow be combined so that interrogation can be directed simultaneously at both of them.

The field version of polygon overlay does this by first computing a new dataset in which the region is partitioned into smaller areas that have uniform characteristics on both field variables. Each area in the new dataset will have two sets of attributes – those obtained from one of the input datasets, together with those obtained from the other. All of the boundaries will be retained, but they will be broken into shorter fragments by the inter-sections that occur between boundaries in one input dataset and boundaries in the other. Note the unusual characteristics of the new dataset shown in Figure 13.19. Unlike the two input datasets, where boundaries meet in a junction of three lines, the new map contains a new junction of four lines, formed by the new intersection discovered during the overlay process. Because the results of overlay are distinct in this way it is almost always possible to discover whether a GIS dataset was formed by overlaying two earlier datasets.

Polygon overlay has different meanings from the field and discrete object perspectives.

With a single dataset that combines both inputs, it is an easy matter to answer all of the queries listed above through simple interrogation. It is also easy to reverse the overlay process – if neighboring areas that share the same land cover class are merged, for example, the result is the land ownership map, and vice versa.

Polygon overlay is a computationally complex operation, and much work has gone into developing algorithms that function efficiently for large datasets. One of the issues that must be tackled by a practically useful algorithm is known as the *spurious polygon* or *coastline weave* problem, an issue discussed in principle in Section 6.3.1. It is almost inevitable that there will be instances in any practical application where the same line on the ground occurs in both datasets. This happens, for example, when a coastal region is being analyzed, because the coastline is almost certain to appear in every dataset of the region. Rivers and roads often form boundaries in many different datasets – a river may function both as a land cover class boundary and as a land ownership boundary, for example. But although the same line is represented in both datasets, its representations will almost certainly not be the same. They may have been digitized from different maps, subjected to different manipulations, obtained from entirely different sources (an air photograph and a topographic map, for example), and subjected to different measurement errors. When overlaid, the result is a series of small slivers. Paradoxically, the more care one takes in digitizing or processing, the worse the problem becomes, as the result is simply more slivers, albeit smaller in size.

In two vector datasets of the same area there will almost certainly be instances where lines in each dataset represent the same feature on the ground.

Table 13.1 shows an example of the consequences of slivers, and how a GIS can be rapidly overwhelmed if it fails to anticipate and deal with them adequately. Today, a GIS will offer various methods for dealing with the problem, the most common of which is the specification of a *tolerance*. If two lines fall within this distance of each other, the GIS will treat them as a single line, and not create slivers. The resulting overlay contains just one version of the line, not two. But at least one of the input lines has been moved, and if the tolerance is set too high the movement can be substantial, and can lead to problems later.

Overlay in raster is an altogether simpler operation, and this has often been cited as a good reason to adopt raster rather than vector structures. When two raster layers are overlaid, the attributes of each cell are combined according to a set of rules. For example, suppose the task is to find all areas that belong to owner A and have land use class B. Areas with these characteristics would be assigned a value, perhaps 1, and all other areas would be assigned a value of 0. Note the important difference between raster and vector overlay: in vector overlay there

Table 13.1 Numbers of polygons resulting from an overlay of five datasets (all are representations of fields). Dataset 1 is a representation of a map of soil capability for agriculture, datasets 2 through 4 are land use maps of the same area at different times (the probability of finding the same real boundary in more than one such map is very high), and dataset 5 is a map of land capability for recreation. The final three columns show the numbers of polygons in overlays of three, four, and five of the input datasets (one acre equals roughly 0.4 hectare)

Acres	1	2	3	4	5	1 + 2 + 5	1 + 2 + 3 + 5	1 + 2 + 3 + 4 + 5
0–1	0	0	0	1	2	2640	27566	77346
1–5	0	165	182	131	31	2195	7521	7330
5–10	5	498	515	408	10	1421	2108	2201
10–25	1	784	775	688	38	1590	2106	2129
25–50	4	353	373	382	61	801	853	827
50–100	9	238	249	232	64	462	462	413
100–200	12	155	152	158	72	248	208	197
200–500	21	71	83	89	92	133	105	99
500–1000	9	32	31	33	56	39	34	34
1000–5000	19	25	27	21	50	27	24	22
>5000	8	6	7	6	11	2	1	1
Totals	88	2327	2394	2149	487	9558	39188	90599

is no rule for combination, and instead the result of overlay contains all of the input information, rearranged and combined so that it can be used to respond to queries and can be subjected to analysis. Figure 13.20 shows an example of raster overlay.

Raster overlay is simpler, but it produces a fundamentally different kind of result.

13.4.4 Spatial interpolation

Spatial interpolation is a pervasive operation in GIS. Although it is often used explicitly in analysis, it is also used implicitly, in various operations such as the preparation of a contour map display, where spatial interpolation is invoked without the user's direct involvement. Spatial interpolation is a process of intelligent guesswork, in which the investigator (and the GIS) attempt to make a reasonable estimate of the value of a field at places where the field has not actually been measured. Spatial interpolation is an operation that makes sense only from the field perspective. The principles of spatial interpolation are discussed in Section 5.5; here the emphasis is on practical applications of the technique, and commonly used implementations of the principles.

Spatial interpolation finds applications in many areas:

● In estimating rainfall, temperature, and other attributes at places that are not weather

stations, and where no direct measurements of these variables are available.

● In estimating the elevation of the surface in between the measured locations of a DEM.

● In *resampling* rasters, the operation that must take place whenever raster data must be transformed to another grid (Figure 7.9).

● In contouring, when it is necessary to guess where to place contours in between measured locations.

In all of these instances spatial interpolation calls for intelligent guesswork, and the one principle that underlies all spatial interpolation is the Tobler Law – "all places are related but nearby places are more related than distant places" (Section 5.2). In other words, the best guess as to the value of a field at some point is the value measured at the closest observation points – the rainfall *here* is likely to be more similar to the rainfall recorded at the nearest weather stations than to the rainfall recorded at more distant weather stations. A corollary of this same principle is that in the absence of better information, it is reasonable to assume that any field exhibits relatively smooth variation – fields tend to vary slowly, and to exhibit strong positive spatial autocorrelation, a property of geographic data discussed in Section 5.3.

Spatial interpolation is the GIS version of intelligent guesswork.

In this section two commonly used methods of spatial interpolation are discussed: inverse

(A)   (B)

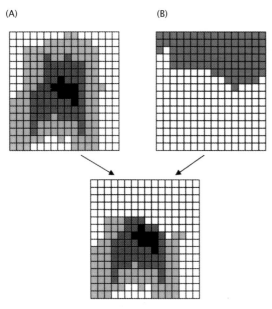

Figure 13.20 The raster overlay case, resulting in a new dataset that applies a set of rules to the input datasets, and is irreversible. The two input datasets are maps of (A) travel time from the urban area (see Figure 13.16) and (B) county (red indicates County X, white indicates County Y). The output map identifies travel time to areas in County Y, and might be used to compute average travel time to points in that county in a subsequent step. This operation is not reversible

distance weighting (IDW), which is the simplest method; and Kriging, a popular statistical method that is grounded in the theory of regionalized variables and falls within the field of *geostatistics*.

13.4.4.1 Inverse distance weighting (IDW)

IDW is the workhorse of spatial interpolation, the method that is most often used by GIS analysts. It employs the Tobler Law by estimating unknown measurements as weighted averages over the known measurements at nearby points, giving the greatest weight to the nearest points.

More specifically, denote the point of interest as x, and the points where measurements were taken as x_i, where i runs from 1 to n, if there are n data points. Denote the unknown value as $z(x)$ and the known measurements as z_i. Give each of these points a weight d_i, which will be evaluated based on the distance from x_i to x. Figure 13.21 explains this notation with a diagram. Then the weighted average computed at x is:

$$z(\mathbf{x}) = \sum_i w_i z_i \bigg/ \sum_i w_i$$

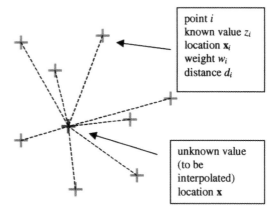

Figure 13.21 Notation used in the equations defining spatial interpolation

In other words, the interpolated value is an average over the observed values, weighted by the w's.

There are various ways of defining the weights, but the option most often employed is to compute them as the inverse squares of distances, in other words (compare the options discussed in Section 5.5 in connection with the nature of geographic data):

$$w_i = 1/d_i^2$$

This means that the weight given to a point drops by a factor of 4 when the distance to the point doubles (or by a factor of 9 when the distance trebles). In addition, most software gives the user the option of ignoring altogether points that are further than some specified distance away, or of limiting the average to a specified number of nearest points, or of averaging over the closest points in each of a number of direction sectors (Figure 13.22). But if these values are not specified the software will assign default values to them.

> IDW provides a simple way of guessing the values of a field at locations where no measurement is available.

IDW achieves the desired objective of creating a smooth surface whose value at any point is more like the values at nearby points than the values at distant points. If it is used to determine z at a location where z has already been measured it will return the measured value, because the weight assigned to a point at zero distance is infinite, and for this reason IDW is described as an *exact* method of interpolation because its interpolated results honor the data points exactly (an *approximate* method is allowed to deviate from

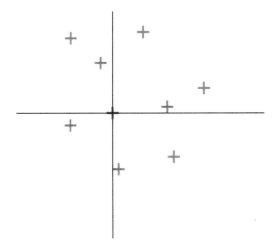

Figure 13.22 Selection of points for IDW interpolation using sectors. In this case four sectors are defined, and only the closest points in each sector are used (green crosses)

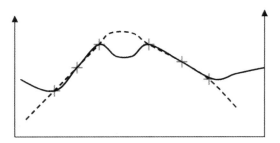

Figure 13.23 Potentially undesirable characteristics of IDW interpolation. This set of six data points clearly suggests a hill profile (dashed line). But in areas where there is little or no data the interpolator will move towards the overall mean (solid line)

the measured values in the interests of greater smoothness, a property which is often useful if deviations are interpreted as indicating possible errors of measurement, or local deviations that are to be separated from the general trend of the surface).

But because IDW is an average it suffers from certain specific characteristics that are generally undesirable. A weighted average that uses weights that are never negative must always return a value that is between the limits of the measured values – no point on the interpolated surface can have an interpolated z that is more than the largest measured z, or less than the smallest measured z. Imagine an elevation surface with some peaks and pits, but suppose that the peaks and pits have not actually been measured, but are merely indicated by the values of the measured points. Figure 13.23 shows a cross-section of such a surface. Instead of interpolating peaks and pits as one might expect, IDW produces the kind of result shown in the figure – small pits where there should be peaks, and small peaks where there should be pits. This behavior is often obvious in GIS output that has been generated using IDW. A related problem concerns extrapolation: if a trend is indicated by the data, as shown in Figure 13.23, IDW will inappropriately indicate a regression to the mean outside the area of the data points.

IDW interpolation may produce counter-intuitive results in the areas of peaks and

pits, and outside the area covered by the data points.

In short, the results of IDW are not always what one would want. There are many better methods of spatial interpolation that address the problems that were just identified, but the ease of programming of IDW and its conceptual simplicity make it among the most popular. Users should simply beware, and take care to examine the results of interpolation to ensure that they make good sense.

13.4.4.2 Kriging

Of all of the common methods of spatial interpolation it is Kriging that makes the most convincing claim to be grounded in good theoretical principles. The basic idea is to discover something about the general properties of the surface, as revealed by the measured values, and then to apply these properties in estimating the missing parts of the surface. Smoothness is the most important property, and it is operationalized in Kriging in a statistically meaningful way. There are many forms of Kriging, and the overview provided here is very brief. Interested readers are encouraged to read the excellent treatment in the GIS text by Burrough and McDonnell (1998), or one of the general introductions to geostatistics (e.g. Isaaks and Srivastava 1989).

There are many forms of Kriging, but all are firmly grounded in theory.

Suppose we take a point **x** as a reference, and start comparing the values of the field there with the values at other locations at increasing distances from the reference point. If the field is smooth (if

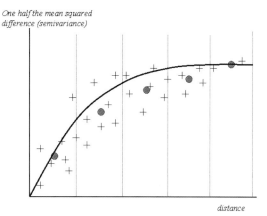
One half the mean squared difference (semivariance)

distance

Figure 13.24 A semivariogram. Each cross represents a pair of points. The solid circles are obtained by averaging within the ranges or *bins* of the distance axis. The solid line is the best fit to these five points, using one of a small number of standard mathematical functions

the Tobler Law is true, that is, if there is positive spatial autocorrelation) the values nearby will not be very different – $z(\mathbf{x})$ will not be very different from $z(\mathbf{x}_i)$. To measure the amount, we take the difference and square it, since the sign of the difference is not important: $(z(\mathbf{x}) - z(\mathbf{x}_i))^2$. We could do this with any pair of points in the area.

As distance increases, this measure will likely increase also, and in general a monotonic (consistent) increase in squared difference with distance is observed for most geographic fields (note that z must be measured on a scale that is at least interval, though *indicator Kriging* has been developed to deal with the analysis of nominal fields). In Figure 13.24, each point represents one pair of values drawn from the total set of data points at which measurements have been taken. The vertical axis represents one half of the squared difference (one half is taken for mathematical reasons), and the graph is known as the *semivariogram* (or *variogram* for short – the difference of a factor of two is often overlooked in practice, though it is important mathematically). To express its contents in summary form the distance axis is divided into a number of ranges or *buckets*, as shown, and points within each range are averaged to define the heavy points shown in the figure.

This semivariogram has been drawn without regard to the *directions* between points in a pair. As such it is said to be an *isotropic* variogram. Sometimes there is sharp variation in the behavior in different directions, and *anisotropic*

semivariograms are created for different ranges of direction (e.g. for pairs in each 90 degree sector; see Figure 13.22).

An anisotropic variogram asks how spatial dependence changes in different directions.

Note how the points of this typical variogram show a steady increase in squared difference up to a certain limit, and that increase then slackens off and virtually ceases. Again, this pattern is widely observed for fields, and it indicates that difference in value tends to increase up to a certain limit, but then to increase no further. In effect, there is a distance beyond which there are no more geographic surprises. This distance is known as the *range*, and the value of difference at this distance as the *sill*.

Note also what happens at the other, lower end of the distance range. As distance shrinks, corresponding to pairs of points that are closer and closer together, the semivariance falls, but there is a suggestion that it never quite falls to zero, even at zero distance. In other words, if two points were sampled a vanishingly small distance apart they would give different values. This is known as the *nugget* of the semivariogram. A non-zero nugget occurs when there is substantial error in the measuring instrument, such that measurements taken a very small distance apart would be different due to error, or when there is some other source of local noise that prevents the surface being truly smooth. Accurate estimation of a nugget depends on whether there are pairs of data points sufficiently close together. In practice the sample points may have been located at some time in the past, outside the user's control, or may have been spread out to capture the overall variation in the surface, so it is often difficult to make a good estimate of the nugget.

The nugget can be interpreted as the variation among repeated measurements at the same point.

To make estimates using Kriging we need to reduce the semivariogram to a mathematical function, so that semivariance can be evaluated at any distance, not just at the midpoints of bins as shown in Figure 13.24. In practice this means selecting one from a set of standard functional forms, and fitting that form to the observed data points to obtain the best possible fit. This is shown in the figure. The user of a Kriging function in a GIS will have control over the selection of distance ranges and functional forms, and whether a nugget is allowed.

Finally, the fitted semivariogram is used to estimate the values of the field at points of interest. As with IDW, the estimate is obtained as a weighted combination of neighboring values, but the estimate is designed to be the best possible given the evidence of the semivariogram. In general nearby values are given greater weight, but unlike IDW direction is also important: a point can be *shielded* from influence if it lies behind another point, since the latter's greater proximity suggests greater importance in determining the estimated value, whereas relative direction is unimportant in an IDW estimate. The process of maximizing the quality of the estimate is carried out mathematically, using the precise measures available in the semivariogram. Readers interested in the mathematical details should consult the references given earlier.

Kriging responds both to the proximity of sample points and to their directions.

Unlike IDW, Kriging has a solid theoretical foundation, but it also includes a number of options (e.g. the choice of the mathematical function for the semivariogram) that require attention from the user. In that sense it is definitely not a *black box* that can be executed blindly and automatically, but instead forces the user to become directly involved in the estimation process. For that reason GIS software designers will likely continue to offer several different methods, depending on whether the user wants something that is quick, despite its obvious faults, or better but more involving on the part of the user.

13.4.5 Density estimation and potential

Density estimation is in many ways the logical twin of spatial interpolation – it begins with points, and ends with a surface. But conceptually the two approaches could not be more different, because one seeks to estimate the missing parts of a field from samples of the field taken at data points, while the other creates a field from discrete objects.

Figure 13.25 illustrates this difference. The two datasets in the diagram look identical from a GIS perspective – they are both sets of points, with locations and a single attribute. But one shows sample measurements from a field, and the other shows the locations of discrete objects. In the discrete object view there is nothing between the objects but empty space – no missing field to be filled in through spatial interpolation. It would make no sense at all to apply spatial interpolation to a collection of discrete objects – and no sense at all to apply density estimation to samples of a field.

Density estimation makes sense only from the discrete object perspective, and spatial interpolation only from the field perspective.

Density estimation has many different roots, but is amply summarized in several texts, in particular that by Silverman (1986), and in discussions of spatial analysis (e.g. Bailey and Gatrell 1995). Although it could be applied to any type of discrete spatial object, it is most often applied to the estimation of point density, and that is the focus here. The most obvious example is the estimation of population density, and that example is used in this discussion, but it could be equally well applied to the density of different

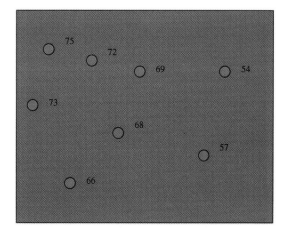

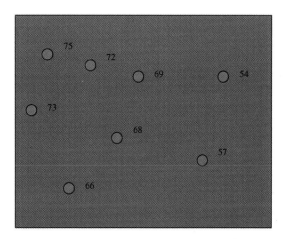

Figure 13.25 Two identical datasets with sharply different meanings. (A) a field of atmospheric temperature measured at eight irregularly spaced sample points. (B) eight discrete objects representing cities, with associated populations in thousands. Spatial interpolation makes sense only for (A), and density estimation only for (B)

(A) 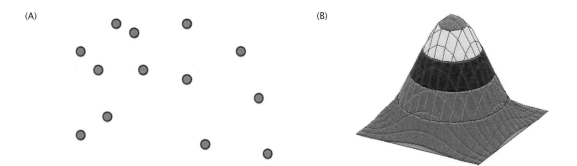 (B)

Figure 13.26 (A) A collection of point objects, and (B) a kernel function. The kernel's shape depends on a distance parameter – increasing the value of the parameter results in a broader and lower kernel, and reducing it results in a narrower and sharper kernel. When each point is replaced by a kernel and the kernels are added the result is a density surface whose smoothness depends on the value of the distance parameter

kinds of diseases, or animals, or any other set of well-defined points.

Consider the continent of Australia. One way of defining its population density is to take the entire population, and divide by the total area – on this basis the 1996 population density was roughly 2.38 per sq km. But we know that Australia's settlement pattern is very non-uniform, with most of the population concentrated in five coastal cities (Brisbane, Sydney, Melbourne, Adelaide, and Perth). So if we looked at the landscape in smaller pieces, such as circles 10 km in radius, and computed population density by dividing the number of people in each circle by the circle's area, we would obtain very different results depending on where the circle was centered. So in general, population density at a location, and at a spatial resolution of d might be defined by centering a circle at the location, and dividing the total population within the circle by its area. Using this definition there are an infinite number of possible population density maps of Australia, depending on the value selected for d. And it follows that there is no such thing as *population density*, only population density *at a spatial resolution of d*. Note the similarity between this idea and the previous discussion of slope – in general, many geographic themes can only be defined rigorously if spatial resolution is made explicit, and much confusion results in GIS because of our willingness to talk about themes without at the same time specifying spatial resolution.

> Density estimation with a kernel allows the spatial resolution of a field of population density to be made explicit.

The theory of density estimation formalizes these ideas. Consider a collection of point objects, such

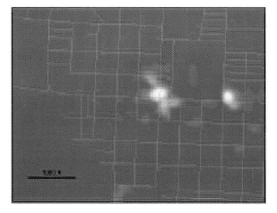

Search radius = 200 feet

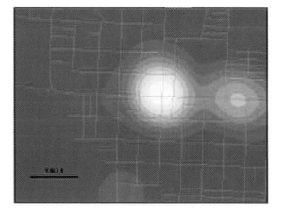

Search radius = 1000 feet

Figure 13.27 Density estimation using two different distance parameters in the respective kernel functions, showing the generally smoother and less peaked nature of the surface that results from the larger distance parameter

as those shown in Figure 13.26. The surface shown in Figure 13.26(B) is an example of a *kernel function*, the central idea in density estimation. Any kernel function has an associated length measure, and in the case of the function shown, which is a Gaussian distribution, the length measure is a parameter of the distribution: we can generate Gaussian distributions with any value of this parameter, and they become flatter and wider as the value increases. In density estimation, each point is replaced by its kernel function, and the various kernel functions are added to obtain an aggregate surface, or field of density. If one thinks of each kernel as a pile of sand, then each pile has the same total weight of one unit. The total weight of all piles of sand is equal to the number of points, and the total weight of sand within a given area, such as the area shown in the figure, is an estimate of the total population in that area. Mathematically, if the population density is represented by a field $\rho(x, y)$, then the total population within area A is the integral of the field function over that area, that is:

$$P = \int_A \rho \, dA$$

A variety of kernel functions are used in density estimation, but the form shown in Figure 13.26 is perhaps the commonest. This is the traditional bell curve or Gaussian distribution of statistics, and is encountered elsewhere in this book in connection with errors in the measurement of position in two dimensions (Section 15.1.2.1). By adjusting the width of the bell it is possible to produce a range of density surfaces of different amounts of smoothness. Figure 13.27 contrasts two density estimations from the same data, one using a comparatively narrow bell to produce a complex surface, and the other using a broader bell to produce a smoother surface.

Questions for Further Study

1. Did Dr John Snow actually make his inference strictly from looking at his map? What information can you find on the Web on this issue? (try www.jsi.com)
2. How exactly do driving directions given by humans differ from GIS-generated ones? Compare the directions you would give with those given by www.mapquest.com for a route that is familiar to you.
3. What is conditional simulation, and how does it differ from Kriging? Under what circumstances might it be useful (a possible source on this

question is the chapter by Englund in Goodchild M F, Steyaert L T, and Parks B O, *Environmental Modeling with GIS*. New York: Oxford University Press).
4. You are given a map showing the home locations of the customers of an insurance agent, and asked to construct a map showing the agent's market area. Would spatial interpolation or density estimation be more appropriate, and why?

Online Resources

NCGIA Core Curricula (www.ncgia.ucsb.edu/pubs/core.html):
Core Curriculum in GIScience, Section 2.1.2.1 (Simple Algorithms for GIS I: Intersections of Lines), Section 2.1.2.2 (Simple Algorithms for GIS II: Operations on Polygons), Section 2.1.2.3 (The Polygon Overlay Operation), and Section 2.14.2 (Exploratory Spatial Data Analysis, Robert Haining and Stephen Wise)
Core Curriculum in GIS, 1990, Units 5 (Raster GIS Capabilities), 14 (Vector GIS Capabilities), 15 (Spatial Relationships in Spatial Analysis), 32 (Simple Algorithms I: Intersection of Lines), 33 (Simple Algorithms II: Polygons), 34 (The Polygon Overlay Problem), 40 (Spatial Interpolation I), 41 (Spatial Interpolation II)
ESRI Virtual Campus courses (campus.esri.com):
Turning Data into Information, by Paul Longley, Michael Goodchild, David Maguire, and David Rhind (Modules ''Query and Measurement'' and ''Transformations and Descriptive Summaries'')
Introduction to ArcView GIS, by ESRI (includes: lessons on ''Selecting Map Features'' and ''Measuring Distance and Area'' in the Module ''Querying Data in ArcView''; lessons on ''Finding Features Nearby and Within'', ''Finding Intersecting Features'', ''Finding Features and Joining their Attributes'', and ''Advanced Geoprocessing'' in the Module ''Analyzing Spatial Relationships using ArcView''; and lessons on ''Creating Themes from Shapefiles'', ''Creating Event Themes from Coordinates Files'', and ''Address Geocoding'' from the Module ''Creating Your Own Data in ArcView'')
Introduction to ArcView 3D Analyst, by ESRI (lessons on ''Analyzing Paths and Profiling Lines'', ''Analyzing Area and Volume'', and ''Analyzing Visibility'' in the Module ''Using ArcView 3D Analyst Analysis Tools'')

Excellent additional information on John Snow is available at the Web site of John Snow Inc: www.jsi.com

Chapter 37, Spatial hydrography and landforms (Band L E)

Chapter 40, The future of GIS and spatial analysis (Goodchild M F and Longley P A)

Reference Links

Maguire D J, Goodchild M F, and Rhind D W (eds) 1991 *Geographical Information Systems: Principles and applications*. Harlow, UK: Longman (Text available online from 'Links to Big Book 1' at www.wiley.com/gis and www.wiley.co.uk/gis).

Chapter 21, The functionality of GIS (Maguire D J, Dangermond J)

Chapter 23, Cartographic modelling (Tomlin C D)

Chapter 25, Developing appropriate spatial analysis methods for GIS (Openshaw S)

Chapter 26, Spatial decision support systems (Densham P J)

Longley P A, Goodchild M F, Maguire D J, and Rhind D W (eds) 1999 *Geographical Information Systems: Principles, techniques, management and applications*. New York: John Wiley.

Chapter 16, Spatial statistics (Getis A)

Chapter 17, Interactive techniques and exploratory spatial data analysis (Anselin L)

Chapter 18, Applying geocomputation to the analysis of spatial distributions (Openshaw S, Alvanides S)

Chapter 19, Spatial analysis: retrospect and prospect (Fischer M M)

Chapter 34, Spatial interpolation (Mitas L, Mitasova H)

References

Bailey T C and Gatrell A C 1995 *Interactive Spatial Data Analysis*. Harlow, UK: Longman Scientific and Technical.

Burrough P A and McDonnell R A 1998 *Principles of Geographical Information Systems*. New York: Oxford University Press.

Isaaks E H and Srivastava R M 1989 *Applied Geostatistics*. New York: Oxford University Press.

Mandelbrot B 1977 *Fractals: Form, Chance, and Dimension*. San Francisco: Freeman.

Mandelbrot B 1983 *The Fractal Geometry of Nature*. San Francisco: Freeman.

Martin D J 1989 Mapping population data from zone centroid locations. *Transactions of the Institute of British Geographers* **NS 14**: 90-97.

Openshaw S, Charlton M, Wymer C, and Craft A 1987 A Mark I geographical analysis machine for the automated analysis of point data sets. *International Journal of Geographical Information Systems* **1**: 335-358. (www.tandf.co.uk/journals)

Silverman B W 1986 *Density Estimation for Statistics and Data Analysis*. New York: Chapman and Hall.

Tomlin C D 1990 *Geographic Information Systems and Cartographic Modeling*. Englewood Cliffs, New Jersey: Prentice Hall.

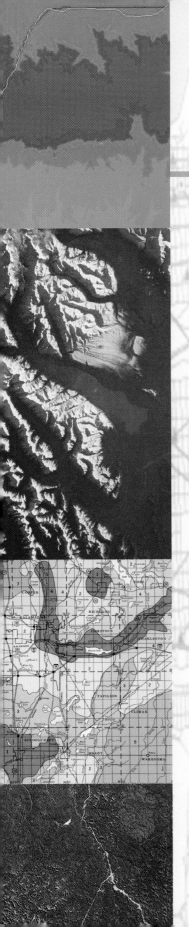

ADVANCED SPATIAL ANALYSIS

14

This second of two chapters on spatial analysis continues into more advanced territory, and focuses on three areas: descriptive summaries, optimization, and hypothesis testing. These methods are all conceptually more complex than those in Chapter 13, but are equally important in practical applications of GIS.

The chapter begins with a brief discussion of data mining. Descriptive summaries attempt to capture the nature of geographic distributions, patterns, and phenomena in simple statistics that can be compared through time, across themes, and between geographic areas. Optimization techniques apply much of the same thinking, and extend it, to help users who must select the best locations for services, or find the best routes for vehicles, or a host of similar tasks. Hypothesis testing addresses the basic scientific need to be able to generalize results from a small study to a much larger context, perhaps the entire world.

Learning objectives

After reading this chapter you will know about:

● Data mining, a new form of analysis that is enabled by GIS and by vast new supplies of data;

● The concept of summarizing a pattern in a few simple statistics;

● Methods that support decisions by enlisting GIS to search automatically across thousands or millions of options;

● The concept of a hypothesis, and how to make inferences from small samples to larger populations.

14.1 More Spatial Analysis

The methods introduced in this chapter are generally more complex than those of Chapter 13, and use more advanced conceptual frameworks. The previous chapter began with simple queries and measurement, but in this chapter the discussion turns to summaries, optimization methods, and inferential methods. Many date from well before the era of GIS, and the advent of geographic databases and cheap computing technology has turned what were once esoteric methods buried in the academic literature into practical ways of solving everyday problems.

The advent of large databases and fast computing has also led to new ways of thinking about spatial analysis, and many of these are captured in the term *data mining*. Surely, the thinking goes, there must be interesting patterns, anomalies, and truths buried in the masses of data being collected and archived every day by humanity's vast new investment in information technology. Cameras now capture images of the traffic on freeways that are precise enough to read license plates. Every use of a credit card creates a record in a database that includes the location of use, along with information on what was bought, and having the name of the user allows that information to be linked to many other records (Figure 14.1). If we only had suitable computer methods, the thinking goes, we could program software to scan through these vast databases, looking for things of interest. Many of these might be regarded as invasions of privacy, but others could be very useful in spotting criminal activity. Data mining has become a major area of application in business, where it is used to detect anomalies in spending patterns that might be indicative of a stolen credit card – for example, a card owned by a resident of California, USA is suddenly used for a series of large purchases in a short period of time in the New York area, in types of stores not normally frequented by the card owner. The *geographic* aspect of this evidence – the expenditures in a small, unusual area – is one of the most significant.

> Data mining is used to detect anomalies and patterns in vast archives of digital data.

Data mining techniques could be used to watch for anomalous patterns in the records of disease diagnosis, or to provide advanced warnings of new outbreaks of virulent forms of influenza. In all of these examples the objective is to find *patterns*, and *anomalies* that stand out from the normal in an area, or stand out at a particular point in time when compared with long-term

Figure 14.1 Massive geographic databases can be built from records on the use of credit cards

averages. In recent years surprising patterns in digital data have led to the discovery of the ozone hole over the Antarctic, the recall of millions of tires made by the Firestone-Bridgestone Company, and anticipation of numerous outbreaks of disease.

14.2 Descriptive Summaries

14.2.1 Centers

The topics of generalization and abstraction were discussed earlier in Chapter 7 as ways of reducing the complexity of data. This section reviews a related topic, that of numerical summaries. If we want to describe the nature of summer weather in an area we cite the *average* or *mean*, knowing that there is substantial variation around this value, but that it nevertheless gives a reasonable *expectation* about what the weather will be like on any given day. The mean (the more formal term) is one of a number of measures of *central tendency*, all of

Figure 14.2 The search for patterns in massive datasets, such as this image of the Earth's surface, is known as data mining

which attempt to create a summary description of a series of numbers in the form of a single number. Another is the *median*, the value such that one half of the numbers are larger, and one half are smaller. Although the mean can be computed only for numbers measured on interval or ratio scales, the median can be computed for ordinal data. For nominal data the appropriate measure of central tendency is the *mode*, or the commonest value (for definitions of nominal, ordinal, interval, and ratio see Section 3.4, and for appropriate

measures of central tendency for cyclic data see Mardia and Jupp 2000).

The spatial equivalent of the mean would be some kind of center, calculated to summarize the positions of a number of points. Early in US history the Bureau of the Census adopted a practice of calculating a center for the US population. As agricultural settlement advanced across the West in the 19th century, the repositioning of the center every 10 years captured the popular imagination. Today, the movement west has slowed, and by the next census may even have halted.

The mean of a set of numbers has several properties. First, it is calculated by summing the numbers and dividing by the number of numbers. Second, if we take any value *d* and sum the squares of the differences between the numbers and *d*, then when *d* is set equal to the mean this sum is minimized (Figure 14.3). Third, the mean is the point about which the set of numbers would balance if we made a physical model such as the one shown in Figure 14.4, and suspended it.

These properties extend easily into two dimensions. Figure 14.5 shows a set of points on a flat plane, each one located at a point (x_i, y_i) and with weight w_i. The *centroid* or *mean center* is found by taking the weighted average of the *x* and *y* coordinates:

$$\bar{x} = \sum_i w_i x_i \Big/ \sum_i w_i$$

$$\bar{y} = \sum_i w_i y_i \Big/ \sum_i w_i$$

It also is the point that minimizes the sum of squared distances, and it is the balance point.

The centroid is the most convenient way of summarizing the locations of a set of points.

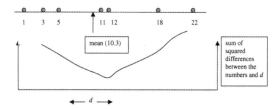

Figure 14.3 The mean minimizes the sum of squared differences with a set of numbers. The upper diagram shows seven numbers. The lower diagram shows how the sum of squared differences between these numbers and *d* is minimized when *d* is the mean value

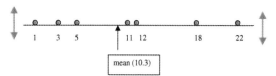

Figure 14.4 The mean is also the balance point, the point about which the distribution would balance if it were modeled as a set of equal weights on a weightless, rigid rod

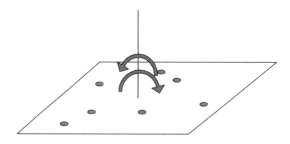

Figure 14.5 The centroid or mean center replicates the balance-point property in two dimensions – the point about which the two-dimensional pattern would balance if it were transferred to a weightless, rigid plane and suspended

Just like the mean, the centroid is a useful summary of a distribution of points. Although any single centroid may not be very interesting, a comparison of centroids for different sets of points or for different times can provide useful insights. Figure 14.6 shows the movements of the centroids of different types of land use in a typical North American city, and nicely summarizes the changes that have occurred since the late 19th century, when all types of land uses were densely mixed, to the patterns of today now that zoning controls have produced an effective segregation, particularly of residential and industrial land uses.

The property of minimizing functions of distance (the square of distance in the case of the centroid) makes centers useful for different reasons. Of particular interest is the location that minimizes the sum of distances, rather than the sum of squared distances, since this could be the most effective location for any service that intended to serve a dispersed population. The point that minimizes total straight-line distance is known as the *point of minimum aggregate travel* or MAT. Historically there have been numerous instances of confusion between the MAT and the centroid, and the US Bureau of the Census used to claim that the Center of Population had the MAT property, even though their calculations were based on the equations for the centroid.

The centroid is often confused with the point of minimum aggregate travel.

There is no simple mathematical expression for the MAT, and instead its location must be found by a trial-and-error process of *iteration*, in which an initial guess is successively improved using a suitable algorithm. There is also an interesting way of computing the MAT using a physical analog, an experiment known as the *Varignon*

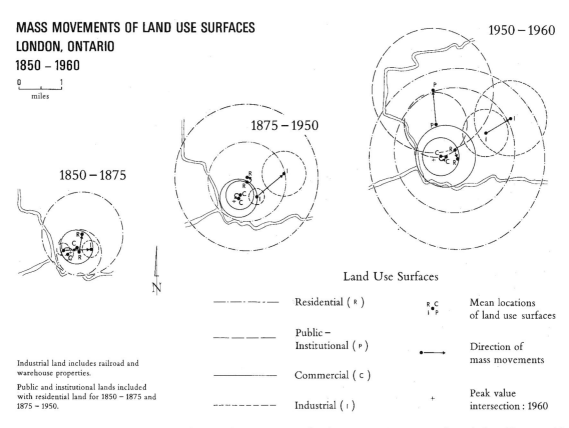

MASS MOVEMENTS OF LAND USE SURFACES
LONDON, ONTARIO
1850 – 1960

0 _____ 1
miles

1950 – 1960

1875 – 1950

1850 – 1875

N

Industrial land includes railroad and
warehouse properties.

Public and institutional lands included
with residential land for 1850 – 1875 and
1875 – 1950.

Land Use Surfaces

Residential (R)

Public –
Institutional (P)

Commercial (c)

Industrial (I)

R   C
I   P
Mean locations
of land use surfaces

Direction of
mass movements

+   Peak value
intersection : 1960

Figure 14.6 An example of the use of centroids to summarize the changes in point patterns through time. The centroids of four land use classes are shown for London, Ontario, Canada, from 1850 to 1960. Circles show the associated dispersions of sites within each class. Note how the industrial class has moved east, remaining concentrated, while the commercial class has remained concentrated in the core, and the residential class has dispersed but remained centered on the core. In contrast the institutional class moved to a center in the northern part of the city. (Source: The Great Lakes Geographer, Dept of Geography, University of Western Ontario)

Frame (see Box 14.1). There is an intriguing paradox associated with the experiment. Imagine that one of the points must be relocated further away from the MAT but in the same direction. The force pulling on the string remains the same, in the same direction, so the position of the knot does not move. But how could the solution to a problem of minimizing distance not be affected by a change in distance? It is straightforward to show with a little application of calculus that the experiment is indeed correct, despite what appears to be a counter-intuitive result – that the location of the MAT does indeed depend only on the weights and the directions of pull, not on the distances. The same property occurs in the one-dimensional case: the point that minimizes distance along a line is the median, and the median's location is not affected by moving points as long as no point is moved to the opposite side of the median.

This example serves to make a very general point about spatial analysis and its role in human-computer interaction. In the introduction to Chapter 13 the point was made that all of these methods of analysis work best in the context of a collaboration between human and machine, and that one benefit of the machine is that it sometimes serves to correct the misleading aspects of human intuition. The MAT, and more complex problems that are based on it, illustrates this principle very well. Intuition would have suggested that moving one of the weights in the Varignon Frame experiment would have moved the knot, even though the movement was in the same direction. But theory says otherwise. Humans can be very poor at guessing the answers to optimization problems in space, and the results of machine analysis can often be sharply different from what we expect.

Box 14.1 The Varignon Frame experiment

This experiment provides a way of finding the point of minimum aggregate travel (MAT). Because it uses a physical model to solve the problem rather than a digital representation we can think of it as an *analog* computer. Suppose we want to find the best location for a steel mill, to minimize the total costs of shipping raw materials to the plant (iron ore, limestone, and coal) and the total costs of shipping the product to markets. Each source of raw materials or market is represented by a point, with a weight equal to the cost per km of shipping the required amount of material. Then the total shipping cost is the product of these weights times distances to the plant, summed over all sources and markets.

Take a piece of plywood, and sketch a map on it, locating the sources and markets (see Figure 14.7). At each of these points drill a hole, and hang a weight on a string through the hole, using the appropriate weight (shipping cost of the required amount of material per km). Bring all of the strings together into a knot. When the knot is released to find its own position the result is the MAT, though friction in the holes may create some error in the solution.

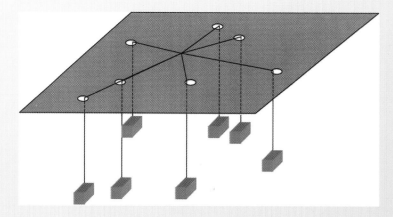

Figure 14.7 The Varignon Frame experiment, an analog computer for the point of minimum aggregate travel (MAT)

The MAT is one instance of a type of spatial analysis in which the objective is not insight so much as *design* – it is a *normative* or *prescriptive* method. In the 1920s the Soviet Union supported the Mendeleev Centrographical Laboratory whose objective was to identify the centers of various kinds of economic activity in the USSR, as the most suitable sites for the state's central planning functions. For example, the center of piano manufacture would be the best place to locate the state planning office for piano production.

Many methods of spatial analysis are used to make design decisions.

In Section 13.3.1 it was noted that simple metrics of distance, such as the length of a straight line, have the disadvantage that they ignore the effects of networks on travel. The approach to the MAT described here also assumes straight-line travel. Comparable approaches based on networks are often better suited to practical problem-solving, and some examples of these are discussed in Section 14.3.

All of the methods described in this and the following section are based on plane geometry, and simple x, y coordinate systems. If the curvature of the Earth's surface is taken into account, the centroid of a set of points must be calculated in three dimensions, and will always lie under the surface. More useful perhaps are versions of the MAT and centroid that minimize distance over the curved surface (the minimum total distance, and the minimum total of squared distances, respectively, using great circle distances, Figure 4.11). See Figure 14.8.

14.2.2 Dispersion

Central tendency is the obvious choice if a set of numbers must be summarized in a single value,

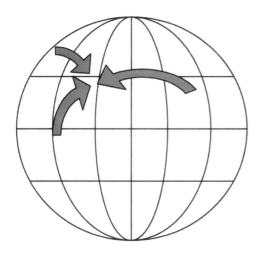

Figure 14.8 Minimizing distance over the curved surface of the Earth

but what if there is the opportunity for a second summary value? Here the measure of choice for numbers with interval or ratio properties is the *standard deviation*, or the square root of the mean squared difference from the mean:

$$s = \sqrt{\sum_i (x_i - \bar{x})^2 / n}$$

where n is the number of numbers, s is the standard deviation, x_i refers to the ith observation, and $\bar{x}$ is the mean of the observations. In weighted form the equation becomes:

$$s = \sqrt{\sum_i w_i (x_i - \bar{x})^2 \Big/ \sum_i w_i}$$

where w_i is the weight given to the ith observation. The *variance*, or the square of the standard deviation (the mean squared difference from the mean) is often encountered, but it is not as convenient a measure for descriptive purposes. Standard deviation and variance are considered more appropriate measures of dispersion than the *range* (difference between the highest and lowest numbers) because as averages they are less sensitive to the specific values of the extremes.

The standard deviation will also be encountered in Chapter 15 in a different guise, as the root mean squared error (RMSE), a measure of dispersion of observations about a true value. Just as in that instance, the Gaussian distribution provides a basis for generalizing about the contents of a sample of numbers, using the mean and standard deviation as the parameters of a simple bell curve. If data follow a Gaussian distribution, then approxi-

mately 68% of values lie within one standard deviation of the mean; and approximately 5% of values lie outside two standard deviations.

These ideas convert very easily to the two-dimensional case. A simple measure of dispersion in two dimensions is the *mean distance from the centroid*. In some applications it may be desirable to give greater weight to more distant points. For example, if a school is being located, then students living at distant locations are comparatively disadvantaged. They can be given greater weight if each distance is squared, such that a student twice as far away receives four times the weight. This property is minimized by locating the school at the centroid.

Mean distance from the centroid is a useful summary of dispersion.

Measures of dispersion can be found in many areas of GIS. The breadth of the kernel function of density estimation (Section 13.4.5) is a measure of how broadly the pile of sand associated with each point is dispersed. RMSE is a measure of the dispersion inherent in positional errors (Section 15.1.2.1).

14.2.3 Histograms and pie charts

Histograms (bar graphs) and pie charts are two of the many ways of visualizing the content of a geographic database. A histogram shows the relative frequencies of different values of an attribute by ordering them on the x axis and displaying frequency through the length of a bar parallel to the y axis (Figure 14.9). Attributes should have interval or ratio properties, although ordinal properties are sufficient to allow the values to be ranked, and a histogram based on ordinal data is sometimes useful. A pie chart is most useful for nominal data, and is used to display the relative frequencies of distinct values, with no necessity for ranking. Pie charts are also useful in dealing with attributes measured on cyclic scales (Box 3.3). Both take a single attribute and organize its values in a form that allows quick comprehension.

As noted in the discussion of queries and interrogation in Section 3.4, it is often helpful if their contents of histograms and pie charts can be linked. For example, the user might be able to select a value on the x axis of the histogram, and to have the records whose attributes exceed that value flagged automatically in both the map and table views.

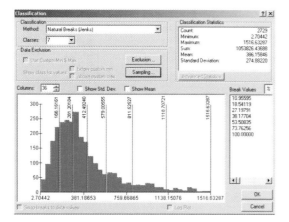

Figure 14.9 A bar graph or histogram showing the number of features with values of a certain attribute. In the case, the length of a street segment is the attribute, and 2729 such objects in the database are shown as a bar graph. The commonest length is about 435 m, a typical length in a US city, and over 300 segments are roughly of this length. The longest segment is over 1500 m (note the summary statistics provided by this ESRI ArcMap function) (Source: ESRI)

14.2.4 Scatterplots

The previous three sections looked at descriptive summaries of a single set of objects – points – with associated attributes that were treated as weights. But much of the power of GIS lies in its ability to compare sets of attributes – often thought of as the process of overlaying layers – in order to explore *vertical* relationships rather than *horizontal* ones (the term *vertical* is often used in GIS to refer to this comparison of attributes, despite the confusion with the true vertical or *z* dimension). These types of relationships were introduced in Section 5.7, and regression was presented there as a way of describing them.

Scatterplots are useful visual summaries of relationships between attributes.

The scatterplots introduced in Section 5.7 provide very good summaries of the relationships between two variables, by displaying the values of one attribute plotted against the other. If both sets of attributes belong to the same objects (e.g. the average income of a census tract, and the percent of the population with university education of the same census tracts), then the construction of a scatterplot is straightforward. If both are attributes of raster datasets, then the scatterplot is built by comparing the datasets pixel by pixel, with one point on the scatterplot for every pixel.

But if the attributes come from different sets of vector objects that do not coincide in space the problem is decidedly more complex, and one that GIS is particularly good at addressing.

Box 14.2 shows a typical example of the problems associated with correlating attributes for objects that do not coincide spatially. In this case both the number of objects and their locations differ between the two datasets, and several options are available for comparing them. A simple expedient is to use spatial interpolation to map both variables as complete fields. Then it would be possible to conduct the analysis in any of three ways:

● by interpolating the second dataset to the locations at which the first attribute was measured;

● by the reverse, i.e. interpolating the first dataset to the locations at which the second attribute was measured; or

● interpolating both datasets to a common geometric base, such as a raster.

In the third case, note that it is possible to create a vast amount of data, by using a sufficiently detailed raster.

Methods of spatial interpolation make it easy to invent geographic data, a potentially dangerous practice.

In essence, all of these options involve manufacturing information, and the results will depend to some extent on the nature and suitability of the method used to carry out the spatial interpolation. We normally think of a relationship that is demonstrated with a large number of observations as stronger and more convincing than one based on a small number of observations. But the ability to manufacture data upsets this standard view. It makes most sense in such situations to interpolate to a number of points that is within the range of the number observed, in other words, bounded by the larger and smaller datasets.

14.2.5 Spatial dependence

The concept of spatial dependence has been encountered at many different points in the previous chapters, which is fitting considering its central role in GIS and the Tobler Law. As noted in Section 5.2, a world without spatial dependence would be an impossibility: there would be no basis for compressing atomic geographic facts into practical forms of representation, no basis for spatial interpolation, and indeed no basis for life. Hell might be a world without spatial

Box 14.2 Comparing attributes when objects do not coincide spatially

GIS users often encounter situations where attributes must be compared, but for different sets of objects. Figure 14.10 shows a study area, with two sets of points, one the locations where levels of ambient sound were measured using recorders mounted on telephone poles, the other the locations of interviews conducted with local residents to determine attitudes to noise. We would like to know about the relationship between sound and attitudes, but the locations and numbers of cases are different. Three solutions are discussed in the text, all using spatial interpolation. The figure shows an acceptable solution, to interpolate to a regular grid of points. The number of grid points should be between the two numbers of cases – in this case, between 10 and 15 (12 are shown).

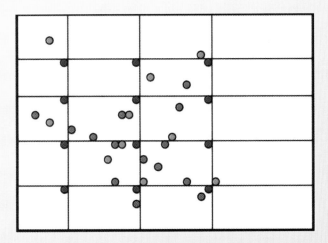

Figure 14.10 Coping with the comparison of two sets of attributes when the respective objects do not coincide. In this instance attitudes regarding ambient noise have been obtained through a household survey of 15 residents (green dots), and are to be compared to ambient noise levels measured at 10 observation points (brown dots). The solution is to interpolate both sets of data to 12 comparison points (blue dots), using the methods discussed in Section 13.4.4

dependence, since it would be impossible to live there in any practical or meaningful way.

Spatial dependence can be measured at any spatial resolution.

Spatial dependence is inherently scale specific. A dataset can exhibit positive spatial dependence at one scale, and negative spatial dependence at another. For example, the chessboard shown in Figure 5.2(A) and discussed in Section 5.3 has negative spatial dependence between squares, but positive spatial dependence within squares. The Moran and Geary indices introduced in Section 5.6.2 compare attributes of neighboring zones, but as noted there it is possible by making appropriate modifications to the weights matrices to compare more distant pairs of zones, and thus

to measure spatial dependence at different scales. For example, we could construct a weights matrix where weights are 1 between pairs of squares that are at least three squares apart, and zero otherwise.

By contrast, the semivariogram that was introduced in Section 13.4.4 in connection with spatial interpolation and Kriging measures and summarizes spatial dependence at all scales, by comparing the attributes at all pairs of points, and plotting one half of the average squared difference against distance, for different ranges of distance.

Spatial dependence is a very useful descriptive summary of geographic data, and a fundamental part of its nature. The semivariogram of a raster dataset describes how difference increases with distance, and whether difference ceases to increase beyond a certain range. By computing semivariograms in different directions it is possible to determine whether a dataset displays marked anisotropy, or distinct behaviors. The Geary and Moran indices are useful summary measures of the degree to which high or low values tend to cluster together, or whether they are distributed in apparent randomness.

14.2.6 Fragmentation and fractional dimension

Figure 14.11 shows a type of dataset that occurs frequently in GIS. It might represent the classes of land cover in an area. In fact it is a map of soils, and each patch corresponds to an area of uniform

class, and is bounded by areas of different class. A landscape ecologist or forester might be interested in the degree to which the landscape is fragmented: is it broken up into small or large patches, are the patches compact or contorted, and how long are the common boundaries between the patches? Fragmentation statistics have been designed to provide the numerical basis for these kinds of questions, by measuring relevant properties of datasets.

An interesting application of this kind of technique is to deforestation, in areas like the Amazon Basin. Figure 14.12 shows a sequence of images of part of the basin from space, and the impact of clearing and settlement on the pattern of land use. The impact of land use changes like

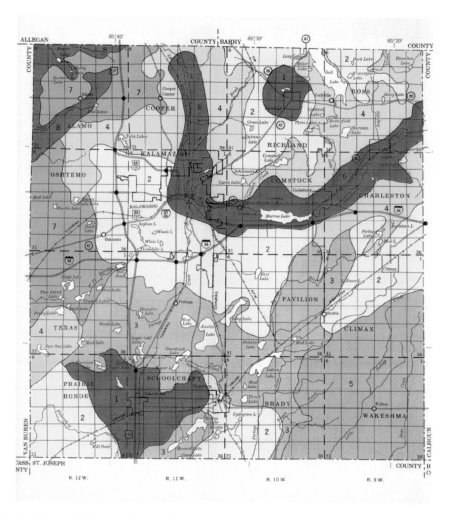

Figure 14.11 A soil map of Kalamazoo County, Michigan, USA. Soil maps such as this one show areas of uniform soil class separated by sharp boundaries (Source: US Department of Agriculture)

these depends on the degree to which they fragment existing habitat, perhaps into pieces that are too small to maintain certain species such as major predators.

(A)

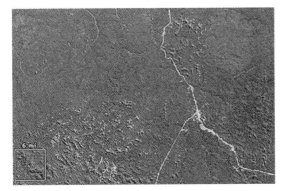

(B)

(C)

Figure 14.12 Three images of part of the state of Rondonia in Brazil, for (A) 1975, (B) 1986, and (C) 1992, showing the acceleration of deforestation. Fragmentation statistics can be used to summarize the shapes of areas, their relative sizes, and relative positions (Source: USGS)

An obvious measure of fragmentation is simply the number of patches. If the representation is vector this will be the number of polygons, whereas if it is raster it will be computed by comparing neighboring cells to see if they have the same class. But an area can be subdivided into a given number of patches in a vast number of distinct ways, so additional measures are needed. One simple measure is the average shape of patches, calculated using the shape measures discussed in Section 13.3.2. A somewhat more informative indicator would be a measure of average shape for each class, since we might expect shape to vary by class – the urban class might be more fragmented than the forest class.

The size distribution of patches might also be a useful indicator. Average patch size is uninformative, since it is simply the total area divided by the number of patches, and is independent of their shapes, or of the relative abundances of large and small patches. But a histogram of patch size would show the relative abundances, and can be a useful basis for inferences.

Summary statistics of the common boundaries of patches are also useful. The concept of fractals was introduced in Section 5.8 as a way of summarizing the relationship between apparent length and level of geographic detail, in the form of the fractional dimension D. Lines that are smooth tend to have fractional dimensions close to 1, and more contorted lines have higher values. So fractional dimension has become accepted as a useful descriptive summary of the complexity of geographic objects, and tools for its evaluation have been built into many widely adopted packages. FRAGSTATS is an example – it was developed by the US Department of Agriculture for the purposes of measuring fragmentation using many alternative methods.

14.3 Optimization

The concept of using spatial analysis for design was introduced earlier in Section 14.2.1, and several of the methods described in that section were shown to have useful design-oriented properties. This section includes discussion of a wider selection of so-called *normative* methods, or methods developed for application to the solution of practical problems of design.

By way of illustration, Box 14.3 shows the overlay method of McHarg, first described as a method to be executed by hand using simple transparencies in his highly influential book *Design with Nature* (McHarg 1969), and later computerized in several early GIS. This is a prime example of the power of GIS to support spatial decisions – a *spatial decision support system*.

The methods discussed in this section fall into several categories. The next sub-section discusses methods for the optimum location of points, and extends the method introduced earlier for the MAT. The second sub-section discusses routing on a network, and its manifestation in the traveling salesman problem (TSP). The final sub-section examines the selection of optimum paths across continuous space, for locating such facilities as power lines and highways, and for

routing military vehicles such as tanks. The methods also divide between those that are designed to locate points and routes on networks, and those designed to locate points and routes in continuous space without respect to the existence of roads or other transportation links.

14.3.1 Point location

The MAT problem is an instance of location in continuous space, and finds the location that

---

Box 14.3 Ian McHarg, landscape architect, and the overlay method

Ian McHarg was born in Glasgow, Scotland, in 1920. He served in the British Army in World War II, went to Harvard in 1946, and moved to the University of Pennsylvania in 1954 to start a program in landscape architecture. His life up to the early 1990s is described in detail in his autobiography, published in 1996 (*A Quest for Life*, published by John Wiley & Sons), but he continues to be very active. He is best known for his highly influential book *Design with Nature* (1969), which laid out the principles of a process of design based on an understanding of ecology, hydrology, and other environmental sciences. The book is his "personal testament to the power and importance of sun, moon, stars, the changing seasons, seed time and harvest, clouds, rain and rivers, the oceans and the forests … essential partners to survival and with us now involved in the creation of the future". Although his methods were originally applied by hand, geographic information systems provided exactly the right tools to apply McHarg's methods to large and complex projects.

Figure 14.13 Ian McHarg, landscape architect

Ian McHarg passed away March 5, 2001, and his absence will be felt by all who knew him and benefitted from his contributions.

Figure 14.14 shows the basic concept of overlay pioneered by McHarg (see Section 13.4.3 for details of polygon overlay).

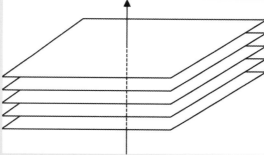

Figure 14.14 McHarg's overlay method, in its original manual form. Each of the five transparencies represents a map portraying one criterion and its importance in a specific decision, such as the location of a power line across a county. The transparency is darkest where the power line would create the greatest negative impact. When the transparencies are stacked above a light source, the areas where most light penetrates are most suitable for the power line. The process is easily operationalized, and made more rigorous and objective, in a GIS

minimizes total distance with respect to a number of points. The analogous problem on a network would involve finding that location on the network that minimizes total distance to a number of points, also located on the network, using routes that are constrained to the network. Figure 14.15 shows the contrast between continuous and network views of the problem, and Section 9.2.3.3 discusses data models for networks.

A very useful theorem first proved by Louis Hakimi reduces the complexity of many location problems on networks. Figure 14.15 shows a typical basis for network location. The links of the network come together in *nodes*. The weighted points are also located on the network, and also form nodes. For example, the task might be to find the location that minimizes total distance to a distribution of customers, with all customers aggregated into these weighted points. The weights in this case would be counts of customers. The Hakimi theorem proves that for this problem of minimizing distance the only locations that have to be considered are the nodes: it is impossible for the optimum location to be anywhere else. It is easy to see why this should be so. Think of a trial point located in the middle of a link, away from any node. Think of moving it slightly in one direction along the link. This moves it towards some weights, and away from others, but every unit of movement results in the same increase or decrease in total weighted distance. In other words, the total

distance traveled to the location is a linear function of the location along the link. Since the function is linear it cannot have a minimum mid-link, so the minimum must occur at a node.

In Section 14.2 the MAT in one dimension was shown to be the location such that half of the weight is to the left and half to the right, in other words, the measure of central tendency for ordinal data discussed in Section 14.2 and known as the *median*. The MAT problem on a network is consequently known as the *1-median* problem. The *p-median* problem seeks optimum locations for any number *p* of central facilities such that the sum of the distances between each weight and the *nearest* facility is minimized. A typical practical application of this problem is in the location of central public facilities, such as libraries, schools, or agency offices, when the objective is to locate for maximum total accessibility.

Many problems of this nature have been defined for different applications, and implemented in GIS. While the median problems seek to minimize total distance, the *coverage* problems seek to minimize the *furthest* distance traveled, on the grounds that dealing with the worst case of accessibility is often more attractive than dealing with average accessibility. For example, it may make more sense to a city fire department to locate so that a response is possible to *every* property in less than five minutes, than to worry about minimizing the *average* response time. Coverage problems find applications in the location of emergency facilities, such

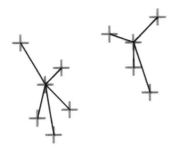

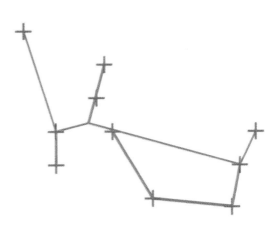

Figure 14.15 Search for the best locations for two sites (blue crosses) to serve dispersed customers (red crosses). On the left the problem is solved in continuous space, with straight-line travel. In continuous space there is an infinite number of possible locations for the sites. On the right it is solved on a network, where Hakimi's theorem states that only junctions (nodes) in the network and places where there is weight need to be considered, making the problem much simpler

Figure 14.16 GIS can be used to find locations for fire stations that result in better response times to emergencies

as fire stations, where it is desirable that every possible emergency be covered within a fixed number of minutes of response time, or when the objective is to minimize the worst-case response time, to the furthest possible point.

Location-allocation involves two types of decisions – where to locate, and how to allocate demand for service.

All of these problems are referred to as *location–allocation* problems, because they involve two types of decisions: where to *locate*, and how to *allocate* demand for service to the central facilities. A typical location–allocation problem might involve the selection of sites for super-markets. In some cases the allocation of demand to sites is controlled by the designer, as it is in the case of school districts when students have no choice of schools. In other cases allocation is a

matter of choice, and good designs depend on the ability to predict how consumers will choose among the available options. Models that make such predictions are known as *spatial interaction models*, and their use is an important area of GIS application in market research. For further reading on location-allocation problems and spatial inter-action models see Ghosh and Rushton (1987) and Fotheringham and O'Kelly (1989).

14.3.2 Routing problems

Point location problems are concerned with the design of fixed locations. Another area of optimization is in routing and scheduling, or decisions about the optimum tracks followed by vehicles. A commonly encountered example is in the routing of delivery vehicles. In these examples there is a base location, a depot that serves as the origin and final destination of delivery vehicles; and a series of stops that need to be made. There may be restrictions on the times at which stops must be made. For example, a vehicle delivering home appliances may be required to visit certain houses at certain times, when the residents are home. Vehicle routing and scheduling solutions are used by parcel delivery companies, school buses, on-demand public transport vehicles, and many other applications.

At the root of all routing problems is the *shortest path*, the path through the network between a defined origin and destination that minimizes distance, or some other measure based on distance, such as travel time. Attributes associated with the network's links, such as length, travel speed, restrictions on travel direction, and level of congestion are often taken into account. Many people are now familiar with the routine solution of the shortest path problem by Web sites like MapQuest.com, which solve many millions of such problems per day for travelers (see Section 1.5.3). They use standard algorithms developed decades ago, long before the advent of GIS. The path that is strictly shortest is often not suitable, because it involves too many turns or uses too many narrow streets, and algorithms will often be programmed to find longer routes that use faster highways, particularly freeways. Routes in Los Angeles, USA, for example, can often be caricatured as 1) shortest route from origin to nearest freeway, 2) follow freeway network, 3) shortest route from nearest freeway to destination, even though this route may be far from the shortest.

The simplest routing problem with multiple destinations is the so-called *traveling salesman problem* or TSP. In this problem there are a number of places that must be visited in a tour from the depot, and the distances between pairs of places are known. The problem is to select the best tour out of all possible orderings, in order to minimize the total distance traveled. In other words, the optimum is to be selected out of the available tours. If there are n places to be visited, including the depot, then there are $(n - 1)!$ possible tours (the symbol ! indicates the product of the integers from 1 up to and including the number, known as the number's *factorial*), but since it is irrelevant whether any given tour is conducted in a forwards or backwards direction the effective number of options is $(n - 1)!/2$. Unfortunately this number grows very rapidly with the number of places to be visited, as Table 14.1 shows.

A GIS can be very effective at solving routing problems because it is able to examine vast numbers of possible solutions quickly.

The TSP is an instance of a problem that becomes quickly unsolvable for large n. Instead, designers adopt procedures known as *heuristics*, which are algorithms designed to work quickly, and to come close to providing the best answer, while not guaranteeing that the best answer will be found. One not very good heuristic for the traveling salesman problem is to proceed always to the closest unvisited destination, and finally to return to the start. Many spatial optimization problems, including location-allocation and routing problems, are solved today by the use of sophisticated heuristics (Box 14.4).

There are many ways of generalizing the TSP to match practical circumstances. Often there is more than one vehicle, and in these situations the division of stops between vehicles is an important decision variable – which driver should cover which stops? In other situations it is not essential that every stop be visited, and instead the problem is

one of maximizing the rewards associated with visiting stops, while at the same time minimizing the total distance traveled. The is known as the *orienteering* problem, and it matches the situation faced by the driver of a mobile snack bar, who must decide which of a number of building sites to visit during the course of a day. ESRI's ArcLogistics is an example of a package built with a GIS that is designed to solve many of these everyday problems using heuristics.

14.3.3 Optimum paths

The last example in this section concerns the need to find paths across continuous space, for linear facilities like highways, power lines, and pipelines. Locational decisions are often highly contentious, and there have been many instances of strong local opposition to the location of high-voltage transmission lines (which are believed by many people to cause cancer, although the scientific evidence on this issue is mixed). Optimum paths across continuous space are also needed by the military, in routing tanks and other vehicles, and there has been substantial investment by the military in appropriate technology. They are used by shipping companies to minimize travel time and fuel costs in relation to currents. Finally, continuous-space routing problems are routinely solved by the airlines, in finding the best tracks for aircraft across major oceans that minimize time and fuel costs, by taking local winds into account. Aircraft flying the North Atlantic typically follow a more southerly route from west to east, and a more northerly route from east to west.

A GIS can find the optimum path for a power line or highway, or for an aircraft flying over an ocean given prevailing winds.

Practical GIS solutions to this problem date back to the work of McHarg (Box 14.3), and its implementation in GIS software. They are normally solved on a raster. Each cell is assigned a *friction* value, equal to the cost or time associated with moving across the cell in the horizontal or vertical directions. The user selects a *move set*, or a set of allowable moves between cells, and the solution is expressed in the form of a series of such moves. The simplest possible move set is the *rook's case*, named for the allowed moves of the rook (castle) in the game of chess. But in practice most software uses the *queen's case*, which allows moves in eight directions from each cell rather than four. Figure 14.18 shows the two cases. When a diagonal move is made the associated cost or time must be multiplied by a

Table 14.1 The number of possible tours in a traveling salesman problem

Number of places to visit	Number of possible tours
3	1
4	3
5	12
6	60
7	360
8	2520
9	20160
10	181440

Box 14.4   Routing service technicians for Schindler Elevator

Schindler Elevator Corporation is a US leader in elevator maintenance. It uses a GIS to plan efficient routes for preventative maintenance technicians (Figure 14.17). Service work for each technician is scheduled in daily routes, with emphasis on geographically compact areas of operation to allow quick response in case of emergency calls (people stuck in elevators need a quick response). Each building is visited periodically so that the technician can inspect, grease, clean, and perform periodic maintenance tasks. Additionally, visits are constrained by customers who, for example, do not want elevators out of service during the busiest times.

Schindler's GIS, named Planning Assistant for Superintendent Scheduling (PASS), went into operation in mid-1999. It creates optimum routes, based on multiple criteria, for a number of technicians assigned to one of Schindler's 140 offices in the USA. As well as assigning visits for multiple technicians, based on maintenance periods and customer preference, it is able to consider many other factors such as technician skill sets and permissible work hours (including overtime), legally binding break times, visit duration, and contract value. Much of this information is stored in Schindler's central SAP/R3 Enterprise Resource Planning (ERP) system, which must be queried to obtain the data before each routing operation.

The system has had a major impact on Schindler's business, saving them several millions of dollars annually (for an initial outlay of around $1 million). It has also allowed them to improve overall operational efficiency by restructuring offices based on routing requirements.

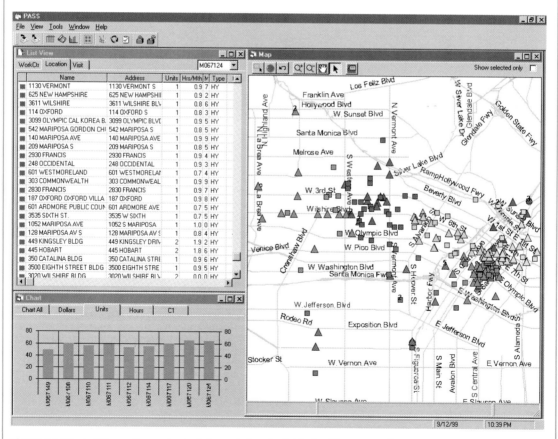

Figure 14.17   A daily routing and scheduling solution for Schindler Elevator. Each symbol on the map identifies a service stop, at the addresses listed in the table on the left (Source: Schindler Elevator Corporation)

factor of 1.414 (i.e. $\sqrt{2}$) to account for its extra length.

Friction values can be obtained from a number of sources. Land cover data might be used to differentiate between forest and open space – a power line might be given a higher cost in forest because of the environmental impact (felling of forest), while its cost in open space might be allocated based on visual impact. Elaborate methods have been devised for estimating costs based on many GIS layers and appropriate weighting factors, and interested readers should refer to the text by Massam (1980) for many useful and practical examples. Real financial costs might be included, such as the costs of construction or of land acquisition, but most of the costs are likely to be based on intangibles, and consequently hard to measure effectively.

Economic costs of a project may be relatively easy to measure compared with impacts on the environment or on communities.

Given a cost layer, together with a defined origin and destination and a move set, the GIS is then asked to find the least-cost path. If the number of cells in the raster is very large the problem can be computationally time-consuming, so heuristics are often employed. The simplest is to solve the problem hierarchically, first on a coarsely aggregated raster to obtain an approximate corridor, and then on a detailed raster restricted to the corridor.

There is an interesting paradox with this problem that is often overlooked in GIS applications, but is nevertheless significant, and provides a useful example of the difference between interval and ratio measurements (Section 3.4). Figure 14.19 shows a typical solution to a routing problem. Since the task is to minimize the

sum of costs, it is clear that the costs or friction values in each cell need to be measured on interval or ratio scales. But is interval measurement sufficient, or is ratio measurement required – in other words, does there need to be an absolute zero to the scale? Many studies of optimum routing have been done using scales that clearly did not have an absolute zero. For example, people along the route might have been surveyed, and asked to rate impact on a scale from 1 to 5. It is questionable whether such a scale has interval properties (are 1 and 2 as different on such a scale as 2 and 3, or 4 and 5?), let alone ratio properties (there is no absolute zero, and no sense that a score of 4 is twice as much of anything as a score of 2).

But why is this relevant to the problem of route location? Consider the path shown in Figure 14.19. It passes through 456 cells, and includes 261 rook's case moves and 195 diagonal moves. If the entire surface were raised by adding 1.0 to every score, the total cost of the optimum path would rise by 536.7, allowing for the greater length of the diagonal moves. But the cost of a straight path, which requires only 60 rook's case and 216 diagonal moves, would rise by only 365.4. In other words, changing the zero point of the scale can change the optimum, and in principle the cost surface needs to be measured on a ratio scale. Interval and ordinal scales are not sufficient, despite the fact that they are frequently used.

14.4 Hypothesis Testing

This last section reviews a major area of statistics – the testing of hypotheses and the drawing of inferences – and its relationship to GIS and spatial

(A)

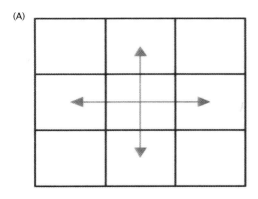

(B)

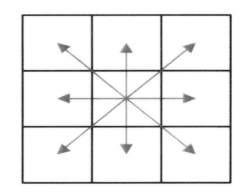

Figure 14.18 The rook's case (A) and queen's case (B) move sets, defining the possible moves from one cell to another in solving the problem of optimum routing across a friction surface

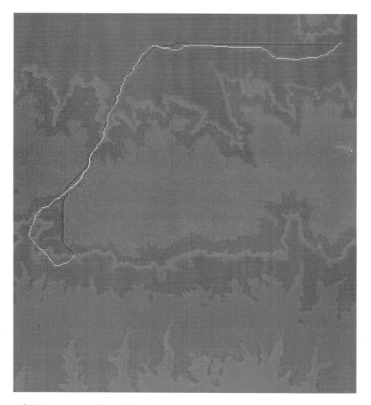

Figure 14.19 Solution of a least-cost path problem. The white line represents the optimum solution, or path of least total cost, across a friction surface represented as a raster, and using the queen's case move set (eight possible moves). Brown denotes the highest friction. The blue line is the solution when cells are aggregated in blocks of 3 by 3, and the problem is solved on this coarser raster. (Courtesy: Ashton Shortridge)

analysis. Much work in statistics is *inferential*: it uses information obtained from samples to make general conclusions about a larger population, on the assumption that the sample came from that population. The concept of inference was introduced in Section 5.4, as a way of reasoning about the properties of a larger group from the properties of a sample. At that point several problems associated with inference from geographic data were raised, and this section revisits and elaborates on that topic, and discusses the particularly thorny issue of hypothesis testing.

For example, suppose we were to take a random and independent sample of 1000 people, and ask them how they might vote in the next election. By *random and independent*, we mean that every person of voting age in the general population has an equal chance of being chosen, and that the choice of one person does not make the choice of any others – parents, neighbors – more or less likely. Suppose that 45% of the sample said they would support George W. Bush. Statistical theory then allows us to give 45% as the best estimate of

the proportion who would vote for Bush among the general population, and also allows us to state a *margin of error*, or an estimate of how much the true proportion among the population will differ from the proportion among the sample. A suitable expression of margin of error is given by the 95% confidence limits, or the range within which the true value is expected to lie 19 times out of 20. In other words, if we took 20 different samples, all of size 1000, there would be a scatter of proportions, and we expect that 19 out of 20 of them would lie within these 95% confidence limits. In this case a simple analysis using the *binomial distribution* shows that the 95% confidence limits are 3% – in other words, the true proportion lies between 42% and 48% 19 times out of 20.

This example illustrates the *confidence limits* approach to inference, in which the effects of sampling are expressed in the form of uncertainty about the properties of the population. An alternative, very commonly used in scientific reasoning, is the *hypothesis-testing* approach. In this case our objective is to test some general statement about the population – for example, that 50% will support Bush in the next election (and 50% will support the other candidate – in other words, there is no real preference in the electorate). We take a sample, and then ask whether the evidence from the sample supports the general statement. Because there is uncertainty associated with any sample, unless it includes the entire population, the answer is never absolutely certain. In this example and using our confidence limits approach, we know that if 45% were found to support Bush in the sample, and if the margin of error was 3%, it is highly unlikely that the true proportion in the population is as high as 50%. Alternatively, we could state the 50% proportion in the population as a *null hypothesis* (we use the term *null* to reflect the absence of something, in this case a clear choice), and determine how frequently a sample of 1000 from such a population would yield a proportion as low as 45%. The answer again is very small; in fact the probability is 0.0008. But it is not zero, and its value represents the chance of making an error of inference, of rejecting the hypothesis when in fact it is true.

> Methods of inference reason from information about a sample to more general information about a larger population.

These two concepts – confidence limits and inferential tests – are the basis for statistical testing, and form the core of introductory statistics texts (e.g. Rogerson 2001). There is no point in reproducing those introductions here, and the reader is simply referred to them for discussions of the standard tests, the *F*, *t*, χ^2, etc. The focus here is on the problems associated with using these approaches with geographic data, in a GIS context. Several problems were already discussed in Section 5.7 as violations of the assumptions of inference from regression. The next section discusses the general issues, and points to ways of resolving them.

14.4.1 Hypothesis tests on geographic data

Although inferential tests are standard practice in much of science, they are very problematic for geographic data. The reasons have to do with fundamental properties of geographic data, many of which were introduced in Chapter 5, and others have been encountered at various stages in this book.

First, many inferential tests propose the existence of a population, from which the sample

has been obtained by some well-defined process. We saw in Section 5.4 how difficult it is to think of a geographic dataset as a sample of the datasets that might have been. It is equally difficult to think of a dataset as a sample of some larger area of the Earth's surface, for two major reasons.

First, the samples in standard statistical inference are obtained independently. But a geographic dataset is often *all there is* in a given area – it *is* the population. Perhaps we could regard a dataset as a sample of a larger area. But in this case the sample would not have been obtained randomly – instead, it would have been obtained by systematically selecting all cases within the area of interest. Moreover, the samples would not have been independent. Because of spatial dependence, which we have understood to be a pervasive property of geographic data, it is very likely that there will be similarities between neighboring observations.

> A GIS project often analyzes all the data there are in a given area, rather than a sample.

Figure 14.20 shows a typical instance of sampling topographic elevation. The value obtained at B could have been estimated from the values obtained at the neighboring points A and C – and this ability to estimate is of course the entire basis of spatial interpolation. We cannot have it both ways – if we believe in spatial interpolation, we cannot at the same time believe in independence of geographic samples, despite the fact that this is a basic assumption of statistical tests.

Finally, the issue of spatial *heterogeneity* also gets in the way of inferential testing. The Earth's surface is highly variable, and there is no such

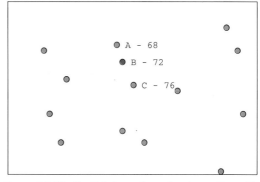

Figure 14.20 Spatial dependence in observations. The value at B, which was measured at 72, could easily have been interpolated from the values at A and C – any spatial interpolation technique would return a value close to 72 at this point

thing as an average place on it. The character-istics observed on one map sheet are likely to be substantially different from those on other map sheets, even when the map sheets are neighbors. So the census tracts of a city are certainly not acceptable as a random and independent sample of all census tracts, even the census tracts of an entire nation. They are not independent, and they are not random. Consequently it is very risky to try to infer properties of *all* census tracts from the properties of all of the tracts in any *one* area. The concept of sampling, which is the basis for statistical inference, does not transfer easily to the spatial context.

> The Earth's surface is very heterogeneous, making it difficult to take samples that are truly representative of any large region.

Before using inferential tests on geographic data, therefore, it is advisable to ask two fundamental questions:

- Can I conceive of a larger *population* that I want to make inferences about?
- Are my data acceptable as a *random* and *independent* sample of that population?

If the answer to either of these questions is *no*, then inferential tests are not appropriate.

Given these arguments, what options are available? One strategy that is sometimes used is to discard data until the proposition of independence becomes acceptable – until the remaining data points are so far apart that they can be regarded as essentially independent. But no scientist is happy throwing away good data.

Another approach is to abandon inference entirely. In this case the results obtained from the data are descriptive of the study area, and no attempt is made to generalize. This approach, which uses local statistics to observe the *differences* in the results of analysis over space, represents an interesting compromise between the nomothetic and idiographic positions outlined in Section 1.3, and there is an excellent overview in Chapter 5 of the text by Fotheringham *et al* (2000). Generalization is very tempting, but the heterogeneous nature of the Earth's surface makes it very difficult. If generalization is required, then it can be accomplished by appropriate experimental design, by replicating the study in a sufficient number of distinct areas to warrant confidence in a generalization.

Finally, a large amount of research has been devoted to devising versions of inferential tests that cope effectively with spatial dependence and spatial heterogeneity. Software that implements these tests is now widely available, and interested readers are urged to consult the appropriate sources. An excellent comprehensive source is Spacestat (see www.spacestat.com).

14.5 Conclusion

This chapter has covered many of the more advanced techniques of spatial analysis that are available in GIS. The last section in particular raised some fundamental issues associated with applying methods and theories that were developed for non-spatial data to the spatial case. Spatial analysis is clearly not a simple and straightforward extension of non-spatial analysis, but instead raises many distinct problems, as well as some exciting opportunities. The two chapters on spatial analysis have only scratched the surface of this large and rapidly expanding field.

Questions for Further Study

1. Besides being the basis for useful measures, fractals also provide interesting ways of simulating geographic phenomena and patterns. Browse the Web for sites that offer fractal simulation software, or investigate one of many commercially available packages. What other uses of fractals in GIS can you imagine?
2. Parks and other conservation areas have geometric shapes that can be measured by comparing park perimeter length to park area, using the methods reviewed in this and the previous chapter. Discuss the implications of shape for park management, in the context of (a) wildlife ecology and (b) neighborhood security.
3. What exactly are *multicriteria* methods? Examine one or more of the methods in the Eastman chapter referenced below, summarizing the issues associated with (a) measuring variables to support multiple criteria, (b) mixing variables that have been measured on different scales (e.g. dollars and distances), (c) finding solutions to problems involving multiple criteria.
4. Every point on the Earth's surface has an antipodal point – the point that would be reached by drilling an imaginary hole straight through the Earth's center. Britain, for example, is approximately antipodal to New Zealand. If one third of the Earth's surface is land, you might expect that one third of all of the land area would be antipodal to points that are also on land, but a quick look at an atlas will

show that the proportion is actually far less than that. In fact, the only substantial areas of land that have antipodal land are in South America (and their antipodal points in China). How is spatial dependence relevant here, and why does it suggest that the Earth is not so surprising after all?

5. (This question requires some knowledge of calculus, specifically the ability to differentiate a simple equation.) The Varignon Frame appears to present a paradox – how can the point that minimizes distance not be affected when distances are changed by moving points directly towards or away from the optimum? Write down the objective function of the problem, differentiate it and set the derivative to zero, and hence prove that the paradox is indeed true. Using the same approach, prove that the centroid minimizes the sum of distances squared.

Online Resources

NCGIA Core Curricula (www.ncgia.ucsb.edu/pubs/core.html):
    Core Curriculum in GIScience, Sections 2.14.1 (Spatial Decision Support Systems, Jacek Malczewski), and 2.14.6 (Artificial Neural Networks for Spatial Data Analysis, Suchi Gopal)
    Core Curriculum in GIS, 1990, Units 57 (Decision Making Using Multiple Criteria), 58 (Location-Allocation on Networks), 59 (Spatial Decision Support Systems), and 74 (Knowledge-Based Techniques)
ESRI Virtual Campus courses (campus.esri.com):
    Turning Data into Information, by Paul Longley, Michael Goodchild, David Maguire, and David Rhind (Modules "Transformations and Descriptive Summaries" and "Optimization and Hypothesis Testing")
    Introduction to ArcView 3D Analyst, by ESRI (includes the Module "Using Avenue with ArcView 3D Analyst"; and also lessons on "Creating 3D Graphic's", "Analyzing TIN Surfaces", "Creating a Surface Fly-by", and "Integrating ArcView 3D Analyst and ArcView Spatial Analyst" in the Module "Advanced Techniques in ArcView 3D Analyst".)
    Introduction to ArcView Network Analyst, by ESRI
    Working with ModelBuilder, by ESRI

Reference Links

Maguire D J, Goodchild M F, and Rhind D W (eds) 1991 *Geographical Information Systems: Principles and applications*. Harlow, UK: Longman (Text available online from 'Links to Big Book 1' at www.wiley.com/gis and www.wiley.co.uk/gis).
    Chapter 24, Spatial data integration (Flowerdew R)
    Chapter 25, Developing appropriate spatial analysis methods for GIS (Openshaw S)
    Chapter 26, Spatial decision support systems (Densham P J)
    Chapter 27, Knowledge-based approaches in GIS (Smith T R, Jiang Y)
Longley P A, Goodchild M F, Maguire D J, and Rhind D W (eds) 1999 *Geographical Information Systems: Principles, techniques, management and applications*. New York: John Wiley.
    Chapter 16, Spatial statistics (Getis A)
    Chapter 18, Applying geocomputation to the analysis of spatial distributions (Openshaw S, Alvanides S)
    Chapter 19, Spatial analysis: retrospect and prospect (Fischer M M)
    Chapter 20, Location modelling and GIS (Church R L)
    Chapter 35, Multi-criteria evaluation and GIS (Eastman J R)
    Chapter 36, Spatial tessellations (Boots B)
    Chapter 40, The future of GIS and spatial analysis (Goodchild M F, Longley P A)

References

Cliff A D and Ord J K 1973 *Spatial Autocorrelation*. London: Pion.

Fotheringham A S, Brunsdon C, and Charlton M 2000 *Quantitative Geography: Perspectives on spatial data analysis*. London: Sage.

Fotheringham A S and O'Kelly M E 1989 *Spatial Interaction Models: Formulations and applications*. Dordrecht: Kluwer.

Ghosh A and Rushton G (eds) 1987 *Spatial Analysis and Location–Allocation Models*. New York: Van Nostrand Reinhold.

Lam N S N and De Cola L 1993 *Fractals in Geography*. Englewood Cliffs, New Jersey: PTR Prentice Hall.

Mandelbrot B B 1983 *The Fractal Geometry of Nature*. San Francisco: Freeman.

Mardia K V and Jupp P E 2000 *Directional Statistics*. New York: John Wiley.

Massam B H 1980 *Spatial Search: Application to planning problems in the public sector*. Oxford: Pergamon.

McHarg I L 1969 *Design with Nature*. Garden City, New York: Natural History Press.

Rogerson P 2001 *Statistical Analysis in Geography*. London: Sage

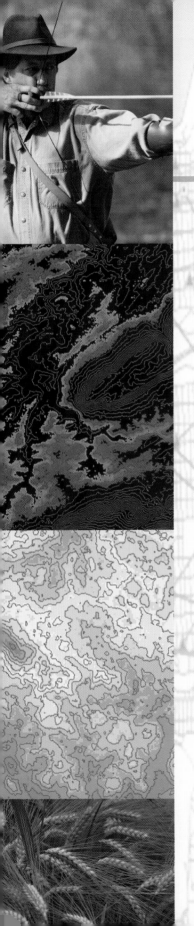

UNCERTAINTY, ERROR, AND SENSITIVITY

15

Data in a GIS can be subject to measurement error, out of date, excessively generalized, or just plain wrong. The purpose of this chapter is to review uncertainty from a technical perspective, and to identify the ways in which GIS practitioners can succeed in living with it, by minimizing its effects. The first part of the chapter looks at uncertainty using the traditional tools of the statistician. This is a very powerful approach, but limited in its applicability for many reasons. The second section looks at how statistical models can be used to understand the impacts of uncertainties on GIS products, through methods known as error propagation. The third section takes a very different tack, by adopting the conceptual frameworks of fuzzy logic and subjective probability, and showing how these methods have attracted significant attention in the GIS community because of their intuitive appeal and generality. The chapter ends with some general guidelines for living with uncertainty.

Learning objectives

On reaching the end of this chapter you will know about:

- Statistical tools for describing and modeling error and uncertainty;

- Techniques for predicting the effects of uncertain data on the results of GIS analysis and decision making;

- Fuzzy logic, a framework for acquiring uncertain data;

- Methods for coping with uncertainty in GIS practice.

15.1 Statistical Models of Uncertainty

Scientists have developed many widely used methods for describing errors in observations and measurements, and these methods may be applicable to GIS if we are willing to think of databases as collections of measurements. For example, a digital elevation model consists of a large number of measurements of the elevation of the Earth's surface. A map of land use is also in a sense a collection of measurements, because observations of the land surface have resulted in the assignment of classes to locations. Both of these are examples of observed or measured attributes, but we can also think of location as a property that is measured.

> A geographic database is a collection of measurements of phenomena on or near the Earth's surface.

This section deals with two basic types of measurement, and the errors to which they are subject. Measurements which take the form of classes, as in the land use example, are one type, and measurements on continuous scales, as in the elevation example, are the other. In the more exact terminology introduced in Box 3.3 these are nominal and interval/ratio scales respectively. The next two sections discuss these two types of error in detail.

15.1.1 Nominal case

Classes are an instance of nominal data (Box 3.3), data whose values serve only to distinguish an instance of one class from an instance of another, or to identify an object uniquely. If classes have an inherent ranking they are described as ordinal data, but for the purposes of simplicity in this chapter the ordinal case will be treated as if it were nominal.

Consider a single observation of nominal data – for example, the observation that a single parcel of land is being used for agriculture (this might be designated by giving the parcel Class A, Agriculture, as its value of the "Land Use Class" attribute). For some reason, perhaps related to the quality of the aerial photography being used to build the database, the class may have been recorded falsely as Class G, Grassland. A certain proportion of parcels that are truly Agriculture might be similarly recorded as Grassland, and we can think of this in terms of a probability, that parcels that are truly Agriculture are falsely recorded as Grassland.

Table 15.1 shows how this might work for all of the parcels in a database. Each parcel has a true class, defined by accurate observation in the field, and a recorded class as it appears in the database. The whole table is described as a *confusion matrix*, and instances of confusion matrices are commonly encountered in applications dominated by class data, such as classifications derived from remote sensing or aerial photography. The true class might be determined by ground check, which is inherently more accurate than classification of aerial photographs, but much more expensive and time-consuming.

Ideally all of the observations in the confusion matrix should be on the principal diagonal, in the cells that correspond to agreement between true class and database class. But in practice certain classes are more easily confused than others, so certain cells off the diagonal will have substantial numbers of entries.

A useful way to think of the confusion matrix is as a set of rows, each defining a vector of values. The vector for row i gives the proportions of cases in which what appears to be Class i is actually Class 1, 2, 3, etc. Symbolically, this can be represented as a vector $\{p_1, p_2, \ldots, p_j, \ldots, p_n\}$, where n is the number of classes, and p_j represents the proportion of cases for which what appears to be class i according to the database is actually Class j.

There are several ways of describing and summarizing the confusion matrix. If we focus on one row, then the table shows how a given class in the database falsely records what are actually different classes on the ground. For example, Row A shows that of 106 parcels recorded as Class A in the database, 80 were confirmed as Class A in the field, but 15 appeared to be truly

Table 15.1 Example of a misclassification or confusion matrix. A grand total of 304 parcels have been checked. The rows of the table correspond to the land use class of each parcel as recorded in the database, and the columns to the class as recorded in the field. The numbers appearing on the principal diagonal of the table (from top left to bottom right) reflect correct classification

	A	B	C	D	E	Total
A	80	4	0	15	7	106
B	2	17	0	9	2	30
C	12	5	9	4	8	38
D	7	8	0	65	0	80
E	3	2	1	6	38	50
Total	104	36	10	99	55	304

Class D. The proportion of instances in the diagonal entries represents the proportion of correctly classified parcels, and the total of off-diagonal entries in the row is the proportion of entries in the database that appear to be of the row's class but are actually incorrectly classified. For example, there were only nine instances of agreement between the database and the field in the case of Class C. If we look at the table's columns, the entries record the ways in which parcels that are truly of that class are actually recorded in the database. For example, of the 10 instances of Class C found in the field, nine were recorded as such in the database and only one was misrecorded as Class E.

The columns have been called the *producer's* perspective, because the task of the producer of an accurate database is to minimize entries outside the diagonal cell in a given column, and the rows have been called the *consumer's* perspective, because they record what the contents of the database actually mean on the ground: in other words, the accuracy of the database's contents.

Users and producers of data look at misclassification in distinct ways.

For the table as a whole, the proportion of entries in diagonal cells is called the *percent correctly classified* (PCC), and is one possible way of summarizing the table. In this case 209/304 cases are on the diagonal, for a PCC of 68.8%. But this measure is misleading for at least two reasons. First, chance alone would produce some correct classifications, even in the worst circumstances, so it would be more meaningful if the scale were adjusted such that 0 represents chance. In this case, the number of chance hits on the diagonal in a random assignment is 76.2 (the sum of the row total times the column total divided by the grand total for each of the five diagonal cells). So the actual number of diagonal hits, 209, should be compared with this number, not 0. The more useful index of success is the *kappa index*, defined as:

$$\kappa = \frac{\sum_{i=1}^{n} c_{ii} - \sum_{i=1}^{n} c_{i.}c_{.i}/c_{..}}{c_{..} - \sum_{i=1}^{n} c_{i.}c_{.i}/c_{..}}$$

where c_{ij} denotes the entry in row i column j, the dots indicate summation (e.g. $c_{i.}$ is the summation over all columns for row i, that is, the row i total, and $c_{..}$ is the grand total), and n is the number of classes. The first term in the numerator is the sum of all the diagonal entries (entries for which the row number and the column number are the same). To compute PCC we would simply divide this term by the grand total (the first term in the denominator). For kappa, both numerator and denominator are reduced by the same amount, an estimate of the number of hits (agreements between field and database) that would occur by chance. This involves taking each diagonal cell, multiplying the row total by the column total, and dividing by the grand total. The result is summed for each diagonal cell. In this case kappa evaluates to 58.3%, a much less optimistic assessment than PCC.

The second issue with both of these measures concerns the relative abundance of different classes. In the table, Class C is much less common than Class A. The confusion matrix is a useful way of summarizing the characteristics of nominal data, but to build it there must be some source of more accurate data. Commonly this is obtained by ground observation, and in practice the confusion matrix is created by taking samples of more accurate data, by sending observers into the field to conduct spot checks. Clearly it makes no sense to visit every parcel, and instead a sample is taken. Because some classes are commoner than others, a random sample that made every parcel equally likely to be chosen would be ineffective, because too much data would be gathered on common classes, and not enough on the relatively rare ones. So instead, samples are usually chosen such that a roughly equal number of parcels are selected in each class (see Section 5.4). Of course these decisions must be based on the class as recorded in the database, rather than the true class. Sampling strategies that focus separately on each class are said to be *stratified* by class (see Section 5.4 for a discussion of various approaches to spatial sampling and their effects).

Sampling for accuracy assessment should pay greater attention to the classes that are rarer on the ground.

Parcels represent a relatively easy case, if it is reasonable to assume that the land use class of a parcel is uniform over the parcel, and class is recorded as a single attribute of each parcel object. But as we noted in Sections 5.2 and 5.3, more difficult cases arise in sampling natural areas, for example in the case of vegetation cover class, where parcel boundaries do not exist. Figure 15.1 shows a typical vegetation cover class map, and is obviously highly generalized. If we were to apply the previous strategy, then we would test each area to see if its assigned vegetation cover class checks out on the ground. But unlike the parcel case, in this example the boundaries

Figure 15.1 An example of a vegetation cover map. Two strategies for accuracy assessment are available: to check by area (polygon), or to check by point. In the former case a strategy would be devised for field checking each area, to determine the area's correct class. In the latter, points would be sampled across the state and the correct class determined at each point. (NW = Northwest, CW = Central West, SW = Southwest, GV = Great Valley, CaR = Cascade Range, MP = Modoc Plateau, SN = Sierra Nevada, DMoj = Desert (Mojave), and DSon = Desert (Sonoran).) (Source: Frank Davis)

between areas are not fixed, but are themselves part of the observation process, and we need to ask whether they are correctly located. Error in this case has two forms: misallocation of an area's class, and mislocation of an area's boundaries. In some cases the boundary between two areas may be fixed, because it coincides with a clearly defined line on the ground; but in other cases, the boundary's location is as much a matter of judgment as the allocation of an area's class. Burrough and Frank (1996) discuss many of the implications of uncertain boundaries in GIS.

> Errors in land cover maps can occur in the locations of boundaries of areas, as well as in the classification of areas.

In such cases we need a different strategy, that captures the influence both of mislocated

boundaries and of misallocated classes. One way to deal with this is to think of error not in terms of classes assigned to areas, but in terms of classes assigned to points. In a raster dataset, the cells of the raster are a reasonable substitute for individual points. Instead of asking whether area classes are confused, and estimating errors by sampling areas, we ask whether the classes assigned to raster cells are confused, and define the confusion matrix in terms of misclassified cells. This is often called *per-pixel* or *per-point* accuracy assessment, to distinguish it from the previous strategy of *per-polygon* accuracy assessment. As before, we would want to stratify by class, to make sure that relatively rare classes were sampled in the assessment. Congalton and Green (1999) provide a comprehensive overview of accuracy assessment, with a focus on data derived from remote sensing.

15.1.2 Interval/ratio case

The second case addresses measurements that are made on interval or ratio scales (Box 3.3). Here, error is best thought of not as a change of class, but as a change of value, such that the observed value x' is equal to the true value x plus some distortion δx, where δx is hopefully small. δx might be either positive or negative, since errors are possible in both directions. For example, the measured and recorded elevation at some point might be equal to the true elevation, distorted by some small amount. If the average distortion is zero, so that positive and negative errors balance out, the observed values are said to be *unbiased*, and the average value will be true.

> Error in measurement can produce a change of class, or a change of value, depending on the type of measurement.

Sometimes it is helpful to distinguish between *accuracy*, which has to do with the magnitude of δx, and *precision*. Unfortunately there are several ways of defining precision in this context, at least two of which are regularly encountered in GIS. Surveyors and others concerned with measuring instruments tend to define precision through the performance of an instrument in making repeated measurements of the same phenomenon. A measuring instrument is precise according to this definition if it repeatedly gives similar measurements, whether or not these are actually accurate. So a GPS receiver might make successive measurements of the same elevation, and if these are similar the instrument is said to be precise. Precision in this case can be measured by the variability among repeated measurements. But it is possible that all of the measurements are approximately 5 m too high, in which case the measurements are said to be biased, even though they are precise, and the instrument is said to be inaccurate. Figures 15.2 and 15.3 illustrate this meaning of precise, and its relationship to accuracy.

The other definition of precision is more common in science generally. It defines precision as the number of digits used to report a measurement, and again it is not necessarily related to accuracy. For example, a GPS receiver might measure elevation as 51.3456 m. But if the receiver is in reality only accurate to the nearest 10 cm, three of those digits are spurious, with no real meaning. So although the precision is one ten thousandth of a meter, the accuracy is only one tenth of a meter. Box 15.1 summarizes the rules that are used to ensure that reported measure-

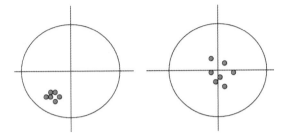

Figure 15.2 The term *precision* is often used to refer to the repeatability of measurements. In both diagrams repeated measurements have been taken of the same position, represented by the center of the circle. In (A) successive measurements have similar values (they are precise), but show a bias away from the correct value (they are *inaccurate*). In (B), precision is lower but accuracy is higher

Figure 15.3 Accuracy and precision identify different aspects of a measuring instrument's ability to hit a target exactly

Box 15.1   Good practice in reporting measurements

Here are some simple rules that help to ensure that people receiving measurements from others are not misled by their apparently high precision.

1. The number of digits used to report a measurement should reflect the measurement's accuracy. For example, if a measurement is accurate to 1 m then no decimal places should be reported. The measurement 14.4 m suggests accuracy to one tenth of a meter, as does 14.0, but 14 suggests accuracy to 1 m.
2. Excess digits should be removed by rounding. Fractions of one half and above should be rounded up, fractions below one half should be rounded down. The following examples reflect rounding to two decimal places:

    14.57803 rounds to 14.58
    14.57397 rounds to 14.57
    14.57999 rounds to 14.58
    14.57499 rounds to 14.57

3. These rules are not effective to the left of the decimal place – for example, they give no basis for knowing whether 1400 is accurate to the nearest unit, or to the nearest hundred units.
4. If a number is known to be exactly an integer or whole number, then it is shown with no decimal point.

ments do not mislead by appearing to have greater accuracy than they really do.

> To most scientists, precision refers to the number of significant digits used to report a measurement, but it can also refer to a measurement's repeatability.

In the interval/ratio case, the magnitude of errors is described by the *root mean square error* (RMSE), defined as the square root of the average squared error, or:

$$\left[\sum \delta x^2 / n \right]^{1/2}$$

where the summation is over the values of δx for all of the n observations. The close relationship between RMSE and the standard deviation was noted in Section 14.2.2. Although RMSE involves taking the square root of the average squared error, it is convenient to think of it as approximately equal to the average error in each observation, whether the error is positive or negative. The US Geological Survey uses RMSE as its primary measure of the accuracy of elevations in digital elevation models, and published values range up to 7 m.

Although the RMSE can be thought of as capturing the magnitude of the average error, many errors will be greater than the RMSE, and many will be less. It is useful, therefore, to know how errors are *distributed* in magnitude: how

many are large, how many are small. Statisticians have developed a series of models of error distributions, of which the commonest and most important is the Gaussian distribution, otherwise known as the error function, the "bell curve", or the Normal distribution (the distribution has already been mentioned in Section 14.2.2 in connection with the measurement of dispersion). Figure 15.4 shows the curve's shape. If observations are unbiased, then the mean error is zero (positive and negative errors cancel each other out), and the RMSE is also the distance from the center of the distribution (zero) to the points of inflection on either side, as shown in the figure. To take the example of a 7 m RMSE on elevations in a USGS digital elevation model, if error follows the Gaussian distribution, this means that some errors will be more than 7 m in magnitude, some will be less, and also that the relative abundance of errors of any given size is described by the curve shown. 68% of errors will be between −1.0 and +1.0 RMSEs, or −7 m and +7 m. In practice many distributions of error do follow the Gaussian distribution, and there are good theoretical reasons why this should be so.

> The Gaussian distribution predicts the relative abundances of different magnitudes of error.

To emphasize the mathematical formality of the Gaussian distribution, its equation is shown below. The symbol σ denotes the standard

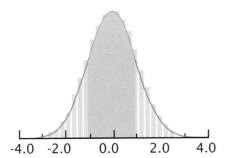

-4.0   -2.0   0.0   2.0   4.0

Figure 15.4 The Gaussian or Normal distribution. The height of the curve at any value of *x* gives the relative abundance of observations with that value of *x*. The area under the curve between any two values of *x* gives the probability that observations will fall in that range. The range between −1 standard deviation and +1 standard deviation is in blue. It encloses 68% of the area under the curve, indicating that 68% of observations will fall between these limits

deviation, μ denotes the mean (in Figure 15.4 these values are 1 and 0 respectively), and exp is the exponential function, or "2.71828 to the power of". Scientists believe that it applies very broadly, and that many instances of measurement error adhere closely to the distribution, because it is grounded in rigorous theory. It can be shown mathematically that the distribution arises whenever a large number of random factors contribute to error, and the effects of these factors combine additively, that is, a given effect makes the same additive contribution to error whatever the specific values of the other factors. For example, error might be introduced in the use of a tape measure because some observers consistently pull the tape very taut, and others do not. If this differential stretching always contributes the same amount of error (e.g. +1 cm, or −2 cm), then this contribution to error is said to be additive.

$$f(x) = \frac{1}{\sigma\sqrt{2\pi}} \exp\left[-\frac{(x-\mu)^2}{2\sigma^2}\right]$$

We can apply this idea to determine the inherent uncertainty in the locations of contours. The US Geological Survey routinely evaluates the accuracies of its digital elevation models (DEMs), by comparing the elevations recorded in the database with those at the same locations in more accurate sources, for a sample of points. The differences are summarized in an RMSE, and in this example we will assume that errors have a Gaussian distribution with zero mean and a 7 m RMSE. Consider a measurement of 350 m.

According to the error model, the truth might be as high as 360 m, or as low as 340 m, and the relative frequencies of any particular error value are as predicted by the Gaussian distribution with a mean of zero and a standard deviation of 7. Now consider what this means for the location of the 350 m contour. Figure 15.5 shows where the contour lies according to the DEM – this figure has been generated simply by running the thick line between values on either side of 350 m. But if we take error into account, using the Gaussian distribution with an RMSE of 7 m, it is no longer clear that a measurement of 350 m lies exactly on the 350 m contour. Instead, the truth might be 340 m, or 360 m, or 355 m. Figure 15.6 shows the implications of this in terms of the location of the contour. 95% of errors would put the contour within the colored zone. In areas colored red the observed value is more than 350 m, but the truth might be 350 m; in areas colored green the observed value is less than 350 m, but the truth might be 350 m. There is a 5% chance that the true location of the contour lies outside the colored zone entirely.

15.1.2.1 Positional error

Quality is an important topic in GIS, and there have been many attempts to identify its basic dimensions. The US Federal Geographic Data Committee's various standards list five components of quality: attribute accuracy, positional accuracy, logical consistency, completeness, and lineage. Definitions and other details on each of these and several more can be found in the text by Guptill and Morrison (1995), and on the FGDC's Web pages (www.fgdc.gov). This section focuses on positional accuracy.

Data quality standards adopted by the US Federal Geographic Data Committee specify five components of quality.

In the case of measurements of position, it is possible for every coordinate to be subject to error. In the two-dimensional case, a measured position (x', y') would be subject to errors in both *x* and *y*, specifically, we might write $x' = x + \delta x$, $y' = y + \delta y$, and similarly in the three-dimensional case where all three coordinates are measured, $z' = z + \delta z$. The *bivariate Gaussian distribution* describes errors in the two horizontal dimensions, and it can be generalized to the three-dimensional case. Normally, we would expect the RMSEs of *x* and *y* to be the same, but *z* is often subject to errors of quite different magnitude, for example in the case of determinations of position using GPS. The bivariate Gaussian distribution

Figure 15.5 Plot of the 350 m contour for the State College, Pennsylvania, USA topographic quadrangle. The contour has been computed from the US Geological Survey's digital elevation model for this area. (The area shown is approximately 5 km across.) (Source: Hunter G J and Goodchild M F 1995 Dealing with error in spatial databases: a simple case study. *Photogrammetric Engineering and Remote Sensing* **61**: 529–37)

also allows for correlation between the errors in x and y, but normally there is little reason to expect correlations. Figure 15.7 shows the bivariate Gaussian distribution applied to errors in position.

Because it involves two variables, the bivariate Gaussian distribution has somewhat different properties from the simple (univariate) Gaussian distribution. Sixty-eight percent of cases lie within one standard deviation for the univariate case (Figure 15.4). But in the bivariate case with equal standard errors in x and y, only 39% of cases lie within a circle of this radius. Similarly, 95% of cases lie within two standard deviations for the univariate distribution, but it is necessary to go to a circle of radius equal to 2.15 times the x or y standard deviations to enclose 90% of the

bivariate distribution, and 2.45 times standard deviations for 95%. For a standard GPS receiver (see Section 10.2.2.2), approximately 90% of error distances were less than 100 m under selective availability (the deliberate corruption of the signal available to civilian users prior to May 2000), and 90% are less than approximately 30 m when selective availability is turned off (selective availability was turned off, we hope permanently, in May 2000).

National Map Accuracy Standards often prescribe the positional errors that are allowed in databases. For example, the 1947 US National Map Accuracy Standard specified that 95% of errors should fall below 1/30 inch (0.85 mm) for maps at scales of 1:20 000 and finer (more detailed),

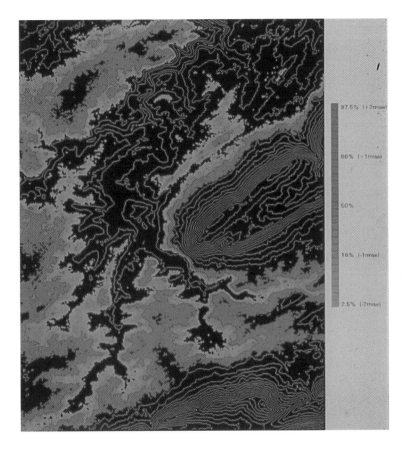

Figure 15.6 Uncertainty in the location of the 350 m contour in the area of State College, Pennsylvania, generated from a US Geological Survey DEM with an assumed RMSE of 7 m. According to the Gaussian distribution with a mean of 350 m and a standard deviation of 7 m, there is a 95% probability that the true location of the 350 m contour lies in the colored area, and a 5% probability that it lies outside (Source: as for Figure 15.5)

and 1/50 inch (0.51 mm) for other maps (scales coarser – less detailed – than 1:20 000). A convenient rule of thumb is that positions measured from maps are subject to errors of up to 0.5 mm at the scale of the map. Table 15.2 shows the distance on the ground corresponding to 0.5 mm for various common map scales.

A useful rule of thumb is that features on maps are positioned to an accuracy of about 0.5 mm.

15.1.3 The spatial structure of errors

The confusion matrix, or more specifically a single row of the matrix, along with the Gaussian distribution, provide convenient ways of describing the error present in a single observation of a

Table 15.2 A useful rule of thumb is that positions measured from maps are accurate to about 0.5 mm on the map. Multiplying this by the scale of the map gives the corresponding distance on the ground

Map scale	Ground distance corresponding to 0.5 mm map distance
1:1250	62.5 cm
1:2500	1.25 m
1:5000	2.5 m
1:10 000	5 m
1:24 000	12 m
1:50 000	25 m
1:100 000	50 m
1:250 000	125 m
1:1 000 000	500 m
1:10 000 000	5 km

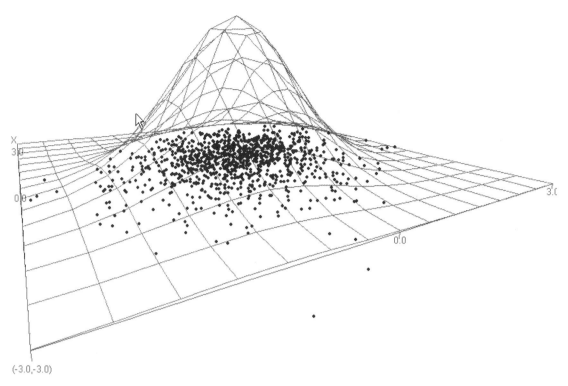

X
3.0

0.0

3.0

0.0

(-3.0,-3.0)

Figure 15.7 The bivariate Gaussian distribution. The surface represents the relative abundance of observations, with the peak occurring at the true location (zero error). The blue dots represent a sample of observations consistent with the distribution – note how the density of blue dots varies in accordance with the height of the surface (Source: P. Darius)

nominal or interval/ratio measurement respectively. When a GIS is used to respond to a simple query, such as "tell me the class of soil at this point", or "what is the elevation here?", then these methods are good ways of describing the uncertainty inherent in the response. For example, a GIS might respond to the first query with the information "Class A, with a 30% probability of Class C", and to the second query with the information "350 m, with an RMSE of 7 m". Notice how this makes it possible to describe nominal data as accurate to a per-centage, but it makes no sense to describe a DEM, or any measurement on an interval/ratio scale, as accurate to a percentage. For example, we cannot meaningfully say that a DEM is "90% accurate".

However, many GIS operations involve more than the properties of single points, and this makes the analysis of error much more complex. For example, consider the query "how far is it from this point to this point?". Suppose the two points are both subject to error of position, because their positions have been measured using GPS units with mean distance errors of 50 m. If the two

measurements were taken some time apart, with different combinations of satellites above the horizon, it is likely that the errors are independent of each other, such that one error might be 50 m in the direction of North, and the other 50 m in the direction of South. Depending on the locations of the two points, the error in distance might be as high as +100 m (see Figure 15.8). On the other hand, if the two measurements were made close together in time, with the same satellites above the horizon, it is likely that the two errors would be similar, perhaps 50 m North and 40 m North, leading to an error of only 10 m in the determination of distance. The difference between these two situations is described in terms of *spatial autocorrelation*, or the inter-dependence of errors at different points in space. Spatial dependence has been discussed at length in several previous chapters, including 3, 5, and 14, and Tobler's Law means that it is almost always present in geographic data. The impact of errors on a property like the distance between two points depends directly on the degree of inter-dependence between their errors.

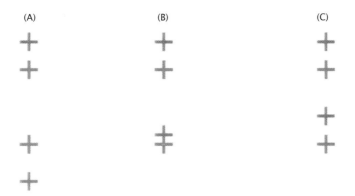

Figure 15.8 Correlations in positional errors and their impacts on the measurement of distance. In each case the red crosses show the true positions, and the green crosses show the measured positions. (A) strong negative correlation (one error positive, one negative), leading to large errors in distance. (B) independence of errors, leading to moderate errors in distance. (C) strong positive correlation, leading to no errors in distance

The spatial autocorrelation of errors can be as important as their magnitude in many GIS operations.

Spatial autocorrelation is also important in errors in nominal data. Consider a field that is known to contain a single crop, perhaps barley (Figure 15.9). When seen from above, it is possible to confuse barley with other crops, so there may be error in the crop type assigned to points in the field. But since the field has only one crop, we know that such errors are likely to be strongly correlated. Spatial autocorrelation is almost always present in errors to some degree, but very few efforts have been made to measure it systematically, and as a

Figure 15.9 Crops are good examples of phenomena that exhibit strong spatial autocorrelation. If an entire field is planted with a single crop, such as barley, then only one point needs to be observed to determine the crop of the entire field

result it is difficult to make good estimates of the uncertainties associated with many GIS operations.

An easy way to visualize spatial autocorrelation and interdependence is through animation. Each frame in the animation is a single possible map, or *realization* of the error process. If a point is subject to uncertainty, each realization will show the point in a different possible location, and a sequence of images will show the point shaking around its mean position. If two points have perfectly correlated positional errors, then they will appear to shake in unison, as if they were at the ends of a stiff rod. If errors are only partially correlated, then the system behaves as if the connecting rod were somewhat elastic.

The spatial structure or autocorrelation of errors is important in many ways. DEM data are often used to estimate the slope of terrain, and this is done by comparing elevations at points a short distance apart. For example, if the elevations at two points 10 m apart are 30 m and 35 m respectively, the slope along the line between them is 5/10, or 0.5. (A somewhat more complex method is used in practice, to estimate slope at a point in the x and y directions in a DEM raster, by analyzing the elevations of nine points – the point itself and its eight neighbors; see Box 13.5).

Now consider the effects of errors in these two elevation measurements on the estimate of slope. Suppose the first point (elevation 30 m) is subject to an RMSE of 2 m, and consider possible true elevations of 28 m and 32 m. Similarly the second point might have true elevations of 33 m and 37 m. We now have four possible combinations of values, and the corresponding estimates of slope range from $(33-32)/10 = 0.1$ to $(37-28)/10 = 0.9$. In other words, a relatively small amount of error in elevation can produce wildly varying slope estimates.

> The spatial autocorrelation between errors in geographic databases helps to minimize their impacts on many GIS operations.

What saves us in this situation, and makes estimation of slope from DEMs a practical proposition at all, is spatial autocorrelation among the errors. In reality, although DEMs are subject to substantial errors in absolute elevation, neighboring points nevertheless tend to have similar errors, and errors tend to persist over quite large areas. Most of the sources of error in the DEM production process tend to produce this kind of persistence of error over space, including errors due to misregistration of aerial photographs. In other words, errors in DEMs exhibit strong positive spatial autocorrelation.

Another important corollary of positive spatial autocorrelation can also be illustrated using DEMs. Suppose an area of low-lying land is inundated by flooding, and our task is to estimate the area of land affected. We are asked to do this using a DEM, which is known to have an RMSE of 2 m. Suppose the data points in the DEM are 30 m apart, and preliminary analysis shows that 100 points have elevations below the flood line. We might conclude that the area flooded is the area represented by these 100 points, or 900×100 sq m, or 9 hectares. But because of errors, it is possible that some of this area is actually above the flood line (we will ignore the possibility that other areas outside this may also be below the flood line, also because of errors), and it is possible that *all* of the area is above. Suppose the recorded elevation for each of the 100 points is 2 m below the flood line. This is one RMSE (recall that the RMSE is equal to 2 m) below the flood line, and the Gaussian distribution tells us that the chance that the true elevation is actually above the flood line is approximately 16% (see Figure 15.4). But what is the chance that *all* 100 points are actually above the flood line?

Here again the answer depends on the degree of spatial autocorrelation among the errors. If there is none, in other words if the error at each of the 100 points is independent of the errors at its neighbors, then the answer is $(0.16)^{100}$, or 1 chance in 1 followed by roughly 70 zeroes. But if there is strong positive spatial autocorrelation, so strong that all 100 points are subject to exactly the same error, then the answer is 0.16. One way to think about this is in terms of *degrees of freedom*. If the errors are independent, they can vary in 100 independent ways, depending on the error at each point. But if they are strongly spatially

autocorrelated, the effective number of degrees of freedom is much less, and may be as few as 1 if all errors behave in unison. Spatial autocorrelation has the effect of reducing the number of degrees of freedom in geographic data below what may be implied by the volume of information, in this case the number of points in the DEM.

> Spatial autocorrelation acts to reduce the effective number of degrees of freedom in geographic data.

15.2 Error Propagation

The examples of the previous section are cases of error propagation, in other words, the effects of known levels of data uncertainty on the outputs of GIS operations. Peter Burrough (Box 15.2) and his group have pioneered much of the research in this area. We have seen how the spatial structure of errors plays a role, and how strong positive spatial autocorrelation can lead to much smaller impacts on estimates of properties such as slope, or area. On the other hand, error propagation can also produce impacts that are surprisingly large, and some of the examples in this section have been chosen to illustrate the substantial uncertainties that can be produced by apparently innocuous data errors.

> Error propagation measures the impacts of uncertainty in data on the results of GIS operations.

In general two strategies are available for evaluating error propagation. The examples in the previous section were instances of *analysis*, where it was possible to obtain a complete description of error effects. A complete analysis is possible in the case of slope estimation, and can be applied in the DEM flooding example. Another example that is amenable to analysis is the calculation of the area of a polygon given knowledge of the positional uncertainties of its vertices.

For example, Figure 15.11 shows a square approximately 100 m on a side. Suppose the square has been surveyed by determining the locations of its four corner points using GPS, and suppose the circumstances of the measurements are such that there is an RMSE of 2 m in both coordinates of all four points, and that errors are independent. Suppose our task is to determine the area of the square. A GIS can do this easily, using a standard algorithm (see Section 13.3). Computers are precise (in the second sense of this chapter, or

Box 15.2 Peter Burrough, physical geographer

Peter Burrough is Professor of Physical Geography at the University of Utrecht in the Netherlands. He sees his major contributions as

"Writing a series of books that have attempted with some success to present an organized overview of the principles of handling spatial data from the point of view of the environmental scientist (physical geographer, soil scientist, ecologist, geologist); dealing with issues of error and uncertainty in data collection and spatial modeling, through the paradigms of statistics and alternatives such as fuzzy sets and fractals; and contributing to a better understanding of how complex, but recognizable spatial phenomena can result from the interaction of multiple processes operating together, but over different scales in space and time".

What has surprised him most about the development of GIS over the past 20 years?

"Two things have surprised me greatly – one is scientific, the other social. The biggest scientific surprise was the discovery that by using simple simulation models in packages like PCRaster, which my group developed at the University of Utrecht [see Box 2.2], one can model the essential aspects of river meandering and delta formation in response to sediment deposition and erosion. I am deeply impressed by the fact that seemingly complex real-world behavior can arise naturally from simple, but iterative mathematics that needs no calculus, and is therefore easy to explain to ordinary people. The social surprise is how conservative most people, including GIS specialists and scientists, are. The consequence is that peer review of research proposals automatically works against adventure. It is probably true that most advances in GIS have come initially, not from brilliant insights into how the world works, but through automating standard tasks. Only when these standard tasks have been automated and procedures have become generally accepted does it seem to penetrate some minds that new possibilities have arisen".

Figure 15.10 Peter Burrough, physical geographer

the sense of Box 15.1), and capable of working to many significant digits, so the calculation might be reported by printing out a number to eight digits, such as 10,014.603 sq m, or even more. But the number of significant digits will have been determined by the precision of the machine, and not by the accuracy of the determination. Box 15.1 summarized some simple rules for ensuring that the precision used to report a measurement reflects as far as possible its accuracy, and clearly those rules will have been violated if the area is reported to eight digits. But what is the appropriate precision?

In this case we can determine exactly how positional accuracy affects the estimate of area. It turns out that area has an error distribution which is Gaussian, with a standard deviation (RMSE) in this case of 200 sq m – in other words, each attempt to measure the area will give a different result, the variation between them having a standard deviation of 200 sq m. This means that the five rightmost digits in the estimate are spurious, including two digits to the left of the decimal point. So if we were to follow the rules of Box 15.1, we would print 10 000 rather than 10 014.603 (note the problem with standard notation here, which does not let us omit digits to the left of the decimal point even if they are spurious, and so leaves some uncertainty about whether the tens and units digits are certain or not – and note also the danger that if the number is printed as an integer it may be interpreted as

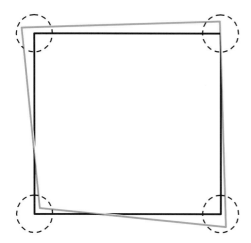

Figure 15.11 Error in the measurement of the area of a square 100 m on a side. Each of the four corner points has been surveyed; the errors are subject to bivariate Gaussian distributions with standard deviations in *x* and *y* of 2 m (dashed circles). The red polygon shows one possible surveyed square (one *realization* of the error model). Distortions are exaggerated

exactly the whole number). We can also turn the question around and ask how accurately the points would have to be measured to justify eight digits, and the answer is approximately 0.01 mm, far beyond the capabilities of normal surveying practice.

Analysis can be applied to many other kinds of GIS analysis, and Heuvelink (1998) discusses several further examples in his excellent text on error propagation in GIS. But analysis is a difficult strategy when spatial autocorrelation of errors is present, and many problems of error propagation in GIS are not amenable to analysis. This has led many researchers to explore a more general strategy of simulation to evaluate the impacts of uncertainty on results.

In essence, simulation requires the generation of a series of realizations, as defined earlier, and it is often called Monte Carlo simulation in reference to the realizations that occur when dice are tossed or cards are dealt in various games of chance. For example, we could simulate error in a single measurement from a DEM by generating a series of numbers with a mean equal to the measured elevation, and a standard deviation equal to the known RMSE, and a Gaussian distribution. Simulation uses everything that is known about a situation, so if any additional information is available we would incorporate it in the simulation. For example, we might know that elevations must be whole numbers of meters, and would simulate this

by rounding the numbers obtained from the Gaussian distribution. With a mean of 350 m and an RMSE of 7 m the results of the simulation might be 341, 352, 356, 339, 349, 348, 355, 350, ...

> Simulation is an intuitively simple way of getting the uncertainty message across.

Because of spatial autocorrelation, it is impossible in most circumstances to think of databases as decomposable into component parts, each of which can be independently disturbed to create alternative realizations, as in the previous example. Instead, we have to think of the entire database as a realization, and create alternative realizations of the database's contents that preserve spatial autocorrelation. Figure 15.12 shows an example, simulating the effects of uncertainty on a digital elevation model. Each of the three realizations is a complete map, and the simulation process has faithfully replicated the strong correlations present in errors across the DEM.

15.3 Fuzzy Approaches

So far the approach has been statistical, that is, uncertainty has been conceptualized as a statistical process, in which outcomes depend on probabilities. We have assumed that it is possible to put a straightforward interpretation on statements like "the database indicates that this field contains wheat, but there is a 0.17 probability that it actually contains barley". Here are two possible interpretations:

1. If 100 randomly chosen people were asked to make independent assessments of the field on the ground, 17 would determine that it contains barley, and 83 would decide it contains wheat.
2. Of 100 similar fields in the database, 17 actually contained barley when checked on the ground, and 83 contained wheat.

Of the two you probably find the second more acceptable, because the first implies that people cannot correctly determine the crop in the field.

These interpretations are called *frequentist*, because they are based on the notion that the probability of a given outcome can be defined as the proportion of times the outcome occurs in some real or imagined experiment, when the number of tests is very large. But while this is reasonable for classic statistical experiments, like tossing coins or drawing balls from an urn, the geographic situation is different: there is only one

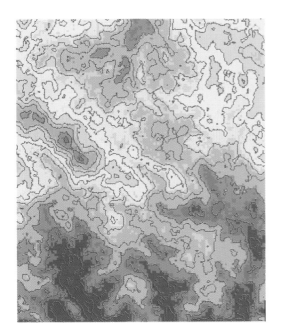
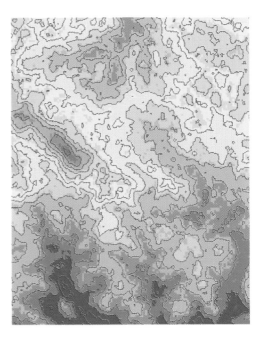
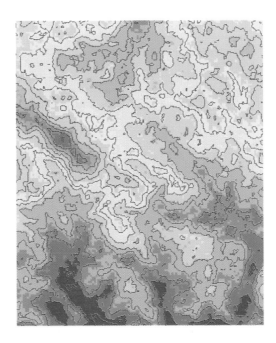

Figure 15.12 Three realizations of a model simulating the effects of error on a digital elevation model. The three datasets differ only to a degree consistent with known error. Error has been simulated using a model designed to replicate the known error properties of this dataset – the distribution of error magnitude, and the spatial autocorrelation between errors. (Courtesy: Ashton Shortridge)

field with precisely these characteristics, and one observer, and in order to imagine a number of tests we have to invent more than one observer, or more than one field (the problems of imagining larger populations for some geographic samples were discussed in Section 14.4).

In part because of this problem, many people prefer the *subjectivist* interpretation of probability, that it represents a judgment about relative likelihood that is not the result of any frequentist experiment, real or imagined. Subjective probability is similar in many ways to the concept of fuzzy sets, and the latter framework will be used here to emphasize the contrast with frequentist probability.

Suppose we are asked to examine an aerial photograph to determine whether a field contains wheat, and we decide that we are not sure. However, we are able to put a number on our degree of uncertainty, by putting it on a scale from 0 to 1. The more certain we are, the higher the number. Thus we might say we are 0.90 sure it is wheat, and this would reflect a greater degree of certainty than 0.80. This degree of belonging to the class *wheat* is termed the *fuzzy membership*, and it is common though not necessary to limit memberships to the range 0 to 1. In effect, we have changed our view of membership in classes, and abandoned the notion that things must either belong to classes or not belong to them – in this new world, the boundaries of classes are no longer clean and crisp, and the set of things assigned to a set can be fuzzy.

> In fuzzy logic, an object's degree of belonging to a class can be partial.

One of the major attractions of fuzzy sets is that they appear to let us deal with sets that are not precisely defined, and for which it is impossible to establish membership cleanly. Many such sets or classes are found in GIS applications, including land use categories, soil types, land cover classes, and vegetation types. Classes used for maps are often fuzzy, such that two people asked to classify the same location might disagree, not because of measurement error, but because the classes themselves are not perfectly defined and because opinions vary (see Section 6.2). As such, mapping is often forced to stretch the rules of scientific repeatability, which require that two observers will always agree. Box 15.3 shows a typical extract from the legend of a soil map, and it is easy to see how two people might disagree, even though both are experts with years of experience in soil classification. Other examples of these issues were discussed in Chapter 6.

Figure 15.13 shows an example of mapping classes using the fuzzy methods developed by A-Xing Zhu of the University of Wisconsin-Madison, USA, which take both remote sensing images and the opinions of experts as inputs. There are three classes, and each map shows the fuzzy membership values in one class, ranging from 0 (darkest) to 1 (lightest). In Figure 15.13 the result of converting to *crisp* categories, or *hardening*, is also shown – to obtain this map, each pixel is colored according to the class for which it has the highest membership value.

Fuzzy approaches are attractive, because they capture the uncertainty that many of us feel about the assignment of places on the ground to specific categories. But researchers have struggled with

---

Box 15.3 Fuzziness in classification: description of a soil class

Following is the description of the Limerick series of soils from New England, USA (the type location is in Chittenden County, Vermont), as defined by the National Cooperative Soil Survey. Note the frequent use of vague terms such as ''very'', ''moderate'', ''about'', ''typically'', and ''some''. Because the definition is so loose it is possible for many distinct soils to be lumped together in this one class – and two observers may easily disagree over whether a given soil belongs to the class, even though both are experts. The definition illustrates the extreme problems of defining soil classes with sufficient rigor to satisfy the criterion of scientific repeatability.

"The Limerick series consists of very deep, poorly drained soils on flood plains. They formed in loamy alluvium. Permeability is moderate. Slope ranges from 0 to 3 percent. Mean annual precipitation is about 34 inches and mean annual temperature is about 45 degrees F.

Depth to bedrock is more than 60 inches. Reaction ranges from strongly acid to neutral in the surface layer and moderately acid to neutral in the substratum. Textures are typically silt loam or very fine sandy loam, but lenses of loamy very fine sand or very fine sand are present in some pedons. The weighted average of fine and coarser sands, in the particle-size control section, is less than 15 percent."

(A)

(B)

(C)

(D)

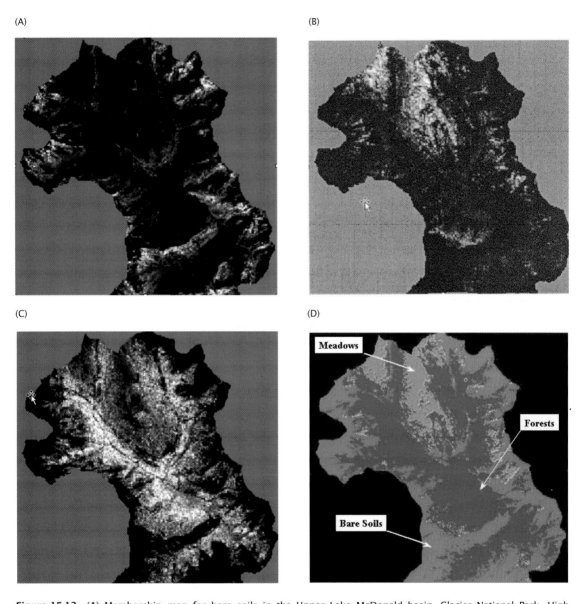

Figure 15.13 (A) Membership map for bare soils in the Upper Lake McDonald basin, Glacier National Park. High membership values are in the ridge areas where active colluvial and glacier activities prevent the establishment of vegetation. (B) Membership map for alpine meadows. High membership values are on gentle slopes at high elevation where excessive soil water and low temperature prevent the growth of trees. (C) Membership map for forest. High membership values are in the middle to lower slope areas where the soils are both stable and better drained. (D) Spatial distribution of the three cover types from hardening the membership maps. (Source: A-Xing Zhu)

the question of whether they are more *accurate*. In a sense, if we are uncertain about which class to choose then it is more accurate to say so, in the form of a fuzzy membership, than to be forced into assigning a class without qualification. But that does not address the question of whether the fuzzy membership value is accurate. If Class A is

not well defined, it is hard to see how one person's assignment of a fuzzy membership of 0.83 in Class A can be meaningful to another person, since there is no reason to believe that the two people share the same notions of what Class A means, or of what 0.83 means, as distinct from 0.91, or 0.74. So while fuzzy approaches make sense at an intuitive level, it is more difficult to see how they could be helpful in the process of communication of geographic knowledge from one person to another.

15.4 Living with Uncertainty

It is easy to find examples to illustrate how uncertainty can be a problem in GIS, and much more difficult to find positive steps to deal with it. So this last section of the chapter attempts to do that, by summarizing our current ability to come to grips with the inevitability of uncertainty, and to make sure that the results of GIS analysis are as meaningful, accurate, and reliable as possible.

> Discovering the effects of uncertainty is much easier than finding effective ways of living with it.

First, since there can be no such thing as perfectly accurate GIS analysis, it is essential to acknowledge that uncertainty is inevitable. It is better to take a positive approach, by learning what one can about uncertainty, than to pretend that it does not exist. To behave otherwise is unconscionable, and can also be very expensive in terms of lawsuits, bad decisions, and the unintended consequences of actions (see Chapter 17).

Second, GIS analysts often have to rely on others to provide data, through government-sponsored mapping programs like those of the US Geological Survey or the UK Ordnance Survey, or commercial sources. Data should never be taken as the truth, but instead it is essential to assemble all that is known about the quality of data, and to use this knowledge to assess whether the data are actually fit for use. Metadata (Section 7.5) are specifically designed for this purpose, and will often include assessments of quality. When these are not present, it is worth spending the extra effort to contact the creators of the data, or other people who have tried to use them, for advice on quality. Never trust data that have not been assessed for quality, or data from sources that do not have good reputations for quality.

Third, the uncertainties in the outputs of GIS analysis are often much greater than one might expect given knowledge of input uncertainties, because many GIS processes are highly non-linear. Other processes dampen uncertainty, rather than enhance it. Given this, it is important to gain some impression of the impacts of input uncertainty on output. An easy way to do this is through sensitivity analysis, a process of deliberately distorting inputs to see the effects of such distortions on outputs. For example, suppose we suspect that a DEM has errors of up to 2 m in each of its elevation values. The DEM is being used to investigate viewsheds, or the areas that can be seen from a given viewpoint. The analysis is run using the data, and the viewshed is mapped (Figure 15.14). We suspect that uncertainty has some impact on the result, but have no way of knowing how much impact.

A sensitivity analysis might proceed as follows. First, generate a new raster layer consisting of random values, scaled to roughly match the known uncertainties. We might do this by assigning each cell either $+2$ or -2. Then add this layer to the DEM, to simulate the variation due to uncertainty. Repeat the analysis with this new distorted DEM, and evaluate the viewshed. We could do this repeatedly, but a small number of simulations are probably sufficient to see the approximate impacts of uncertainty. By assigning the random values independently we have ignored the spatial autocorrelation that almost certainly exists, and there are more sophisticated methods that allow it to be incorporated into the simulation.

Fourth, rely on multiple sources of data whenever you can. It may be possible to obtain maps of an area at several different scales, or to obtain several different vendors' databases. Raster and vector datasets are often complementary (e.g. one could combine a remotely sensed image with a topographic map). Digital elevation models can often be augmented with spot elevations, or GPS measurements.

Finally, be honest and informative in reporting the results of GIS analysis. It is safe to assume that

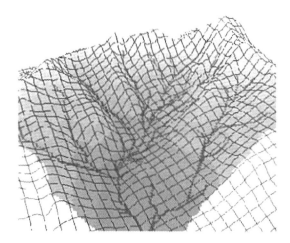

Figure 15.14 A watershed extracted from a DEM using a GIS algorithm. If the DEM is subject to errors, these will be reflected in uncertainties about the watershed boundary. Sensitivity analysis and simulation provide ways of evaluating these effects (Source: Band 1999)

GIS designers will have done little to help in this respect – results will have been reported to high precision, with more significant digits than are justified by actual accuracy, and lines will have been drawn on maps with widths that reflect relative importance, rather than uncertainty of position. It is up to you as the user to redress this imbalance, by finding ways of communicating what you know about accuracy, rather than relying on the GIS to do so. It is wise to put plenty of caveats into reported results, so that they reflect what you believe to be true, rather than what the GIS appears to be saying. As someone once said, when it comes to influencing people "numbers beat no numbers every time, whether or not they are right", and the same is certainly true of maps. Mark Monmonier has written an excellent book on the art of lying with maps (Monmonier 1996).

Questions for Further Study

1. Find out about the five components of data quality used in GIS standards, from the information available at www.fdgc.gov. How are the five components applied in the case of a standard mapping agency data product, such as the US Geological Survey's Digital Orthophoto Quarter-Quadrangle program (search the Web for the appropriate documents)?
2. Mapping projects that produce classifications (e.g. production of land cover maps from remote sensing) often take a standard of 85% for accuracy – in other words, only 15% of tested points or areas should be found to be wrongly classified. Is this a reasonable criterion?
3. How accurate is the Global Positioning System? If you have access to a receiver, conduct a simple experiment by repeatedly measuring the coordinates of the same location many times, over a period of several days, and record the results (compare Figure 15.7). You may also have access to very accurate measurements of the coordinates of certain points (e.g. geodetic control monuments, or very accurate surveys of an area). How do errors change through time – are they correlated, and over what time intervals? How does their distribution in space compare with the bivariate normal distribution?
4. What tools do GIS designers build into their products to help users deal with uncertainty? Take a look at your favorite GIS from this perspective. Does it allow you to associate metadata about data quality with datasets? Is there any support for propagation of uncertainty? How does it determine the number of significant digits when it prints numbers? What are the pros and cons of including such tools?

Online Resources

NCGIA Core Curricula (www.ncgia.ucsb.edu/pubs/core.html):
    Core Curriculum in GIScience, Sections 2.10 (Handling Uncertainty, edited by Gary Hunter), 2.10.1 (Managing Uncertainty in GIS, Gary Hunter), 2.10.2 (Uncertainty Propagation in GIS, Gerard Heuvelink), 2.10.3 (Detecting and Evaluating Errors by Graphical Methods, Kate Beard), 2.10.4 (Data Quality Measurement and Assessment, Howard Veregin)
    Core Curriculum in GIS, 1990, Units 45 (Accuracy of Spatial Databases) and 46 (Managing Error)
ESRI Virtual Campus course (campus.esri.com):
    Turning Data into Information, by Paul Longley, Michael Goodchild, David Maguire, and David Rhind (Module 'Uncertainty, Error and Sensitivity')

Reference Links

Maguire D J, Goodchild M F, and Rhind D W (eds) 1991 *Geographical Information Systems: Principles and applications*. Harlow, UK: Longman (Text available online from 'Links to Big Book 1' at www.wiley.com/gis and www.wiley.co.uk/gis).
    Chapter 12, The error component in spatial data (Chrisman N R)
Longley P A, Goodchild M F, Maguire D J, and Rhind D W (eds) 1999 *Geographical Information Systems: Principles, techniques, management and applications*. New York: John Wiley.
    Chapter 12, Data quality parameters (Veregin H)
    Chapter 13, Models of uncertainty in spatial data (Fisher P F)
    Chapter 14, Propagation of error in spatial modelling with GIS (Heuvelink G B M)
    Chapter 15, Detecting and evaluating errors by graphical methods (Beard M K and Buttenfield B P)
    Chapter 37, Spatial hydrography and landforms (Band L)
    Chapter 45, Managing uncertainty in GIS (Hunter G J).

References

Burrough P A and Frank A U 1996 *Geographic Objects with Indeterminate Boundaries.* London: Taylor and Francis.

Congalton R G and Green K 1999 *Assessing the Accuracy of Remotely Sensed Data: Principles and practices.* Boca Raton, Florida: Lewis Publications.

Guptill S C and Morrison J L 1995 *Elements of Spatial Data Quality.* Oxford: Elsevier.

Heuvelink G B M 1998 *Error Propagation in Environmental Modelling with GIS.* London: Taylor and Francis.

Monmonier M S 1996 *How to Lie with Maps.* Chicago: University of Chicago Press.

GIS AND MANAGEMENT

16

In GIS, things work successfully because they are managed. Because geography is ubiquitous, use of GIS typically includes material drawn from different sources and GIS can potentially underpin all decisions within and between organizations. Succeeding in these circumstances is not easy. Anyone who does succeed has managed the environment appropriately. For these reasons, effective GIS implementation requires understanding of management issues.

At a high level, the same management issues and needs exist in government, business, and academia. This becomes ever more the case with the growth of the Knowledge Economy (KE), to which GIS contributes powerfully. The KE, and management that operates within it, is affected by the growth of networking (such as the Internet) and other technological changes. But many business principles remain unchanged, e.g. understanding the business drivers and how GIS can contribute to those most important locally. Technology alone is only one contributor to success.

Learning objectives

After reading this chapter, you will understand:

- Government, academia, and commerce as "businesses";

- The interaction between these sectors inside GIS;

- GIS and the growth of the Knowledge Economy and knowledge industries;

- The impact of the Internet and the Web upon GI and management;

- GIS as a business in its own right;

- Business drivers;

- Management approaches to GIS.

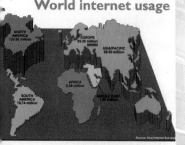

alla, report the Australian Bureau of Statistics° In
rnet users are between the ages of 20 and
999, there were 1.4 million users accessing the
veys°

World internet usage

etnam officially connected to the Internet in 1997
rst six months° In March 1999 there were 1.3
sia° A recent survey suggests that 81 per
male° In March 1999, Nua Internet Surveys report-

16.1 Government, Academia and Commerce Are All "Businesses"

Most of us live in a capitalist society. Our tangible wealth and physical well-being – and hence freedom to do many things – come increasingly from the exploitation of knowledge or skills. Much of this exploitation is done by business and is derived from scientific or technological innovation. Increasingly, we see the breakdown of previously discrete sectors as commercial considerations impinge more widely and there is convergence between the activities and operations of commerce and industry, government, the not-for-profit sector, and academia. Increasingly everyone is concerned with explicit goals (many of which share similar traits) and with ''knowledge management'' (see Box 16.1). Moreover, in the GIS world at least, we also see significant overlap in functions – for instance, both government and the private sector are producers of geographic information (GI). For these reasons, we use ''business'' as a single term to describe the corporate activities in all three sectors. Accordingly, we think of the GIS world as being driven by organizational or individual objectives, using raw material (data or information: see Section 1.2), tools (GIS) and human capital (skills and approaches) to create new products or services (which may be other information or knowledge) and exploit them. This view is reflected in the nature of people being appointed to senior posts: for instance, the head of Britain's national mapping organization appointed in 2000 was formerly employed in commerce (see Box 16.2).

This commonality between the different sectors should not be exaggerated. For example in publicly owned commercial organizations in the USA, pleasing the stockholders every quarter is a commonplace requirement. The lack of profitability and resulting low price of shares in Smallworld – a pioneer in object orientation – led to its purchase by General Electric in Fall 2000. The different drivers in each sector are set out later in this chapter, in Table 16.2. But every organization now has to listen and respond to customers, clients, or fellow stakeholders. Every organization has to listen to citizens whose power can sometimes be mobilized successfully against even the largest corporations. Every organization has to plan strategically and deliver more for less input. Everyone is expected to be innovative and deliver new products or services much more frequently than in the past. Everyone has to act and be seen to be acting within the law, regulatory frameworks, and conventions: these provide far-reaching constraints. And everyone

has to be concerned with risk minimization, knowledge management, and protection of the organization's reputation.

All organizations and their employees are subject to similar needs and incentives, though what is most important varies locally.

The relationship between the different sectors also shifts over time. In the 1960s, 1970s, and even 1980s, much of the available GIS software was produced by government or by individuals. Only with the arrival of significant commercial enterprises – and hence competition – has GIS become a global reality with the number of users perhaps ten thousand times greater today than it was in the 1970s. Very little general purpose and widely used software is now produced outside the commercial sector, the exception being the Idrisi software from Clark University (see Sections 1.4.3.2 and 8.5.7).

16.2 Why is Management Important in GIS?

GIS has two obvious relationships to management. The first is that it can help us to manage many projects so as to produce more effective, more efficient, more equitable, or more predictable outcomes. An example of this is the exploration of toxic waste sites which minimize hazard to the community whilst not inducing large additional transport cost or pollution burdens on the taxpayer. Management of projects is needed everywhere, as Box 16.3 shows – a classic example of how GIS implementations can evolve to take advantage of advances in technology, organizational developments, and serendipity. Good managers are able to turn these situations to their advantage.

The second relationship between GIS and management is that all GIS projects themselves need to be managed. This includes everything from the specification of needs through to the selection and procurement of the tools, and the development of staff skills through to the operations of the GIS team (see Chapter 18).

Management is all about making conscious, desired changes to the world. Luck and idiocy on the part of one's competitors only plays a small part in what happens. Science and technology alone – however good – are not sufficient conditions for success: for many years, the Soviet Union had more scientists and technologists than the rest of the world combined. Microsoft did not succeed simply because they produced smart software. It succeeded because the organization

Box 16.1   Mapping disadvantage: the need to know

The Prince's Trust – a UK charity – gives new opportunities to 14 to 30 year olds, reaching out to those who need help most. It helps young people develop confidence, learn new skills, get into work, and start businesses. Some 11,000 people – staff and volunteers – help the Trust achieve its objectives.

Identifying where the biggest problems exist and hence targeting resources to help is of crucial importance for the Trust. After reviewing other work on deprivation and disadvantage, they carried out a major research study. The starting point was to identify a range of measures related to disadvantage amongst the young and how these limited success in later life. The 14 measures were identified by longitudinal research, i.e. analysis of details of individuals over many years. These included being involved in crime as a perpetrator or victim, dependency on state benefits, and not gaining qualifications at school or being absent from it.

The Trust commissioned work on identifying which areas in England and Wales scored highly on these criteria. Figure 16.1 shows the map of education disadvantage scores for the London boroughs whilst Table 16.1 shows how the aggregate score was built up from the three individual education criteria. Thus Islington is in the top (worst) 10% of areas in England in terms of examination results (getting three "blobs"), in the worst 25% for absenteeism (two "blobs"), and in the 10% most disadvantaged in terms of the proportion of children getting free school meals (three "blobs").

The GIS work helped the Prince's Trust to improve targeting of its resources. The results were kept simple in order to convey the problems to politicians and lay people. And the study also showed that, though young people in all high scoring areas needed help, very different policy interventions are needed in different areas.

Shows young people's experience of education.
Combines data on qualifications, absenteeism, and entitlement to free school meals.

AT LEAST 8 • THE TOP 10%   4 OR 5 • BETWEEN 10-25%   2 OR 3 • BETWEEN 25-50%   0 OR 1 • BETWEEN 50-100%

Figure 16.1   Areas of disadvantage in Longon for young people on education criteria, based on a study for the Prince's Trust, a UK charity

Box 16.1   Continued

Table 16.1   Educational attainment in London

london education BOROUGH	percent of 15 year-olds gaining less than 5 GCSEs (A to G)	the proportion of authorised and unauthorised absence from school	percent of the school population entitled to free school meals	borough total
Islington	•••	••	•••	••••••••
Greenwich	••	••	•••	••••••••
Hackney	•••	•	•••	•••••••
Haringey	•••	•	•••	•••••••
Kensington and Chelsea	•••		•••	•••••••
Southwark	••		•••	••••••
Westminster, City of	•••		•••	••••••
Camden	•	•	•••	•••••
Lambeth	•	•	•••	•••••
Tower Hamlets	•	•	•••	•••••
Hammersmith and Fulham		•	•••	••••
Lewisham		•	•••	••••
Wandsworth	••		••	••••
Barking and Dagenham	•	••	•	••••
Newham		•	•••	••••
Waltham Forest	•	•	••	••••
City of London			•••	•••
Brent			••	••
Ealing			••	••
Hounslow	•		•	••
Merton	•	•		••
Croydon			•	•
Enfield			•	•
Hillingdon	•			•
Kingston upon Thames	•			•

••• THE TOP 10%	•• BETWEEN 10-25%	• BETWEEN 25-50%	NO • BETWEEN 50-100%	– NO DATA AVAILABLE

had great management and smart people. It had a vision of what it wanted to achieve, ideas on how to do it, plus superb marketing skills. It also had the ability to reprioritize and reformulate plans and activities as often as necessary. Box 16.4 shows some examples of where bosses of organizations got things wrong. Bill Gates recovered; Ken Olson did not. How the operation and the people within it are enthused, trained, and developed, how a culture of high expectations is established and changing plans early – all these management roles make the difference between success and, at best, mediocrity.

We began Chapter 1 in this book with the assertion that geography is a factor in almost everything (Section 1.1). The results of many decisions have consequences which are geographically varied and existing geographic variations influence key decisions. It is obvious therefore that GIS, decision-making, and management collectively have ramifications for the entire population of each and every country. Management is about achieving desired ends through people and with the use of tools such as GIS – and irrespective of the sector. "People skills" cannot sensibly be learned in a vacuum, outside a context of tools and organizational culture. For this reason, this chapter deals "in the round" with many of the elements relevant to good management and how management impinges on

Box 16.2   Vanessa Lawrence: national mapping agency Chief Executive

Vanessa Lawrence (Figure 16.2) is Director General and Chief Executive of Ordnance Survey, Britain's national mapping agency. She is a Geography graduate with an MSc in remote sensing, image processing, and applications. After university Vanessa joined the Longman group, where she had a role to commission and be in charge of the management and profitable development of textbooks and reference books for the higher education market in GIS and Geography, etc. One of the books she commissioned was the very successful two-volume GIS reference work by Maguire *et al* (1991). A 1989–1990 scholarship enabled her to travel around the world to consider the rapid development of GIS and related fields. She was then appointed to be the Technical Director of GeoInformation International, a Pearson company and new venture specializing in providing a wide range of information to the GIS community.

Vanessa went to Ordnance Survey from Autodesk, the world's fourth largest PC software company. She joined that firm in 1996 and headed the team that established Autodesk's GIS business in parts of Europe, the Middle East, and Africa. She helped to develop one of the world's largest geographic information systems, a project that created the electoral area structure for the 1999 South African General Election.

At Ordnance Survey, Vanessa heads the government department that produces and maintains a wide range of maps and computer data for business, leisure, and academic uses. She is also advisor to the UK Government on all issues relating to mapping, geographic information, and surveying. Ordnance Survey was the first national mapping agency in the world completely to digitize all its maps. Computer data forms around 75% of Ordnance Survey's business.

"The most surprising thing is the sheer speed at which electronic positioning information is becoming available to customers who wouldn't naturally think of themselves as map users" says Vanessa, pointing out that at least 80% of business information has some kind of geographic content. "That shows just how readily data can be sorted or accessed through a GIS. Quite often, people using information facilities on the Internet or the latest generation of mobile phones may not even realize that a GIS is underpinning their service. A vast range of work in both public and private sectors already relies on Ordnance Survey data," she says. "My key task at Ordnance Survey is to ensure that it becomes a fully-fledged e-business, offering continuous Web access to all our data in customer-friendly formats".

Figure 16.2   Vanessa Lawrence, Chief Executive

GIS and how GIS may impinge on management. We deal with the detailed aspects of how to create and run a GIS in Chapter 18. Here we take a high-level view of how good management makes things happen, either within a GIS facility or by using it for organizational ends.

All organizations need to be responsive to the needs of customers and other stakeholders and must demonstrably be effective. This is achieved through good management, rather than luck or individual genius.

Most GIS management is not about dealing with technical specialists (though they can either be very easy or very difficult to deal with because of their technical obsessions). For every expert software developer in the GIS industry, there are at least 10 expert applications people in the user

Box 16.3   Evolving GIS at Kruger National Park (KNP), South Africa

The vision of those working in the Park is that all South Africans can have a joint stake and stewardship in one of Africa's most important ecological resources. GIS is playing an important role in the evolution and realization of this vision.

KNP has been active in GIS since the early 1990s. In the early days GIS software products from four vendors were used to accomplish different specialist or niche products in the areas of ecological analysis and facilities management. A review of these projects highlighted the need to create a central data repository.

In 1996 a UNIX workstation GIS was purchased and a serious attempt was made to gather, verify, convert, and store all relevant baseline data layers in a central location called GISCENTRAL. Data layers were collected on natural themes (landscapes, land types, geology, rivers, and drainage channels), operational themes (management blocks, veldt condition assessment plots, and burn blocks), and artificial (human) themes (roads, bridges, dams, rest camps and other visitor facilities, entrance gates, and boundaries). This high-end GIS was quite well used by GIS professionals in the park and by graduate and postgraduate academic visitors.

At the same time, KNP also purchased several desktop GIS software packages. These proved to be of great interest to non-computer specialists, who could download pre-built data layers onto their laptops, and view and analyze them with their local software.

A GIS seminar was held in 1997 at the Park to inform top management and users on the progress and future direction of GIS in the Park. One year later there was extensive restructuring of KNP and other national parks. At about the same time, developments in Internet technology offered new opportunities to spread GIS out to a dispersed user population. Almost without realizing it, the combination of new technology, the restructuring, and some informal lobbying opened the door to an Internet GIS implementation. Today KNP has several Internet applications providing distributed access to geographic data and functions. For example, visitors can view and query live maps (Figure 16.3) and even add information about game sightings and then query the locations of, say, buffalo on a particular day.

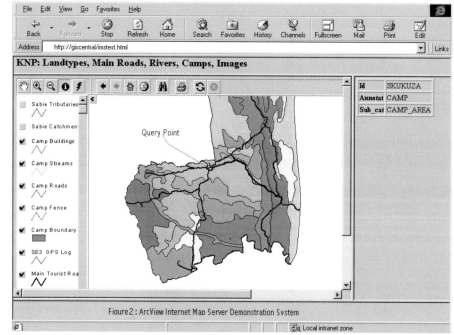

Figure 16.3   Internet GIS at Kruger National Park, South Africa (A) Land types, main roads, rivers, and camp sites in the Park (Source: South Africa National Parks) (Cntd.)

Box 16.3 Continued

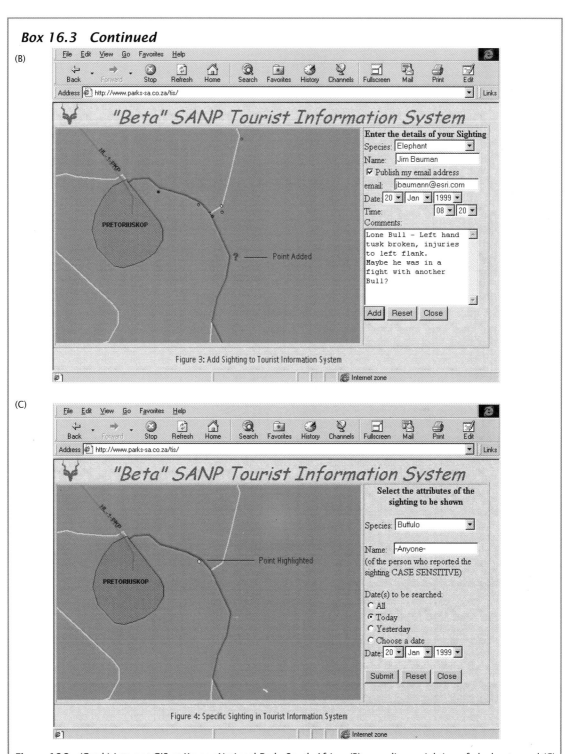

(B)

Figure 3: Add Sighting to Tourist Information System

(C)

Figure 4: Specific Sighting in Tourist Information System

Figure 16.3 (Cntd.) Internet GIS at Kruger National Park, South Africa; (B) recording a sighting of elephants; and (C) searching for sightings of buffaloes (Source: South Africa National Parks)

Box 16.4 Some technological visions which proved wrong

"Everything that can be invented has been invented". Commissioner, US Office of Patents 1899
"There is a world market for maybe five computers". IBM chairman 1943
"There is no reason why anyone would want a computer in their home". Ken Olson, President of DEC
    (then the world's second largest computer manufacturer) 1977
"640k [kilobytes] ought to be enough for anybody". Bill Gates 1981

enterprises and, for each of them, there seems to be at the very least 10 users, i.e. a ratio of 1:10:100 or even greater (see Section 8.6). There is then much more chance that you will be a GIS manager than a software developer. As a result, your success will come from dealing with all categories of people, requiring the use of different languages or techniques to communicate and to convince the different communities. Beyond that, the context in which you manage is rarely constant for long so we now describe the changing world of knowledge creation and exploitation – a world in which GIS is prospering.

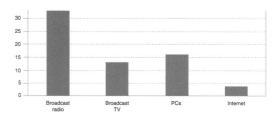

Figure 16.4 Increased speed of up-take of innovations, shown by the years needed for it to be accepted by 50 million Americans. (Source: US Department of Commerce)

16.3 The Knowledge Economy, Knowledge Management, and GIS

Different people mean different things by the term "Knowledge Economy". For some, this exists when citizens have been empowered by laws and by access to technologies which, in combination, enable access to information from many different sources. The information enables us to develop new ideas or govern ourselves more effectively and harmoniously, keeping the legislature in check. For others, the term means a society where enterprise can flourish through individuals or companies creating new or improved products – all on the basis of early knowledge of new ideas or products from elsewhere. In practice, these overlap – but sometimes may also conflict, as we see later.

There is a gathering pace world-wide in commercial developments through the creation of new products, processes, and services and in the rapidity of their dissemination and acceptance on a national or world-wide basis (see Figure 16.4). Inventive entrepreneurs in Hong Kong take for granted the truth of the phrase "If it works, it's obsolete". The life cycle of new products, services, and even of knowledge itself is becoming ever shorter under competition which is ever more global. The financial industries exemplify this most clearly: in 1980, the average daily turnover in foreign exchange was $180 million; twenty years later it was $1.5 trillion. This

and many other business developments have been greatly facilitated – by scientific and technological developments, especially in the information and communication technologies (ICT) – but these are not the only factor and technology has not always done what was anticipated (Box 16.5).

This escalation is most rapid in the knowledge industries where substantial national and business revenues may be at stake through licensing of new processes or products (Figure 16.5). It is widely agreed that these industries will be the key to future national and business success. As an example, the World Bank's *1998 World Development Report* said:

"For countries in the vanguard of the world economy, the balance between knowledge and resources has shifted so far towards the former that knowledge has become perhaps the most important factor determining the standard of living.... Today's most technologically advanced economies are truly knowledge-based."

Earlier in this book we have identified "knowledge" as the highest form (apart from wisdom) in a hierarchy of descriptions (see Section 1.2). Knowledge does not arise simply from having access to large amounts of information. Rather, it can be considered as information to which value has been added by interpretation based on a particular context, experience, and purpose. Thus while knowledge may exist in any sector of human activity, its volume, nature, and the forms of it needed to succeed will vary between applications. The knowledge

Box 16.5   Technology, globalization, and unintended consequences

Globalization is not only a factor of technology: it has been driven as much by world-wide trade liberalization and the removal of restrictions on cross-border capital movements. Moreover technology effects are not always predictable or benign: Tenner (1997) has highlighted the unintended consequences of many new technological innovations, some very harmful, and the lower impact than was originally anticipated in many cases. One example is that deaths from hurricanes in the USA have dropped from six thousand a year in 1900 to a few dozen now thanks in large measure to radar and other tools to track and predict storms. Yet insurance payouts have soared – $7.3 billion after Hurricane Andrew – in part because people felt safer in building in hurricane-prone areas.

requirements in designing mobile communications will be different to those needed to print a newspaper, for example. On this definition, GIS people are likely to be part of the knowledge industries – but only if they are required to innovate and have skills to develop and exploit tacit as well as codified knowledge.

Our definition is conceptually clear and not industry-specific. But not everyone uses it. There are some good reasons for this – for example, there is a need for simple cross-national and cross-industry comparisons of the "knowledge industries". Without international comparisons, management of economies is difficult. For this purpose, the Organization for Economic Co-operation and Development (OECD) defines knowledge industries by industrial sector – a less than perfect, yet simple, approach. To OECD, the knowledge-based industries are made up of knowledge-based services and high-tech industry. Knowledge-based services are taken as tele-communications, computer and information services, finance, insurance, royalties, and other business services. The high technology industries

are defined as aerospace, computers and office equipment, pharmaceuticals, and radio, TV, and communications equipment. On these definitions, the entirety of GIS is clearly a part of these "knowledge industries" and the "Knowledge Economy"!

> Irrespective of the detailed definition, those active in GIS are working in the knowledge industries and contributing to the Knowledge Economy.

Since knowledge – however defined – is becoming so crucial, it is no surprise that investors increasingly recognize the growing importance of knowledge assets in the way they value firms. Box 16.6 demonstrates one approach to valuing knowledge in commercial enterprises. However, intangible assets occur at least as frequently in government-based organizations though valuing the equity in such cases is even more complex. It is also obvious that intangible assets may be negative (e.g. where brand image is destroyed by loss of public acceptance of the quality or safety of products or services provided by the government body or commercial firm). Intangible assets are present in extreme form in academia where the poaching of one key member of staff or a team may destroy the reputation of the organization in that field overnight.

16.4   Information, the Currency of the Knowledge Economy

In this section, we discuss several different aspects of the role and characteristics of information in general and geographic information (GI) in particular. We start with how information underpins decision-making – the basis of many GIS installed in the past.

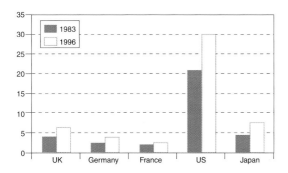

Figure 16.5   Knowledge Economy revenues from overseas earnings on royalties and license fees (US$bn). (Source: UK Competitiveness White Paper 1998 Copyright UK Government, Department of Trade and Industry)

Box 16.6   How do we place a value on knowledge?

In fast-growing sectors like biotechnology and computer software, including some parts of GIS, a large part of the value of the company resides in the knowledge embodied in its patents and in its staff. Sveiby (1997) pointed out the huge growth in the difference between Sun's stock market value and its book value in 1995–96 because of the announcement of Java; this represented a major increase in its intangible assets. He categorized the different types of intangible assets as below.

Visible equity (book value)	Intangible assets (stock price premium)		
Tangible assets minus visible debt	**External structure** Brands, customer and supplier relations	**Internal structure** The organization: management, legal structure, manual systems, attitudes, R&D, software	**Individual competence** Education, experience

16.4.1   Information and knowledge – the basis for good decision-making?

Until now, we (and many others) have assumed that having knowledge and information gives power and capability (Box 16.7). The individual who knows more than his or her colleague can therefore make better decisions. As authors, we hold to the view that better information often *does* permit better decisions to be made. But the reality is usually much more complicated, especially in GIS. In part this is because GIS is rarely the *only* important input to decision-makers. Many decisions require the balancing of conflicting objectives and judgments about the trade-offs, as known in the light of available knowledge. Thus the role of "raw information" in shaping the running of the world can easily be exaggerated, as Brown and Duguid (2000) have shown: they demonstrate clearly that managers are not simply information processors, with the "right" decisions simply falling out of an analytical process. They illustrate this by citing five commonplace yet erroneous assumptions about management and information which were well-known as long ago as 1967:

● *That more information leads to better decisions.* In reality, most managers increasingly suffer from information overload, especially of irrelevant information. Visualization techniques (Chapter 12) can help to reduce the problem greatly but rarely cure it completely.

● *That managers need the information they want.* Managerial "wish lists" often do not match to what is actually needed to run the business.

● *That managers are able to model the decision they wish to make.* It is unusual for a clear model to exist which identifies the relevant variables and their inter-relationships.

Box 16.7   The importance of "facts"

There is a great academic argument about what constitutes "facts" and to what extent these are universally true or are context-dependent (see Section 17.1.5). Some individuals however have brooked no such argument! In *Hard Times*, Charles Dickens has one of his characters say:

"Teach these boys nothing but facts. Facts alone are wanted in life. Plant nothing less and root out everything else. You can only form the minds of reasoning animals upon Facts ... Stick to Facts, Sir."

- *That managers do not need to understand the information system.*
- *That information systems lead to enhanced communication within the organization.* In practice, the internal structures of the organization may vitiate this potential advantage. Too much information may also clog up responses. For example, communications between the public and the UK Prime Minister rose twenty-fold in 5 years in the late 1990s. This was due partly to extended use of e-mail. It caused major staffing and other problems, leading to public dissatisfaction with the responses and a sense of exclusion – the exact opposite of what was intended.

Chapter 2 introduced the application-driven nature of much GIS. But if we look from an institutional perspective, successful use of GIS requires an understanding of the aims, objectives, and ways of operating of the organization and a limiting of activities to what is most essential. It also suggests that the potential benefits of GIS for managers of big projects or even for corporate bosses are:

- Providing "factual information" about the location of resources, human or natural (*a facilitation process*).
- Computing derived "facts" (see Box 16.7), e.g. fastest routes through a network, population within 5 miles of a toxic waste dump, change since last year (*a facilitation process, producing new information*).
- Selection, compression, and visualization of information to facilitate decision-making (*information filtration*).
- The search for regularity and possible causality: looking for patterns and correlates of geographic distributions, e.g. does spending on certain goods coincide with particular ethnic groups or distance between shops and home? Chapter 2 gives more examples (*information filtration, leading to evidence*).
- Creation of added value by linkage of information from different sources. In general, such linkage extends the range of applications possible – with two information sets for a given region, we have one combination; with 20 sets, we have over a million combinations in total (*information fermentation, leading to more information, evidence, and possibly knowledge*).
- Predicting future events which are geographically distributed (*evidence or knowledge-based operations*).

Despite the over-hyping of the role of information in much decision-making, it is clear that information *is* crucial to the success of the Knowledge Economy. This seems even more obvious when we take the Internet into account. Many people have claimed – from the Chief Executive of Oracle onwards – that this "changes everything" (see also Section 1.4.3.1). Indeed in the Preface to this book, we have espoused a different kind of education and learning – one which is predicated upon ready (and hence near-ubiquitous, immediate, and low cost) access to information via the Internet and the Web.

In practice, the move to a Knowledge Economy is a bit more sophisticated, unfinished, and unpredictable than that, as the battles over copyright protection of information make clear (see Chapter 17). GIS plays only a modest role in them: the main battle-grounds are in meetings of the World Intellectual Property Organization, in the US Congress, in parliaments around the world, and in courts where claims of software and music piracy are matched by arguments that copyright owners are trying to establish monopolies in perpetuity.

So far as the economics of information are concerned however, the respected information economists Shapiro and Varian (1999) have argued that none of the principles is changed by the advent of a new delivery channel (the Internet). They claim that there is no need for a new set of principles to guide business strategy and public policy in the Internet age. Have you read, they ask rhetorically, the literature on differential pricing, bundling, signalling, licensing, lock in, or network economics? Have you studied the history of the telephone system or the battles between IBM and the Justice Department (they might have added between that department and Microsoft)? "Technologies change, economic laws do not" they concluded.

Is this really true, especially for geographic information or GI? To answer this we first have to examine the characteristics of information in general and GI in particular – and to recall that most geographic information has hitherto been created by governments. Thus both governments – subject to political and citizen pressure – and private sector entrepreneurs have a vital interest in this area. Moreover, government interests are not everywhere the same. All this is somewhat different from other situations, e.g. in the case of pharmaceuticals, government is not directly involved in production of products; multiple drugs produced by different multinationals sometimes compete and government's role is largely to regulate the safety and ensure that citizens are not being held to ransom by extortionate charging. In GI by contrast, governments provide much of the content. The commercial "manufacturing sector" is presently modest in size; many existing products originate within government and value is typically added by the private sector. In addition, there are substantial benefits to be gained from having one geographic framework (see Section 17.3.2.2) and having as few duplicate coverage datasets as possible.

16.4.2 The characteristics of information

We can think of the very same information being used for two different purposes and thus being valued very differently by the same people in different circumstances:

- *For consumption*. Individuals will decide how much value they assign to it based upon their valuation of pleasure, time saving, or some other metric and their awareness of the uses and the potential benefits of the information.

- *As a factor of production*, where the information is used as part of a good or service. The end user of that good or service will make his or her decision on the uses to which it can be put, its availability to others (it will tend to be more valuable if it is not available to others, such as competitor organizations) and the ease of substitutability. One obvious example is the use of the same geographic information in making planning decisions in government and for business purposes in commerce. Some decisions will place a high social value on the information whilst others will give it a high economic value. Actually measuring these values is rather difficult so discussion thus far has typically been at a very high conceptual level (but see Box 18.1 on cost–benefit analysis).

It is crucial to appreciate that information, from a management perspective, has a number of unusual characteristics as a commodity. In particular, it does not wear out through use (though it may diminish in value as time passes; census of population data from 10 years ago is less reliable for current policy-making than that from last year). It is usually argued that information is in general a public good but several types of public good can be distinguished. These distinctions are important because implicitly they may influence policy discussion. A "pure" public good has very specific characteristics:

- The marginal cost of providing an additional unit is zero. Thus copying a small amount of GI adds effectively nothing to the total cost of production.

- Use by one individual does not reduce availability to others ("non-rivalry"). This characteristic is summarized in the famous Thomas Jefferson quotation that: "He who receives an idea from me, receives instruction himself without lessening mine; as he who lights his taper at mine, receives light without darkening me."

- Individuals cannot be excluded from using the good or service ("non-excludability").

In practice, however, information is actually an optional public good, in that – unlike defense – it is possible to opt to take it or not; you do not have to make use of US Geological Survey data. Moreover, the accessibility and cost of the systems to permit use of information influence whether, in practice, it is a public good in any sense. Since only 0.05% of the world's population has ready access to GIS tools, it can scarcely be argued that geographic information in digital form is yet a universal public good in the traditional sense. Finally – to be pedantic – it may also be best to define information as a quasi-public good since it may be non-rival but its consumption can in certain circumstances be excluded and controlled. The business cases and vast investments of a number of major GI purveyors such as Space Imaging Inc. are based on this proposition; if everything they produced could be copied for free and re-distributed at will by anyone, their business would be untenable. Thus the pecuniary value of information may well depend on restricting its availability whilst its social value may be enhanced by precisely the opposite approach.

> Geographic information has traditionally been seen as a special type of public good but its characteristics differ in different countries and in different sectors of the economy.

What is certainly true however is that information in digital form – however much it costs to collect or create in the first instance and even to update – can be copied and distributed via the Internet at near-zero marginal cost. In general, information is costly to produce but cheap to reproduce. Its widespread availability has many positive (and some negative) externalities, i.e. benefits (or disbenefits) which arise far beyond the original creator, as described for GI in the next section. It is also an "experience good" which consumers find hard to value unless they have used it before. There is no morality about information itself: its source (especially GI) can often readily be disguised, for example by changing the representation of roads from single lines to cased (double) ones, changing their color, adding some sinuosity, or computing tables of distance between places.

16.4.3 The characteristics of GI

GI has the standard characteristics of information described above but also some others which mark it out as distinctive. For instance, one characteristic of most of the many different markets for GI is their gross immaturity compared with those of other types of information – and certainly with those for other commodities. There are several reasons for this situation. The first is the expectations of many customers that GI is a free good (provided free or at very low cost and without restriction on its use) as well as a public good, since it has been provided historically by governments to taxpayers.

Beyond that, certain detailed geographic information has some of the characteristics of a natural monopoly. This effectively means that the first person with the infrastructure wins because others can never justify duplication of facilities. Such monopolies are obviously antipathetic to competition, which demands multiple choice. The obvious parallel for such GI is with telephone lines before the advent of cellular telephones. For instance, assume that only one organization has a complete set of information for every house in the country, that it is unwilling to share it freely and that the majority of that organization's costs are sunk ones (i.e. they have been spent and cannot be recovered). In such circumstances – not uncommon even now in some parts of the world with government-held information – it is unlikely that the collection and duplicate sale of information can be made competitive. Nor may competition be in the national interest so far as the "geographic framework information" is concerned (see Section 17.3.2) because it can give rise to deleterious externalities (see below). The creation of multiple but different framework datasets will guarantee the creation of artifacts in any computer analysis using combinations of datasets related to each of the different frameworks (e.g. Figure 6.11). Putting a value on such externalities is difficult, especially since the costs are visited on those at the end of supply chains and are often only discovered long after analyses of the data have led to mistaken conclusions. This provides great opportunities for lawyers.

> Some geographic information has the characteristics of a natural monopoly.

A third factor in making GI special is the relative lack of transiency of much traditional geospatial data, which makes it analogous to "reference books". A typical dataset derived from topographic maps may only change by 1% or less per year. Satellite imagery of the same area may change much more because of land cover change, imperfect conflation of images sensed at different times, and the effects of the particular classifier used – but any given business may very well be completely uninterested in this change for its own particular purposes and one purchase may therefore be adequate.

> Everyone sharing one "geographic framework" avoids many problems.

Another characteristic is that related to users. For example, the skills of the user and the available software determine how much can be done with the data. In addition, the value of GI to different users varies by many orders of magnitude. If use of GI can improve efficiency by 0.1% in some target marketing or utility company performance (Chapter 2) it may easily repay all investment but that is unlikely to be true in many other cases. Knowledge of all this on the part of the information producers is variable at present. A practical consideration which often determines the value of GI is the extent of linkage with other GI that is possible – and hence the extent of "value added". Because other data exist does not mean, however, that their existence is known about or that the data are actually available. To add to the complications, we know relatively little about quality of many GI sets and hence of their *fitness for purpose*. This is because of the difficulty of defining quality in a sound way. Sometimes even knowing everything does not help very much because there are no alternatives – there is a legacy of long-lasting geographic information. Moreover, we do not have a good idea of whether data quality matters legally for there have been relatively few tests as yet of liability involving GI in the courts.

It is also worth understanding that the present GI marketplace does not meet the specification by economists of an *efficient market*. In particular, there is far from perfect awareness of the market on the part of the purchasers, and there are major market distortions – varying by country – due to subsidies, legal constraints, public perceptions, etc. and the existence of substantial externalities which further render the existing markets sub-optimal in terms of classical economics. Externalities arise where:

● production of a good by one "agent" imposes costs on or delivers benefits to other producers or consumers; or

● consumption by one individual imposes costs on or delivers benefits to other consumers or users.

It will be seen, therefore, that a pure public good is a special form of externality; provision for one group of users will benefit other potential users because it is not possible to exclude them. Pollution of water, of air, and by noise are classic examples of external costs; refuse collection, education, and public health are classic examples of external benefits. It is an important feature of markets where externalities are present that output levels resulting from free market provision will not be optimal.

The views taken of GI also differ by sector. Much of the above is largely of theoretical interest to those commercial data suppliers working mainly in the short term. To survive, they can (and often must) treat the provision of information as much more akin to that of any other commodity than can government. The information provider in the latter sector is not only constrained by rules outlawing exclusive deals, differential pricing on a major scale, etc.: he or she is expected to behave to the highest standards of equity, probity, propriety, and consistency – and to demonstrate that this is the case, rather than getting the best bottom line.

GI is a tradeable commodity and a strategic resource. Its supply and use are governed by public policy, contractual, financial, and other considerations.

16.4.4   The Internet, World Wide Web, and GI

The Internet has been hailed as a distribution channel, a communications tool, a market place, and an information system – all in one. It alters almost everything managers do, from finding suppliers to coordinating projects to collecting and managing customer and client data. If this is

all true, surely it is also of crucial importance for GIS and GI?

The answer is mostly yes – but there are some caveats. Consider first the proposition that, if Internet sales allow airlines to fill seats that would otherwise be empty, the price of tickets required to meet fixed costs (and of breaking even) falls. In addition, the Internet will have enabled more people to travel to more places than hitherto. This is generally seen as a good thing. But this really only applies to goods in finite supply (e.g. airline seats) and which are "perishable" (i.e. are valueless after take-off). It says very little about the GI situation or the role of the Internet and the World Wide Web (as a facilitator of use of the Internet) in it. Moreover, use of the Internet is only more efficient if it displaces other activities. Reading material online is more efficient than printing and transporting material physically. But ordering books by the Internet from Europe and having them air-freighted from the USA is scarcely energy-efficient! Finally, a 2000 US Department of Commerce study showed that households with annual incomes over $75,000 are 20 times more likely to have Internet access than others – the so-called digital divide. It is clear from all this that the Internet is not yet a universal benefit and suffers from some unrealistic expectations (Box 16.8).

So far as GIS/GI users and managers are concerned, there are however a number of real or apparent advantages of the Internet. These are:

● The ease of setting up and using "information location tools" (using clearinghouses or geolibraries: see Section 7.5 and Chapter 19).

● The possibility to preview or "taste" simplified versions of the information to check its suitability (essential in what is for many an "experience good" – see Section 16.4.2).

● The capacity for mass customization, i.e. tailoring to the needs of the individual through use of automated tools and hence segmenting the

---

Box 16.8   Some Internet hype

There is much nonsense spoken about the Internet. The extravagant claims of MIT's Media Lab staff are well known. For example, Nicholas Negroponte said "[thanks to the Internet, the children of the future] are not going to know what nationalism is"; his colleague Michael Dertouzos has claimed that digital communications will bring "computer-aided peace ... which may help to stave off future flare ups of ethnic hatred and national breakups". All of this is reminiscent of earlier quotes: a Victorian enthusiast said "it is impossible that old prejudices and hostilities should longer exist, while such an instrument has been created for the exchange of thought between all the nations of the Earth". He spoke in 1858 and referred to the advent of the first transatlantic telegraph cable.

market to a huge degree and permitting differential pricing.

- Its ability to transfer data at very low cost.
- The ability to transfer costs to the users from the producers. Instead of paying for staff to deal with orders, package and dispatch material, etc., use of the Internet requires that users spend their own time on explicit selection from a menu of options and invest in the communication costs involved.
- The reduction in search time for those who do not live near to good libraries and the immediacy of obtaining the results (even if the quality of the source material is currently far lower than in traditional, peer-reviewed book form).
- The seamless way in which information may be obtained, i.e. it is not usually necessary in well-designed systems to have to write letters, send FAXs, get angry on the telephone, and cope with the information which is itself supplied in some other medium.
- If charging is implemented, it can be done through simple, low-cost operations like use of credit cards (as in Terraserver; see Section 19.3.2).
- Initial supply at least of the information may be tracked and conditions of use enforced on the initial user through his or her inescapable agreement to terms and conditions of use before the information is downloaded (as for software).
- The astonishing and rapidly growing familiarity of and use by many people of the Internet. This trades upon network externalities (see Section 17.3.2.2 and Figure 16.4).

The Internet facilitates improved marketing, supply, and formatting of GI.

In the GI case, the greatest of all benefits come from customized mass production – where subsets of the data are selected to meet particular needs – and from the "added value" which is potentially obtained by linking datasets together. In the latter case we create something new and get it almost for nothing. But the benefit is not because of the Internet *per se* but rather because of the digital form in which that information is held, together with its geographic identifiers. The Internet thus is a "simplifier" of problems rather than a necessary condition for extracting added value. More importantly, in a world where fast answers are expected, the Internet and World Wide Web may make the delays so short that they are for the first time tolerable for managers and users.

The implications of all this for GI may well be that the Internet encourages its wider use, irrespective of its source or charging level. The Internet does *not* differentiate between public and private sector information supply. In a world where both sectors make information available through the Internet, user preferences will be shaped by the quality and range of the products, the marketing, ease of access and use in combination with other products, technical support, brand image of the supplier, and perhaps price.

16.5 GIS as a Business and as a Business Stimulant

Thus far, we have treated GIS as a factor in achieving business success in commercial, government, or academic organizations. Just how big is the GIS business – indeed, how can we measure it? First let us consider the spending on the purchase of software, hardware needed to run it, data, and human resources (people). Truly comparable "supply-side" figures are hard to come by since published accounts of different firms often group activities in different ways reflecting their particular organizational structure. Not surprisingly, firms portray their figures in the best possible light. Figure 16.6 shows how global GIS revenues are said to have increased according to Daratech, a consultancy specializing in measuring market performance. These figures should be treated with some caution but they show that software sales alone have more than doubled in five years.

An alternative approach is to consider one company and break down its revenues and other indicators of success such as licenses. ESRI – perhaps the largest GIS company world-wide – is a difficult company to investigate since it is privately owned and does not have to make its accounts public. But the authors of this book have an inside track to some of its records! These show that the company has many more than three-quarters of a million licensees for its different products. It is estimated that this translates to about half a million active users at any one time. ESRI gross incomes have risen between 15% and 25% annually over the past decade, with the financial year 2000 gross revenues being about $400 million; if the revenues of the partly owned ESRI "franchises" in other countries are added in, the total revenue *comfortably* exceeds half a billion US dollars. Another indicator of growth has been the numbers attending their annual conference: Figure 16.7 shows how this has changed from 23 attendees in 1981 to about 10 000 in 2000 – in

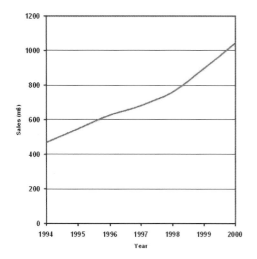

Figure 16.6 The global GIS software market ($m) in 2000, according to Daratech Inc. (Copyright: Daratech)

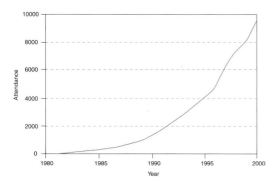

Figure 16.7 Attendance at the annual ESRI User conference since 1981 – a surrogate for interest in GIS as a whole (Source: ESRI)

many ways an allegory for GIS as a whole. The company also estimates that their activities leverage at least 15 times the revenues they get in terms of staff, hardware, training, and other expenditures on the part of the users. All of this suggests that ESRI alone generates expenditures of between $5 billion and $6 billion annually.

If the world population had grown at the same rate as attendance at ESRI conferences since 1981, the global population would now be 1890 billion people, instead of 6 billion!

Figure 16.8 shows the proportions of the total global GIS software market achieved by various different players – their market share – again according to Daratech. If we assume each has the same leverage factor as ESRI, the total expenditure

on GIS and related activities world-wide cannot be much less than $15–20 billion. Depending on how wide is the definition of GIS, it could be much higher still (e.g. if the finances of both commercial and military satellites are included). Given the number of ESRI licenses, those of other commercial firms, use of "free software" provided by governments and other parties plus the illegal, unlicensed copying and use of commercial software which still occurs, the likely number of active GIS professional users must be well over three million people world-wide. At least double that number of individuals will have had some direct experience of GIS and perhaps an order of magnitude more people (i.e. well over 10 million) will have heard about it and perhaps used passive Web services such as local mapping.

Other ways of quantifying activities are to consider the expenditures of public bodies, all of which – other than security agencies – are generally reported. In the European Union for example, the civilian national mapping agencies alone in total spend between $1 and $1.5 billion whilst land registration and other map and GI-using public sector bodies expend several times this amount. World-wide, this figure can be multiplied by perhaps 10 or more. The military expenditure in this area is very considerable: the US National Imagery and Mapping Agency (NIMA) has a publicly stated budget of about $1 billion. How much of all this is on GIS and GI is partly a matter of definition. But, from all this "demand side" information, we can get some approximate confirmation of the supply side estimates in the previous paragraphs.

How much other economic benefit this GIS activity underpins is not clear for few studies have been done and disentangling the interactions is very difficult. One such study by consultant

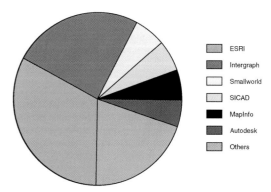

Figure 16.8 GIS software global market share of the major vendors in 2000 according to Daratech. (Copyright: Daratech)

economists has nevertheless suggested that the $150 million annual expenditure on updating the national mapping of Britain supports about one thousand times that sum in the Gross Value Added to the economy. In no sense is the connection absolute: other sources of information – or guesswork – might be used if national Ordnance Survey mapping were not available. Nevertheless, even if wrong (either way) by a factor of three this does demonstrate the scale of the GI/GIS business considered on a wide definition. We suspect that the contribution that GI (and GIS) makes across the world varies greatly but there are no figures to confirm this suspicion. We can however be confident that the GIS community is significant and is becoming more so every year.

16.6 Business Drivers

We have now considered the different sectors using and contributing to GIS, some of the inter-actions between them, the nature of our "raw material", and the changing nature of the economy thanks partly to new technology. It is now time to explore the things that force or encourage individuals and organizations – business drivers. This is essential if we are ever to manage them to successful outcomes.

Understand motives and you understand why many things happen.

Table 16.2 is a simple tabulation of the drivers (motives) that shape business decisions. It must be stressed that all aspects of this classification are simply illustrative. For example, the nature, extent, and abilities of the public sector vary

Figure 16.9 A categorization of how firms in the GIS arena have actually operated. Virtually no firm operates in all three areas of activity though some operate at different positions along any one of the three edges

widely between countries. In France, for example, government departments may own shares in private sector firms.

No one organization is present in all aspects of GIS, even if that term is restricted to cover only software and its application. Figure 16.9 is a categorization of how firms in the GIS arena – and other arenas – have actually operated. It turns out to be rather difficult to operate equally well in any two of the three vertex areas: it often causes internal prioritization problems and difficulties in relationships with business partners who see this as competition. The consequence is that any one individual will only see the whole picture if he or she has worked in different kinds of organizations.

16.7 Some Basics of Leadership and Management

By now, the reader should have come to conclude that GIS is non-trivial. Lots of people are involved, some are making money from it, some simply want to achieve organizational objectives, there are many interactions between the different players, large sums of money may hang on the out-turns, and the rate of change in the knowledge industries (of which GIS is part) is high and going higher. Hopefully this will have led to the conclusion that good management is essential. We now turn to what *is* good management and how this is manifested in the GIS world.

Management is an unfortunate term – for many, it implies a dull, routine job ensuring processes are followed and production targets are met. But today's manager is required to anticipate future oppor-tunities and possible disasters and take action appropriately. He or she has to keep up-to-date with changes in organizational aims and mores – and shape them if possible. Keeping up-to-date in all the factors that impinge on them is crucial. And they must be able to persuade their colleagues and staff to give of their best and targets must be achieved. Such pro-active management is more properly titled leadership.

Management is not "rocket science" – almost anyone can be competent at it with good training, enough practice, an understanding of "the big picture", and a modicum of understanding of people. But it is not a trivial pastime – far from it. Few people are good at it all the time. Getting good results normally requires unremitting effort, intelligence, and an ability to welcome criticism and adapt behavior appropriately. Arrogance and good management abilities are not often found together.

Table 16.2 Some business drivers and typical responses. Note that some of the responses are common to different drivers

Sector	Selected business drivers	Possible response	GIS example
Private (publicly quoted body)	Create bottom line profit and return part of it to shareholders. Build tangible and intangible assets of firm. Build brand awareness	Get first mover advantage; create or buy best possible products, hire best (and most aggressive?) staff. Invest as much as needed in good time; ensure effective marketing and "awareness raising" by any means possible; reduce internal cost base	Purchase and exploitation by Autodesk of MapGuide software and subsequent development – one of earliest web mapping tools. Engagement of ESRI in collaboration with educational sector since c.1980 – leading to 80%+ penetration of that market and most students then becoming ESRI software-literate
	Minimize risk	Set up risk management procedures, arrange partnerships of different skills with other firms, establish secret cartel with competitors (illegal)/gain *de facto* monopoly. Establish tracking of technology and of the legal and political environment. Beware of anything which may adversely affect the organization's "brand image" – one of the key business assets. Avoid partnerships with "fly by night operations", with no understanding of GIS/GI or with individuals with poor reputations. Know what is coming via network of industry, government, and academic contacts	Typically, GIS firms will partner with other information and communication technology organizations (often as the junior partner) to build, install, or operate major information technology (IT) systems. Many GIS software suppliers have partners who develop core software, build value-added software to sit "on top", act as system integrators, resellers, or consultants. Data creation and service companies often establish partnerships with like bodies in different countries to create pan-national seamless coverage
	Get more from existing assets	"Sweat assets", e.g. by finding new markets which can be met from existing data resources, re-organized if necessary, or by selling staff time as consultants	Target marketing: use data on existing customers to identify like-minded consumers and then target them using geodemographic information systems (see Section 2.3.3) Create new markets, e.g. target telecom services at existing Internet Service Provider (ISP) customers
	Create new business	Anticipate future trends and developments – and secure them	Go to GIS (e.g. ESRI) conferences and network. Monitor developments, e.g. via competitors' staff advertisements. Buy out start-ups with good ideas (e.g. ESRI and the Spatial Database Engine or SDE)

Table 16.2 Continued

Sector	Selected business drivers	Possible response	GIS example
Government	Justify actions to politicians so as to get funding from taxes	Identify why and to what extent proposed actions will impact on policy priorities of government and lobby as necessary for tax appropriations. Check that no other bodies in government are proposing similar initiative. Obtain political champions for proposed actions. Ensure politicians will become heroes if this succeeds	Creation of US National Spatial Data Infrastructure (NSDI: see Chapter 19) and equivalents in other countries. Impact of President Clinton's Executive Order 12906 Activities of Center for Geographic Information in North Carolina – funded initially on project-by-project basis Vice-President Gore's support for Digital Earth (Section 19.3.3.1)
	Provide good value for money (VfM) to taxpayer	Demonstrate that what is done is actually done effectively (meeting specification, on time, and within budget) and efficiently (via benchmarking of services provided against other organizations)	Constant reviews of VfM of British government bodies (including Ordnance Survey, Geological Survey, etc.) over 10 years and comparison with private sector providers of services. National performance reviews of government (including GIS users) in USA
	Respond to citizens' needs for information	Identify and foster this; set up/encourage delivery infrastructure; set laws which make some information availability mandatory	Setting up of US national GI clearinghouse under Executive Order 12906 (Section 19.2.3); equivalent developments elsewhere (Australia, Canada, Netherlands, UK). Prime Ministerial target in UK for 25% of all government services delivered online by 2003. In GI, response has been to make government information readily available via the Web and encourage feedback by e-mail.
	Act equitably and with propriety at all times	Ensure that all citizens and organizations (clients, customers, suppliers) are treated identically and that government processes are transparent, well-known, and are followed strictly	Treat all requests for information equally (unless a pricing policy is in operation and this permits dealing with high value transactions as a priority) Put suitable material (i.e. that which does not breach privacy guidelines) on Web but also ensure material is available in other form for citizens without access to the Internet
Academic	Establish high research reputation	Carry out, publish, and disseminate results of new research, advancing the field Get research grants from prestigious sources, run conferences attended by top class participants, form informal "club"	Focus on GIS/GI things that business does not do by way of research or cannot talk about for fear of losing competitive advantage, i.e. fundamental research (often technical), "soft, human-focused" research, or policy-relevant research

Continued

Table 16.2 Continued

Sector	Selected business drivers	Possible response	GIS example
Academic (cntd)		Win prizes	Set up research centers, get funding to attract graduate students and other researchers. Publish in top journals in the GIS-related field (Box 1.4), including some (but not too many) in journals outside of field. Talk to people in different disciplines to get new ideas. Attend many conferences; accept keynote speech invitations
	Establish reputation for teaching excellence	Provide top class educational experience for students (increasingly important). Alumni are the best form of advertising in the long term! Build international links to anticipate globalization of learning Differentiate courses from others in ways suited to the local market needs.	Recognize that GIS principles and technology are near-identical world-wide so: Set up consortium to create curriculum materials (e.g. NCGIA Core Curriculum – see Section 17.2.1) and/or deliver identical courses world-wide (e.g. UNIGIS model) and/or develop state-of-the-art courses which can be run on a residential (e.g. the University of London's M.Sc. in GISc) or distance learning basis (e.g. City University's MGI course) Pro-actively seek overseas students to build international base
	Obtain funding for research and teaching	External funding partly creates success and prestige in many science fields – less so in some social or human sciences. You need to out-shine your equivalents. Money from the best sources (peer-reviewed academic or science sources) normally gives more prestige than that from industry or government!	Get ideas from what others are doing/have done elsewhere and adapt and develop these for your national context. Conferences, research seminars are the most fruitful form of such personal development (notably the Spatial Data Handling conferences for technical work; vendor conferences are often helpful to keep up-to-date with commercial and applied research developments and help to identify lacunae in commercial enterprises). Face-to-face meetings are essential for dealing initially with key people. Build up and celebrate international network of research and teaching contacts, publishing with *simpatico* top quality people overseas

Management is not "rocket science" but neither is it easy.

16.7.1 Bear traps for GIS managers

Successful management can take many forms; there is no simple invariant guide to success. But there are some basic principles to be followed, as set out in any standard textbook. In addition, there are some other matters to be factored in, both generally and in GIS. These are:

● *People cause more problems than technology.*

● *Things change increasingly rapidly* – both the technology and the user expectations.

● *Complexity is normal but some of it is self-imposed* – and all organizations tend to believe they are unique and complicated.

● *Uncertainty is always with us.*

● *Inter-dependence is inescapable.* Creation and implementation of a GIS often has far wider ramifications than originally imagined and hence impacts on many more people, at least indirectly.

● *Users often have very imprecise ideas of what they need* and typically confuse such needs with their "wants".

● *Detailed specifications of needs are usually seen as necessary* to hold people (contractors, managers, etc.) to account. But these then complicate (and make expensive) changes to a project.

● *There are considerable differences in national or even regional cultures.* These usually have significant impacts in running projects in other countries. The globalization of GIS, often using terminology or facilities emanating from the USA, has partly disguised this effect but it is of fundamental importance.

16.7.2 What kind of management for what kind of GIS?

There is a down side to GIS' capacity to under-pin new services and create new products. Operational constraints ensure you generally have to take data "as they come" from other people (see Section 17.3.2). Their classification or level of accuracy may not be what you would like in an ideal world (see Chapter 5). There is also a risk in that the results may be provably wrong – and lead to humiliation or being sued (being proved wrong is much easier than being proved right in GIS, just as in most scientific endeavors: see Chapters 6 and 15 on uncertainty). Taking information from multiple

sources can lead to later conflict over the ownership of products if prior agreements have not been forged. And few GIS managers have avoided at least one project with cost over-runs: these can come about even in the best-run projects since many GIS analyses are non-trivial and may have to be repeated with different algorithms to test the stability of the results.

But running a GIS is even more complicated than that. As Campbell (1999) has pointed out forcefully, the idea that GIS can be implemented and run simply by following "cookbook" instructions and that success will then inevitably follow is quite wrong. It is contrary to all management experience in projects based on the implementation of high technology. There are many GIS projects which have failed because the managers followed such an approach. In practice, the managerial approach has to be sensitive to the institutional context, organizational objectives, and the existing staff. Campbell argues there are three different managerial perspectives relevant to IT in general and GIS in particular. These can be described as technological determinism, managerial rationalism, and social interactionism. The first two remain the norms in GIS implementation.

Successful GIS implementation and use requires more than just technical solutions.

Technological determinism is a Utopian approach which stresses the inherent technical merits of an innovation. The approach can be recognized in those GIS projects which are defined in terms of equipment and software. It is usually sold to potential users on the basis that "There Is No Alternative (TINA); this will do everything you need and has lots of intangible benefits because of the capacity of the technology". Relatively little emphasis is placed on the human and organizational aspects compared with fine-tuning the software and meeting the detailed technical specification.

Managerial rationalism assumes much of the importance of new IT systems but recognizes the importance of human and organizational factors. It stems from thinking in the 1940s and onwards which took as the starting point that "rational approaches" – involving defined rules and procedures – should be followed by management and staff. The role of users is typically to receive the benefits, although modest attempts to engage them in review of the system may be undertaken.

Finally, so far as Campbell (1999) is concerned, comes social interactionism. This starts with the recognition of uncertainty on how organizations really work and a belief that knowledge (see

Section 1.2) and culture within the institution have big impacts on the success of IT projects. In this view decision-making is an interactive, iterative, and often fast-changing process between individuals and groups – both within the organization and without – which are sometimes in conflict and sometimes in collaboration. Success comes from placing stress on the organizational and user acceptance and use of the technology, rather than on the intrinsic merits of the system.

We observe that many GIS projects are still driven by technical people seduced by the excitement of a technical tool of huge potential. In many cases, they seem to have no appreciation of their modest skills in management, trusting in technological determinism or managerial rationalism to carry them through.

16.8 Discussion

In this chapter we have seen that GIS is now nothing more or less than an increasingly useful part of management's weaponry. Since everyone is a manager at some stage, this is important. GIS is used to support competition with other players where that is the nature of the industry, i.e. it exists to obtain comparative advantage for one group as compared with others. On the other hand, it is used as a means of evaluating policy and practice options in both the public and private sectors. And it is used as an information sharing device where that is deemed desirable.

But good management is far from easy and general prescriptions need to be mapped into local circumstances. The business environment faced by different organizations manifestly differs on many counts. It varies by virtue of the level of operations (e.g. international, national, or local), sector (private/public – government of different levels/not for profit), by the level of competition (formal and informal – as occurs in many governments), by business culture (e.g. willingness to collaborate with other bodies), and other factors. Some environments are much more conducive to success than others: for instance, the constraints of working with poorly skilled staff, on out-of-date equipment, and to impossible deadlines are non-trivial. As always, it is the interaction of the various elements of the business environment and the incentive structures and operating constraints that pose the greatest challenges for management in general and for GIS in particular.

Questions for Further Study

1. Write down – without looking it up on your intranet or in paper documentation – the five or six key strategic aims of your organization. Are the most important drivers in the organization in accord with these aims?
2. For your own organization or selected part of it, draw up a (short) list of the functions it carries out in support of its strategic aims. What GI are collected to help the organization achieve these aims? How is that information converted into useful knowledge? Where are decisions actually taken? Who are the key people involved? What difference has GIS/GI made?
3. Assume that tomorrow you are to take over responsibility for five demoralized and underfunded staff in a GIS group in academia, business, or government. What would be your first actions? How do you decide whether someone is doing a good job and what do you do if they are not?
4. What are the special characteristics of information as a commodity (as compared, for example, with baked beans)?

Online Resources

Sites that enlarge on arguments made in this chapter:
www.ntia.doc.gov/ntiahome/speeches/ntca120198.htm
www.princes-trust.org.uk/whatsframes.htm
www.whereonearth.com/gijs/flash_index.htm

Reference Links

Longley P A, Goodchild M F, Maguire D J, and Rhind D W (eds) 1999 *Geographical Information Systems: Principles, techniques, management and applications*. New York: John Wiley.
Chapter 44 Institutional consequences of the use of GIS (Campbell H J)

References

Brown J S and Duguid P 2000 *The Social Life of Information*. Cambridge, MA: Harvard Business School Press.

Reeve D E and Petch J R 1999 *GIS Organizations and People: A socio-technic approach.* London: Taylor and Francis.

Shapiro C and Varian H R 1999 *Information Rules: A strategic guide to the network economy.* Cambridge, Mass.: Harvard Business School Press.

Sveiby K E 1997 *The New Organizational Wealth.* San Francisco: Berrett-Koehler.

Tenner E 1997 *Why Things Bite Back: Technology and the revenge of unintended consequences.* New York: Vintage Books.

G-BUSINESS: GIS ASSETS, CONSTRAINTS, RISKS, AND STRATEGIES

17

Recent years have seen a big increase in the value – however measured – of "g-business" in commerce, government, not-for-profit organizations and even academia. This has been built on the use of a range of business assets. Here we concentrate on two assets only – ownership of GI and the staff and their GIS skills. The first has been highly contentious and the second is becoming ever more important and may become formalized through professional-type accreditation.

In addition to exploiting assets, successful businesses of all kinds have to circumvent or live with constraints and manage the risks inherent in their operations. We summarize here common GIS-related constraints and risks and show how they illustrate the need for a GIS strategy and implementation plan.

Learning objectives

After reading this chapter, you will know about:

- The protection of innovations;

- The impact of the Internet on g-business;

- The availability, pricing, quality, and ownership of GI;

- Skills of existing GIS people as a constraint;

- The nature of existing GIS education and future needs;

- Legal constraints on GIS operations;

- Risk management and GIS strategy.

17.1 GI as a Business Asset and the Law

The law touches all of our activities and there is a geography to the law. Here we consider only the law in relation to ownership of intellectual property and how this varies between individual jurisdictions both across the world and even within one country (e.g. between States in the USA). We also indicate how the advent of the Internet has thrown some aspects of the law into turmoil.

The law touches all of our activities and there is a geography to the law.

Given this heterogeneity and turmoil, none of what follows should be taken as other than general discussion of the issues. In no sense does this chapter – or this book as a whole – offer legal advice. Our aim is to highlight the issues. The law is often a morass. It is vital that anyone with management responsibilities working in GIS seeks local, professional legal advice at an early stage wherever the law might be involved in a problem.

We started (Section 1.4.3) by noting that GIS embraced networks, hardware, software, data, people, and procedures (see also Section 8.1). In all of these, individuals or organizations may create something better than has been done before. In such circumstances, it is a normal reaction to consider how to achieve the greatest benefit from a given investment of time and other resources. The outcomes of this creativity may be made generally and freely available – either because that is "good science" or because wider availability will benefit the owners more. In many commercial operations, however, protection of one's own creative works (e.g. through licensing) is the very basis of operations. This is the essence of the commercial view of the Knowledge Economy (see Chapter 16). Governments have long recognized the need for protection of investment – witness the Berne Convention Treaty of 1886, subsequently modified on several occasions, and signed by most countries. Another example is the US government's attempts in the late 1990s to protect illegal copying of American popular music by Chinese factories.

Copyright is a universally agreed form of protecting investment, but copyright owners can opt not to enforce it.

At the other end of the spectrum, there is clearly a societal interest in ensuring that innovations diffuse widely, rather than being held to ransom by any unscrupulous monopolist. The law attempts to balance this right to exploit results of creativity and of investment against the need to foster society-wide gains. Most typically this is done by limiting the time of protection enjoyed by an innovation, such as for patents for new drugs or copyright on published material. A variety of means of legal protection exist, through legislation on copyright, designs, patents, trademarks, and trade secrets. Because of its nature, most of the material in GIS has been protected by copyright legislation but patenting of software and processes is becoming somewhat more common and threatens to change the world of GIS. Lawyers will certainly benefit from this increasingly litigious GIS world.

Patenting is a process impinging increasingly on the IT industry generally and GI in particular (Box 17.1). Patents have traditionally been given for a new manufactured good. Some have been awarded for software but a recent development has been their award for processes. The reality is that patenting is much more widely applicable than is commonly supposed and is possible for anything which is useful, novel, and non-obvious – including business methods. One early attempt to patent a GI-related process was that by employees of ICL, the one-time UK computer giant now owned by Fujitsu. These individuals set out in the early 1970s to patent cartographic digitizing through manual line following methods. They were forced to withdraw the application by their superiors. In mid-2000, Etak – a pioneer in car guidance systems – was awarded four patents relevant to GIS. These were on a method for storing map data in a database using space-filling curves, a system for tracking an object using an electronic map and a GPS receiver, for path-finding computation, and for caching for the path-finding computation. Another example of a GIS-related patent is shown in Box 17.1. Firms owning patented processes commonly seek royalties from others who use arguably identical approaches. In such cases, the choices are either to give in – avoiding legal costs but paying for something which may have been developed independently – or fight the case in court arguing independent development and prior use. The latter course ties up senior management, incurs legal costs, and may end in failure and awards of punitive damages.

Society has ordained that there must be a balance between the right to benefit from one's innovations and protection against long-term monopolies.

Box 17.1 Interactive automated mapping patent

Patent Number: 5,214,757
Name: Interactive Automated Mapping
Issuing office: United States Patent Office
Date of Patent: May 25,1993
Inventors: Thad Mauney, Aglaia C. F. Kong, Douglas B. Richardson, all of Yellowstone County, Mont.
Primary Examiner: Gary V. Harkcom
Attorney, Agent, or Firm: Dorsey & Whitney

Abstract: An automated, fully transportable mapping system utilizes position information gathered from a Global Positioning Satellite (GPS) capture program to create new maps or annotate existing maps contained in a Geographic Information System (GIS) database in real time. In addition, the present invention displays position information captured by GPS in real-time, enabling users to track the path on which they are traveling. Attributes related to the position information may also be entered in real time, and are stored in a file for subsequent inclusion in a GIS database.

The rest of the patent document describes the specific inventions claimed, the technical field, background art (context) and detailed description of the preferred embodiment.

17.1.1 The rise of the commercial sector

If this all seems remarkably focused on the private sector aspects of GIS, remember that things have changed rapidly. Two decades or less ago such commercial activities were very small in relation to governmental ones. By and large, the modest number of firms involved around the world were tiny in size and were restricted to creating software for sale. They operated alongside a do-it-yourself software world in which individuals or small groups in government created their own software and gave it away to anyone who asked for it.

Now, operational software is almost entirely created by commercial enterprises which have also branched out into service provision, consultancy, and data collection. The advent of desktop and Internet GIS has been made possible by the commercial sector. Data are often encapsulated with software to give that particular software a market advantage. In parallel, the creation of universal standards is being propelled through *de facto* ones used by the major software vendors. It is no exaggeration therefore to say that the commercial world is now probably the prime driver of the world-wide GIS industry (rather than GI Science or academia). Government's role is increasingly to regulate commerce, foster competition, and ensure that the core information infrastructure is available to all on equitable terms (though not necessarily free). Academia's role is that of creating suitably skilled and motivated people, to advance research in areas which commercial and government labs find unattractive – and challenge orthodoxy.

The commercial sector is now the prime driver of the GIS industry.

17.1.2 The government as a trader in information

Commercial imperatives are readily understood. But the practice of selling information is not restricted to commerce: many governments are also seeking to recover some of the heavy up-front costs of information (especially GI) collection. They typically do this through charging for use of an asset which has been funded over many years by the taxpayer and/or, in some cases, by users.

Much of the change has been brought about through governments world-wide reviewing their roles, responsibilities, and taxation policies and reforming their public services as a consequence. In many cases this is driven by a changed view of the role of the state (Osborne and Gaebler 1992, Foster and Plowden 1996). Fundamental reviews and subsequent reforms have taken place in New Zealand, Australia, the Nordic countries, the UK, the USA, and elsewhere. These have led to some dramatic changes in what government does and how it does it. In some countries the privatization and commercialization concepts of the 1990s are now being re-visited and different models, sometimes under the flag of "the Third Way", are being explored.

In certain cases, the imperative for these reviews and reforms has been financial. In Europe, the Maastricht Treaty forced the reduction of government deficits sharply as a prelude to the creation of a common currency. In others, the imperative was ideological – that the state should do less and the private sector should do more. The reform of government has led to two approaches in regard to GI. The first has been the outsourcing of production: thus for example the US Geological Survey agreed to get at least half of its mapping and imagery produced under contract by the private sector. The other has been simultaneously to force economies within government and get the user to pay for the bulk of GI. It is worth stressing that this is nothing new in principle: almost all government organizations have historically made *some* charge for data, including Federal Government bodies in the USA. The only issue is for what elements users should be charged, e.g. the whole product or only the costs of documentation and distribution. The use of the Internet has changed the situation significantly. It potentially facilitates the widening of access to information whilst reducing government's costs, rather than being a device for raising revenue.

The advantages and disadvantages of a "charging for everything" approach and of the "free access" one are summarized in Box 17.2 though these are subject to much argument. Charging has led to a reduction in wasteful expenditure, focusing of activities on high priority areas, and fostering of engagement with the private sector. In Britain, for instance, the ongoing costs of running the national mapping organization in the late 1990s were almost totally met from revenues raised from users. Moreover, the main elements of the database were guaranteed never to be more than six months out-of-date. Figure 17.1 shows how the proportion of funding arising from the users rose

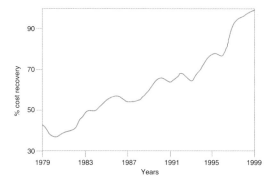

Figure 17.1 Cost recovery by Ordnance Survey, Britain's national mapping agency, by charging users for digital information, paper maps and services. From 2000 onwards OS has to operate profitably. (Source: OS Annual Reports)

over many years. That said, the policy also has some downsides, as suggested in Box 17.2.

17.1.3 The impact of the Internet

The Internet has had some significant but also some quite subtle impacts on the GI industry and on government (see Section 16.4.4). Here we examine only the impact of the Internet on GI as a business asset. However, since the industry is relatively small, immature, and low profile compared with many others, we use developments in the music industry as an analogy (see Box 17.3). This is not a perfect analogy since governments – one way or another – are a vital part of the supply chain in GI. Nevertheless the music industry illustrates what may come about and neatly illustrates the issues.

What are the implications of all this for GI? It is clear that, if material is sufficiently popular and has a high consumption value (see Section 16.4.2), Napster-like facilities should surely grow up. The only reason that this has not occurred in GI is because the market is not yet sufficiently large, because GI is still not a consumer "must have", and because the identical product is not required everywhere (the geography of Peoria or Santa Barbara is not as universally coveted as a Madonna record). But street mapping on the Internet has become widely available and used; it is typically funded through advertising on Web sites and keying sales outlets (e.g. of restaurants) to the maps and to the observer's own location, i.e. by adding value to the basic map. We can conclude that the lack of intrinsic value of the map is because of the commodification of that type of GI, i.e. there are competing multiple products, none of which is unique, and so prices fall. The Internet facilitates commodification once multiple products exist. In addition, topographic maps or the geographic framework more generally (see Section 17.3.2.1) are a necessary but not sufficient requirement for many users. Value adding (e.g. see Box 17.6) is crucial to generating significant revenues in GI.

All of the above assumes that markets for GI are homogeneous. That is not true and is becoming even less so. Different applications require GI datasets to be classified in different ways and held at different levels of spatial resolution – there is little point in working with national-level data when studying local problems. So we may very well see increasing market differentiation through the creation of parallel datasets – those for specific local, niche use (inevitably more expensive) and those of more global, mass use (and hence cheaper).

Box 17.2   The advantages and disadvantages of the "user pays" and "free access" philosophies (and some contrary views)

Arguments for cost recovery from users

- charging which reflects the cost of collecting, checking, and packaging data actually measures "real need", reduces market distortions, and forces organizations to establish their real priorities (*but not all organizations have equivalent purchasing power, e.g. utility companies can pay more than charities; differential pricing invites legal challenge*);
- users exert more pressure where they are paying for data and, as a consequence, data quality is usually higher and the products are more "fit for purpose";
- charging minimizes the problem of subsidy by taxpayers of users – about a 20 000 (taxpayers) to 1 (user) ratio globally according to Section 16.5 (*but some users are acting on behalf of the populace, such as local governments*);
- empirical evidence shows that governments are more prepared to part-fund data collection where users are prepared to contribute meaningful parts of the cost. Hence full data coverage and update is achieved more rapidly (*but once the principle is conceded, government usually seeks to raise the proportion recovered inexorably*);
- it minimizes frivolous or trivial requests, each consuming part of fixed resources (*but this rationale is now less relevant given the user-driven accesses to the Internet in some countries*);
- it enables government to reduce taxes;

Arguments for dissemination of data at zero or copying cost

- the data are already paid for, hence any new charge is a second charge on the taxpayer for the same goods (*but taxpayers are not the same as users – see above*);
- the cost of collecting revenue may be large in relation to the total gains (*not true in the private sector – why should it be so in the public sector?*);
- maximum value to the citizenry comes from widespread use of the data through intangible benefits or through taxes paid by private sector added-value organizations (*this contention is neither provable or disprovable because of the many factors involved. Holding it or the contrary view is probably ideological*);
- the citizen should have unfettered access to any information held by his/her government (*a matter of political philosophy*).

Box 17.3   Fraud or fair dealing on the Internet? The Napster case

In mid-2000, a legal furore grew over the activities of Napster, an online music distribution service. Napster instituted "peer to peer" copying capabilities, enabling individuals to copy commercially published music put up on the Web – 20 million of whom did so in the first year of operations. The music industry took the company to court. Napster's defense was that its users do not infringe copyright because they make only personal use of the information and use the information in ways that count as "fair use", such as sampling, under the law (see below). In any case, they argued, Napster itself should not be held liable for any copyright infringement – just as Internet service providers are not held responsible for illegal content so long as they do not officially know about it. The judge hearing the case disagreed, saying there is a big difference between making a copy for friends and making it available for millions to copy. The legal process grinds on but it seems likely that the original judge's decision will be upheld in the "final analysis".

But this does not mean that the "problem" will go away. Even after Napster was originally closed down, users switched in greatly increased numbers to a variety of other sites offering copy-cat services. This may be a latter-day parallel to the original fierce Hollywood opposition to video-recording of films. By change of its business model, Hollywood now earns more from videos than from the original film screenings. Napster may trigger the same response in the music world.

17.1.4 Plus ça change, plus c'est la même chose ...

None of this is new. Exactly the same problems caused by new technology and its impact on ownership rights have arisen at many different moments in history. Box 17.4 illustrates this perfectly.

The same problems may arise not from new technology but from changes of public policy. Such policy is made and re-made in the light of contemporary societal views and political fashions. There is no "right policy" in many fields of human endeavor. A prime example of this is given in Box 17.5.

Virtually all governments engage and have engaged in some form of cost recovery from GI products but the level of charging varies considerably. Even within one country, the policy and practice can vary between different public bodies. National information supply policies have also varied considerably over time.

17.1.5 The big information ownership questions

It is best to cover the practical aspects of GI ownership and protection in a problem-oriented way. Frequently asked questions and brief answers follow.

Can geographic data, information, evidence, and knowledge be regarded as property?

The answer to this is normally yes, at least under certain administrations. The situation is most obviously true in regard to data or information – either under copyright or the European Database Directive (see below). Who owns the data is sometimes difficult to define unequivocally, notably in personal data. An example is bills at a supermarket which have been paid by credit card

and subsequently geocoded to generate aggregate purchasing characteristics for people in a ZIPcode or postcode area.

Can geographic "facts" (c.f. representations or interpretations of them) properly be protected? What are geographic "facts"?

Copyright protection under law is a widespread, even ubiquitous, matter though its form differs significantly. Even in one country, for instance, it can vary: in the UK copyright is for 70 years from the death of the author but Crown (government) copyright extends only for 50 years from publication. There is little argument about copyright where it is manifestly based on great originality and creativity, e.g. in relation to a new popular song or painting. Rarely however is the situation as clear-cut, especially in the GIS world where some information is widely held to be in the public domain, some is regarded as "facts", and some underpins either the commercial viability or the public rationale of government organizations.

In the USA, the Supreme Court's ruling in the famous *Feist* case has been widely taken to mean that factual information gathered by "the sweat of the brow" – as opposed to original, creative activities – is not protectable by copyright law. Telephone numbers are widely regarded as facts and names and addresses as "geographic facts". Despite this, several jurisdictions in the USA and elsewhere have found ways to protect compilations of facts, provided that these demonstrate creativity and originality. The real argument is about which compilations are sufficiently original to merit copyright protection.

Outside of the USA, the "facts are not protectable" view is much less common. In the UK, for instance, the Copyright Designs and Patents Act of 1988 has been widely used to give copyright protection to compilations of (what may be regarded as) facts. The concept of Crown

---

Box 17.4   The Aeolian case

A new technology has set the music industry on its ear. Threats of suits and countersuits are exchanged like a litigious call-and-response. Congress holds hearings, and much official fretting ensues. From lawyers to musicians to listeners, everyone is asking: can copyright law cope with this innovation?

The year is 1909, not 2000, and the company is named Aeolian, not Napster. Aeolian, which manufactures rolls for a newfangled invention called the player piano, is roiling the music business. Composers consider its technology a threat to their vitality, and as various interests jockey to control the distribution and sale of music, Congress is called upon to bail the industry out. Today, nearly 100 years and a few million MP3 downloads later, Congress finds itself confronting a similar predicament.

Glenn Otis Brown 2000 Copyright goes old school. *Wired On-line*, 27 July

Box 17.5 Swings and roundabouts of public policy on GI

Following the policies of the British government of the day, Ordnance Survey took action to prevent "free riding" or illegal use of government's maps at least as early as 1817. They expressly warned anyone infringing their copyright that this would lead to legal action (see Figure 17.2). In the 1930s world of mapping, however, public policy changed to a "charge only for ink and paper" basis; in the 1960s it changed back again to charging users a fraction of the total costs of operations. In 1999 the move to a Trading Fund status effectively led to OS having to become profitable and thus meet full costs of updating the mapping and the interest on capital.

TRIGONOMETRICAL SURVEY OF GREAT BRITAIN

It having been represented to the Master-General and Principal Officers of His Majesty's Ordnance, that certain mapsellers and others have, through inadvertence or otherwise, copied, reduced, or incorporated into other works and published, parts of the "Trigonometrical Survey of Great Britain," a work executed under the immediate orders of the said Master-General and Board, the said Master-General and Board have thought proper to direct, that public notice be given to all mapsellers and others, cautioning them against copying, reducing, or incorporating into other works and publishing, all or any

part of the said "Trigonometrical Survey," or of the Ordnance maps which may have been or may be engraven therefrom.

    "Every offender after this notice given, will be "proceeded against according to the provisions of "the Act of Parliament made for the protection "of property of this kind."

By order of the Board,

R.H.Crew, Secretary

Office of Ordnance, 24th February 1817

The London Gazette, Saturday March 1, 1817, number 17225, page 498

Figure 17.2 The earliest known warning against infringing GI copyright: the warning printed by the Board of Ordnance in the *London Gazette* of 1 March 1817

---

copyright, held in all information produced by the UK central (equivalent to US Federal) government, has been ubiquitously used to protect that information. This has been to enable the accuracy, integrity, and official status to be protected, along with revenue generation where it is believed appropriate. Even if this copyright protection were to be challenged, all data creators in Europe now have a 10 year protection of their "sweat of the brow" efforts under the Database Directive of 1996 which has had to be incorporated into national law in all 15 European Union (EU) countries. Such protection may be extended if the database is substantially refreshed within the period (as would occur within databases of a "bundle" of road and related attribute data). Such geographic differences in the law have potential geographic consequences for commercial GI activity.

But what is a geographic "fact"? Onsrud – a distinguished surveyor and a lawyer – claimed in

his 1993 paper, as quoted by Cho (1998: 208), that: "Facts, algorithms, physical truths and ideas exist for everyone's use. It is difficult to argue that the outline of a building, the bounds of a land parcel, or a line of constant elevation on a map ... are expressions of originality. Any other person or sensor attempting to represent these facts would have little choice but to do so in much the same way". This of course completely neglects the cartographer's art and, in particular, generalization and multiple geographic representations (see Chapter 7).

The Onsrud interpretation is untenable. In the first instance, representations of "facts" can incorporate considerable originality, whether cartographic, photographic, or otherwise. In the second instance, many geographic features are fuzzy in some measure: ask different people to draw their interpretation of the Appalachians or the Mid-West and you will get different answers. In the third place, if it were true, the entire business case for commercial high resolution satellites – at least in so far as they are funded from and the business is based within the USA – is directly challenged (see below).

Given that governments still generate most geographic information, should and can this be protected?

This is a matter of huge debate, both between and within countries. For instance, the World Meteorological Organization consists of representatives of governments from around the world. It was disrupted in 1995-6 by an internal argument about which information shared between its members was in the public domain and which was exploitable commercially. Crudely expressed, the USA said all such information should be available in the public domain and would cut off provision of information from their satellites if the rest of the world did not fall in line. The Europeans argued otherwise. A compromise resulted whereby "core data" were put in the public domain and "added value" data derived from some additional modeling was deemed commercially exploitable under licenses.

Is information collected directly by machine, such as a satellite sensor, protectable?

In the Onsrud world (see above), automated sensing should not be protectable by copyright because it contains no originality or creativity. For organizations which have invested hundreds of millions of dollars in setting up the data collection and distribution systems and forged partnerships with other businesses on the basis of likely shared profits, this would not be good news!

Clearly this interpretation of copyright law is not the view taken by the major players, such as Space Imaging Inc. (see Section 19.3.2), since they include copyright claims for their material and demand acceptance of this before selling products to firms or individuals. But they may have fewer problems than many other GI suppliers. This is because there are at present only a few suppliers of satellite imagery. The cost of getting into the business is high; thus any re-publication or re-sale of an image could be tracked back to source relatively easily. In addition, there is some evidence that some of the key customers (notably the military) are interested primarily in near-real time results. Ten year old imagery is only of limited interest to a few customers, chiefly for change detection (see below).

In any event, imagery up to 10 years old would be protectable under the EU's Database Directive, at least within the European Union area itself. Whether this would be enforceable elsewhere in the world will depend upon emerging legal regimes to control selling over the Internet.

What is the half-life of geographic information i.e. over what period does it lose half its value?

"Half-life" – analogous to the rate of decay of radioactive isotopes – is an apparently attractive concept. If shown to be meaningful, it could help with investment decisions, decisions on whether to defend court actions, and much else. For example, if GI loses half its market value in, say, one year, seeking to protect it through copyright enforcement for the standard 50 or 70 years may well be absurd. "Transiency of use profiles" have been used to set charging or access controls in the past by the US Federal Government in relation to Landsat data.

Unfortunately, the half-life varies greatly, both by information type and by application. Some GI are, for most purposes, much more ephemeral than others. Thus the bulk of the value of meteorological data is much more transient than that of geology. The moral for information providers – if they are only information providers – is that they should build transiency into their data wherever they can persuade customers to accept it and impose a leasing arrangement, rather than make one-off sales. Curiously, however, there is increasing evidence that data of supposed short commercial "half-life" may well have longer term commercial value. One British firm seems to have found a valuable market for historical information on sites which might possibly be contaminated through previous industrial land uses (Box 17.6). Moreover, cyclic transiency is becoming increasingly important: as one example, information derived from Censuses of Population declines in value as it ages in the 10 years between censuses. Yet its value rises again at the next Census since it forms the benchmark by which change is computed. In practice, then, the value of the same information differs hugely to different people and for different applications.

How can tacit geographic and process knowledge – such as that held in the heads of employees and gained by experience – be protected?

Exploiting tacit knowledge is becoming a serious business problem from at least two possible points of view. The first is how to value an organization and its employees (Section 16.3). Traditionally this has been done in terms of its profitability or its turnover. Such approaches worked well for much manufacturing industry and labor-intensive service organizations (such as banks until a few years ago). Yet if it makes nothing and employs only a few bright people who one day might have an idea which changes the world, how is that organization – such as a GIS vendor or service provider – to be valued?

The second reason for the importance of this topic is that "know how" is a foot-loose element. Suppose you have created a brilliant team which has created a wholly new form of GIS, far superior to the existing systems which have normally to cope with the legacies of their antecedents. Suppose further that the lead designer is seduced away by a large software vendor. Though no code is actually transported, the essence of the software and the lessons learned in creating the first version would greatly facilitate the creation of a competitor. This situation has happened in GIS/GI organizations from the earliest times. Such dilemmas are not restricted to the private sector. Losing key staff is always a major problem.

So what if anything can be done of this kind? The only formal way to protect tacit knowledge is to write some appropriate obligations into the contract of all members of staff – and be

prepared to sue the individual if he or she abrogates that agreement. In practice this is not always straightforward for other laws like the European Union Human Rights Directive place constraints on what "restraints to trade" may be imposed, especially if the power of the parties making the contract are judged unequal. Keeping your staff busy and happy is a better solution!

How can you prove theft of your data or information?
It is now quite common to take other parties to court for alleged theft of your information. One of the authors previously led an organization which successfully sued the Automobile Association (AA) in the UK for illegal copying, amending, and reproducing (in millions of copies) portions of Ordnance Survey (OS) maps. This was despite the terms, conditions of use, and charges for such

---

Box 17.6   A value-adding GI organization

A prime example of the way in which GIS and GI can be used to provide a valuable service and generate business is the case of Landmark Information Group. This claims to be the UK's leading supplier of digital mapping, property, and environmental risk information. The organization has forged data supply agreements and partnerships with a number of government departments and Agencies plus statutory and non-statutory bodies.

Landmark's business is founded on two principles. The first is that all property-related investment decisions are affected by potential environmental risks and liabilities. The second is that combining and consolidating environmental data from a wide variety of sources not only improves the availability of data and makes them more convenient to access, it also creates new types of information services which could provide new solutions to old problems. In other words, it creates added value. On this basis, a wide range of nationally complete information has been assembled from different sources and georeferenced so it fits together. These include:

- 230 000 current and 500 000 historical Ordnance Survey (OS) digital maps dating back to the 1850s, all in computer form, together with a record including address and coordinates for every address in the country
- A century-long set of Trade Directory Entries and Business Census data indicating sites with potentially contaminative uses, plus other such sites identified by analysis of the OS maps
- Areas at risk of coastal or riverine flooding
- Natural hazard areas (e.g. those liable to coal mining subsidence, radon-affected, and groundwater vulnerability areas)
- Planning consents and site information (e.g. discharge consents into streams and rivers, hazardous substance consents; registered landfill sites; registered waste transfer, treatment, or disposal sites; enforcement and prohibition notices; and prosecutions brought against organizations under various environmental regulations)
- Air Pollution Control records, i.e. authorizations

From all this information integrated within their GIS, Landmark offers a variety of services tailored to different markets – to environmental professionals, real estate professionals, property lawyers, local governments, utility companies, and individual house owners. The business has grown from its origins in 1995 to a $20 million turnover business in 2000. A sample of one part of the standard report for house owners on the potential environmental hazards surrounding their home is shown in Figure 17.3; the full report is produced for $50.

Box 17.6 Continued

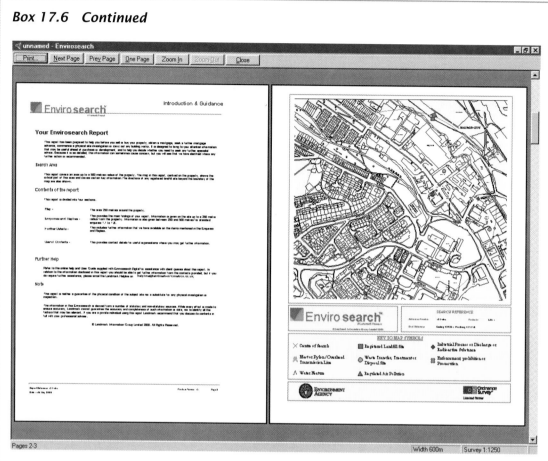

Figure 17.3 Excerpt from a Landmark report showing possible sources of environmental contamination around a specified house, based on the collation of information from different sources. (Copyright: Landmark Information Group and the Crown)

mapping being published and well-known to the AA: the case was finally concluded in 2001, when OS won $30 million.

The legal process in such circumstances is an adversarial one. It is therefore crucial to have good evidence to substantiate one's case and to have good advice on the laws as they exist and have been previously interpreted by courts. How then do you prove that some GI is yours and how do you demonstrate that any differences which actually exist do not demonstrate a different provenance? This is all in the context where digital technology makes it easy to copy or scan GI, disguise its "look and feel", and reproduce products to very different specifications, perhaps generalized.

The solution is both pro-active and reactive. In

the first instance, the data may be watermarked in both obvious and non-visible ways. Thus a crest in 5% gray – like a watermark in paper – embedded in the data file can be removed by duplicitous individuals but serves as some level of warning (and as advertising where it is not removed). A series of groups of small numbers of colored pixels scattered apparently randomly throughout a raster dataset ("salting") can be effective. Proof acceptable to courts requires a good audit trail within one's own organization to demonstrate that these particular pixels were established by management action for that purpose.

In addition to watermarking, finger-printing can also be used. For example, at least one major US commercial mapping organization adds occasional fictitious roads to its road maps for this purpose; fictitious names are included in address lists for the same reason. In areas subject to temporal change (like tides or vegetation change), it is very unlikely that any two surveys or uses of different aerial photographs could ever have produced the same result in detail. The final fingerprint category is that of the in-house mistake: if this is not catastrophic, it may be managerially wise to nurture it for legal support at some future date when it turns up in a rival's map or data file! When all this is added to the idiosyncratic hand of a cartographer it should, in the right hands, be enough to demonstrate a good case though rather different techniques may be called for in regard to imagery-based datasets of slow-changing areas.

Who owns information derived by adding new material to source information produced by another party?

Suppose you find that you have obtained a dataset which is useful but which, by the addition of an extra element, becomes immensely more valuable or useful. This is common: in effect, you get something for nothing by adding geographic information together. The number of data combinations rises very rapidly as you add a few more "layers" – two layers gives one combination, 20 layers gives 190 pairs and over 1 million combinations in total. The more combinations, the wider the range of problems which may potentially be tackled.

> The first law of GIS says you get something for nothing by bringing together GI from different sources and using it in combination.

So who owns and can exploit the results in derivative works? The answer is both you and the originator of the first dataset. Without his or her input, yours would not have been possible. This is true even if the original data are only implicit – if for instance you trace off only a few key features and dump all the rest of the original. Moreover, you may be in deep trouble if you combine your data with the original data and omit to ask permission to do so of the originator. In countries, typically in mainland Europe, where "moral rights" exist the author has the right to integrity (i.e. to object to any distortion or modification of the work which might be prejudicial to his or her honor). He or she also has the right to attribution as the author, to decide when a work will be published, and the right to withdraw a work from publication.

How should I price GI for sale or licensing?

This is a big topic, covered for information in general by Shapiro and Varian (1999). As we have stressed earlier, GI is more complicated than most information partly because of the typical government role as initial supplier and sometimes sharer of the revenues. In general however, those seeking to benefit from owning such intellectual property should recall that information is costly to produce but cheap to reproduce; that once the first copy of an information good has been produced, most costs are sunk and cannot be recovered; that multiple copies can be produced at roughly constant and near-zero unit costs; and that there is no natural limit to the number of additional copies which can be made. This all suggests that there are only three sustainable structures for an information market. The first is the dominant firm model (e.g. Microsoft). By virtue of its size and scale economies it enjoys a cost advantage over its rivals. The second is the differentiated product market with a number of firms producing the same kind of information but in different varieties. This is easy in GI where spatial resolution and generalization may be tuned. It is also the most common structure in the information goods market in general. The third structure is where governments are involved in the market and have some control of it due to a natural monopoly or statutory protection.

Given all this but ignoring the last special case above, the information owner should:

● Price by perceived value, rather than cost of production.

● Achieve cost leadership through economies of scale and scope.

● Differentiate the products, exploiting the infinitely re-useable digital nature of the data, i.e. create multiple versions of the same product differing in accuracy, appearance, and currency and create different products by linking different datasets together and extracting "new" information goods. Do not let your information become a commodity, i.e. where everyone is selling the same good. In that case, fierce price competition will decide who wins.

● Personalize the pricing as well as the product (this has the incidental advantage for governments selling information goods that it enables them to avoid legal traps for equity of treatment for all customers, irrespective of whether they are rich utilities or poor citizens). One way to achieve this is to group prices (e.g. for all local governments) via some kind of sector-wide deal or Service Level Agreement, thereby permitting bulk discounts quite legally.

● Seize first mover advantages (i.e. get into the market first and establish brand image, customer base, etc.), but also be in continuous revolution in terms of exploiting new technology and tools. For the market leader in a commodity-like product area, Shapiro and Varian (1999) offer the advice to avoid greed. For example "limit pricing" sets prices just low enough to discourage competitors from investing to join the market since they will never make a suitable return on their employed capital. They also urge the business manager to "play tough" – a new potential competitor should know that you are willing to lower your prices aggressively if necessary.

● Understand your many markets and their current profitability by a variety of means and use promotions to measure demand, e.g. through regional pricing. And anticipate what is likely to change this situation!

17.1.6 Conclusions on GI as a business asset

GI as an asset is a tricky, immature thing. To make a business out of it requires ingenuity and careful targeting of niches. A major problem is that GI is still seen as a low value good by many people used to getting it free from governments. It helps your business enormously if you can add value to the raw datasets provided by government or others (see Box 17.6).

GI is a tricky asset. Adding value is the best way of becoming a successful niche player.

17.2 GIS People as a Business Asset

In many businesses, especially those forming part of the Knowledge Economy (Section 16.3), staff are usually seen as the prime business asset and the factor differentiating one organization from another. The supply of GIS experts and GIS-literate people is also a crucial factor in determining the rapidity of uptake of the existing facilities and their successful use. The nurturing and management of such scarce and high value individuals is therefore an important matter for enlightened self-interest on the part of all organizations. Setting direction, providing encouragement, incentivization and reward, fostering personal development, re-positioning staff to suit changing circumstances, reprimanding and firing them is what managers exist to do, both inside and outside of the GIS community.

What are the skills and other characteristics of those working in or with GIS? We actually know very little about the three million or so people using GIS world-wide at present. Personal experience indicates that many first generation GIS people were trained in a big range of initial disciplines and came to GIS through a job requirement: as a result, excellent cross-fertilization of ideas and viewpoints occurred. But what has been happening since GIS took root widely within educational institutions in the 1990s?

One of the few surveys carried out of GIS staff was that by *Spatial News* in mid-2000. In response to a Web-based questionnaire, some 645 US individuals returned answers to 13 questions. The answers to this survey are probably misleading since in no sense are the respondents a statistically representative sample. That said, there are enough interesting features in the results to make them worth reporting (but suspect). The overall profile of the working GIS professionals who responded was as follows:

The average period the GIS professional has spent in the industry is about 4 years though a few have been around for over 20. Just under 2 out of 5 of these professionals work for government and just over half work in the private sector, largely in privately owned firms (rather than publicly listed corporations). Half of all these people work in enterprises with more than 100 staff; 15% work with less than 10 colleagues. Eight out of 10 GIS professionals who responded have a degree and a third of them have a Masters. Their average salary was about $44,000 though a few were earning over $75,000. Nearly 6 out of 10 thought they were underpaid. Only 22% thought they were definitely keeping pace with the "cutting edge technology" and three out of 10 knew they were falling behind significantly. Nearly half of the respondents saw "GIS technical and analytical skills" as the most desired skills in the GIS marketplace whilst another quarter saw "GIS programming" skills in this way. A similar number thought they would not be working for the same employer in 5 years time.

In summary, most people in the survey think of themselves as skilled technical people but in constant need of updating. Revealingly, few of the questions touched upon anything beyond these issues, i.e. the results were pre-determined by the questions. The historical origins of GIS as a technical tool looking for problems has led to many GIS-trained employees being highly skilled technically but unable to deal with ease with business issues or even with the user community outside of that one in which they were first employed. This situation calls for improvement.

> Most GIS industry people think of themselves as technical experts. This over-strong concentration is not in our long-term interests.

17.2.1 Existing GIS education and training

There are at least three reasons why we would expect global similarities in GIS curricula and educational approach. The first of these is the pervasive influence of the software vendors whose systems have harmonized GIS terminology and concepts. The second is the effect of the GIS Core Curriculum created by a team led from the University of California Santa Barbara in 1990. This was made cheaply available world-wide and has been used in well over 100 universities; it has set a norm for ways of thinking about and delivering GIS education. Finally, there have been a number of subsequent attempts to create international consortia delivering GIS education material in different countries. The most notable of these is perhaps the UNIGIS Consortium, with nodes in 14 countries.

Increasingly, however, Web-based material is enabling many universities to deliver courses to individuals irrespective of where they are physically located (though the value of face-to-face meetings is increasingly recognized as being necessary even for a distance learning course). We recognized in the Preface to this book that GIS education as traditionally practiced anticipated the widespread acceptance of "Mode 2" approaches to learning. Thus learning in GIS is typically achieved in substantial measure by "doing", rather than by reproducing "facts" given in lectures. That "doing" is best guided or supported by a tutor or mentor. Such tutoring can be provided in a rote-learning situation by a software wizard. More complex educational issues, where judgment is concerned, require a human being to act in this capacity.

Forer and Unwin (1999) have summarized much of what had happened in the past then debated the "proper" nature of GIS education. At least 2000 courses are now run annually world-wide in which GIS is a major element (Figure 17.4). We can summarize the types of courses as shown in Table 17.1 below.

17.2.2 Continuing Professional Development (CPD)

CPD has been defined by the UK national GIS "umbrella body", the Association for Geographic

Figure 17.4 Students graduate with a degree in GIS – prepared for the needs of the "real world"? (Courtesy: Richard Bailey)

Information (AGI), as "the systematic maintenance, improvement and broadening of knowledge and skill, and the development of personal qualities necessary for the professional and technical duties throughout an individual's working life". Of necessity, it is continuous (given the rate of change in working life), focused on professional or organizational matters, broad-based, and structured in relation to an individual's own development plan and the needs of his or her organization.

Successful CPD in GIS requires a thorough understanding of market needs. Basic technical training in GIS may be similar world-wide. But the requirements of different groups in employment will differ hugely. In addition, the previous experience of those in each group is much more heterogeneous than those at school or university undergraduate level.

There is obvious need for accreditation of such GIS-related courses and two approaches exist – to do it through existing professional or statutory bodies (as for accountants, surveyors, or those in

Table 17.1 A simple summary of GIS/GI-related training and education courses. The distinction drawn between the two delivery modes is of course less clear-cut in practice

	Delivery mode	
	Face-to-face courses	"Distance learning"
Software training courses	Commonplace – all vendors provide this themselves or via licensed partners. Can be very remunerative	Becoming ever more important and valuable, e.g. ESRI Virtual Campus
Software development and customization	Varies greatly. Fundamental training is provided by university Computer Science courses. Vendors may provide specific courses, e.g. Smallworld's courses on Magik, their macro language	
School-level education	GIS was formerly part of the UK government's Core Curriculum for schools (it was deleted when time pressures occurred but mapping remains). Elsewhere the school curriculum is typically defined locally	
Undergraduate level education	GIS is taught in many hundreds of undergraduate programs around the world but in many instances as modules in degree courses in Geography and non-geography courses (e.g. forestry or environmental sciences). Only a few universities run undergraduate courses entirely in GIS (e.g. Kingston University, UK)	The UNIGIS consortium has set up a distance learning course with a common "pick and mix" curriculum now used by universities around the world. This benefits from some face-to-face teaching
Postgraduate level university education	Numerous Masters programs exist (e.g. the long-established course at University College London).	Some of these programs can be taken in a distance mode but a few have been designed from the outset for that, e.g. those at City University, London and Pennsylvania State University, USA
Short courses for professionals (i.e. Continuous Professional Development or CPD)	This is a matter of rapidly growing importance and is covered separately	An increasing number of these courses are given by distance-based means

the insurance community) or to set up a scheme administered centrally by individuals or bodies experienced in GIS and GI. Both have obvious advantages and drawbacks. In the UK, a pioneering CPD scheme has been launched by the AGI. The principles are that this is voluntary (at present), low in cost, requires commitment from the individual to undertake at least 20 units of professional development in each of three years of the cycle, and a record of the activities is monitored. The model is thus one of a modest, personally managed program of professional development. This is to be achieved by participation in a wide range of activities, including "on the job training", attendance at formal courses, seminars, conferences, and workshops, writing papers, giving presentations, and acquiring other key skills (such as in business studies, management training, and language skills). Under the AGI scheme, the courses or materials required can be produced by any body which chooses to create them; no authentication or formal accreditation of these is presently carried out though some quality assurance mechanism is clearly desirable in the medium term.

17.2.3 Professional accreditation in GIS

There is an attraction in having properly certified or accredited learning or training. This is most successful where safety-critical activities are involved, e.g. in Engineering. Globalization is leading to globally acceptable certification. At present, however, there is no known professional certification of GIS/GI courses in a way analogous to those in other professions. But there are areas of GIS where professional skills are being exercised, where its exploitation is not simply a matter of pressing buttons, and where the results may impact strongly on people's lives. In such circumstances, the whole question of judgment, quality of work, and liability (see Section 17.3.1) comes into play.

Attempts have been made both at the national and international levels to introduce such certification. Within the USA, there are a number of bodies actively considering such professional certification. Certification committees have been formed by at least five national societies or state organizations. The need for such certification remains controversial: a recent URISA President proclaimed herself in favor of the definition and adoption of a code of ethics by GIS practitioners but not professional certification – as recommended by her colleagues.

At the international level there has been parallel activity. Perhaps surprisingly, this has come through the International Standards Organization. The role of ISO is to develop voluntary international standards covering all technical fields. It works by application of a strict formal procedure which has been developed to ensure that approved standards have consensus from all national bodies involved. One of ISO's 3000 technical committees (TC), number 211- Geographic Information/Geomatics, is concerned with standardization in the field of digital geographic information (see Section 10.4.1). The Canadian representatives on ISO/TC211 put forward a proposal in early 1998 for a new work item on "Geographic Information Science/Geomatics – Qualifications and Certification of Personnel". The initial motivation for this proposal arose from concerns of foreign aid agencies who were having difficulties assessing the qualifications of consultants seeking funding for work to be carried out in less developed countries.

The Canadian proposal, known as TC211 document N573, laid out an ambitious scope of work which included:

- Developing a system and plan for the qualification and certification, by a central independent body, of candidate institutions, programs, and personnel in the field of Geographic Information Science or Geomatics.

- Defining the boundaries between GIScience/Geomatics and other related disciplines and professions.

- Specifying the technologies and tasks pertaining to GIScience/Geomatics.

- Establishing skill sets and competency levels for technologists, professional staff, and management in the field.

Nine countries, including the USA, voted against the N573 proposal on the grounds that it is inappropriate for a technical standards organization to determine professional credentials. Nevertheless, the proposal was approved and the new work item introduced into the ISO/TC 211 program. At the time of writing it is impossible to predict the outcome of this enterprise. It may come to nothing or it may sway the entire professional accreditation of GIS people world-wide.

17.2.4 Fixing the shortcomings of current GIS education and learning

The "survey" of GIS staff characteristics quoted earlier plus other evidence suggests that a technical fixation is still current in GIS. This creates a self-perpetuating role for GIS as a pursuit for those "with clever fingers": the ISO/TC211 proposal confirms this belief. It is no surprise therefore that relatively few GIS people make it to the higher levels in major organizations (Box 17.7 presents an exception). If we wish to have GIS skills and approaches more deeply involved in high level decision-making, we need to modify the nature of our education and learning.

The reality – and one both espoused and noted throughout this book – is that GIS is now a much "broader church" than simply a collection of technical expertise. There appears to be a significant mis-match between the long-term advantage and the nature and volume of present education and training facilities for GIS. In a world where there are far more users and recipients of GIS work than software specialists, most existing GIS courses are too techno-centric and consistent world-wide. In particular, they neglect the business issues in Chapters 16 to 19 of this book.

Clearly not all courses and learning need to address all these issues. Some will be particularly relevant to CPD or other groups. The introduction of locally related legal, cultural, and application-related elements – as well as buttressing the global technical, business, and management issues – will, however, be to the benefit of GIS practitioners, the discipline, and business (used in our wide sense) alike.

17.3 Constraints on GIS Use

There are many detailed constraints on the use of GIS. These include software capability; the lack of

Box 17.7 Karen Siderelis: Chief Geographic Information Officer

Karen Siderelis (Figure 17.5) is Chief Geographic Information Officer with the US Geological Survey. In this newly established position, she formulates and implements plans, policies, and strategies to manage the scientific information and information infrastructure of the USGS. Key aspects of Karen's job are maintaining strong working relationships with Federal, State, local, and private sector organizations and providing leadership in the coordination of scientific information.

The US Geological Survey is an organization of roughly 10 000 employees housed in about 200 offices throughout the United States. Its annual budget is approximately $1 billion. It is the nation's primary provider of knowledge and information about issues in the natural sciences and is steeped in a tradition of discipline-based applied and basic science including geology, hydrology, geography, and biology. USGS also has been the Federal Government's primary agency responsible for mapping the nation's surface and subsurface features. It is evolving toward an integrated natural science and information agency focused on understanding and solving complex societal problems. GIS and related technologies have been and will continue to be crucial to the success of the Survey.

From 1977 until 2000 Karen Siderelis was employed by the North Carolina Center for Geographic Information and Analysis (CGIA) and served as Director since 1981. CGIA is an agency in the Office of the Governor that is responsible for statewide coordination of geographic information and for operating a GIS service Center.

Figure 17.5 Karen Siderelis, Chief Geographic Information Officer

Karen's key contributions over the last 20 years have been in the area of spatial data infrastructure – creating collaborative, multi-participant environments for sharing and using geographic information. Through her work at the North Carolina CGIA, she led the development of one the most advanced Statewide spatial data infrastructures in the United States and has been an active advocate for the National Spatial Data Infrastructure (NSDI: see Section 19.2). In her new role at USGS, she serves as a champion for the NSDI, helps integrate scientific information with other spatial data, and encourages scientific rigor in the management of spatial data and technology.

When asked what has pleased her most about the development of GIS over the last 20 years, Karen says: "I find that GIS increasingly is being *relied upon* to ensure the safety and health of the public – in natural disasters, crime, pollution, disease, etc. Rarely a week goes by that I am not alerted to a new and innovative application of GIS that is related to public safety. However, we are not advancing quite as quickly in building spatial data infrastructure to support and to integrate these important applications. I expect the infrastructure to catch up with the applications in the near future!"

necessary human skills; an inadequate model of the organization, poor understanding of its business needs, and lack of specification of the desired outcomes; plus a lack of will or ability to partner with other bodies working in related areas (see Chapter 19); and poor leadership and management. These are covered in different sections of this book. Here we specifically consider the law and the availability of GI itself as major constraints on GIS operations. The impact of each

will vary depending on where you live or operate and through time – for many of these constraints are mutating constantly.

17.3.1 The law and GIS

Central to many aspects of use of GIS is the legal framework under which it is operated (see Box 17.8). One aspect of it – the ownership and

Box 17.8   Some legal factors that managers using GIS have to think about

During our careers in GIS, we can expect to have to deal with at least some of the following manifestations of the law:

● human rights laws
● fair trading (or anti-trust) laws
● copyright and other intellectual property rights (IPR) laws
● data protection laws
● public access laws such as freedom of information legislation
● public procurement laws or regulations
● legal liability laws
● employment law
● health and safety law
● statutory or other authority for public agencies to involve in trading or related activities

exploitation of geographic information and intellectual property rights (IPR) – has already been treated separately above.

The law touches everything. Be prepared.

There is a geography of the law. The legal framework varies in significant respects from country to country. For example, the Swedish tradition of open records on land ownership and many personal records goes back to the 17th Century but much less open frameworks exist even in other countries of Europe. Box 17.9 summarizes the situation and the way in which commerce has to adjust to the legal framework of each country in which it operates, though increasingly, supra-national bodies like the World Trade Organization and the European Union force

Box 17.9   Globalization of commerce and the law

"Commerce is global. Law, for the most part, is not ... never has that truth caused greater uncertainty than now, as global business clashes with local and national law in the borderless new world of electronic commerce ... all the uncertainty is good for lawyers but not for the growth of e-commerce... The threshold question is: whose laws apply? When an American woman slips and falls in a hotel in Italy, can she sue in New Jersey, just because the hotel advertises globally on the Internet? When a Christmas tree catches fire in an American home and causes a fatality, can the store where it was purchased sue the Hong Kong manufacturer in a US court just because the HK company has a presence in cyberspace?

In both cases, the US courts said 'no'. But the question of jurisdiction remains unresolved, both internationally and within the US: what consumer laws, contract law, privacy laws and other laws apply to e-commerce applications? Where does a transaction take place? How will conflicts in law be determined?

US courts have begun to sketch out a tentative standard for jurisdiction, determined by the degree of inter-activity at the web site, and especially by whether a sale took place ... if the company sells goods directly to a Michigan resident through the web, then his home state's laws would probably apply. The question of foreign jurisdiction is even murkier. According to a draft EU Electronic Commerce Directive, the law of the consumer's country would apply in any case where a purchase was made through a web site ... In the end, it comes back to the same question: how can business be borderless when the law is not? No one yet has the answers."
Financial Times, 23 December 1999

greater homogeneity in certain laws and regulations. Such matters affect all commercial transactions between organizations in different countries and may affect governments through trade and other activities. Despite all that, other matters are still national in scope – especially the creation and maintenance of "official" (government-produced) geographic information.

First let us deal with those items of the law where GIS is not unique in any way. Human rights laws set out the framework of citizens' rights within the state. In some cases, such as the European Convention on Human Rights, these frameworks have implications for GIS through the treatment of privacy in regard to data protection (see below and Box 17.10). Fair trading (or anti-trust) laws have only affected us indirectly via anti-trust actions (e.g. against Microsoft). It is entirely conceivable that monopolistic practice in regard to some widely needed geographic information or to some hugely dominant software might be actionable in the courts; so far, this has not occurred. All sizeable bodies have procurement policies designed to ensure the organization benefits from competition, avoids fraud, and sources materials or services from reputable suppliers. These will strongly affect the procurement of GIS hardware, software, data, and consultancy services, especially in the European Union (EU) where trade-bloc-wide procurement laws exist. Finally – so far as the generality of laws is concerned – employment law covers a wide variety of topics. These include terms and conditions of employment, equal opportunities, sexual discrimination, and harassment, etc. In

general this and health and safety law have exactly the same impacts in GIS as in the rest of the ICT industry.

Privacy however is another matter. The loss of privacy is one of the downsides of new technology, at least for those nations which care passionately about it. Data protection laws are important for GIS, at least in so far as the GI concerned relates in some way to named or otherwise identifiable individuals. The use of postal addresses provides some form of identification for many purposes. Curry (1999) has argued that GIS has played an important role in undermining privacy. He has been concerned about the bringing together of geographic information and personal information, leading to unprovably correct simulations of the characteristics of human individuals or households and their subsequent use in marketing and elsewhere. He has argued that this is unacceptable and immoral. He, along with civil rights campaigners, favors an "opt in" approach. In general terms, the industry favors "opt out" provisions – the data subject must say he or she wants to be excluded from having their records used by firms. US Federal laws provide few constraints upon such actions though there have been many proposals to improve matters. In contrast, strong laws exist in Europe. This dichotomy has led to US/Europe conflict; it was apparently resolved with EU approval of "safe harbor" treatment of the USA. Firms in that country which publicly commit themselves to a set of data protection principles acceptable to the EU can trade in information in Europe. Despite these laws and agreements, there is

---

Box 17.10   Privacy is a lost cause?

The Economist of 1 May 1999 summarized the situation as follows:

"Remember, they are always watching you. Use cash when you can. Do not give your phone number, social security number or address, unless you absolutely have to. Do not fill in questionnaires or respond to telemarketeers. Demand that credit and data marketing firms produce all information they have on you, correct errors and remove yourself from marketing lists. Check your medical records often. If you suspect a government agency has a file upon you, demand to see it. Block caller ID on your phone, and keep your number unlisted. Never use electronic toll-booths on roads. Never leave your mobile phone on – your movements can be traced. Do not use store credit or discount cards. If you must use the Internet, encrypt your email, reject all 'cookies' and never give your real name when registering at web sites. Better still, use someone else's computer. At work, assume that calls, voice mail, email and computer use are all monitored.

. . . Anyone who took these precautions would merely be seeking a level of privacy available to all 20 years ago . . . . Yet . . . all these efforts to hold back the rising tide of electronic invasion will fail . . . Faced with the prospect of [privacy] loss, many might prefer to eschew even the huge benefits that the new information economy promises. *But they will not, in practice, be offered that choice*" [authors' emphasis].

much evidence of a diminution in privacy (see Box 17.10).

Information access laws encode the citizen's rights to access government information. The US Freedom of Information Act (FOIA), the Reduction of Paperwork Act, and other laws implement a distinctive Federal Government information policy. That government sees distribution of data it holds at the cost of dissemination (or less), allied to no restriction being placed on its further distribution, as a matter of principle. This applies to GI as much as any other data. The only major exceptions to this policy are where material has been supplied by foreign governments under specific agreements permitting internal business use (e.g. data for military purposes; much of this is of course GI). Such information may not be passed on and copyright remains with the supplier (see Section 19.3.1). Again, there is a geography of the law. The US Freedom of Information Acts differ from equivalents in other countries (e.g. Australia and Canada) which have FOIAs enshrined in the rights of citizens and the obligations imposed on data creators and users.

Liability is a creation of the law to support a range of important social goals, such as avoidance of injurious behavior, encouraging the fulfillment of obligations established by contracts, and the distribution of losses to those responsible for them. It is a huge and complicated issue. Liability in data, products, and services related to GIS will in many cases be determined by resort to contract law and warranty issues. This assumes that gross negligence has not occurred when the situation can be far worse and individuals as well as corporations can be liable for damages. Onsrud (1999) gives a general overview of this in regard to GIS, using GIS scenarios to illustrate the law of contract and tort, though he stresses that this describes only the situation in the USA (and at the time he wrote). He also stressed that other liability burdens may also arise under legislation relating to specific substantive topics such as intellectual property rights, privacy rights, anti-trust laws (or non-competition principles in a European context), and open records laws.

Clearly all this is a serious issue for all GIS practitioners. Consider, for instance, the following realistic scenario. You create some software and assemble some data from numerous sources in your spare time and for your own interest. A friend admires the outcome and asks for a copy: you are glad to oblige, knowing he or she is doing some important work on behalf of the local community. Decisions are made using these tools which subsequently are shown to be founded on results which are in error because of poor programming and inadequate, error-prone data.

You get sued by the community, by your friend – and by data suppliers who are annoyed that you have ripped off their data and, by implication, also brought them into disrepute. Minimizing losses for users of geographic software and data products, and reducing liability exposure for creators and distributors of such products is achieved primarily through performing competent work and keeping all parties informed of their obligations. If you have a contract to provide software, data, services, or consultancy to a client, that contract should make the limits of your responsibility clear. Simple disclaimers rarely count for much.

17.3.2 Availability of geographic information as a constraint to GIS use

If GI is needed but is not available, someone will be faced with a substantial cost in creating or buying it. Without such "rocket fuel", vendors have much more difficulty in selling software. And politicians who require speedy guidance on what to do after natural disasters (see Section 2.3.4.2) or quick answers in regard to a policy issue simply will not be able to use GIS. It follows therefore that data existence, availability, and ease and cost of access have major impacts on the entire GIS enterprise and on its future potential.

17.3.2.1 Types and sources of GI

First however it is crucial to understand the sources and characteristics of different types of geographic information. Table 17.2 sets out a simple classification of the general situation, cross-tabulating information source and type. Since "area coverage data" and "human individual data" are infinite in scope, we will not deal with them here.

Framework information are those which underpin all GIS operations. They form the base or template on which other (area coverage) GI are assembled. There are two main reasons why a spatial framework is essential in almost all GIS operations: it provides both spatial context and spatial structure. Thus the framework may be used to rectify distortions in, or sub-set and re-orientate, satellite imagery. The second reason follows from the fact that geographic information is typically collected by many different agencies, usually within the boundaries of national territories; and most of these agencies have historically collected data primarily to suit their own local purposes. Thus, if the benefits of added value are to be obtained from data integration within a GIS (see Box 17.6), there must be a mechanism (or spatial "key") to register the

Table 17.2 A classification of GI

	Geographic framework information	Area coverage information	Human individual information
Government(s)	Traditionally almost all produced by federal/national or local governments through one agency in each government. In some countries, some of this is now being updated by commercial enterprises (e.g. TIGER files in US)	A huge variety of data, often produced by many different agencies. Many of these have been created to meet local/institutional needs and are hence often incompatible in some form.	Many datasets held as part of government administrative processes (e.g. social security, which *could* be linked to other data via address or social security identifier); generally this is not legal. Rarely if ever are the individual level data made available but area aggregate data may be made available (e.g. from population censuses)
Private sector	Produced very little wide-area data in this category until the late 1990s other than aerial photography for government. Now some major commercial products, based on satellite imagery (see Section 19.3.2) exist though some of them partially originate within government	Until the 1980s, very few commercial developments. Now growing in importance, sometimes through adding value to poorly described or formatted government data (designed for internal use in many cases).	Many individual data held through customer records. Since the 1970s, national geodemographic datasets have been created and used for many in-house purposes. Now many simple datasets (e.g. phone and address lists) exist which can be merged easily with area classifications to impute family characteristics

different sets of information together in their correct relative positions. In general, only the geographic framework can provide this ubiquitous data linkage key through the use of coordinates or (sometimes) via implicit geographical coding (e.g. ZIP or postcodes).

Geographic framework information underpins all use of GIS.

What constitutes the framework varies in different countries. In the USA, the Federal Geographic Data Committee's strategic plan (Section 19.2.1) defines the information content of the US framework as consisting of:

geodetic control (control points, datums, etc.)	digital orthoimagery (see Box 17.11)
elevation	transportation
hydrography	the geography of
public land cadaster information	governmental/ administrative units

Similar definitions – including placenames – would in practice occur in many countries. Most typically, the framework is manifested in user terms through

topographic mapping – the "topographic template". Such mapping will normally contain all this information but partly in implicit form (e.g. geodetic control).

Important differences do exist between national frameworks. These are suited for different types of application, were created in response to different legal and governmental environments, and have persisted in a historic form for different periods. Typically, only one or two public sector organizations are responsible for providing the overall national framework (as in Britain, France, the Nordic countries, India, Japan, and New Zealand). But, in countries with federal governments, the typical situation is that the national government produces relatively low resolution materials (see Box 17.11) and local and more detailed versions of the framework are produced by other (e.g. state or local) governments (as in Australia, Germany, and the USA). Detailed parts of the framework may be unavailable for reasons of state security or lack of investment. Moreover, the vast bulk of framework information is presently still only available in paper map or coordinate list form. To date, the private sector has been mainly

Box 17.11   The framework created by the US national mapping agency, US Geological Survey (USGS)

These materials form part of the US National Mapping Program (NMP). The same material is usually available in paper and digital forms. The NMP incorporates the documented needs of 40 Federal agencies and the 50 States, which are solicited and analyzed as part of a continuous requirements assessment process. More details are available in the Online Resources Web addresses.

Topographic Quadrangle Maps. The USGS's most familiar product is the 1:24 000 scale map. This is the primary scale of data produced. National coverage is also produced at scales of 1:50 000, 1:100 000, 1:250 000 and 1:2 000 000. The same information is available in two digital forms. Digital Line Graphs (DLGs) are vector files containing line data, such as roads and streams, digitized from the maps. DLG's offer a full range of attribute codes and are topologically structured. Digital Raster Graphics are scanned images of the USGS topographic maps.

Digital Elevation Models. These comprise terrain elevations at regularly spaced horizontal intervals. DEMs are developed from stereo models or digital contour line files derived from the topographic quadrangle maps.

Orthophotoquads. These are distortion-free aerial photographs that are formatted and printed mainly as standard 7.5 minute, 1:24 000-scale quadrangles. The standard is a black-and-white or color-infrared 1-meter ground resolution quarter quadrangle (3.75-minute) image.

Satellite Images. Digital data available includes Advanced Very High Resolution Radiometer (AVHRR) images, and Landsat Thematic Mapper (TM) and Multi-Spectral Scanner (MSS) data.

Geographic Names Information System (GNIS). This contains information about physical and cultural geographic features in the United States, including the Federally recognized name and location. It forms the official national record of domestic geographic names information.

involved in marketing some of this framework data or as contractors in its collection (but see Table 17.2).

The vast bulk of geographic information now in existence – as measured by topic and thematic detail, and probably still by data volume – has been collected by and for government. This situation is changing for two reasons. First, GPS and other data collection technologies enable individuals or organizations to collect locational information in real time and directly in computer form. Second, the market for commercially produced datasets is increasing rapidly, encouraging niche players to participate. But users still need "spatial context" – the framework – for use of their own data. In principle this may be created in the long term by the conflation of GPS datasets collected by many parties. This is however a non-trivial task and the widespread availability of "framework data" from a small number of reputable providers renders it pointless in many cases.

17.3.2.2   Information consistency and quality

Does use of a single framework matter? In general, competition is a good thing: it provides greater user choice and drives down prices. In the case of framework geographic information, however, this is less obviously true. Three types of external benefits arise from the use of a common geographic framework:

● Ensuring consistency in the *collection* of information (producer externalities). Inconsistency can raise the costs of using data and limit the range of applications. It is, for example, important that all the emergency services use the same framework.

● Providing users *access* to the same information (network externalities). Network benefits exist for any user of software or data being widely used by others; these have been given as an explanation for the high market share enjoyed by Microsoft.

● Promoting *efficiency* of decision-making (consumer externalities). An example is pollution control where access to consistent information allows pressure groups to be more effective in influencing government policy and in monitoring activities.

Finally, the issue of information quality is a key one for management use of and reliance on GIS. We have already covered it in earlier, especially theoretical and technical, sections on uncertainty (see Chapters 6 and 15). The important point here is that there are no agreed mechanisms for certifying quality of geographic information. Nor are there presently any independent bodies for carrying out the task. This is not surprising – the specifications of many datasets compiled across the world vary enormously, the creators often guard their independence jealously, and there are many trade-offs in deciding how to calibrate such a multi-faceted characteristic. But it presents a real problem for a GIS manager: how do you know that datasets proffered by agencies A, B, and C are "fit for purpose" for the specific job that needs to be done? Sometimes the answer is obvious – a dataset may not contain all the required themes or be at such an absurdly low resolution that its use is nonsensical. Other times a dataset may be the only one available even if far from perfect for the application. If this is the case, how safe is it to proceed with its use? The end result of this uncertainty is that great reliance is placed upon institutional longevity and accountability as surrogates for high quality and dependability – both are attributes of government bodies. This "safety first" approach applies in acquisition of both geographic information and software. The approach is entirely understandable as a risk reduction strategy (see Section 17.4).

> The framework's quality is generally publicly defined and well known and its governmental source is liable to stay in operation to update and guarantee it.

17.3.2.3 Information availability and accessibility

Suppose the desired information exists – and is known to exist via some "clearinghouse" or on-line catalogue (Section 19.2.3). That existence and its availability and accessibility are, however, sometimes very different matters. Much existing geographic information is simply not available in many parts of the world and numerous conditions are placed upon its use in others. For example, there are many parts of the world – especially areas of tension such as certain national boundaries like the borders of India and Pakistan – where strict limits to availability exist. These limitations are typically set by the local military and politicians, especially at resolutions more detailed than, say, that represented by 1:500 000 scale mapping. Other access constraints exist in information held by commercial enterprises to suit their competitive advantage. In most cases where such data *are* made available, strict conditions of use apply, e.g. restriction to internal business purposes only. Finally, as we saw earlier (Section 17.1.2), national data policies on information trading by government may also constrain information accessibility and "usability" via pricing mechanisms.

17.4 Risk

Risk identification and management is fundamental to the success of any organization. Failure to control it can lead to financial disaster, to public humiliation for the organization and the loss of jobs, especially of senior personnel.

Central to the notion of risk is accountability. It is essential on a practical basis to "lock in" your superiors in the project, obtaining their agreement to the strategy and key operational decisions. Failure to do so may well mean your demise – a near-certainty if things go wrong. Senior managers must receive a formal GIS proposal for their consideration and formal acceptance. This will normally be developed in several stages – perhaps the floating of some initial ideas, supported by evidence of what has been achieved elsewhere and tangible and intangible benefits (see Chapter 18), followed by a vision and strategy for getting there and finally by detailed business and implementation plans.

That is the generality. More particularly, introduction of GIS can increase institutional risk through failure to meet core operational requirements or suffering major cost over-runs. Alternatively, intelligently used GIS can reduce business risks for the organization as a whole. The different types of risk need to be considered separately (Table 17.3). In general, risks are greater when GIS are being used for wholly new activities within organizations than when such systems are being used to do existing things better or cheaper, especially if they are introduced incrementally.

It is important to realize that some risks are inherently unpredictable in detail though their aggregate likelihood can sometimes be insured against (e.g. via professional indemnity insurance). Perhaps the most striking and global examples involve money, such as the 40% depreciation of the Mexican peso in the week of 19 December 1994. Any external (e.g. US) organization trading in Mexico (e.g. in selling GIS software or services) and getting paid in pesos could have been ruined by this event.

Equivalent, externally imposed, high risk events include major natural disasters. It is worth noting that GIS is now a major contributor to the estimation of risk for the world's largest insurers, in tasks ranging from estimates of credit-worthiness of individuals to monitoring the distribution and frequency of natural hazard events (Section 2.3.4.2). More directly related to GIS however is the impact of major technological change: like everyone else, the authors of this book failed to anticipate the impact of the World Wide Web on use of the Internet. Many other GIS developers

Table 17.3 A classification of risk faced by organizations

Major risk category	Sub-category	GIS-related example
Business risk	Product/service risk	Implementation of system over-runs financial or time budgets. Analyses, outputs, outcomes found flawed or unconvincing due to software inadequacy or staff incompetence
	Economic risk (via change of markets)	No-one wants line maps when up-to-date image maps available?
	Technology risk (out-of-date compared to competitors)	Competitors can achieve the same results (e.g. analyses, maps) faster, better, more cheaply
Event risk	Reputation risk	Erroneous or delayed results lead to collapse in confidence in organization, fanned by the media (e.g. "State denounces GIS firm for unprofessional work . . .")
	Legal risks	As above, leading to legal action to recover damages and/or costs
	Disaster risks	Software incapable of doing what promised, fire destroys all facilities including data and results to date
	Regulatory or policy shift risk	Government decrees that information policy has changed or that all materials contributing to end results have to be in public domain
	Political risk	Elected official supports competitor
Financial risk	Credit, market, liquidity, operational, price, delivery, etc. risks and fraud	Other normal business risks – all must be guarded against

and operators were caught with "old style" systems which were expensive to operate, maintain, and run by comparison with what was rendered possible by the WWW.

17.4.1 Managing risks in GIS through investment and risk appraisals

Managing risk within GIS projects requires at least two appraisals, carried out at different stages. These may overlap with – and be part of – the business case for a GIS. The appraisals are:

● A sound investment appraisal before commitment is made to proceed.

● A risk appraisal to identify the significant factors which may go wrong in implementation of the project.

The investment appraisal is needed before proceeding. This is because someone above you in the management chain needs to analyze your logic and test it to destruction. Most people – especially techno-enthusiasts – are poor critics of their own ideas, concepts, and proposals. The

opportunity costs of going ahead also need to be exposed. In many organizations, there is only a finite amount of financial resource available for new developments. This limitation may arise from an unwillingness to add to the tax burden (in the public sector) or from restrictions on bank or market support (in the private sector). Such appraisals also need to be done when money is not a major problem: the most precious commodity in many businesses is senior management skills and it may be necessary to use these to take forward or to rescue another project. In short, the GIS business case is a necessary but not sufficient case for proceeding – a more wide-ranging investment appraisal is needed and this can rarely be done by GIS staff alone. It needs at least to be sanctioned by the finance director or his/her equivalent.

Managing risk requires both an investment appraisal and a risk appraisal – you only go ahead with a GIS project if the likely returns are big enough, if you are sure you can achieve them, and if there is no better use of internal resources.

An investment appraisal needs to cover at least:

● *The predicted costs* of people, materials, hardware and software, training, and other organizational costs insofar as they can be quantified (e.g. disruption through changes to building occupancy), all broken down by time.

● *The predicted benefits* (see Section 18.1). Wherever possible these must be quantified in money terms and sum to more than the costs – "hard nosed" managers are much more likely to approve a project on the basis of tangible benefits than on a mixture of these and intangibles. That said, the intangible benefits often become substantial in successful projects and become disbenefits in unsuccessful ones. In all cases these must be related to institutional aims and priorities. Thus it may be possible to justify something which is more expensive than the *status quo* if it delivers some other explicitly valued outcomes. As one example, one British government agency found that its land registration transactions took an elapsed time of about 30 days – but that the actual time that staff spent doing work on each application totaled little more than the same number of minutes. The only way to effect a step-change in the performance decreed by central government was to install a major IT system, incorporating a GIS, to automate and change the processes and change work structures. This was adopted despite the very significant increase in short and medium term costs.

● *Predicted costs and benefits of the status quo*, i.e. the "do nothing option".

● *Sensitivity analyses* of the impacts of variations in the plan on the business case. Such variations could include cost over-runs, failure to achieve the predicted benefits in the predicted time, etc.

● *Conclusions* on the basis of all the above, preferably agreed individually with as many of the key players as possible before the formal consideration of the appraisal.

The role of the risk assessment is to anticipate the most likely and significant risks and work out what to do about them in advance, in so far as that is possible. Such anticipation of risk requires a formal process and people charged to identify and manage it. No project should start without a table of potential risks, classified by probability and anticipated magnitude of impact (which may be very different) and by steps to be taken to reduce the probability and/or minimize the impact. Moreover, the risk assessment is iterative: after every major event or even periodically, the process shown in Figure 17.6 must be repeated.

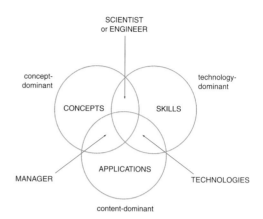

Figure 17.6 The iterative nature of risk assessment and management

17.5 Exploiting the Possibilities and Avoiding the Risks: the Role of the GIS Strategy

Strategies set direction, based upon the organization's aims and an analysis of its strengths, weaknesses, opportunities, and threats (a SWOT analysis). It will however be rare for any organization to have a free-standing GIS strategy. In the first instance, it will normally have to fit within the organization's overall strategy and business plan. It will also need to be supported by other strategies with which it will be associated, such as the finance, human resources, information systems and services, and estates strategies – and possibly others such as external relations/publicity and international ones. Moreover, it may simply be embedded, even buried, within the Information Technology, Systems, or Services strategy. We think that is often a mistake because of the cross-cutting nature of GIS and (especially) GI across many levels and (potentially) most of the organization's activities – see why we think "spatial is special" in Section 1.1.

Despite all that, here we treat the GIS strategy as a free-standing entity for illustrative purposes. The particular form of that strategy will vary depending on how GIS is seen within the organization (see Table 17.4). Finally, strategies are useless without an implementation plan, senior management commitments to proceed, and milestones at which progress is monitored. Most of these issues are dealt with in more detail in Chapter 18.

17.5.1 Getting to the GIS strategy and the implementation plan

Organizations routinely achieve things through standard processes. A top-down process view of

Table 17.4 Different types of GIS operations seen from a management perspective

Type of GIS operation	Purpose	Characteristics
Individual GIS operation – desktop or Internet-based	To provide access to information and modeling capabilities for one individual	Can be highly effective and tailored to specific needs but idiosyncratic, leading to massive duplication of resource if multiple people do it and inconsistent use (hence costs) of tools, data, advice
Departmental project	To complete one project (e.g. carry out visualization analysis of proposed power station or an economic impact analysis)	Often done on "quick and dirty" basis – capacity (e.g. tools, staff) may be bought in if project is short term. Result is a product (rather than an on-going service). Usually highly focused: data quality only has to be as good as required for this application. May lead to gross waste if further applications of same GIS discovered later
Departmental application (i) operational department	GIS tools used to support at least one on-going and well-defined business function (e.g. creating and maintaining tax maps, land records, planning permissions)	Advantages are clear focus and relatively simple system design and data stream analysis. Facility may well be best managed by department responsible for this core function. Corporate approval and support may or may not be needed, depending on policies, risk assessment. On-going support will however be essential
Departmental application (ii) strategy department	GIS used to explore "what if" questions and implications and interactions of development ideas	Much more difficult to design since questions to be tackled are inherently unpredictable. Fit to other strategy elements (e.g. finance estimates of cash flow, capital requirements) may have to be done through knitting together of outputs from multiple systems via spreadsheets
Corporate enterprise facility	System used in multiple departments and data shared across all of them (but with different departments updating different elements of corporate database)	Potentially highly valuable – can be designed to meet overall business imperatives, minimizes duplication of effort across common needs, provides consistent information, facilitates good long-term corporate and IT planning, permits added value from creation of data linkages and new applications. Can be very difficult to make successful if aim is to meet needs of whole organization. Requires identification of relatively stable needs and also much flexibility in tools. It also needs a culture in GIS support areas which values challenge, together with excellent communications and shared values across whole organization. Such a system can become a severe constraint on institutional action if business needs change markedly.

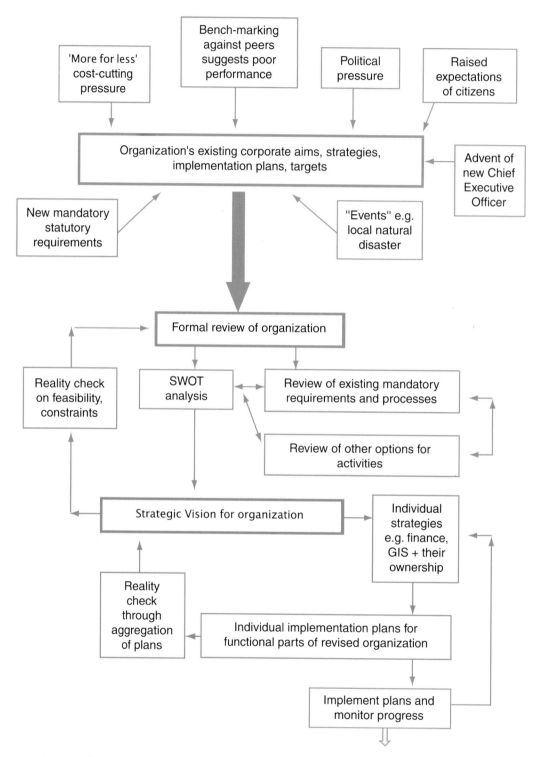

Figure 17.7 External and internal pressures on a hypothetical local government organization and consequential moves to a corporate and a GIS strategy and implementation plans

how to get to a GIS strategy is shown in Figure 17.7. To aid realism, we use a hypothetical example of a mid-size local government, with perhaps a thousand or so employees. Things sometimes happen in a top down way but, in practice, GIS strategies are often partly grown from the "grass-roots".

Getting to a GIS (and corporate) strategy requires the following analyses and actions:

● A statement of the current situation, agreed by all the key players as an accurate summary. This needs to cover the reason why the organization exists and how well it is achieving those aims (based largely on customer, client, or "owner's" views). It needs also to describe any major shortcomings and the causes of them. In addition, it needs to summarize the organization's culture and values and present structure and operational management. And it needs a process map of the organization. All of this needs to be as dispassionate as possible – which is why consultants are often called in to help.

● A statement of the organization's strengths, weaknesses, opportunities, and the threats to it. From this comes recognition of the particular constraints that need to circumvented.

● How existing activities fit to the strategic vision of the corporate body.

● A statement of the changes necessary and a comparison of these with the financial, staffing, and other constraints plus an assessment of their ramifications. These may involve staff numbers and skills, disposition of staff, change of software, the need to acquire or create additional data/information, or technology.

● Obtaining "buy in" from senior managers and the staff concerned.

Steering an organization is both necessary and difficult – at least if it is done on the basis of rationality and through teams, rather than the exercise of autocracy. The market position, capability, resources, and existing structures of the body – whether academic, governmental, not-for-profit, or commercial – must be considered. The consequences of plans and the motivations of those propounding them need to be considered, along with the risks involved and means of minimizing them. The need to "keep the show on the road" whilst making changes is normally essential, as is the need to justify actions to stakeholders in the organization. And all of this is typically carried out in a climate of uncertainty about external developments and much else. It is a non-trivial exercise. Every GIS project is also difficult – for the same reasons.

Questions for Further Study

1. Who owns the geographic information you use? What rights do you have to it? How would you protect your organization's GI against misuse by competitors?
2. If you were being educated for your own job, what would be the ideal components of a GIS course? Is distance-based learning for GIS a satisfactory approach? If so, for what types of personal development is it best suited?
3. Suppose tomorrow morning you receive a writ for damages from a commercial firm for whom you have carried out consultancy. It alleges your results – on which they acted – are now demonstrably false and you did not take due care and attention. What do you and your lawyer do first? How do you "cover your back"?
4. What are the main risks that might befall your GIS project over the next 18 months? What steps will you take to ensure the problems do not occur or, if they do, to obviate them or contain their effects?
5. How would you go about creating a new GIS strategy for your organization?

Online Resources

The NCGIA Core Curricula provide examples of how GIS courses might be organized. In addition these sections are particularly relevant to this chapter's content (www.ncgia.ucsb.edu/pubs/core.html)
Core Curriculum in GIScience, 3.2.1 (Public Access to Geographic Information, Albert Yeung), 3.5 (Teaching GIS, David Unwin), 3.5.1 (Curriculum Design for GIS, David Unwin), 3.5.2 (Teaching and Learning GIS in Laboratories, David Unwin)
Core Curriculum in GIS, 1990, Unit 70 (Legal Issues)
Information as an asset
www.sims.berkeley.edu/resources/infoecon
www.echo.lu

Structure and content of a Masters degree course designed to be taken either by face-to-face or distance learning *and* incorporating business and management issues as well as technical ones (see especially the WebCT pages)
www.soi.city.ac.uk/mgi

The multinational UNIGIS consortium
www.unigis.org

Information on the progress of professional certification for GIS in the USA

cem.uor.edu/users/kemp/certification

Terms of Reference for ISO Technical Committee TC211 on Geographic Information/Geomatics, concerned with standardization in the field of digital geographic information and professional certification

www.statkart.no/isotc211

Some sources of geographic framework and other GI, each with links to other sites

www.usgs.gov
www.fgdc.gov
www.ngdf.org.uk
www.cerco.org.fr
www.ign.fr
www.ordsvy.gov.uk

Value-adding industry

www.landmark-information.co.uk
www.upmystreet.com

Reference Links

Maguire D J, Goodchild M F, and Rhind D W (eds) 1991 *Geographical Information Systems: Principles and applications*. Harlow, UK: Longman (Text available online from 'Links to Big Book 1' at www.wiley.com/gis and www.wiley.co.uk/gis).

Epilogue (Rhind D W, Goodchild M F, Maguire D J)

Longley P A, Goodchild M F, Maguire D J, and Rhind D W (eds) 1999 *Geographical Information Systems: Principles, techniques, management and applications*, New York: John Wiley.

Chapter 46 Liability in the use of GIS databases (Onsrud H J)

Chapter 54 Enabling progress in GIS and education (Forer P, Unwin D)

Chapter 55 Rethinking privacy in a geocoded world (Curry M R)

References

Cho G 1998 *Geographic Information Systems and the Law*. New York: John Wiley.

Foster C and Plowden F 1996 *The State under Stress*. Buckingham: Open University Press.

Masser I 1998 *Governments and Geographic Information*. London: Taylor and Francis.

Onsrud H 1993 *Law, Information Policy and Spatial Databases*. NCGIA Working Paper. Orono, Maine: NCGIA.

Osborne D and Gaebler T 1992 *Reinventing Government*. Reading, Mass.: Addison-Wesley.

Shapiro C and Varian H R 1999 *Information Rules*. Cambridge, Mass.: Harvard Business School Press.

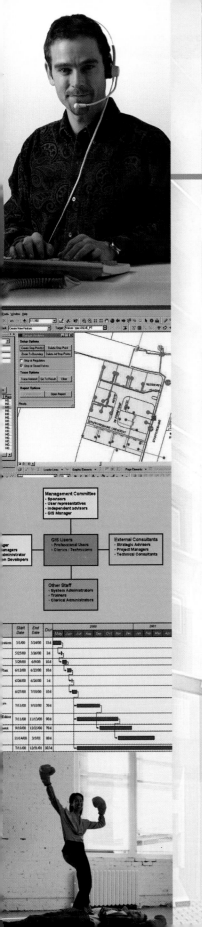

OPERATIONAL ASPECTS OF GIS

18

For many users, purchasing a GIS is their first experience of procuring a large computer system. Large GIS require great care and attention. Methodologies have been developed to choose, implement, and manage operational GIS.

Choosing GIS involves four key stages: analysis, specification and evaluation of alternatives, and implementation of the system.

Implementing GIS requires consideration of issues such as planning, support, communication, resource management, and funding. The five key dimensions to managing an operational GIS are support for customers, operations, data management, application development, and project management.

Learning objectives

After reading this chapter, you will understand:

- How to choose the right GIS;

- Key GIS implementation issues;

- How to manage an operational GIS effectively with limited resources and ambitious goals;

- Why GIS projects fail – some pitfalls to avoid and some useful tips about how to succeed;

- The roles of staff members in a GIS project.

18.1 Introduction

This chapter is concerned with the practical aspects of managing an operational GIS. It is embedded deliberately in the midst of high-level management concepts: success comes from combining strategy and implementation. It is the role of management in GIS projects to ensure that operations are carried out effectively and efficiently, and that a healthy, sustainable GIS can be maintained. GIS tasks must be completed on time, within budget, and according to quality standards. The three key operational aspects of this are choosing, implementing, and managing a GIS.

Successful operational management of GIS requires completion of tasks on time, within budget, and according to quality standards.

Before actually starting to implement a GIS, ask the fundamental question: do I need a GIS? There are many applications where this is obvious (e.g. see Box 16.1). But at the strategic level, there are four general reasons to implement a GIS (Huxhold and Levinsohn 1995, Obermeyer 1999, Tomlinson 1996):

● *Cost reduction.* GIS can replace in part or completely many existing operations, e.g. drafting maps, locating and maintaining customers, and managing land acquisition and disposal more efficiently.

● *Cost avoidance.* Examples of GIS use are in locating facilities away from areas at high risk from natural hazards (e.g. tornadoes, floods, or landslides) and by minimizing delivery routes (e.g. letters, refrigerators, and beer).

● *Increased revenue.* Examples include finding and attracting new customers, making maps for sale and as a tool to support consultants (e.g. facility siting, natural resource conservation, and real estate management).

● *Getting non-tangible (or intangible) benefits.* These benefits are difficult to measure, but they can be very important nonetheless. Examples include making better decisions, use of consistent information, improved public image, reproducible results, and the ability to document the processes and methodology used to solve a problem (see Chapter 17).

GIS projects are similar to many other large IT projects in that they can be broken down into four major life cycle phases (Figure 18.1). These are: business planning (strategic analysis and requirements gathering); system acquisition (choosing and purchasing a system); system

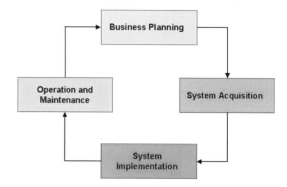

Figure 18.1 GIS project lifecycle stages

implementation (assembling all the various components and creating a functional solution); and operation and maintenance (keeping a system running). These phases are iterative. Over a decade or more, several iterations may occur, often using different generations of GIS technology and methodologies. Variations on this model include prototyping and rapid application development – but space does not permit detailed discussion of them here.

GIS projects comprise four major life cycle phases: business planning; system acquisition; system implementation; and operation and maintenance.

18.2 Choosing a GIS

A general model of how to specify, evaluate, and choose a GIS has been presented by Clarke (1991). Many variations on this model have been used widely and successfully to acquire GIS over the past 20 years or so (see, for example, Heywood *et al* 1998, Huxhold and Levinsohn 1995, and Tomlinson 1996). Clarke's model (Figure 18.2) is based on 14 steps grouped into four stages: analysis of requirements; specification of requirements; evaluation of alternatives; and implementation of a system. Such a process is both time-consuming and expensive. It is really only appropriate for large GIS implementations (contracts over $100 000 in value). For them at least, it is important to have investment and risk appraisals (see Section 17.4).

Choosing a GIS involves four stages: analysis of requirements; specification of requirements; evaluation of alternatives; and implementation of a system.

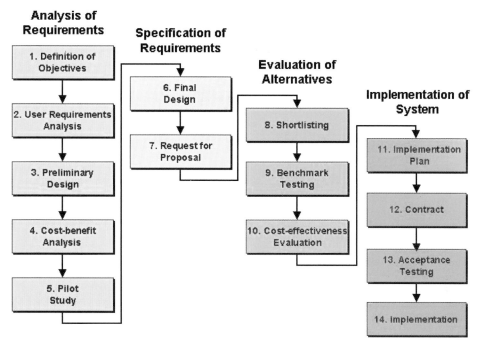

Figure 18.2 General model of the GIS acquisition process (Source: Clarke 1991)

18.2.1 Stage 1: Analysis of requirements

The first stage in choosing a GIS is an iterative process for identifying and refining user requirements, and for determining the business case for acquiring a GIS (see Section 17.4.1). The deliverable for each step is a report that should be discussed with users and management. The results of each report help determine successive stages.

Step 1   Definition of objectives
This is often a major decision for any organization. The process of choosing a GIS begins with development of the organization's strategic plan (Section 17.5) and an outline decision that GIS can play a role in the implementation of this plan. Objectives must be stated in a form understandable to managers. The outcome from Step 1 is a document that managers and users can endorse as a plan to proceed with the acquisition.

Step 2   User requirements analysis
These will determine how the GIS is designed and evaluated. Analysis should focus on what information is being used, who is using it, and how the source is being collected, stored, and maintained. This is a map of existing processes (which may possibly be improved as well as being replicated

by the GIS). The necessary information can be obtained through interviews, documentation, reviews, and workshops. The report for this phase should be in the form of workflows, lists of information sources, and current operation costs. The clear definition of examples of future applications (Figure 18.3), information products (e.g. maps and reports: Figures 17.3 and 18.4), function utilization, and data requirements is essential to successful GIS implementation.

Step 3   Preliminary design
Preliminary design is based on results from step 2. This design will be used for subsequent cost-benefit analysis (Step 4 below) and will enable specification of the pilot study. The four key tasks are: develop preliminary database specifications; create preliminary functional specifications; design preliminary system models; and survey the market for potential systems. Database specifications involve estimating the amount and type of data (see Chapter 11). To help with the definition of functional specifications, there are several published functional lists (e.g. Guptill 1988). Many consultants also maintain checklists and vendors frequently publish descriptions of their systems on their Web sites. The choice of system model involves decisions about raster

APPLICATION

Display zoning map information for a user-defined area.

FUNCTIONS USED IN THE APPLICATION:

Review and prepare zoning changes.

DESCRIPTION OF APPLICATION:

This application uses zoning and related parcel-based data from the database to display existing information related to zoning for a specific area that is defined by the user. The application must be available interactively at a workstation when the user invokes a request and identifies the subject land parcel. The application will define a search area based upon the search distance defined and input by the user, and will display all required data for the area within the specified distance from the outer boundary of the subject parcel.

DATA INPUTS:

User defined: Parcel identifier
Search distance

Database: Zoning boundaries
Zoning dimensions
Zoning codes
Parcel boundaries
Parcel dimensions
Parcel numbers
Street names
Addresses

PRODUCTS OUTPUT:

1. Zoning map screen display with subject parcel highlighted, search area boundaries, search distance, all zoning data, parcel data, street names and addresses.
2. Hard copy map of the above.

Figure 18.3 Sample application definition form (Source: after Huxhold and Levinsohn 1995)

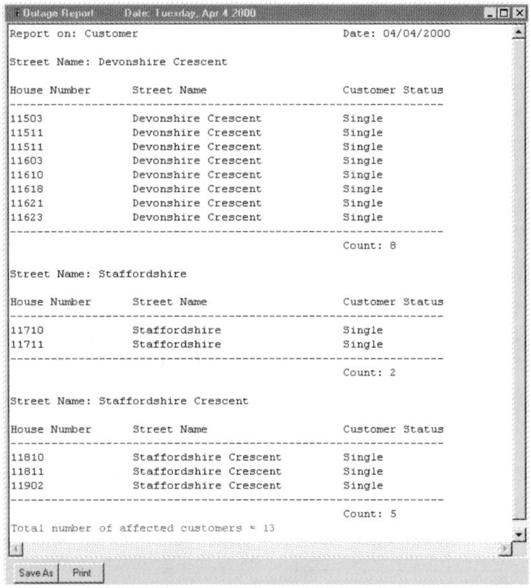
Outage Report Date: Tuesday, Apr 4 2000

Report on: Customer Date: 04/04/2000

Street Name: Devonshire Crescent

House Number	Street Name	Customer Status
11503	Devonshire Crescent	Single
11511	Devonshire Crescent	Single
11511	Devonshire Crescent	Single
11603	Devonshire Crescent	Single
11610	Devonshire Crescent	Single
11618	Devonshire Crescent	Single
11621	Devonshire Crescent	Single
11623	Devonshire Crescent	Single
		Count: 8

Street Name: Staffordshire

House Number	Street Name	Customer Status
11710	Staffordshire	Single
11711	Staffordshire	Single
		Count: 2

Street Name: Staffordshire Crescent

House Number	Street Name	Customer Status
11810	Staffordshire Crescent	Single
11811	Staffordshire Crescent	Single
11902	Staffordshire Crescent	Single
		Count: 5

Total number of affected customers = 13

Save As Print

Figure 18.4 Example report (in this case of outages) that must be created by a newly acquired GIS

and vector data models and system type (see Chapters 8 and 9). Finally, a market survey should be undertaken to assess the capabilities of commercial-off-the-shelf (COTS) systems. This might involve a formal Request For Information (RFI) to a wide range of vendors.

Whether to buy or to build a GIS is a major decision that must be faced (Section 8.4). This occurs at "green field" sites – where no GIS technology has hitherto been used – and at sites where a GIS has already been implemented but is in need of modernization. Today there is a general move in large organizations away from home-grown, or heavily customized, solutions designed for specific projects. These are the so-called stovepipe systems – each has its own technology that covers the full range of tasks. The trend is towards general-purpose COTS solutions. This move has been encouraged by the fact that COTS technologies have ongoing programs of enhancement and maintenance, and because they can be used for multiple projects (Section 17.5).

There is major move in GIS away from building proprietary GIS toward buying COTS solutions.

Step 4 Cost–benefit analysis

Purchase and implementation of a GIS is a non-trivial exercise, expensive in both money and staff resources. It is quite common for organizations to undertake a cost–benefit (also called benefit–cost) analysis to justify the effort and expense, and to compare it against the alternative of continuing with the current data, processes and products – the *status quo* (see Box 18.1, Section 17.4.1 Obermeyer 1999).

Cost–benefit cases are normally presented as a spreadsheet, along with a report that summarizes the main findings and suggests whether the project should be continued or halted. Senior managers then need to assess the merits of this project in comparison with any others competing for their resources (see Section 17.4).

Step 5 Pilot study

A pilot study is a mini-version of the full GIS implementation that aims to test several facets of the project. The primary objective is to test the system design before finalizing the system specification and committing significant resources. Secondary objectives are to develop the understanding and confidence of users and sponsors, to test samples of data if a data capture project is part of the implementation, and to provide a testbed for application development.

Box 18.1   Cost–benefit analysis

Cost–benefit analysis begins with a determination of GIS implementation costs (hardware, software, data, and people) and predicted benefits (improved service, reduced costs, greater use of data, etc.). A value is then assigned to both costs and benefits. When these are summed, the benefits should exceed the costs over the full lifecycle if the project is to proceed. This sounds quite simple and straightforward. In practice, however, cost–benefit is more of an art than a science because there are many complicating factors. For example, it is difficult to assign a value to many activities, and in many traditional public organizations fixed costs (such as staff time, centralized administration and computing facilities) are not taken properly into account. In addition, a positive cost–benefit may not ensure that the project proceeds – that will also depend on whether the organization's cash flow and management time resources are adequate and whether the risks are acceptable (see Section 17.4).

Obermeyer (1999) suggests that the potential costs and benefits of implementing a GIS fall into two categories (Table 18.1): economic (tangible) and institutional (intangible). Unfortunately, the easier to measure economic benefits tend ultimately to be less important in GIS projects than the much more fuzzy institutional benefits. This is especially true of systems that perform decision-support rather than operational tasks (e.g. city planning versus postal delivery routing).

Table 18.1   Simple examples of GIS costs and benefits (Source: after Obermeyer 1999 with additions)

Category	Costs	Benefits
Economic (tangible)	Hardware	Reduced costs (e.g. of staff)
	Software	Greater throughput
	Training	Increased revenues
	New staff or skills	New market services
	Additional space	Expanded market services
	Data	
Institutional (intangible)	Interpersonal shifts	Improved client relationships
	Layoffs of staff	Better decisions
	Staff anxiety	Improved morale
	Neglect of other projects	Better information flow
		Better culture of "achievers"

Although cost–benefit analysis is the best-known method of assessing the value of a GIS, a number of alternatives have been proposed and summarized by Obermeyer (1999). "Cost-effectiveness analysis" compares the cost of providing the same service using alternative means. The "payback period" is derived by dividing the cost of implementing a system by the estimated annual net benefits of using it. The "value-added approach" emphasizes the new things that a GIS allows an organization to perform. In all assessments, account should be taken of the different value of money at different times.

There are surprisingly few readily accessible examples of GIS cost–benefit studies. It is a more difficult and expensive task than it might seem. Many users do not actually bother to carry it out, especially now that GIS is becoming a mainstream science and technology. This attitude is short-sighted because adoption of GIS has long-term consequences – such as staff training needs – which go beyond simple financial expenditure on equipment and software. Many organizations also gain competitive advantage from GIS and do not want to publish their analyses. Moreover, cost–benefit studies are often carried out by consultants who do not want to make their approach public for fear that they will no longer be able to charge for it.

A pilot is a mini-version of a full GIS implementation designed to test as many aspects of the final system as possible.

It is normal to use existing hardware or to lease hardware similar to that which is expected to be used in the full implementation. A reasonable cross-section of all the main types of data, applications, and product deliverables should be used during the pilot, but the temptation must be resisted to try to build the whole system. Users should be prepared to discard *everything* after the pilot if the selected technology or application style does not live up to expectations.

The outcome of a pilot study is a document containing an evaluation of the technology and approach adopted, an assessment of the cost-benefit case, and details of the project risks, and impacts. Risk analysis is an important activity, even at this early stage (see Section 17.4). Assessing what can go wrong can help avoid potentially expensive disasters in the future. The risk analysis should focus on the actual acquisition processes as well as on implementation and operation.

18.2.2 Stage 2: Specification of requirements

The second stage is concerned with developing a formal specification that can be used to solicit and evaluate proposals for the system.

Step 6   Final design

This creates the final design specifications for inclusion in a Request For Proposals (RFP: also called an Invitation To Tender or ITT) to vendors. Key activities include finalizing the database, defining the functional and performance specifications, and creating a list of possible constraints. From these, requirements are classified as mandatory, desirable, or optional. The deliverable is the final design document. This document should provide a clear description of essential requirements – without being so prescriptive that innovation is stifled, costs escalate, or insufficient vendors are able to respond.

Step 7   Request for proposals

The RFP document combines the final design document with the contractual requirements of the organization. These will vary from organization to organization but are likely to include legal details of copyright, intellectual property ownership, payment schedules, procurement timetable, and other draft terms and conditions (see Section 17.3.1). Once the RFP is released to vendors by official advertisement and/or personal letter, a minimum period of several weeks is required for vendors to evaluate and respond. For complex systems, it is usual to hold an open meeting to discuss technical and business issues.

18.2.3 Stage 3: Evaluation of alternatives

Step 8   Short-listing

In situations where several vendors are expected to reply, it is customary to have a shortlisting process. Submitted proposals must first be evaluated, usually using a weighted scoring system, and the list of potential suppliers narrowed down to between two and four. This allows both the prospective purchaser and supplier organizations

to allocate their resources in a more focused way. Short-listed vendors are then invited to attend a benchmark setting meeting.

Step 9   Benchmarking

The primary purpose of a benchmark is to evaluate the proposal, people, and technology of a GIS vendor. Each vendor is expected to create a prototype of the final system that will be used to perform representative tests. The results of these tests are scored by the prospective purchaser. Scores are also assigned for the original vendor proposal and the vendor presentations about their company. Together, these scores form the basis of the final system selection. Unfortunately, benchmarks are often conducted in a rather secretive and confrontational way, with vendors expected to guess the relative priorities (and the weighting of the scores) of the prospective purchaser. Whilst it is essential to follow a fair and transparent process, to maintain a good audit trail, and to remain completely impartial, a more open co-operative approach usually produces a better evaluation of vendors and their proposals.

Step 10   Cost–effectiveness evaluation

Surviving proposals are next evaluated for their cost–effectiveness. This is again more complex than it might seem. For example, GIS software systems vary quite widely in the type of hardware they use, some need additional database management system (DBMS) licenses, customization costs will vary, and maintenance will often be calculated in different ways. The goal of this stage is to normalize all the proposals to a common format for comparative purposes. The weighting used for different parts must be chosen carefully since this can have a significant impact on the final selection. Good practice involves debate within the user community – for they should have a strong say – on the weighting to be used and some sensitivity testing to check whether very different answers would have been obtained if the weights were slightly different. The deliverable from this stage is a ranking of vendors' offerings.

18.2.4 Stage 4: Implementation of system

The final stage is planning the implementation, contracting with the selected vendor, testing the delivered system, and actual use of the GIS.

Step 11   Implementation plan

A structured, appropriately paced implementation plan is an essential ingredient of a successful GIS implementation. The plan commences with identification of priorities, definition of an implementation schedule, and creation of a resource budget and management plan. Typical activities that need to be included in the schedule are installation and acceptance testing, staff training, data collection, and customization. Implementation should be coordinated with both users and suppliers.

Step 12 Contract
An award is subject to final contractual negotiation to agree general and specific terms and conditions, what elements of the vendor proposal will be delivered, when they will be delivered, and at what price. General conditions include contract period, payment schedule, responsibilities of the parties, insurance, warranty, indemnity, arbitration, and provision of penalties and contract termination arrangements.

Step 13 Acceptance testing
This is to ensure that the delivered GIS matches the specification agreed in the contract. Part of the payment should be withheld until this step is successfully completed. Activities include installation, and tests of functionality, performance, and reliability. It is seldom the case that a system passes all tests first time and so provision should be made to repeat aspects of the testing.

Step 14 Implementation
This is the final step at the end of what can be a long road (see also Section 18.3 below). The entire GIS acquisition period can stretch over many months or even longer (see Section 19.5). Activities include training users and support staff, data collection, system maintenance, and performance monitoring. Once the system is successfully in operation, it may be appropriate to publicize its success for enhancement of the brand image or political purposes.

See Box 18.2 for details of a new application requiring integration of four existing ones.

18.2.5 Discussion of the acquisition model

The general model outlined above is widely employed as the primary mechanism for large GIS procurements in public organizations. Although it has many advantages, it also has some significant shortcomings:

● The process is expensive and time-consuming for both suppliers and vendors. A supplier can spend as much as 20% of the contract value on winning the business and a purchaser can spend a similar amount in staff time, external consultancy fees, and equipment rental. This ultimately makes systems more expensive – though competition does drive down cost.

● Because it takes a long time and because GIS is a fast-developing field, proposals can become technologically obsolete within several months.

● The short-listing process requires multiple vendors, which can end up lowering the minimum technical selection threshold in order to ensure enough bidders are available.

● In practice, the evaluation process often focuses undue attention on price rather than the long-term organizational and technical merits of the different solutions.

● This type of procurement can be highly adversarial. As a result, it can lay the foundations for an uncomfortable implementation partnership (see Chapter 19) and often does not lead to full development of the best solution. Every implementation is a trade-off between functionality, time, price, and risk. A full and frank discussion between purchaser and vendor on this subject can generate major long-term benefits.

● Many organizations have little idea about what they *really* need. Furthermore, it is very difficult to specify precisely in any contract exactly how a system must perform. As users learn more, their aspirations also rise – resulting in "feature creep" (the addition of more capabilities) often without an increase in budget. On the other hand, some vendors adopt the strategy of taking a minimalist view of the capabilities of the system featured in their proposal and make all modifications during implementation and maintenance through chargeable change orders. All this makes the entire system acquisition costs far higher than was originally anticipated; the personal consequences for the budget holders concerned can be unfortunate.

As a result of these problems, this type of acquisition model is not used in small or even some larger procurements. A less complex and formal selection method is *prototyping*. Here a vendor or pair of vendors is selected early on using a smaller version of the evaluation process outlined above. The vendor(s) is then funded to build a prototype in collaboration with the user organization. This fosters a close partnership to exploit the technical capabilities of systems and developers, and helps to maintain system flexibility in the light of changing requirements and technology. This approach works best for those procurements – sometimes even some large ones – where there is some uncertainty about the most appropriate technical solution and where the organizations involved are mature, able to control the process, and not subject to draconian procurement rules.

Box 18.2 Domestic energy delivery at Relient Energy, USA

Relient Energy is a domestic electricity and natural gas energy delivery company that services the needs of 4.4 million customers spread over Texas, Arkansas, Louisiana, Oklahoma, and Minnesota. High standards of safety and reliability are essential for energy delivery. This business also requires management of a large amount of outside plant (utility assets such as transformers, meters, and power poles) and generates a very large number of transactions (for example, maintenance work orders and service connections).

Relient Energy was formed by the merger of four energy companies. A key need for the new company was amalgamation of all the existing GIS into a single, distributed system that created significant management challenges. First, there was the decision about which of the existing systems should be chosen for the new company-wide GIS. Even after a detailed technical evaluation there was still a considerable – and at times acrimonious – debate over which legacy system should be selected. Eventually, a new generation version of one of the existing systems was chosen. This required a new design for a large, distributed database for almost 1000 users spread over four main sites. Data had to be converted from the existing systems and loaded into the new database. Marrying data from two different existing data models was both a syntactic and semantic challenge, taking several months. All of this required strong project management capabilities, as well as great diplomatic skills to ensure that the new system matched the new requirements, was delivered on time, and that it was acceptable to all users.

The Relient Energy GIS went into operation in 2000. It is responsible among other things for maintaining an up-to-date inventory of all Relient's assets, for creating maintenance work order maps, and for network analysis operations such as outage analysis (Figure 18.5).

Implementing a GIS is time-consuming and it often changes people's lives, or at least their careers. GIS users build special relationships with GIS vendors and other users of the same technology. It is sometimes even more onerous to re-implement a GIS, requiring data migration, application redevelopment, and learning new technology. The risks involved are obvious.

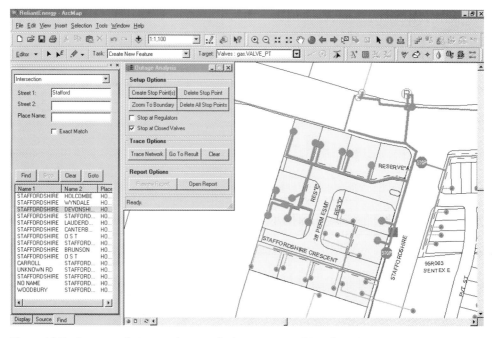

Figure 18.5 Outage analysis is used to see which customers will be affected when gas flow is restricted at a designated stop point (Source: ESRI)

Prototyping is a useful alternative to big linear system acquisition exercises. It is especially useful for smaller procurements where the end result is more uncertain.

18.3 Implementing a GIS

The goal of this section is to provide a checklist of important management issues when implementing a GIS.

Plan effectively
Good planning is essential through the full lifecycle of all GIS projects. Both strategic and tactical (operational) planning are important to the success of a project. Strategic planning involves reviewing overall organization goals and setting specific GIS objectives. Operational planning is more concerned with the day-to-day management of resources. There are several general project management productivity tools available that can be used in GIS projects. Figure 18.6 shows one diagramming tool called a GANTT chart. Several other implementation techniques and tools are discussed in Box 18.3.

Obtain support
If a GIS project is to prosper, it is essential to garner support from all stakeholders. This often requires establishing executive (director-level) leadership support; developing a public relations strategy by, for example, exhibiting key information

products or distributing free maps; holding an open house to explain the work on the GIS team; and participating in GIS seminars and workshops, locally and sometimes nationally.

Communicate with users
Involving users from the very earliest stages of a project will lead to a better system design and will help with user acceptance. Seminars, newsletters, and frequent updates about the status of the project are good ways to educate and involve users. Setting expectations about capabilities, throughput, and turn around at reasonable levels is crucial to avoid any later misunderstandings with users and managers.

Anticipate and avoid obstacles
These include staffing, hardware, software, data-bases, organization/procedures, timeframe, and funding. Be prepared!

Avoid false economies
Money saved by not paying staff a reasonable (market value) wage or by insufficient training is often manifested in reduced staff efficiencies. Furthermore, poorly paid or trained staff often leave through frustration. You cannot prevent this by contractual means (see Section 17.2).

Cutting back on hardware and software costs by, for example, obtaining less powerful systems or cancelling maintenance contracts, may save money in the short term but will likely cause serious problems in the future when workloads

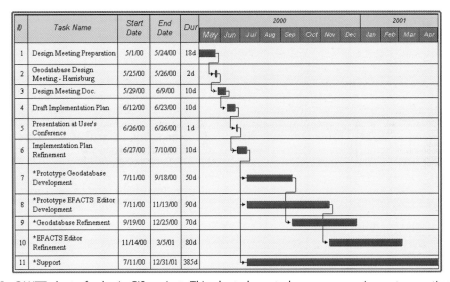

Figure 18.6 GANTT chart of a basic GIS project. This chart shows task resource requirements over time, with task dependencies. This is a relatively simple chart, with a small number of tasks.

Box 18.3 Implementation techniques

The table below summarizes some of the many tools and techniques available for GIS managers to use in implementation projects.

Table 18.2 GIS implementation tools and techniques (after Heywood *et al* 1998 with additions)

Technique	Purpose
SWOT analysis	A management technique used to establish Strengths, Weaknesses, Opportunities, and Threats in GIS implementation (see Section 17.5). The output is a list and narrative.
Rich picture analysis	Major participants are asked to create a schematic/ picture showing their understanding of a problem using agreed conventions. These are then discussed as part of a consensus-forming process.
Demonstration systems	Many vendors and GIS project teams create prototype demonstrations to stimulate interest and educate users/funding agencies.
Interviews and data audits	These aim to define problems and determine current data holdings. The output is a report and recommendations.
Organization charts, system flow charts, and decision trees	These are all examples of flow charts that show the movement of information, the systems used, and how decisions are currently reached.
Data flow diagrams and dictionaries	Charts that track the flow of information and computerized lists of data in an organization.
Project management tools	GANTT charts (see Figure 18.6) and PERT (Program Evaluation and Review Techniques) are tools for managing time and resources.
Object model diagrams	These show the objects to be modeled in a GIS and the relationships between them (see for example Figures 9.18 and 11.3).

increase and the systems get older. Failing to account for depreciation and replacement costs (i.e. by failing to amortize the GIS investment), will store up trouble ahead. The amortization period will vary greatly – hardware may be depreciated to zero value after, say, four years whilst buildings may be amortized over 30 years.

Ensure database quality and security
Investing in the database quality is essential at all stages from creation onwards. Catastrophic results may ensue if any of the updates or (especially) the database itself is lost in a system crash or corrupted by hacking, etc. This requires not only good precautions but also contingency (disaster recovery) plans and trials of them.

Accommodate GIS within the organization
Ideally, organizational procedures should be adapted to promote a GIS, not hinder it. Equally, the GIS must be managed in a way that fits with the organizational aspirations and culture if it is to be a success (see Section 16.7.2). All this is especially a problem because GIS projects often blaze the trail in terms of introducing new technology, interdepartmental resource sharing, and generating new sources of income.

Avoid unreasonable timeframes and expectations
Inexperienced managers often underestimate the time it takes to implement GIS. Good tools, risk analysis, and time allocated for contingencies are

Table 18.3 Percentage distribution of GIS operational budget elements over three time periods (Source: after Sugarbaker 1999)

Budget item	Year 1–2	Year 3–6	Year 6–12
Staff and benefits	30	46	51
Goods and services	26	30	27
Equipment and software	44	24	22
Total	100	100	100

important methods of mitigating potential problems. The best guide to how long a project takes is experience in other similar projects – though the differences between the organizations, staffing, tasks, etc. need to be taken into account.

Funding

Securing ongoing, stable funding is a major task of a GIS Manager. Substantial GIS projects will require core funding from one or more of the stakeholders. None of these will commit to the project without a business case and risk analysis (Sections 17.4 and 17.5). Additional funding for special projects, and from information and service sales, is likely to be less certain. It is characteristic of many GIS projects that the operational budget will change significantly over time as the system matures. The three main components are staff, goods and services, and capital investments. A commonly experienced distribution of costs between these three elements is shown in Table 18.3.

Prevent meltdown

Avoiding the cessation of GIS activities is the ultimate responsibility of the GIS Manager. According to Tomlinson (1996), some of the main reasons for the failure of GIS projects are:

● Lack of executive-level commitment
● Oversight of key participants
● Inexperienced managers
● Unsupportive organizational structure
● Political pressures
● Inability to demonstrate benefits
● Unrealistic deadlines
● Poor planning
● Lack of core funding

18.4 Managing an Operational GIS

Sugarbaker (1999) characterized the many operational management issues throughout the lifecycle of a GIS project as: customer support, operations, data management, and application development and support. Success in any one – or even all – of these does not guarantee project success, but they certainly help to produce a healthy project. Each is now considered in turn. The errors inherent in operational GIS management are described in Box 18.4 and the remit of a strategic GIS manager is described in Box 18.5.

Key management issues for operational GIS are customer support, operations, data management, and application development and support. The successful GIS Manager must achieve success in all of them.

18.4.1 Customer support

In progressive organizations *all* users of a system and its products are referred to as customers. A critical function of an operational GIS is a customer support service. This could be a physical desk with support staff or, increasingly, it is a networked electronic mail and telephone service. Since this is likely to be the main interaction with GIS support staff, it is essential that the support service creates a good impression and delivers the type of service users need. The unit will typically perform key tasks including technical support and problem logging plus supplying requests for data, maps, training, and other products. Performing these tasks will require both GIS analyst-level and administrative skills. It is imperative that *all* customer interaction is logged and that procedures are put into place to handle requests and complaints in an organized and structured fashion. This is both to provide an effective service and also to correct systematic problems.

Customer support is not always seen as the most glamorous of GIS activities. A GIS Manager who recognizes the importance of this function and delivers an efficient and effective service will be rewarded with happy customers. Happy customers remain customers. Effective staff management includes finding staff with the right

Box 18.4 Managing accuracy and error

A clear understanding of the concepts and implications of uncertainty (Chapter 6) and accuracy, error, and sensitivity (Chapter 15) is essential for GIS project participants. Understanding of business risk arising from GIS use and how GIS can help reduce organizational risk is also essential (Chapter 17). This section focuses on the practical aspects relevant to managers of operational GIS.

Organizations must determine how much uncertainty they can tolerate before information is deemed useless. This can be difficult because it is application-specific. An error of 10 m in the location of a building is irrelevant for business geodemographic analysis but it could be critical for a water utility maintenance application that requires digging holes to locate underground pipes. Some errors in GIS can be reduced but at a sometimes considerable cost. Tomlinson (1996) suggested that trying to remove the last 10% of error typically costs 90% of the overall sum. As we concluded in Section 6.5, uncertainty in GIS-based representations is almost always something that we have to live with to a greater or lesser extent. The key issue here is identifying the amount of uncertainty that can be tolerated for a given application, and what can be done at least partially to eliminate it or ameliorate its consequences.

A typology of the errors commonly encountered in GIS is in Chapter 15. Some practical examples of errors in operational GIS are:

● Referential errors in the identity of objects, e.g. a street address could be wrong, resulting in incorrect property identification during an electric network trace.

● Topological errors, e.g. a highway network could have missing segments or unconnected links, resulting in erroneous routing of service or delivery vehicles.

● Relative positioning errors, e.g. a gas station incorrectly located on the wrong side of a dual carriageway road could have major implications for transportation models.

● Absolute errors in the real location of objects in the real world, e.g. tests for whether factories are within a smoke control zone or floodplain could provide erroneous results if the locations are incorrect. This could lead to litigation.

● Attribute errors, e.g. incorrectly entering land use codes would give errors in agricultural production returns to government agencies.

Managing error requires use of quality assurance techniques to identify them and assess their magnitude. A key task is determining the error tolerance that is acceptable for each data layer, information product, and application. It follows that both data creators and users must make analyses of possible errors and their likely effects, based on a form of benefit–cost analysis.

interests and aspirations, rotating GIS Analysts through posts, and setting the right (high) level of expectation in the performance of all staff.

18.4.2 Operations support

Operations support includes system administration, maintenance, security, backups, technology acquisitions, and many other support functions. In small projects, everyone is charged with some aspects of system administration and operations support. But as projects grow beyond five or more staff, it is worthwhile specifically designating someone to fulfil what becomes a core role. As projects become larger, this grows into a full-time function. System administration is a highly technical and mission-critical task requiring a dedicated, properly trained and paid person.

Perhaps more than any other role, clear written descriptions are required for this function to ensure that a high level of service is maintained. For example, large, expensive databases will require a well-organized security and backup plan, to ensure that they are never lost or corrupted. Part of this plan should be a disaster recovery strategy. What would happen, for example, if there were a fire in the building housing the database server or some other major problem?

18.4.3 Data management support

The concept that geographic data are an important part of an organization's critical infrastructure is becoming more widely accepted. Large, multi-user geographic databases use database management system (DBMS) software to allocate resources, control access and ensure long-term usability. DBMS can be sophisticated and complicated, requiring skilled administrators for this critical function.

A database administrator (DBA) is responsible for ensuring that all data meet all of the standards of accuracy, integrity, and compatibility required by the organization. A DBA will also typically be tasked with planning future data resource requirements – derived from continuing interaction with current and potential customers – and the technology necessary to store and manage them. Similar comments to those outlined above for System Administrators also apply to this position.

18.4.4 Application development and support

Although a considerable amount of application development is usual at the onset of a project, it is also likely that there will be an on-going requirement for this type of work. Sources of application development work include improvements/enhancements to existing applications, as well as new users and new project areas starting to adopt GIS.

Software development tools and methodologies are constantly in a state of flux and GIS managers must invest appropriately in training and new software tools. The choice about which language to use for GIS application development is often a difficult one. Consistent with the general movement away from proprietary GIS languages, wherever possible GIS managers should try to use mainstream, open languages that appear to have a long lifetime. Ideally, application developers should be assigned full-time to a project and should become permanent members of the GIS group to ensure continuity (but this often does not occur – see Section 19.5).

18.4.5 Project management

A GIS project will almost certainly have several sub-projects or project stages requiring a structured approach to project management. The GIS Manager may take on this role personally, although in large projects it is customary to have one or more specialist project managers. The role of the project manager is to establish user requirements, to participate in system design and to ensure that projects are completed on time, within budget, and according to an agreed quality plan.

18.4.6 GIS staff

Section 17.2 discussed the key role of staff as assets in all organizations. Several different staff will carry out the operational functions of a GIS. The exact number of staff and their precise roles will vary from project to project. The same staff member may carry out several roles (e.g. it is quite common for GIS administration and application development to be performed by the same technical person), and several staff members may be required for the same task (e.g. there may be many digitizing technicians and application developers). Figure 18.7 shows a generalized view of the main staff roles in medium to large size GIS projects.

All significant GIS projects will be overseen by a management board built up of a senior sponsor (usually a Director or Vice-President), members of the user community, and the GIS manager. It is also useful to have one or more independent members to offer disinterested advice. This group may seem intimidating and restrictive to some. Used in the right way it can be a superb source of continued funding, advice, support, and encouragement.

Day-to-day GIS work typically involves three key groups of people: the GIS Team; the GIS users; and external consultants. The GIS Team comprises the dedicated GIS staff at the heart of the project. The GIS Manager is the team leader. This individual needs to be skilled in project and staff management and have sufficient understanding of GIS technology and the organization's business to handle the liaisons involved. Larger projects will have specialist staff experienced in project management, system administration, and application development.

GIS users are the customers of the system. Other than the leaders of organizations who may rely upon GIS indirectly to provide information on which they make key decisions, there are two main types of user. These are professional users and clerics/technicians. Professional users include engineers, planners, scientists, conservationists, social workers, and technologists who utilize output from GIS for their professional work. Such users are typically well educated in their specific field, but may lack advanced computer skills and knowledge of the GIS. They are usually able to learn how to use the system themselves and can tolerate changes to the service. Clerical and technical users are frequently employed as part of the wider GIS project initiative to perform tasks like data collection, map creation, routing, and service call response. The members of this group typically have limited training and skills for solving *ad hoc* problems. They need robust, reliable support.

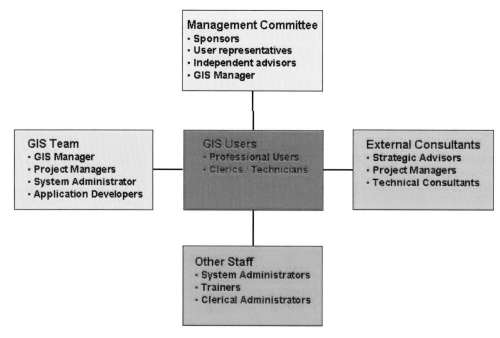

Figure 18.7 The GIS staff roles in a medium to large size GIS project

Box 18.5 *Martin Callingham: GIS manager*

Martin Callingham (Figure 18.8) retired in 2000 from Whitbread plc where he managed Group Market Research and was responsible for both the Spatial Analysis and Direct Marketing teams. He studied musical composition at Trinity College of Music London and then Chemistry at the University of Sheffield, before pursuing a scientific career in research and development in Unilever, the multi-national corporation. He then moved into the area of human perception, finally ending up in a market research company from where he moved to Whitbread. The latter company dates from 1748, has a turnover of over $6 billion, and employs over 100 000 staff. Its origins are as a brewer, commanding somewhat over 17% of the UK market. More recently, however, it has sold off its brewing interests to concentrate on "leisure retailing" – running restaurants, health clubs, and hotels. Apart from its own brands it has the UK franchises to Pizza Hut, TGI Fridays, and Marriott Hotels. This has left the company reduced in size but more focused.

Figure 18.8 Martin Callingham, GIS manager

In Whitbread, Martin pioneered the use of ethnographic research and was instrumental in developing an advanced computer-based culture for the handling of information. He was one of the first commercial users of geodemographics, working with data for 10 000 areas in Britain based on the 1971 Census of Population. Later, he built up the capability of the department to create mathematical models of customer choice. He has pursued an active campaign both within and without the company about the need to combine information of different types and from different sources more intelligently

Box 18.5 *Continued*

than used to be the case. Once PCs became powerful enough to handle large census files Martin set up a Spatial Analysis (see Chapter 14) competency in Whitbread.

As Group Market Research Director, Martin's key contribution over the last ten years has been in the area of raising the corporate awareness of the "spatial dimension". This is much more than the operational and tactical use of GIS in site-selection and evaluation through market area analysis (e.g. Section 2.3.3). It has involved operating not only at these levels but also at the highest strategic levels within the company. For example, when Whitbread was considering merging its 1500 own off-licenses (liquor stores) with those of a competitor, there was concern within Whitbread that the UK regulator, the Office of Fair Trading (OFT), would require a number of shops to be sold – which would render the whole deal pointless. This was because the combined chains might have been construed as creating a series of local monopolies. The work that was requested by a Board Director (at a sensitive time, when the problem was known only to the few people who were working on the deal) confirmed that even under the closest scrutiny the numbers of shops at risk would only be about 30. This meant that the Board could be confident that the deal was not at risk from OFT intervention. The secret in-house project happened and succeeded due largely to Martin's ability and willingness to work at technical and strategic leadership levels within Whitbread.

Another key intervention was during Whitbread's abortive attempt to buy Allied Breweries plc in 1999, when there was a need to convince City of London financiers that there was value in joining the "Big Steak" and "Brewers Fayre" chains together. This was achieved by showing that over 65% of Big Steak outlets were within three miles (5 km) of another outlet in the same chain, and that the same was true of 27% of Brewers Fayre outlets. The implication of the analysis was that, by swapping the brands around in selected locations and putting more Whitbread brands in the Allied sites, consumer choice could be increased and the market grown.

In both of these cases Martin reported directly to a Board Director, who was aware of the importance of the spatial dimension from his previous work on judging the "beer mix" of market segments (see Box 6.1). In both instances the essential work had to be completed in a weekend!

Martin has spoken widely and often provocatively about the cost of geographic data and what he sees as the mindless restrictions that are often put on its use (see Section 17.3.1 on privacy issues). He says: "I am surprised how long it has taken for the ideas behind GIS to spread through the commercial world, and how very limited its ambitions are. I do not think that consultants to the commercial world have been very innovative in the twenty years since Richard Webber [Box 2.1] created the two main geodemographic systems that are still in commercial use today. It is also interesting that the [UK] Office for National Statistics has come around to a similar view, in that they are pricing their data such that resellers will have to truly add value. There are many things that can be done, and more are becoming possible all the time. But I do not feel that there is a sucking force out there making it happen". (See Sections 17.1.2 and 17.1.5 for discussions of data pricing and intellectual property rights issues.)

They may also include staff and stakeholders in other departments or projects that assist the GIS project on either a full- or part-time basis, e.g. system administrators, clerical assistants or software engineers provided from a common resource pool, or managers of other databases or systems with which the GIS must interface. Finally, many GIS projects utilize the services of external consultants. They could be strategic advisors, project managers, or technical consultants able to supplement the available staffing. Although these may appear expensive at first sight, they are often well trained and highly focused. They can be a valuable addition to a project, especially if internal knowledge or resources are limited, and for benchmarking against approaches elsewhere.

They key groups of GIS staff are: the management board; the GIS team (headed by a GIS Manager); the users; external consultants; and various customers.

18.5 Conclusions

Managing a GIS project *is* different from using GIS in decision-making. The first normally requires good GIS expertise and first class project management skills. In contrast, those involved at different levels of the organization's management chain need some awareness of GIS, its capabilities and limitations – scientific and practical – alongside their substantial leadership skills. But our experience is that the division is not clear-cut. GIS project managers cannot succeed unless they understand the objectives of the organization, the business drivers and the culture in which they operate (Chapter 16), plus something of how to value, exploit, and protect their assets (Chapter 17). Equally, decision-makers can only make good decisions if they understand more of the scientific and technological background than they may choose: running and relying on a good GIS service involves much more than the networking of a few PCs running one piece of software. Mutual respect and team working between the users and the experts is at least as crucial in GIS as anywhere. We now move on to consider how such partnerships have operated in GIS to date.

Questions for Further Study

1. List 10 critical GIS project tasks that a GIS manager must perform.
2. Why do GIS projects fail? In addition to the material presented in this text, you might also examine recent conference proceedings, such as those of the Association for Geographic Information www.agi.org.uk.
3. Outline the roles of the main members of a GIS project team.
4. Find example advertisements from GIS magazines or on the web (for example see www.geosearch.com, www.geojobsource.com, or www.find-gis.com) that describe the functions and salaries (see www.sandag.cog.ca.us/sdhost/salary-survey98.html and www.gisjobs.com) of typical GIS staff positions. How do these compare with other industries? And how well are technical GIS staff remunerated as compared with those in management positions?
5. Prepare a new sample GIS application definition form using Figure 18.3 as a guide.

Online Resources

NCGIA Core Curricula (www.ncgia.ucsb.edu/pubs/core.html):
   Core Curriculum in GIScience, Section 3.1 (Making it Work, Hugh Calkins and others)
   Core Curriculum in GIS, 1990, Units 60 (System Planning Overview), 61 (Functional Requirements Study), 62 (System Evaluation), 63 (Benchmarking), 64 (Pilot Project), 65 (Costs and Benefits)
ESRI Virtual Campus course (campus.esri.com):
   Planning for a GIS, by Roger Tomlinson
GIS development guidelines:
   www.archives.nysed.gov/pubs/gis/gisindex.htm

www.gis.about.com/science/gis/cs/implementation/index.htm   General   GIS implementation issues
www.gis.com/data/usingdata/index.html
Using data in GIS
www.geography.about.com/science/geography/library/weekly/aa120197.htm
Interview with GIS Specialist about a City GIS implementation
www.geography.about.com/science/geography/library/weekly/aa080397.htm
Implementing GIS in practice

Reference Links

Maguire D J, Goodchild M F, and Rhind D W (eds) 1991 *Geographical Information Systems: Principles and applications.* Harlow, UK: Longman (Text available online from 'Links to Big Book 1' at www.wiley.com/gis and www.wiley.co.uk/gis).
   Chapter 31, GIS specification, evaluation and implementation (Clarke A L)
   Chapter 33, Managing an operational GIS (Blakemore M J)
Longley P A, Goodchild M F, Maguire D J, and Rhind D W (eds) 1999 *Geographical Information Systems: Principles, techniques, management and applications.* New York: John Wiley.
   Chapter 25, GIS customization (Maguire D J)
   Chapter 41, Choosing a GIS (Bernhardsen T)
   Chapter 42, Measuring the benefits and costs of GIS (Obermeyer N J)
   Chapter 43, Managing an operational GIS. (Sugarbaker L J)
   Chapter 53, Managing a whole economy: the contribution of GIS (Smith Patterson J, Siderelis K)

References

Guptill S C (ed.) 1988 *A Process for Evaluation of Geographic Information Systems.* Federal Interagency Coordinating Committee on Digital Cartography, Technology Exchange Working Group, Technical Report 1. Reston, Virginia: US Geological Survey.
Heywood I, Carver S, and Cornelius S 1998 *An Introduction to Geographical Information Systems.* Harlow: Addison Wesley Longman.
Huxhold W E and Levinsohn A O 1995 *Managing Geographic Information System Projects.* New York: Oxford University Press.
Tomlinson R 1996 *Managing a GIS.* Redlands, California: ESRI.

SUCCESS THROUGH GIS PARTNERSHIPS AT LOCAL, NATIONAL, AND GLOBAL LEVEL 19

Business (as we define it) may operate within commerce, government, or academia. This chapter examines how business collaboration has occurred in GIS and GI, the extent to which it has worked, and the factors that sometimes make it fail. It covers collaborations at the local, national, and the global levels. The significance of leadership by GIS "champions" is highlighted. Ultimately, however, most GIS success comes down to the effectiveness of partnerships and interactions between individual people. Each of these individuals has different perspectives and motivations. A simulation exercise enables the reader to explore how such interactions occur and develop, often in unexpected ways.

Learning objectives

After reading this chapter, you should know about:

- The benefits of collaboration and GIS partnerships of different kinds;
- Some of the pitfalls of partnerships;
- National Spatial Data Infrastructures;
- The need for globally consistent GI;
- Different kinds of projects to create global GI and a Global Spatial Data Infrastructure;
- How organizations and collaborations work in practice.

Box 19.1 GIS Day: an informal global partnership with local outreach

GIS Day is an annual grassroots event which began in November 1999. It is part of the National Geographic Society's Geography Awareness Week, designed to promote geographic literacy in schools, communities, and organizations. GIS users and vendors open their doors to schools, businesses, and the general public to showcase real-world applications of the technology. News of the event is spread by use of the Internet and by advertising. Any organization can host such an event and have it advertised on the Web. Potential visitors can search for the nearest site to themselves. No less than 2400 organizations hosted GIS Day events in more than 91 different countries in 1999 (see Figure 19.1). It was estimated that 2.4 million children and adults were educated on GIS technology through geography on that day.

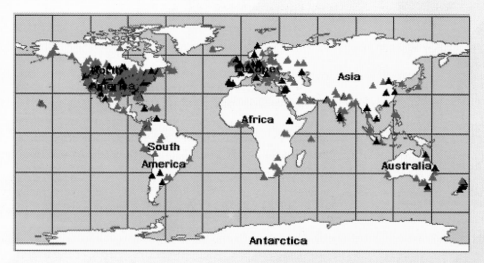

Figure 19.1   The distribution of world-wide GIS Day events in 1999

19.1 Collaborations at the Local Level

Chapters 16 and 17 have emphasized the competitive nature of our world. But there are many occasions in GIS on which we find it best to collaborate or partner with other institutions or people. For instance, data often come from very different people to those with software and no-one has detailed data for many different areas of the world. The need for partnerships applies at the personal, institutional, local area, national, and global levels. We recognized this in the business drivers cited earlier (Section 16.6). The form of the partnerships can range from the highly formal, based on contract, to the informal where participation is entirely voluntary. A good example of the latter is GIS Day (see Box 19.1). Here we look at some major enterprises based on collaboration – including one that failed – to attempt to draw some general conclusions.

Working in partnership is fraught with hazards, yet there is often no alternative. Partnerships can add staff skills, technology, marketing skills, and brand image. They can bring fresh insights on the real user needs, create new products, lead to good publicity, and can lead to cost- and risk-sharing.

The scope for collaboration or partnerships at the local level – typically involving operational matters – varies according to local circumstance. In the USA, for instance, the lack of consistent highly detailed, country-wide GI has fostered many part-nerships to create a single, shared "geographic framework" between different utility companies and local governments – thereby saving much investment and gaining network externalities (see Section 17.3.2.2).

Box 19.1 describes a global but informal partnership between GIS and geography. A second unusual example of local, community use

Box 19.2 *Community use of GIS by the Navajo nation*

Because of the different cultural backgrounds between the authors of this book and those involved in this project, we use the latter's words.

"Dinè CARE is an all-Navajo environmental membership organization, based within the Navajo homeland ... We promote alternative uses of natural resources that are consistent with the Dinè philosophy of Beauty Way. Our main goal is to empower local and traditional people to organize, speak out and determine their own destinies. Membership for us is different from Anglo organizations. It is open to any member of the Navajo Nation who shares our commitment and active participation in the work of protecting our homeland. We do not collect dues or run membership drives. Group membership is a foreign concept ... We live in the communities, our work is very integrated with our communities, and we don't see ourselves as separate from the community. Therefore, all work that takes place is for the good of the community. Our members are individuals and communities from different parts of the 25,000 square mile Navajo Nation spread across Arizona, Utah and New Mexico. They are mostly non-English speaking and live without the conveniences of electricity and telephones in a geographically rural area: 44% of Navajo homes lack plumbing or electricity, while over 80% lack a telephone.

The reality is that our people are doing this work because they have no choice anymore. Many of our traditional people are being discriminated against and exploited on their own lands, simply because their ways are not "progressive" or centered around Anglo notions of economic development (see Figure 19.2).

Our [GIS] project involves working with the Sanostee community developing a restoration plan on a heavily utilized portion of the Navajo Forest. Since 1991, Dinè CARE has sought to defend the forests of the Chuska Mountains and Defiance Plateau from the adverse effects of over 100 years of unmitigated timber cutting. Our current project is mapping the Sanostee region. Although our GIS database is incomplete, we have created some preliminary maps showing vegetation and hydrography of the Sanostee forest landscape. Our data include a 1993 Landsat satellite image ... classified [to] show the most recent view of vegetation land cover. Hydrography, forest stands, roads, GAP, DEM, and boundaries are also included. We are receiving a new, detailed digital soils layer from the Bureau of Indian Affairs ... our goal is to document precisely the condition of the forest and to offer the community a plan for commercial logging alternatives. This includes forest and watershed restoration and regeneration, identification of roads for closure and protection areas based upon endangered species, archeology and sacred sites".

Figure 19.2 From the 1940s through 1971, hundreds of uranium mines were opened throughout the Navajo nation. Dinè families were exposed to radiation. The Shiprock Mountain area contains detritus from mining which has contaminated groundwater. Dinè CARE is much involved in securing compensation for those affected, as well as restoring environmental degradation (Courtesy: Dinè CARE)

Table 19.1 Countries with NSDIs in 2000, after Henry Tom of Oracle Corporation

Argentina	India	Norway
Australia	Indonesia	Pakistan
Canada	Japan	Poland
Colombia	Kiribati	Russian Federation
Cyprus	Macau	South Africa
Finland	Malaysia	Sweden
France	Mexico	United Kingdom
Germany	Netherlands	USA
Greece	New Zealand	
Hungary	Northern Ireland	

is shown in Box 19.2. The particular factors which make such collaborations a success obviously include a high degree of shared interest in the outcomes, mutual trust and shared ownership of what is done, and resources with which to contribute – sometimes only the time commitment of individuals.

19.2 Working Together at the National Level

In some cases, national associations of local bodies have been set up to share knowledge and information and even to act as a coordinating mechanism. The 500 or so local governments in the UK for instance have formed a Geographic Information House to trade in information and have forged a partnership with the Land Registry and other central government partners to create a National Land Information Service.

Beyond such associations of like bodies, we now see truly national partnership initiatives in a number of countries. All of the countries in Table 19.1 have or are implementing partnerships through new forms of National Spatial Data Infrastructure (NSDI). The nature of these initiatives, their title, who is involved, and much else differs from country to country. Despite these differences, there are also striking similarities between NSDIs. This reflects global congruence in the nature of the biggest problems faced, such as duplication of activity and lack of awareness of what already exists.

We concentrate on the US example since this was probably the earliest coherent scheme, it has influenced many others, and it has been underpinned by the leader of the nation state. Web addresses of other NSDIs are given at the end of the chapter. Rhind (1999) has described the early experiences of national spatial data infrastructures.

> Many countries are implementing partnerships through new forms of National Spatial Data Infrastructure. Despite the differences, there are also a number of striking similarities between them.

Box 19.3 Statement of the problem

In the United States, geographic data collection is a multibillion-dollar business. In many cases, however, data are duplicated. For a given piece of geography, such as a state or a watershed, there may be many organizations and individuals collecting the same data. Networked telecommunications technologies, in theory, permit data to be shared, but sharing data is difficult. Data created for one application may not be easily translated into another application. The problems are not just technical – institutions are not accustomed to working together. The best data may be collected at the local level, but they are unavailable to State and Federal government planners. State governments and Federal agencies may not be willing to share data with one another or with local governments. If sharing data among organizations were easier, millions could be saved annually, and governments and businesses could become more efficient and effective.

Public access to data is also a concern. Many government agencies have public access mandates. Private companies and some state and local governments see public access as a way to generate a revenue stream or to recover the costs of data collection. While geographic data have been successfully provided to the public through the Internet, current approaches suffer from invisibility. In an ocean of unrelated and poorly organized digital flotsam, the occasional site offering valuable geographic data to the public cannot easily be found.

Once found, digital data may be incomplete or incompatible, but the user may not know this because many datasets are poorly documented. The lack of metadata or information on the "who, what, when, where, why, and how" of databases inhibits one's ability to find and use data, and consequently, makes data sharing among organizations harder ... If finding and sharing geographic data were easier and more widespread, the economic benefits to the nation could be enormous.

19.2.1 The USA problem

The *Strategy for NSDI* document was initially published by the US Federal Geographic Data Committee (FGDC) in 1994 and revised in April 1997. Its summary of the problem is reproduced in Box 19.3.

19.2.2 Getting to the solution

During the early 1990s, the Mapping Science Committee (MSC) of the United States National Research Council began to investigate the research responsibilities and the future of the then National Mapping Division (NMD) of the US Geological Survey. The MSC coined the phrase "National Spatial Data Infrastructure" and identified it as the comprehensive and co-ordinated environment for the production, management, dissemination, and use of geospatial data. As such, the NSDI was thought of as the totality of the policies, technology, institutions, data, and individuals that were producing and using geospatial data or GI within the United States. A 1994 MSC report a year later urged the use of partnerships in creating the NSDI.

The US Federal Geographic Data Committee (FGDC) adopted the term NSDI to describe a "national digital spatial information resource". Its staff discussed the concept of the NSDI with the Clinton Administration teams which were exploring means to "reinvent" the Federal Government in early 1993. The NSDI was recognized as a means to foster better intergovernmental relations, to empower State and local governments in the development of geospatial datasets, and to improve the performance of the Federal Government. In September 1993, the NSDI was listed as one of the National Performance Review (NPR) initiatives to reinvent that government. Vice-President Gore stated that "In partnership with State and local governments and private companies we will create a National Spatial Data Infrastructure".

One of the primary means of implementing the initiatives arising from the National Performance Review was through Presidential Executive Orders. In April 1994, Executive Order #12906: "Co-ordinating Geographic Data Acquisition and Access: The National Spatial Data Infrastructure" was signed by President Clinton. This directed that Federal agencies carry out certain tasks to implement the NSDI; these tasks were similar to those outlined by the FGDC in its Strategic Plan a month earlier. However, the Executive Order created an environment within which new partnerships were not only encouraged, but required. Though US Presidential Executive Orders are strictly only applicable to Federal

agencies, in this case those agencies were directed specifically to find partners among other levels of government. In addition, the Executive Order had significant effects in increasing the level of awareness about the value, use, and management of geospatial data among Federal agencies specifically. Perhaps more importantly, it raised the political visibility of geographic information collection, management, and use nationally and internationally.

This early political support continued. Speaking in September 1998, Vice President Gore called for stronger efforts nationwide to enhance both the "livability" and economic competitiveness of American communities. The Vice President committed the Administration to expanding its support for the use of geographic information systems technologies and the encouragement of increased public access and sharing of geographic data. The rationale for this was to put "more control, more information, more decision-making power into the hands of families, communities, and regions – to give them all the freedom and flexibility they need to reclaim their own unique place in the world."

19.2.3 What is the NSDI in practice?

The NSDI is defined in the Presidential Executive Order as "the technology, policies, standards, and human resources necessary to acquire, process, store, distribute, and improve utilization of geospatial data". That Order and the FGDC identified three primary areas to promote development of the NSDI. The first was the development of standards, the second improvement of access to and sharing of data by developing a National Geospatial Data Clearinghouse, and the third was the development of the National Digital Geospatial Data Framework. All of these efforts were to be carried out through partnerships among Federal, State, and local agencies, the private and academic sectors, and non-profit organizations – but with the FGDC playing a key role.

The Federal Geographic Data Committee operates as a bureaucracy through a series of subcommittees based on different themes of geospatial data (e.g. soils, transportation, cadastral), each chaired by a different federal agency. Figure 19.3 illustrates the original sub-committees and who led each one. Several working groups have also been formed to address issues on which there is a desire among agencies to coordinate and which cross subcommittee interests (e.g. Clearinghouse, Standards, Natural Resource Inventories).

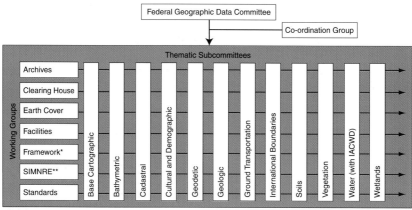

Figure 19.3 The original structure of the Federal Geographic Data Committee sub-committees and lead responsibilities for them

Many of these groups have developed standards for data collection and content, classifications, data presentation, and data management to facilitate data sharing. For example, the Standards Working Group developed the metadata standard, which was formally adopted by the FGDC in mid-1994 and has since been adapted in the light of international and other national developments (see Section 7.5). All of the FGDC-developed standards undergo an extensive public review process. This includes nationally advertised comment and testing phases plus solicitation of comments from state and local government agencies, private sector firms, and professional societies. The NSDI Executive Order mandated that all Federal agencies must use all FGDC-adopted standards.

The second activity area is intended to facilitate access to data, minimize duplication, and assist partnerships for data production where common needs exist. This has been done by helping to advertise the availability of data through development of a National Geospatial Data Clearinghouse. The strategy is that agencies producing geographic information describe its existence with metadata. They then serve those metadata on the Internet in such a way that they can be accessed by commonly used commercial search and query tools (see Section 11.2.1). This has all been rendered practicable by the development of

metadata-creating software within GIS by the larger vendors (Section 7.5). As a result of all this, nearly all Federal agencies, as well as most States and numerous local jurisdictions, have become active users of the Internet for disseminating geographic data. This model does not necessarily assume that GI will be distributed for free. Obtaining some of the datasets requires the payment of a fee, others are free. The Clearinghouse can also be used to help find partners for database development by advertising interest in or needs for GI. The growth in use of the Clearinghouse is shown in Figure 19.4.

The third NSDI activity area is the conceptualization and development of a US digital geospatial framework dataset. This was intended to form the foundation for the collection of other data, minimize data redundancy, and facilitate the integration and use of geospatial data. The Executive Order directed the FGDC to develop a plan for completing initial implementation of the framework by the year 2000 but this has proved somewhat more difficult than the other two activities.

19.2.4 The NSDI stakeholders

The NSDI stakeholders now include public sector bodies, within Federal, State, and local

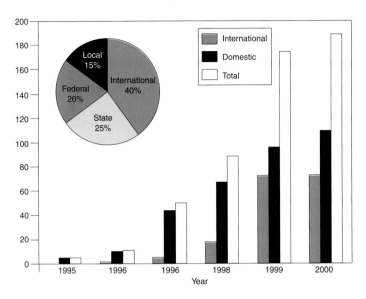

Figure 19.4 US NSDI Clearing House growth in use. The high level of non-US usage is striking. (Source: Henry Tom, Oracle)

government, and the private sector. In general terms, the NSDI is steered by representatives of the following organizations: the National States Geographic Information Council, the National Association of Counties, the Open GIS Consortium, the University Consortium for Geographic Information Science, the National League of Cities, the FGDC, Cooperating State Councils, International City/County Management Association (ICMA), and the Intertribal GIS Council.

Engagement with the private sector was slow to occur. This is now channeled through the Open GIS Consortium, itself a partnership within the NSDI partnership. OGIS is a voluntary membership organization which was initiated to create international standards for geographic information and the means to handle it. The 100 or so strong membership of OGIS is primarily US-based but does include a number of non-US players. This reflects the origins of OGIS and also the US dominance of the GIS marketplace. The power of some of the players involved is considerable (e.g. Lockheed Martin).

19.2.5 Is NSDI a success?

Since the creation of NSDI, many more organizations – from different levels of government and occasionally from the private sector – have formed consortia in their geographic area to build and maintain digital geospatial datasets. Examples include various cities in the US where regional efforts have developed among major

cities and surrounding jurisdictions (e.g. Dallas, Texas), between city and county governments (e.g. San Diego, California), and between State and Federal agencies (e.g. in Utah). The characteristics of these partnerships vary depending on the level of technology development within the partner jurisdictions, on institutional relations, on the funding, and on the type of problems being addressed.

Inevitably, many problems have arisen in such an ambitious project involving huge numbers of people and organizations. Incentivizing different organizations to work together and producing benefits for all the organizations incurring costs is not a simple matter. As indicated above, the inputs of the private sector were initially modest except in delivering metadata software – which they needed to do in any case to meet the needs of their government customers – and steering the creation of standards. The NSDI concept of "bottom up" aggregation of data from many sources to form national datasets is also an intrinsically complex one. The logistics of drawing together data from many thousands of organizations (e.g. US counties) – which vary greatly in resources and inclinations – is daunting. It is also surprising that NSDI has not concerned itself more explicitly with education and training. That has been, is, and will be a continuing constraint on the development and use of GIS and GI (see Section 17.2).

Managing NSDI has attracted two widely different comments. Some have said it is crucial for "no-one to be in charge". Others have seen some management as essential to minimize duplication and foster consistency. FGDC's attempts to lead the whole enterprise have caused something of a backlash. For instance, Federal FGDC members have voiced concerns that its expanded role to address a wider community has impeded the coordination work at the Federal level. Discussions have been held on the creation of a new national organization – a national GeoData Alliance. This is intended to coordinate geographic information activities across all sectors. It is anticipated that a public/private partnership will eventually take over the leadership of the national organization. With a national entity established, the FGDC could return to Federal coordination as its primary role.

Notwithstanding all these complications, NSDI has in its short lifetime generated huge levels of interest in the USA and beyond. Some considerable successes have been achieved, as indicated above. Perhaps its greatest success however has been as a catalyst, acting as a policy focus, publicizing the importance of geographic information, and focusing attention on the benefits of collaboration. This is especially important in a country as large and governmentally complex as the USA.

19.2.6 *Where next and how should NSDI be funded?*

NSDI funding is complicated by its multi-partner, multi-sectoral nature, involving both public interest and private gain, and many potential players. A major report on this topic took a revolutionary approach to the problem (see Online Resources). It argued that the transformation of geographic data into a universally trusted commodity supplied to users throughout the Knowledge Economy is just beginning. Maximizing its value requires not only current, accurate data in a standardized form, but also decision support tools (GIS) and pooling of the data through the Internet and intranets. The report identified the need for capital to correct and "align" inconsistent legacy datasets with the new standards to make them universally transacted commodities. It then suggested how capital can be justified, raised, and used to develop NSDI.

The report's authors envisaged NSDI as a network of data consortia, organized by region, industry, and thematic issue, with its own audience of data providers and users. Also central to their vision is the use of intergovernmental and public/private sector financing. They criticized the current "spare change" method

of NSDI funding and suggested that future finding might be provided through a new National Information Technology Development and Finance Corporation. Privately led, publicly accountable Spatial Information Service Bureaus would be set up with a variety of roles, including certification of data quality. Copyright issues would need to be "clarified" to encourage private sector funding. Public sector bodies would have to do much work to harmonize the format and content of their databases, but these could then be exploited. All this amounts to a very different model for NSDI.

19.3 Nationalism, Globalization, and GIS

We have shown above how the USA has attempted to enhance the quality, utility, accessibility, and awareness of GI. This has been carried out through institutional structures and "umbrella bodies", holding together disparate partners. But GI has hitherto largely been created to suit national needs and is constrained by its historical legacy and the need for continuity through time for comparative purposes. Global GI is currently little more (and sometimes less) than the sum of the national parts and is not readily available. Yet the need for globally or regionally available, easily accessible, and consistent geographic information – especially the framework variety – is increasingly evident. Even though few people currently have management responsibilities which are global, the present shortcomings in GI consistency, availability, and accessibility have severe consequences for many actual and potential users who operate beyond national frameworks. The United Nations Regional Director of Development for the Pacific forcefully argued the need for greater coordination of data collection and supply on a regional and global basis:

"In any area of the world, a number of core data types are required to assess environmental conditions and develop strategies that can lead to long term sustainable human development. For many areas of the world, these data do not exist at the scales required to support these activities. Development of such datasets is labour-intensive and costly and, although these datasets could support a wide variety of applications, no single use can generally justify the full cost of development". Htun (1997)

Global GI is currently little more than the sum of the highly varied national parts and is not readily available.

Box 19.4 A failed international partnership

The International Map of the World or IMW (Thrower 1996) was first proposed in 1891 by Albrecht Penck, the German geomorphologist. His proposal was for a 2200 sheet series of maps at 1:1 million scale, produced to an internationally agreed standard specification. The British Government invited all interested parties to a meeting in London in 1909. It was agreed that the responsibilities for creating the individual map sheets should fall on the respective national mapping organizations. Based on decisions taken at this meeting and confirmed in Paris in 1913, the specification was finalized.

The ultimately sad story of the IMW has some lessons as an international collaborative exercise that never came to fruition, despite the great efforts made. Some 750 map sheets were eventually published though some were not wholly in accord with the standard specification. IMW is now to all intents and purposes a piece of history. The reasons for its ultimate failure were primarily organizational and political ones. There was a lack of commitment of finance by those who agreed to participate. Conflicts in priority with national objectives existed. There was a lack of clearly articulated needs which the IMW was designed to meet and few demonstrations of success from its use. And there was a lack of clear management responsibility for overall success. Finally, there was much duplication of work because of the technologies then available. Much of the same material had also to be created on both national and global map projections.

19.3.1 Global databases

The need for global databases has long been recognized. There are two obvious approaches: to use existing mapping or to create something wholly new from imagery or other sources. Box 19.4 summarizes 60 years of experience in creating the first detailed global map, a common geographic framework world-wide.

Despite the IMW experience, global topographic mapping had been created in various forms by the end of the twentieth century. There are about 11 map series in existence between 1:250 000 and 1:5 million scales which cover all or large areas of the world. The characteristics of the source materials used in the creation of these series differ considerably as does the internal heterogeneity of each series in terms of generalization, resolution, accuracy, and currency.

In the 1990s in particular a number of developments occurred. The US and allied military organizations (see below) created and distributed at low cost the Digital Chart of the World (DCW), now known as VMap Level 0. Derived from the paper Operational Navigation Chart (ONC) 1:1 million scale map of the world, this has been widely used in a variety of applications – including in low price hand-held GPS receivers. Its popularity is despite well-known shortcomings in the quality of the cartographic source material. This project was brought to reality because one organization led it strongly, there was a clear operational need, and financing of it could be justified on that basis. The release of DCW in 1993 led to robust discussions between a number of national mapping organizations (NMO) and the US Defense Mapping Agency (now the US National Imagery and Mapping Agency or NIMA) about infringements of national Intellectual Property Rights. Since then, an agreement has been forged for subsequent VMap products. This ensures that, where NMO-sourced material is used in NIMA products, the providing nation has the option of making this available only for NATO military use and making other arrangements (if desired) for commercial and public dissemination of its material.

Finally, there have been many other plans to build global databases which include mapping – all based on some form of partnership. The most advanced project is that proposed by the Japanese NMO for a new 1:1 million scale global digital map. An International Steering Committee for Global Mapping (ISCGM) has produced a full specification. The committee's proposals involve using a combination of VMap level 0, 30 arc second (i.e. approximately 1 km resolution) Digital Elevation Model data and the Global Land Characteristics Database. Thus ISCGM is using vector and raster data already in the public domain, though not hitherto linked together, to provide the default dataset. It is intended to be replaced on a country-by-country basis wherever NMOs agreed to provide higher resolution data, though the final result would be somewhat inconsistent, especially at national boundaries. Recognizing the inherent difficulties, the ISCGM's specification ensures that, where any difference occurs, both sets of international boundaries will be stored.

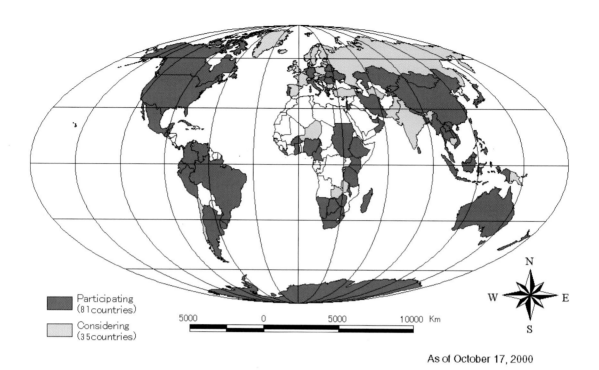

Figure 19.5 The countries which have committed themselves to the Global Mapping Project. (Courtesy: ISCGM)

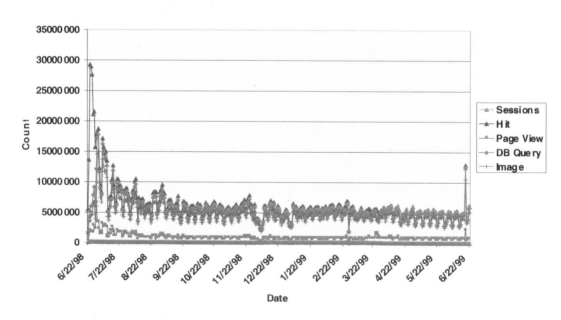

Figure 19.6 Hits on Microsoft's Terraserver System in its first year of operations. Note the right-hand "spike" was caused by an article in *The Economist*. (Courtesy: Microsoft)

Figure 19.5 shows the 81 countries that committed to an involvement in this project in late 2000, though what this means for supply of data in practice is not clear. Rhind (2000) has argued that this is potentially useful but does not go far enough, arguing for a Terraserver-like approach (see below) to 1:250 000 scale mapping.

19.3.2 Global databases from new GIS approaches

Not all global GI has to come from existing mapping. At least four bodies – three of them commercial – are involved in creating or conflating and disseminating wide area or global GI to some common standards. All four involve some form of partnership. The lead organizations are Microsoft, NASA, NavTech, and Space Imaging Inc.

Microsoft's TerraServer began as an experiment to demonstrate the capability of their technology using the largest data files available – imagery provided by SOVINFORMSPUTNIK, and the Landsat images and digital photography provided by US Geological Survey (see Section 7.5.4). The total initial data volume held is large – after compaction, it totals around 2 terabytes (i.e. 10^{12} bytes). In 1999, its first year, the system was sized to cope with 40 million hits per day. Actual traffic was then averaging 7 million hits with peaks of 15 million per day and 6000 concurrent Web connections, equating to 60 000 users per day (see Figure 19.6). These images are accessed online, with different charging mechanisms for the different suppliers' products. High-quality photographic prints can be ordered automatically to arrive by post. A typical charge for a small Soviet image is $8, collected by use of a credit card. That said, coverage is as yet far from global in terms of high resolution imagery (Figure 19.7).

> Not all global GI comes from existing mapping. Various public and private sector partnerships are creating new GI which, in time, will be global in coverage.

Navigation Technologies (NavTech), a global company, has been building databases for over a decade for the intelligent transportation industry

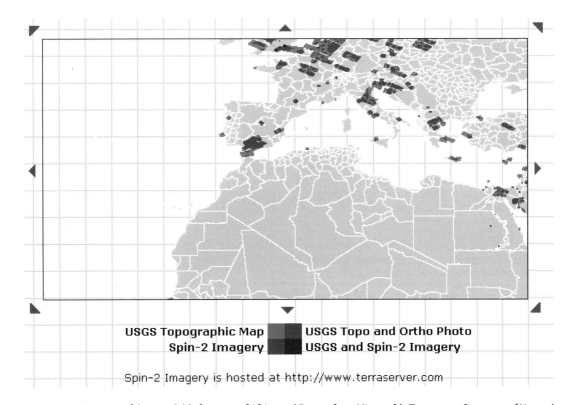

Figure 19.7 Coverage of data available for parts of Africa and Europe from Microsoft's Terraserver System as of November 2000. The coverage of SPIN-2 Russian satellite imagery is sparse in the many parts of the world which were of limited strategic interest to the (former) Soviet Union (Source: www.terraserver.com)

and for in-vehicle navigation. These are built to one world-wide specification. Every road segment in the NavTech database can have up to 150 attributes (pieces of information) attached to it, including information such as street names, address ranges, and turn restrictions. In addition, the database contains Points of Interest information in more than 40 categories. The information has multiple sources and several NMOs have relationships with Navigation Technologies. For example, the Institut Géographique National in France owns shares in NavTech.

NASA launched the $220 million Shuttle Radar Topography Mission in February 2000 (see Figure 3.5(B)). The mission's aim was to obtain the most complete "high resolution" (by the standards of what previously existed) digital topographic database of the Earth. The primary intended use of this data is for scientific research. It did this through use of C-band and X-band inter-ferometric synthetic aperture radars to acquire topographic data between latitudes 60°N and 56°S during the 11-day mission. The resulting grid of heights is scheduled to be available in the public domain within 18 months of the shuttle launch. The data released into the public domain are 1 arc second resolution (i.e. about 30 m).

The first high resolution satellite was launched by Space Imaging Inc. in September 1999. This produces various products, the most detailed being 1 m resolution. Space Imaging also collects and distributes Earth imagery from Canadian, European, Indian, and US sources. The company also delivers a broad array of aerial-derived information products through its Mapping Alliance Program. In essence, Space Imaging is setting out to be a global geoinformation products and services organization, working with a variety of regional partners across the world.

19.3.3 Regional, global, and political approaches to GIS/GI collaboration

The world does not simply consist of independent countries. There are already a number of group-ings which bring together countries for particular purposes – such as the International Standards Organization (ISO), the World Intellectual Property Organization (WIPO), and the World Trade Organ-ization (WTO). These are voluntary groupings bound through signature of treaties or through the workings of a consensus process. But more fundamental groupings bind countries together for a multiplicity of purposes. The prime example is the European Union (EU), containing 360 million people and likely to gain another 140 million by 2010.

The European Commission (EC), the public service of the EU, has been considering a variety of GIS and GI issues since the 1980s. Progress has been slow, complicated by the decentralized, multi-cultural, and multi-lingual nature of decision-making amongst the countries involved and the relatively low priority accorded to GI matters in EC work. Indeed, an attempt to define a GI policy framework for Europe through the GI2000 initiative has failed. There have been some achievements in GIS research terms via the coordination and funding provided by the EC's Joint Research Center in Ispra, Italy. More promising is the incorporation of GI within broader programs. Examples of this include the eEurope and eContent initiatives to foster European industry, and use of new technologies and content by citizens. The EC Green Paper on Public Sector Information, which highlighted the need for much easier access to information by citizens, may have long-running consequences since much information of this type is geographic. Once initiatives to take forward policies are agreed and financed by the EU, they then acquire considerable momentum (see Box 19.5).

On the global stage however, two different approaches to promoting better global GI and GIS have been tried, in addition to the Global Map project (Section 19.3.1). These are the promotion of the Digital Earth scheme and the fostering of a Global Spatial Data Infrastructure. Though arising from different sources, they have some overlaps.

19.3.3.1 Digital Earth

"Digital Earth" formally originated in a 1998 speech by US Vice-President Gore (see Section 3.4 and Box 3.4). He said a pressing problem facing the United States was how to turn the huge amounts of geographic data that were being collected into information that people could use. The challenges can be summarized, with Vice-Presidential quotes, as below:

● There is no effective connection (interoperabil-ity) between the many datasets required to accomplish better management of the Earth's resources, minimization of the problems of nat-ural disasters, etc. "We have islands of oppor-tunity amidst oceans of information chaos."

● There is no common vision to activate coopera-tion among the many elements of the technol-ogy base and the national and international community. "We have landed on the moon, we must now understand our footing on Earth."

Box 19.5 François Salgé: Fostering national and international partnerships

François Salgé is one of the best-known international figures in GIS. From 1979 until 2000 he was employed by Institut Géographique National, the French National Mapping Organization (NMO). In that period, he played a major role in the advent and success of European bodies such as CERCO (the group of 35 European NMOs), MEGRIN (CERCO's commercial arm), EUROGI (the European GI "umbrella body" whose members include national user and vendor bodies) and CEN/TC 287 (see Section 17.2.3).

François' key contributions over the last 20 years have been in the area of geographic information both in France and in Europe. Through his work in IGN, he led the development of the French GI reference data, contributing to shift IGN from a map maker to information provider. He was MEGRIN's first executive director and was president of the European Standards organization (CEN)'s GI technical standardization committee. In addition, he was co-director of the European Science Foundation's GISData program. Throughout the 1980s and 1990s he committed himself to building the European GI infrastructure: its reference data, standards, legal, and research component and policy. This was done through active lobbying of interested parties including the European Commission.

He is now Secretary General of both CNIG and AFIGéO, the French national council of GI and its private sector counterpart, the French not-for-profit association for GI. In this newly established position, he advises the French government on GI policies, formulates related strategies, and coordinates all GI actors in order to plan and implement actions for improving the French Spatial Data Infrastructure and developing the GI economic sector. He also pursues various activities aimed at fostering a European GI policy within EUROGI.

Renowned for his sartorial elegance and his "trade mark" bow tie, Francois is a "visionaire". In the early 1980s, he devised object-based, seamless GI databases when GIS and RDBMS were in their infancy. In the late 1980s he argued strongly for interoperability of GI when transfer standards were high on the agenda. Later he promoted information infrastructure concepts while organizing concerted NMO activity on topics of mutual concern. He is a realist with a penchant for self-derision: he often recalls that "being right too early means being wrong". He is clear that – whilst the technology is sometimes difficult to get right – the attitude and skills of the people involved are crucial to success in GIS, especially in national and international partnerships. François' challenge for the new millennium is "just-in-time information" to support decision-making which necessitates organizing mechanisms for answering decision-makers' questions just when and where they require. His vision is "collaborative GI resources" operating on territories shared between interested parties, such as citizens, the local, national, and European governments, and the private sector.

Figure 19.8 François Salgé, Secretary General, French national GI council.

● There is low awareness and little accessibility by the "common person" to the vast information resources about our planet and its people. "Even rocket scientists find complex applications involving multiple datasets daunting."

Gore's analysis was a damning indictment of what has been achieved to date. His solution was a mechanism for tying together data from multiple,

networked sources and making that integrated resource available visually as a browsable, 3D version of the planet. It would make use of the new broadband Internet and other emerging technologies. He envisaged in the longer term having a complete and consistent digital map of the Earth at 1 m resolution.

Gore legitimated the scheme by arguing that, if successful, "[Digital Earth] will have broad societal and commercial benefits in areas such as education, decision-making for a sustainable future, land-use planning, [and] agricultural and crisis management. [It] could allow us to respond to man-made or natural disasters, or to collaborate on the long-term environmental challenges we face". More than 30 US government agencies that either produce or use GI have subsequently become involved in the initiative.

Despite all the rhetoric, there are tensions between different players. Vice-President Gore stressed the public good benefits and involved Federal Government. The science community agreed, seeing the priorities as environmental protection, disaster management, natural resource conservation, sustainable economic and social development and the improvement of the quality of life. But big players in the private sector have called for government to provide commercial incentives and to make business opportunities available.

19.3.3.2 The Global Spatial Data Infrastructure

The other approach to building world-wide consistent facilities has been to seek to create a

Global Spatial Data Infrastructure (GSDI: Figure 19.9). GSDI is the result of a voluntary coming together of national mapping agencies and a variety of other individuals and bodies. The Steering Committee has defined GSDI as *the broad policy, organizational, technical, and financial arrangements necessary to support global access to geographic information.* The organization has sought to define and facilitate creation of a GSDI by learning from national experiences. As part of this process, it funded a scoping study carried out by Australian consultants. In early 2000 they outlined their conceptual view of how the inputs to decision-making were linked to a SDI and the potential benefits that would arise from having one. They showed that the reality of the benefits was difficult to prove. The fundamental question for any GSDI business case is: what exactly is the product that governments and other agencies will be asked to fund? The consultants said bluntly that "The broad definitions of GSDI are not sufficiently operational (or practical) to build a clear business case around". Key questions to be answered to give credibility to a formal business case for GSDI include:

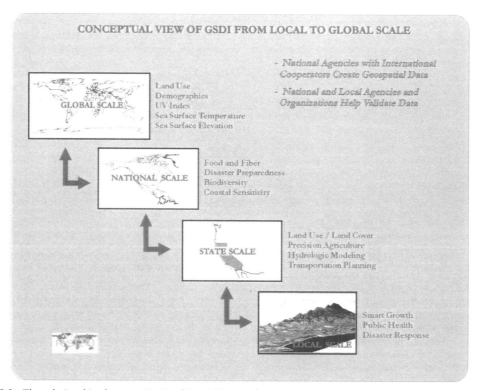

Figure 19.9 The relationships between National Spatial Data Infrastructures in a global federation, as envisaged by the chairman of the Global Mapping project

- Is GSDI a real Global SDI or a federation of national or regional SDI? Figure 19.9, produced by the chairman of the ISCGM (see Section 19.3.1), shows symmetric interacting relationships between SDIs at different levels. How this is to work in practice is a much bigger problem.
- Is GSDI really about encouraging developing countries (or at least those that do not currently place emphasis on geographic data)? In this case, the product may be a suite of policy recommendations and technical assistance approaches that development agencies can apply in their client countries.
- Is GSDI a process, a general framework, or is it a particular product such as a world map or a comprehensive database? Is it to be driven by scientific considerations rather than to "business solutions"?
- Who are the stakeholders?
- What about the demand side of GSDI?

Progress on a GSDI has been slow to date. This is not surprising given the lack of any global government and the complications in building a national, let alone pan-national, NSDI. The real reason is not about inadequate technology but about committed finance, expressed need, and politics.

> Building a Global Spatial Data Infrastructure is incredibly difficult. We cannot agree yet on what it should be and there is little money available. Yet other developments can be influenced by GSDI.

19.4 Political Power and GIS

We have already recorded some startling demonstrations of the effect on GIS/GI of commitments by politicians. Almost all of them have been in the United States: political leadership interest in GI elsewhere has been more muted. The most notable example of all was President Clinton's 1994 promulgation of Executive Order 12906. This was followed up in Vice-President Gore's 1998 address which called for stronger efforts nationwide to enhance the "livability" and economic competitiveness of American communities.

> Clinton and Gore were the best friends GIS has yet had amongst politicians – they embraced the technology and its applications.

There are however many other less well-known examples where GI and GIS have been identified by politicians as central to improving some situation. Perhaps the most striking example is Agenda 21. This agreement was designed to facilitate sustainable development. The manifesto was agreed in Rio in 1992 at the Earth Summit and was supported by some 175 of the governments of the world. It is noteworthy that eight chapters of the Agenda 21 plan dealt with the need to provide GI. In particular, Chapter 40 aimed at decreasing the gap in availability, quality, standardization, and accessibility of data between nations. This was reinforced by the Special Session of the United Nations General Assembly on the Implementation of Agenda 21 held in June 1997. The report of this session includes specific mention of the need for global mapping, stressing the importance of public access to information and international cooperation in making it available.

In practice of course, actions sometimes do not match the fine words of politicians. Nevertheless, it is clear that political support is a crucial requirement for investment if that can only come from governments. Rarely does this political support come from a simple interest in technical matters or even a fascination with geography on the part of the politicians. To obtain support requires fitting the GIS agenda to the existing political agendas, raising awareness, and lobbying.

19.5 Bringing It All Together: How We Actually Operate in the GIS World

Selecting and bringing in a GIS on time, to budget, and meeting all the user expectations is not common. It needs good advice (see Chapter 18) and experience. Following such guidance and previous good practice can help. But what is much more difficult – unless you have been there – is to experience the inter-group dynamics typical in all organizations. This section is designed to provide such experience through simulation of GIS activities in a medium-sized organization. Ideally, to make it work, only the impresario (e.g. the class teacher) should have access to all the briefings. If in places it might seem overly dramatic, it is nevertheless our experience that the real world can be stranger than fiction!

19.5.1 Background: transforming the government of Erehwon through use of GIS

The State of Erehwon has been criticized in the press and others for being ill-prepared for the floods of mid-2001. These led to over 100 deaths and damage of over $500 million; this followed similar ineptness in the face of earlier natural disasters and the ever-declining economic health of the state. Erehwon has been compared very unfavorably with its neighbor, Nerehwon, and other states of the Union. These have made a real success (and publicity coups) from their use of GIS. Nerehwon's GIS is used for emergency planning, for zoning, traffic management, land registration, siting of hazardous materials, school busing, and many other public activities. It is accessed by all the departments of the State government and up-dated by data custodians in different departments.

The new Governor came to office on a wave of anger at the supposed ineptness of her predecessor. She is determined to make Erewhon a leading player in quality, speed, effectiveness, and value for money of public services. As part of this aim and after 3 months in office, she has hired a "hot shot" GIS expert to co-ordinate GIS activities across the State. A Federal Government GIS expert is on hand to provide advice. The plan is that these will collectively get the ideas right. The implementation will then be handled by the State's IT department. A report setting out a vision for the State's use of GIS and what needs to be done to achieve it has been written by the "hot shot" and signed by the Governor. That is summarized in Box 19.6. The prospect of a large investment in GIS has all the usual GIS vendors salivating: all are anxious to keep up to date on developments in Erehwon and influence them wherever possible. Meanwhile, the existing staff are less than impressed: they feel disparaged by the introduction of arrogant (though bright) people from "out of State".

Only the first Act of the exercise is given here; others can be written to take in particular local controversies or problems. The first Act ("everything is possible – and necessary") occurs at an early stage of planning and focuses on meetings between the "hot shot" and IT staff, with involvement from the Governor's advisor (making a cameo appearance), plus the finance, procurement, and other State staff. Preliminary meetings with potential vendors are held to ascertain whether what is being considered is technically feasible. From it must come a plan that is acceptable to the Governor.

It is recommended that the second Act ("getting on with it") is set six months later and is concerned primarily with the final selection of the GIS vendor, negotiation of the complete package, and agreement on roles between in-house teams and the consultants and others – and setting irrevocable milestones for "deliverables". The third and final Act ("rescue the project")

should be set a year later. Milestones have been missed, the Governor is furious, the innocent are being blamed, and a rescue operation is required.

In each case, individual participants are required to seek to follow their briefs, making judgments in the light of discussions and new information. They need to challenge the minutes of the previous meeting if they seem in any way partial. To complicate matters, externally introduced changes of situation and policy occur, usually at crucial moments. Ideally, each Act should be observed by several people, each assigned to concentrate on a small number of actors; the observers should observe different people in each Act. The success of the actors is judged by how much of their brief they achieve (which sometimes requires others to lose). The Governor's advisor and/or the Director of Planning are the judges of overall success for the "business of government". The impresario is required to sum up the situation at the end of each Act and each individual will get personal briefing on his or her own achievements. There is no single, "right" outcome – as in reality.

19.5.2 Act 1: everything is possible – and necessary

Objectives
1. To set up a program for achieving the vision, with timetable of deliverables.
2. To prioritize the goals.
3. Then to have initial discussions with vendors to ascertain their capacity to deliver what is required to time and within the overall budget.
4. To produce a report agreed by State staff on the program, priorities, and risks (and how they are to be managed) for presentation to the Governor by her Advisor.

Dramatis personae in addition to the impresario

Director of Planning for the State	Advisor to the Governor
Director of Finance of the State	Director of State IT group
Chief Information Officer of State	Head of GIS in State IT group
Head of Procurement (works to Finance Director)	Hot shot GIS expert and State's GIS coordinator
Secretary to the Director of Planning (taking a record of each formal meeting)	Federal Government expert in GIS and GI, with involvement in NSDI (see Section 19.2)

Box 19.6 The GIS vision for the state of Erehwon

Erehwon is failing its citizens. Its services are poor compared with other States. They are costly, unfriendly, and ineffective. There is no periodic analysis of needs – we just do what we have always done. The decaying employment base of the State is leading it into a vicious spiral of decay. This must be reversed. I am determined to make Erehwon amongst the leading States. By demonstrating efficient, effective, and focused State operations and a good infrastructure, we will attract new technology industries – the basis of the Knowledge Economy.

Part of our problem is that we have poor management information, especially for "what exists and happens where" across the State. Other States have transformed their effectiveness through their geographic information systems. So must we. We need an integrated and up-to-date database of all the infrastructure, population, and employment sources in the State. We must be able to ask questions that will help us to steer the development of Erehwon. Moreover, we need to be able not only to see the present day in maps and tables. We want to be able to trace historical changes, roll forward current trends, and ask "what if" questions. And the answers must come quickly, accurately, and in whatever form we require. All staff will have access to those data and software tools for which they are authorized – from any desktop in the State government. Citizens will also be able to access much of the information from their homes over the Internet.

The system will need inputs from many, perhaps all, departments. We expect any duplication of activities and data holdings to be weeded out and cost savings to be achieved thereby. If need be, we will re-structure the state government to escape from "silo mentality".

We know this can be done. Others have done it. Skilled consultants and nationally renowned software vendors tell us it is achievable in much less than 18 months if we plan for this and push it forward with energy and commitment from all levels in government. That 18 months is for everything to be working – we expect much quicker wins in some areas. We will neither expect or accept any excuses.

The leadership of the overall project will be placed in the hands of one individual and he or she will have line management for others in different departments as necessary to get the job done. It *will* be done and Erehwon's performance will be radically enhanced.

Georgina Blood
Governor of Erehwon

Sales Director of *Everywhere GIS* (vendor)
Local newspaper reporter specializing in failures in State and local government
Chief Executive of *Tomorrow's GIS* (vendor)

Note: Deputies for state officials may also be allowed to attend meetings. If inadequate numbers of players exist, the roles of the two Heads and the Federal expert may be deleted. The briefings below should be amended as necessary to fit with the gender and number of the players.

Brief for Director of Planning
You have been promoted internally by the new Governor. You know your job is dependent on making her mission a success. Your relationship with the Governor's Advisor is friendly but not close: she seems formidable and has no responsibility for making things actually happen (unlike you). You have grave reservations about the technical people who get excited about the

latest toy without understanding the business imperatives. On the other hand, you think the project budget is too small and the procurement process is likely to take so long that the Governor's objectives will be impossible. Your role is to get a plan that is both acceptable to the Governor and deliverable. Its progress must be monitorable on a weekly basis. To help you, the minutes of formal meetings will be taken by your secretary who is smart, knows the way government works, and is wholly loyal – though last weekend you saw her at a ball game with the "hot shot" GIS advisor ...

Brief for Advisor
You are an old friend of the Governor. She trusts your advice and suspects advice from many officials. Re-election is a long way off but you (and she) recognize that this will only be achieved if the quality and value of what state government delivers is much improved – and this is recognized by voters. Thus the GIS program has *to be seen to work*; who delivers it is unimportant. The press

need to be on-side. You wonder also whether the GIS program could not be used to generate wider benefits, e.g. as a focus for research and investment in the newly established Science Park. Your role is to keep your eye on "the big picture" at the political and economic development level and to show the Governor that the GIS plan will work. So the plan must be a good one. Time however presses ...

Brief for Finance Director of the State

Your predecessor left with the previous Governor, amidst general suspicion that mis-allocation of funds might have occurred. You came from the private sector and are appalled by the sloth, complacency, unwillingness to challenge past practice, poor project management skills, and lack of interest in meeting targets and providing value for money which you find in Erehwon. You have put a plan to the Governor for radical changes in the way in which State government works, including cascading objectives publicly set by the Governor to named officials. Performance-related pay is crucial to make this work. You believe that a fundamental review of programs and departmental contributions to them is long overdue: you suspect that you can make 20% cost reductions at least in the cost of state government *and* get much better performance. The GIS program is small but you are committing time to it for three reasons. It is the first to come up since you have identified what needs to be done across the state, it is important because it touches many departments (some of which seem to serve little useful purpose), and the Governor is obviously enthralled by the idea.

Worryingly, the Governor is not yet committed to your radical approach and her Advisor is difficult to get "on the same wavelength". Your aim is to use the GIS project as a text book example of good management which achieves results for the Governor and which is carried out at low cost and by using external labor and skills where possible.

Brief for Director of State's IT group

You have been in IT for many years and have been responsible for IT strategy, its implementation, and for service delivery. You grew up in a mainframe era but have accommodated to the desktop PC one. You are a strong advocate of formal system specification and use of industry standard tools. But on many occasions you have had to drive the system specification because the users have only a superficial idea of what they want. To make matters worse, users never understand any of the implications of what they say they want and want to make continuing changes in the specification. Your Group gets an

annual budget from the state which in the past has usually been agreed on the basis of the past year plus a few percent increase. Rumors suggest that the Finance Director wishes to introduce zero-based budgeting and contracting out of many activities to the private sector. You fear that the Governor is unimpressed by the steady achievements of your group: she has clearly been misled by vendors about the ease with which IT things can be achieved. Your brief is to ensure sanity prevails in the GIS program, that it integrates properly with other IT projects and takes its turn in the priority queue, that there is enough money allocated, and that it is handled by your experts (using external experts only where necessary).

Brief for Chief Information Officer of State

You have been appointed to bring order out of the anarchy of the present situation. There is little use of information in any corporate sense apart from some of the more basic financial data. Different departments hold their own databases, update these when it suits their own purposes, and are resistant to change. There is no central catalog of databases, even on paper. Duplication of data is rife and its quality is unknown. Much effort is wasted dealing with idiosyncratic queries from citizens and from the Federal Government. Obvious synergies between departments are not being seized, either from ignorance or from protection of empires.

You answer directly to the Chief Administrator of the state and rank on the same level as the Directors (though you have a tiny staff; your method of operating is to be by persuasion or, if that does not work, through the Chief Administrator). The GIS project is uniquely important because of its high visibility and its span across most of the departments and their information/data holdings. It is vital that the project not only delivers something sensible but also forges structures, processes, and relationships to transform the longer-term shared use of information. You have already had one argument with the Director of Finance who questioned your belief that all information should be free – he pointed out that Nerehwon had generated useful income from licensing its data to the private sector and simultaneously reduced frivolous queries. Since he is a powerful figure, you need to avoid direct conflict with him but the "free use of data" practice in the Federal Government may be a useful precedent.

Brief for "hot shot" GIS advisor and coordinator

There is huge potential for GIS to make a difference in Erehwon. The state is so backward that cloning what has been done elsewhere should be easy. The problems are that you do not control the financial or human resources: some of the younger staff are bright but older ones are hayseeds who are terrified of the Governor and fear losing their (long-held) jobs. The local IT people seem to think that there is nothing special about GIS – it is just one more IT application and has to fit in their Neanderthal grand IT plan. Moreover, your last GIS job as a manager of IT in state government was a disaster. You have taken a year out at a university to get up-to-date with new GIS and IT approaches but you suspect that some people in Erehwon know of this past. Your brief is to force something safe but innovative and politically sellable out of a crowd of bureaucrats – and secure much of the credit.

Brief for Head of GIS in State IT group

You work to the Director of IT. He is a wise old chap but hopelessly out of date. He is unaware of half of the impacts of Web-based technology and prefers to operate as if he is still building mainframe systems. You see great advantage in going with truly contemporary technology rather than the 1990s hybrid desk-top/web system of the standard GIS suppliers (who have the problem of supporting compatibility with long-established users). Coincidentally, you play tennis with the son of the Advisor to the Governor. That may be useful.

Apart from your boss's lack of knowledge, you have other challenges. The first is that your boss typically creates teams of a few people working part-time on a project; each person may be in one of a number of groups. You are to "matrix manage" those part-time people on the GIS project. Who is the overall project officer (responsible for the work of vendors, however little they do beyond providing software) is not yet clear – and it should be you. This matrix approach gives staff wide experience but slows development. Finally, the existing PCs and the network are antediluvian and need replacement if the project is to be a success. The project will be a disaster unless there is a proper budget.

Brief for Head of Procurement (works to Finance Director)

You are tasked with getting better value for money through bulk purchasing and standardizing goods purchased. You have discovered that hundreds of suppliers are used in local purchases across the State, resulting in loss of huge potential savings and lots of incompatibility problems. Moreover, you are concerned about some major contracts

placed by your predecessor and your staff. These seem not to have been properly tendered. As a result, you are determined to get a grip on purchasing of goods and services. The GIS project is the first opportunity to demonstrate the benefits of good processes to the public, to your boss, and to other Directors across the State. It is also an opportunity to ensure that you and your colleagues are taken seriously by other departments and help the Director of Finance to ensure that the State's accounts are not qualified by new, hard-line external auditors. Apart from professional pride, qualification of the accounts would be politically unhelpful for the Governor.

You know little about IT and GIS in particular but you "smell trouble" with the different sets of agendas which seem to be operating and the high expectations raised by the GIS project. You must also ensure that the software to be purchased does not leverage large unbudgeted expenditures on hardware. The costs of yearly licensing the software after capitalizing the initial purchase cost do not seem to have been properly considered ...

Brief for Director of Planning's Secretary

You make the record of each formal meeting and chase up action between meetings as necessary. You pride yourself on creating accurate minutes and clear action points as a record. Many times you have made a shambles of a meeting seem good in retrospect. The last thing that your boss wants is to be embarrassed in front of the new people if actions have not been progressed as agreed under his chairmanship. You like working for your boss though he has been getting tetchy, even paranoid, since the new Governor took office. After a recent divorce, you are determined not to work excessive hours and to have some fun. Extended meetings and internal wrangling are not for you. You have just been offered another job.

Brief for Sales Director of *Everywhere GIS*

You are a senior staffer in a large, US-based GIS vendor which has been in business for 20 years. In that time your firm has won business in 30 countries and in 25 US states with its systems. It now has many thousands of customers, some of whom are effectively "locked in" to your systems through the use of proprietary formats, etc. Originally pioneers in many areas of GIS, the firm has been perceived in recent times as being behind the cutting edge. But it has developed a new range of products which will ensure you catch up with the rest in a year or so when it is available (alpha trials have recently had mixed success). The firm's CEO does not want details of these products released yet because of the bugs. He calculates that inertia selling will ensure *Everywhere GIS* can continue to dominate the market because of its installed base. You are less sure given the rising tide of competition. This quarter's sales revenues are lower than last year. If year-end figures are the same there will be a consequent impact on stock price and staff bonuses.

Normally you would not get involved in another sale to "just another" state – and a backward one at that. This should also be an easy sale for your staff. But you have heard on the grapevine that there is likely to be strong support for a local vendor. You also know that the Governor is a close friend of the President and may soon be asked to chair a national GIS initiative ... and your firm needs money for cash flow to roll out the new software.

Brief for Chief Executive Officer (CEO) of *Tomorrow's GIS*

You are a young (30ish) CEO of a 20 person firm which has built a revolutionary GIS using contemporary technology, making no concessions to legacy systems. It is faster and easier to use than anything else on the market and works supremely well in a distributed Internet and intranet environment. It is particularly suited to new users of GIS. You have already won three smallish contracts and this is a particularly important one since your firm is local – you and your colleagues worked in the local university and your firm was supported by seed corn money from the state. Not to win this contract would undermine your credibility elsewhere in the country. And you are getting towards an awkward period where you will need more cash to build the business. You know that you will inevitably be up against *Everywhere GIS* who have the great advantage of many installed sites – likely to influence risk-averse bureaucrats. But your system is much better and is cheaper in terms of resources needed, training, etc. In addition, *Everywhere GIS* now has high overheads, is generally inflexible, and its leaders are loath to do special deals for individual States.

Brief for newspaper reporter

The state of Erehwon is legendary for the poor quality of its government. The new Governor rode to power on a promise to change all that and you are determined to measure progress. Thus far there have been some good appointments of external people – a novel experience – but little improvement is yet visible. You have heard about this pan-government GIS project from some contacts. It is ideal as a measure of whether the

Governor is serious and will succeed. For this reason, you will keep a close watch on it and make use of the contacts you have made over 20 years in the job to get the inside story. You will write a short article for the paper after the forthcoming meetings and will ensure that suitably challenging questions are raised at any public forum. Finally, your editor is not a great fan of the Governor so there will be additional pressure to seek out any project failures and especially to investigate any possible malfeasance. As it happens, a young reporter on the paper is also living with one of the leaders of *Tomorrow's GIS* ...

Brief for Federal Government expert in GIS and GI

You have seen it all, with many years experience in GIS in Federal Government and of collaboration with many States. Erehwon is at least a decade behind the leading States and Federal agencies. Your role is to offer help wherever possible, ensure that the State does things in accord with the tenets of the National Spatial Data Infrastructure and to set up data, information, and knowledge exchange arrangements. But you must do all this without providing any resources (the Federal Government cut-backs have been more severe than expected). You like Erehwon – your wife was born there, your annual holidays are taken in the State, and you can live well there on your DC salary. If they really are serious about GIS you might get a secondment to help them out or even take over the job of that idiot who is their "hot shot" GIS advisor. After the mess he made in his last job ...

19.5.3 The process

Act 1 is best operated by having at least three sets of meetings (but the number of meetings and participants in each one is a matter for the players). The first one should be of the key players – the Directors and the Advisor – whilst a later meeting(s) will involve all the State staff listed. Later still (once the internal story has been sorted out), a group will meet with each of the two contractors to assess the technical feasibility of what is being planned and of the timescales involved. Multiple bilateral meetings between individuals will be needed to prepare for the formal meetings. The citizens of Erehwon also need to kept informed at appropriate times.

For each formal meeting, the most senior person present should take the chair and propose an agenda and objectives for the meeting. Failure to participate in the meeting counts as wimpish behavior, with consequent effects on reputation. The assessors should give individual feedback to each player only after the first meeting in which they participate. Formal meetings should be followed up by minutes recording key points discussed and actions agreed; all those present should receive them. Informal meetings between individuals may or may not be recorded. The newspaper reporter will seek to meet with whoever he thinks appropriate at any time. Some staff and the vendors may wish to call him to ensure he is aware of recent developments . . . .

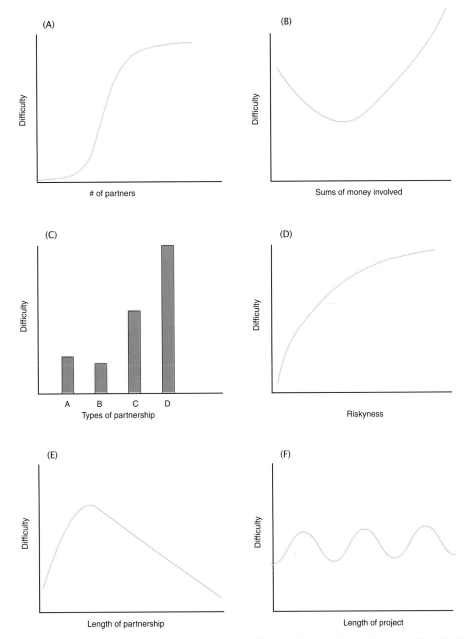

Figure 19.10 How the difficulty of partnering varies with different factors. Key to diagram (C) is A: Private sector partnerships; B: Unpublicized partnerships between government agencies; C: Publicized partnerships between government agencies; D: Public/private sector partnerships with more than three partners

19.6 Conclusions

Change in GIS is endemic – even though some policy change may take us back to a situation commonplace years earlier, thanks¹ to the combination of different events. In this context, nowhere is GIS management more complex than in a partnership environment. Other people's agendas are prone to change as they become affected by things which are not central to your own concern. Figure 19.10 suggests how the level of difficulty in partnerships varies with some different circumstances. In addition, the inter-actions between the different factors can com-plicate things still further. The active involvement of politicians brings high rewards and high risks. If they support you then lose office, their successors may be alienated. If they see advantage in your scheme, they can often make it happen.

Despite all that, there is often no alternative to working in partnership. It can bring vital components – staff skills, technology, marketing skills, and brand image – to add to yours and make a new product or service. It can bring fresh insights on the real user needs and can create superb support and publicity from fellow citizens. It can lead to cost- and risk-sharing. In many GIS and GI matters in particular, there is no real alternative to partnerships because of the way in which public and private sectors each provide some crucial element of the final whole. The probability of success is raised by keeping the objectives clear, ensuring overall responsibility is agreed and adequate resourcing exists, and keeping the numbers of partners to the minimum possible. And remember that the interactions and trust between the people involved will ultimately determine how well the partnership succeeds.

Questions for Further Study

1. Imagine that you are trying to get three local parties – a local government, a utility company, and a commercial supplier of GI – to work together to produce a Web-accessible GI database for free use. What do you imagine would be the problems? How would these change as you move up the spatial scale to the nation, to the region (e.g. the Americas or Europe), and to the world as a whole?
2. Do a one hour search by whatever means you have available locally for a global database in one thematic area (e.g. land use) and note the findings. How many sources of supply have you found? How many organizations were involved, in what sort of partnerships? What is

the relationship between the datasets? What is the internal coherence and quality of what you have found?
3. What do you predict will be the final outcome of the Global Map, Digital Earth, and GSDI projects?
4. How would you "sell" the benefits of GIS and GI to a politician? What would be the things likely to influence him or her? What particular support would you seek?
5. Read an account by any senior politician or an investigative journalist about how things really happen inside government-controlled organizations and partnerships. Are you surprised by what you read?

Online Resources

ESRI Virtual Campus course (campus.esri.com): Protecting Your Investment in Data with Metadata, by George Shirey (to include the lesson "Metadata, the Backbone of the National Spatial Data Infrastructure (NSDI)" in the Module 'Metadata: What's the Big Deal?')

General programs:
Agenda 21:
www.un.org/esa/sustdev/agenda21text.htm
and
www.un.org/ecosocdev/geninfo/sustdev/indexsd.htm
GIS Day:
www.gisday.com/gisday/index.html
Navajo Nation community GIS project:
www.dinecare.indigenousnative.org/gis.html
Global Map project web site:
www1.gsi-mc.go.jp/iscgm-sec/index.html;
General introduction to Digital Earth:
www.develop.larc.nasa.gov
www.digitalearth.gov
www.digitalearth.gsfc.nasa.gov
digitalearth.gsfc.nasa.gov/handouts.html#ppp3
www.develop.larc.nasa.gov/index2.html

Global GI providers or distributors:
NASA shuttle radar mission:
www-radar.jpl.nasa.gov/srtm
Navtech: www.navtech.com
Space Imaging Inc:
www.spaceimaging.com
Microsoft Terraserver:
www.terraserver.microsoft.com
NIMA site: www.nima.mil

Selected National Mapping Organizations:
Australia: www.auslig.gov.au/index.htm
France: www.ign.fr

Great Britain: www.ordsvy.gov.uk/ and
www.ordsvy.gov.uk/literatu/external/oxera99
Japan: www.gsi-mc.go.jp
USA: mapping.usgs.gov
Regional or national umbrella organizations:
www.cerco.org
www.eurogi.org
www.eurogi.org/gsdi/index.html
www.agi.org.uk

Selected local, National, and Global Spatial Data:
Infrastructures:
www.fgdc.gov
www.fgdc.gov/I-Team.html
www.fgdc.gov/publications/reports.html
www.ngdf.org.uk
www.cgdi.gc.ca
www.auslig.gov.au/index_sdi.htm
www.gsdi.org and
www.gsdi.org/canberra/masser.html
www.anzlic.org.au/anzdiscu.htm
www.land.state.az.us/agic/agichome.html
Open GIS Consortium:
www.opengis.org

Reference Links

Longley P A, Goodchild M F, Maguire D J, and Rhind
D W (eds) 1999 *Geographical Information*

Systems: Principles, techniques, management and applications, New York: John Wiley.
Chapter 47 Characteristics and Sources of Framework Data (Smith N, Rhind D)
Chapter 56 National and International Geospatial Data Policies (Rhind D)

References

Htun N (1997) The need for basic map information in support of environmental assessment and sustainable development strategies. In Rhind D W (ed.) *Framework for the World*: Cambridge: GeoInformation International, 111–119.

Rhind D (2000) Current shortcomings of global mapping and the creation of a new geographical framework for the world. *Geographical Journal*, **166**: 295–305.

Thrower N 1996 *Maps and Civilization*. Chicago: University of Chicago Press.

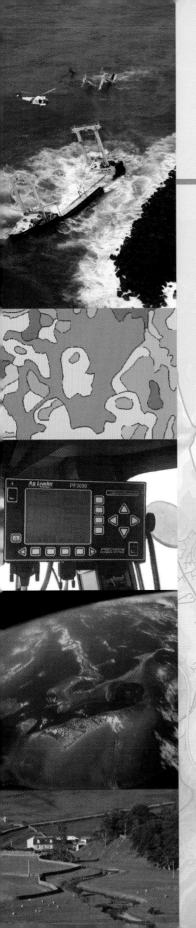

EPILOG

The goal of this final chapter is to revisit and summarize the key ideas introduced in this book and to offer some predictions about the future of GIS as we look forward into the early part of the 21st century. Our journey has encountered a restatement of many existing ideas about GIS in the context of geographic information systems and science. It has also seen the introduction of some new ideas about GIS principles, techniques, and practices. This chapter also reviews the relationship between GISystems, GIScience, and GIStudies -- how GIS is embedded in society. In the final section we look at both pessimistic and optimistic future scenarios for GIS.

Learning objectives

After reading this chapter you will be able to:

● Review the main topics covered in this book, including
  – The differences between GISystems, GIScience, and GIStudies
  – The impact of the Internet on GIS
  – How GIS and management interact
  – How GIS can be applied in a real-world practical context;

● Anticipate the way GIS is likely to develop in the next few years;

● Better plan your involvement with or within GIS.

20.1 Introduction

In *Geographic Information Systems and Science* we have tried to define the content and context of GIS. For the many who are already experienced in the field we have introduced some different ideas and a new way to think about GIS as both systems and science. For those new to GIS we have tried to write a primer that introduces the modern thinking that is becoming increasingly prevalent. There is no question that GIS in the 21st Century is a valuable field of endeavor and that progress is accelerating on several fronts. Allied to heightened awareness, uptake is increasing, cost-benefit cases are even more demonstrable, critical analyses of the use of GIS are being published, there are fundamental improvements in scientific methodologies, there is a realization that management is a vital ingredient in a successful project (technology is necessary, but not sufficient), and ever more ingenious and eclectic applications are appearing every week.

This Epilog closes the book with two major parts. The first reviews some of the main themes that have run throughout the book: the complementary roles of systems, science, and studies; the role of the Internet in redefining GIS over the past few years; the importance of management in a field that remains largely technology-led; and the applied nature of GIS in the real world. The second section offers some views on what the future is likely to hold for GIS. Here we suggest possible optimistic and pessimistic scenarios for the world of GIS.

20.2 Revisiting Some Major Themes

20.2.1 Systems, science, and studies

In Chapter 1 we identified geographic information science (GIScience) as the set of fundamental issues arising from the use of geographic information systems (GISystems). That definition is helpful, because it suggests that people using GIS will inevitably ask questions about such topics as scale, accuracy, and the relationship between humans and computers. Questions like these can only be answered through systematic scientific inquiry, and access to major branches of knowledge like spatial statistics and cognitive science. As Chapters 8–15 in particular demonstrate, GISystems provide the tools to collect, manage and display geographic data very efficiently and effectively. The implication here is clearly that *geographic information systems support and drive science* (Figure 20.1).

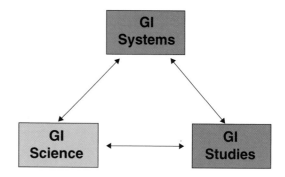

Figure 20.1 The inter-relationship between GI Systems, Science, and Studies

Geographic information systems support and drive science.

On the other hand, many of the issues raised by GIS have been around for centuries and humans have already accumulated vast stores of relevant theory and technique, as we first noted in Section 1.6 and as will have become apparent throughout this book. Cartographic design principles, so important for effective communication using GIS (Chapter 12), originated long before the advent of either digital computers or GIS. This suggests an alternative view of the relationship between science and systems: *geographic information systems implement the storehouse of knowledge known as geographic information science.* The term *geographic information science* was not coined until the 1990s, but from this perspective the act of coining it reflected a need to reprioritize old science to be more relevant to new systems, and to motivate new research in old areas of science to solve problems that systems have brought to the fore. The story of the representative fraction, and the difficulties of moving it into a digital world covered in Section 3.7, illustrates how new technology has created a need for cartographers to rethink a problem that they may long have thought solved.

From this perspective GISystems mine GIScience, and the future of GISystems depends on a constant flow of new ideas from GIScience. GIScience extends well beyond the immediate set of issues motivated by today's GISystems, into areas such as cognitive science and information science, where tomorrow's discoveries may make possible a new generation of GISystems that bear little relationship to today's designs. This new perspective, of GIScience predating GISystems, is much more ambitious than the reverse one, and concerned with a much longer historical perspective.

Geographic Information Science drives the development of Systems.

But this argument misses one very important element – the need to reflect on GISystems, their impact on society, and their appropriate place within it. We noted in Section 1.6 that any technology reflects in part the nature of the society that invents and uses it, and in turn influences the way that society evolves. GIStudies is not about the technology itself, or about the science that makes that technology possible, but about how that technology is embedded in a societal context. Careful reflection is a vital part of the role of the academy and GIStudies is a vital part of the academy's relationship to GISystems. As the title suggests, and as Section 1.6 shows, this book has tended to focus more on the scientific and technical aspects of GISystems and GIScience than on the reflective aspects of GIStudies. We note the now considerable and expanding literature expressing a critical perspective on GIS and its role in science and society (Pickles 1995, Curry 1998). An important perspective raised here is that applied science creates value, which allows increased consumption. Appraisals of GIS in society are about the redistribution of the fruits of scientific activity. This is a book about production, not consumption, but we acknowledge that means have ends.

GIStudies is concerned with the way systems and science interact with society.

However important it may seem, information technology is inevitably limited in its ability to represent the complexity of the real world, or to address and solve its problems. Understanding its limitations has been one of the themes of this book, notably in the chapters on representation, uncertainty, and management. But sometimes even the most skeptical of GISystems users may fail to see the wood for the trees. It is important, therefore, that GIStudies continue to attract people who are less immersed in the technology and can think deeply about the meaning of GIS, its role in society, and about GIS weaknesses and limitations.

20.2.2 GIS and the Internet

A second theme of this book has been the role of the Internet in redefining GIS, a topic that was introduced in Section 1.4.3.1 and discussed critically in Section 16.4.4. The Internet met a need that none of the previous information communication technologies had been able carry out effectively:

- A set of straightforward protocols that could accommodate immense increases in activity.
- A decentralized structure that required no top-down investment or control and made it easy to connect to the network.
- Compatibility with the Unix operating system that had become very popular in the late 1980s, largely supplanting many proprietary systems.
- The ability to send any kind of information through the same network.

Even so, it was not until the advent of the World Wide Web (WWW), and the release of the first easy-to-use browser (Mosaic) in 1993 that the Internet really entered the public consciousness. Today, of course, fortunes have been made (and lost), and the Internet and WWW have fuelled what many claim to be the longest period of sustained economic growth in history.

What is the real impact of the Internet on GIS? First, and perhaps most importantly, it has shifted the vision of GIS and its basic role. GIS was invented in the 1960s to perform user tasks that could not be performed manually, or could only be performed manually at much greater cost and with much less accuracy. In the Internet world GIS has become a means of communicating geographic information between users. It is the means whereby one part of an organization provides its input to another, the means whereby a group makes its knowledge available to other groups, and the means whereby a group attempts to persuade others of the appropriateness of its views.

The Internet was by a long way the single biggest external stimulus to GIS in the 1990s.

To support this, the Internet provides easy access to distributed systems and databases. A simple click is all that is needed to obtain a map or dataset from a remote site, or to submit data for analysis by a remote server. Simple clicks allow the user to navigate from one site to another following hyperlinks. Browser software, combined with the HTTP (hypertext transmission protocol) communication standard, allows users to see information organized in many different forms, using color, sound, and animation. This is not the world of the 1960s, when the only means to interact with a computer was through a teletype machine that printed in black and white at 10 characters per second. It is not even the world of the mid-1990s where handling geographic information meant access to a powerful and expensive desktop workstation and a reasonable understanding of technology.

Second, the Internet is driving a reorganization of how GIS work is done. The old world of data producers, software producers, and users who either created their own data and software or obtained them on magnetic media through the mail has given way to a more diverse world in which Web sites provide access to both data and the services that use data to create useful information. In this new world users are both consumers and producers of data and both data and software flow freely or are sold for profit over the Internet. As the communication model suggests, this is an interactive new world of senders and receivers, empowered by software. WWW browsers have fundamentally redefined how users interact with computers, by making sharing easy, and by supporting a rich world of multiple media. In Section 1.5.3 we even went as far as to define yet another use of the GIS TLA (three letter acronym) – geographic information service. A GIService is a 24 by 7 (24 hours a day, 7 days a week) operation that provides access to geographic data and processing capabilities. Such services are already becoming available over the Internet. They provide services for flood risk mapping, lifestyle classification, site suitability analysis, address geocoding, mapping and routing (Figure 1.16). We envisage considerably more activity in this new GIS area in the next few years.

GIS will see the introduction of many new on-line GIServices in the next few years.

Third, the Internet is having a profound impact on the significance of geographic space. It appears at first glance to make geography irrelevant, since it makes no difference whether the site supplying your map request is next door or several thousand miles away. Many Internet addresses say very little about the geographic location of the sender (e.g. domains such as .edu and .com), and a service such as Altavista automatically searches the entire Internet, without any bias towards sites that are in close geographic proximity. Only recently have substantial efforts begun to add more geographic content to the Internet. Restaurant owners are clearly more interested in promoting their sites to nearby markets and the same goes for any service that operates in a geographic context, such as local government or political representation. This process is likely to grow in importance, as the WWW becomes swamped by its own growth (do users really need to search the 30+ million sites on Internet today to find a nearby restaurant?).

This trend to a more geographically based line of Internet services is also in line with recent developments in the area of wireless services, as we saw in Section 8.5.3. The new generation of wireless phones is already to some extent Internet-enabled, offering the ability to retrieve and display short e-mail messages and even simple maps. New location-based services are able to identify nearby restaurants and retailers, based on techniques that determine the location of the user's device.

The idea of the "death of distance" brought about by the emergence of new communications technology, such as the Internet, is a gross simplification of reality. The geography of our world remains a variegated one. Where we start from, how we behave, and how we as individuals and societies prosper are all strongly influenced by differences in the contemporary geographies in which we operate. Contemporary geographies comprise a complex of natural, business, technological, educational, legal, and social environments (see Box 20.1 and Figure 20.2). By our actions we also help to shape that geography. We believe that GIS can help us to reshape the world, to enhance the prosperity and quality of life for many of the world's inhabitants. This is why we are involved with geographic information systems and science.

20.2.3 GIS and management

A third key thrust of this book has been the relationship between GIS and management. We have already discussed the problems of undue focus on technology in the context of the debate about systems and science. Similar comments apply to the relationship between science, technology, and management. There are two important dimensions to GIS and management: the management of GIS and the use of GIS for management.

GIS projects are just like all other projects – to be successful they need to be well managed. In Chapters 16–19 we present the many ways in which management can help contribute to the success or failure of GIS projects. These include working to clearly defined and known objectives, casting objectives in the form of projects of no more that 12 months duration, estimating and minimizing risks, and communicating effectively with co-workers and users. Although this will be entirely obvious to many, it is still surprising how many GIS projects fail because of lack of good management. It seems that the appeal of the technology, the excitement of a new scientific discovery, and the thrill of applying GIS in new

Box 20.1 Two views of the importance of geography

"The world has never been a level playing field, and everything costs." David Landes 1998, *The Wealth and Poverty of Nations: Why some are so rich and some are so poor*. New York: Little, Brown and Company.

"The main proposition of 'The New Geography' is at once both startling and obvious. Notwithstanding the 'death of distance' caused by modern communications and by the portability of 'new economy' businesses along with creators of their intellectual property, place has never been more important. Freed from older constraints of location – nearness to raw materials, markets, or pools of cheap labour – new businesses will go wherever they think their highly educated and well-rewarded workers will be most attracted by the quality of life. Such sought-after people have become, in effect, sophisticated consumers of place, be they high tech 'nerdistans' in Raleigh, North Carolina, elite rural "Valhallas", or revived metropolitan regions such as Manhattan, Seattle or the inner lakeshore of Chicago.

Apart from the bleak prospects that follow from this for places that are aesthetically challenged ... Mr Kotkin draws attention to ... rising inequality ... In the heart of Silicon Valley, East Palo Alto and Palo Alto, divided by ... Highway 101, share a name but little else. The former district, with its largely poor, Latino population, has one of the highest murder rates in the nation; the second, a home to venture capitalists and net entrepreneurs, the highest house prices. In many ways the new geography is depressingly familiar". *Economist* review of Kotkin J 2000 *The New Geography*. New York: Random House.

areas all combine to overshadow the willingness to plan properly, provide adequate resources, and be realistic about timescales. We can all learn from our mistakes and we advise all would be GIS managers to read about failed GIS projects such as the English Police system (Openshaw *et al* 1990) and the London Ambulance Control system (cited by Reeve and Petch 1999).

> Poor management remains the number one reason why GIS projects fail.

Over the past 10 years many have begun to realize that a significant number of the world's problems are inherently geographic in terms of both their causes and their impacts. At the same time, investigation has revealed that crosscutting inter-disciplinary science and social science is required in order to seek and implement solutions. We argue that global warming, the hole in the ozone layer, the AIDS epidemic, desertification, inner city social unrest, and other massive problems can benefit from using GIS to manage data and support investigation. GIS can help integrate data from multiple sources, record how events and process change over space and time, support the evaluation of alternative explanatory models, and communicate results in an effective way.

> GIS can play a significant role in helping to resolve global problems.

GIS is increasingly a useful part of management's weaponry. It can and does provide competitive advantage (e.g. in assessing store location, reducing customer complaints from electricity outages, and modeling the propensity of steep slopes to fail in times of high rainfall). It is a useful means of evaluating policy and practice options in both the public and the private sectors. GIS also has an important role to play in communication and information sharing, and in public policy formulation and implementation.

The GIS Manager of today needs to be concerned not just with science and technology, but also a wide range of management issues. We saw in Chapter 17 that these include managing people and business assets, as well as an awareness of legal issues such as intellectual property, data protection, ownership, and copyright. We will return to these topics again later in this chapter. GIS Managers need to be able to create and implement complex information systems as discussed in Chapter 18.

20.2.4 GIS applications

The fourth distinctive characteristic of this book is the focus on GIS applications – how GIS can be used to solve substantive real-world problems. GIS theory means nothing until it is put into practice. As all GIS practitioners well know, GIS theories and models always have to be adapted to the constraints imposed by the real world: data, time, computer power, and human resources are never infinite.

Section 2.1.1 entitled "One day of life with GIS" demonstrated clearly how GIS pervades many aspects of everyday life and that it has enormous practical value. GIS has the potential to be employed wherever people seek answers to geographic questions, however large or small. The applications of GIS are many and we have tried to present a representative fraction of possible applications as a series of Boxes throughout this book. They range from those that deal with space, at the sub-architectural scale (Box 10.3 "Preserving medieval art with GIS in Bucharest, Romania"), through a range of geographic scales from the local (Box 16.3 Evolving GIS at Kruger National Park, South Africa), and the national (Box 8.5 Using GIS in Honduras hurricane disaster rescue and relief) to the global (Box 19.1: GIS Day: A global but informal partnership).

> GIS has been applied in thousands of different application areas. It is the commonality of the underlying systems, science, and studies that links them all together.

In Chapter 2 we chose four examples that seem to us to illustrate something of the range of GIS applications. Transportation logistics (Section 2.3.4) is concerned with the efficient movement of goods and services over networks – dispatching technicians to repair gas leaks, delivering letters and parcels, collecting dead animals, and hauling lumber from a mill to a distribution warehouse are all representative of the problem domains. In this application field the focus is very much on conserving resources and reducing costs.

The history of environmental and natural resources applications is as old as GIS itself. GIS allows us to measure, monitor, model, and manage the environment better at a time when it is under severe pressure from human development (Section 2.3.5). It is perhaps in this application area that GIS has been used most for modeling.

Government users were among the first to discover the value of GIS. Today governments throughout the world are trying to improve the quality of their products, processes, and services, by using resources more efficiently. Governments also use GIS to incorporate public values in decision making, to deliver services in equitable ways, and to represent the views of citizens by working with elected officials. Tax assessment, as described in Section 2.3.2.2, is a classic local government GIS application.

GIS has also been employed widely in the business sector for business and service planning (Section 2.3.3). Many models of consumer

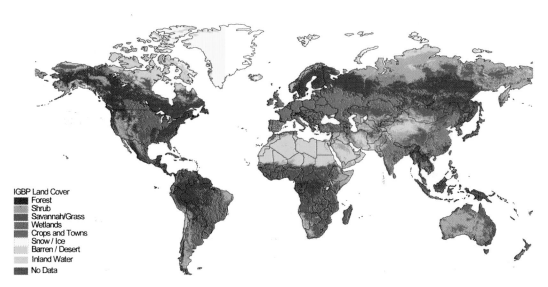

IGBP Land Cover
- Forest
- Shrub
- Savannah/Grass
- Wetlands
- Crops and Towns
- Snow / Ice
- Barren / Desert
- Inland Water
- No Data

Figure 20.2 The pressures on the global environment have changed dramatically as *Homo sapiens* have spread and multiplied. Humans have either ploughed or paved over almost a quarter of the land. If pasture is taken into account, humanity can be said to have transformed about half the land on the planet for its own uses. The geographic distribution of land cover is shown in this global map assembled from Advanced Very High Resolution Radiometer data by NASA and the US Geological Survey.

activity and facility location are based on composite indicators for small areas. The term "geodemographics" has been coined to describe this group of GIS applications. The list of specific applications includes store location, predicting consumer spending, branch rationalization, and decisions about which goods to stock.

20.3 GIS Futures

Throughout this book we have seen some of the many ways in which GIS has been used to represent geographic reality. However, as we saw in Chapter 2, the final goal of GIS usage is rarely descriptive "what is" or "where is" activity. Rather the goal is to use GIS for prescriptive "why is" and "what if" scenarios, and probably the most important test of a GIS is its ability to forecast (Table 20.1). We have come a long way in our descriptions of the role of science, through software, on society. In this section we look towards different scenarios for GIS itself. Just as GIS forecasts are characterized by uncertainty, so the future of GIS is itself uncertain.

Figure 20.3 illustrates the way in which the uncertainty of outcomes is often represented. The activity under study (e.g. rates of consumer spending, demand for water, erosion of slopes, deaths from a disease, or growth in GIS usage) has been observed to grow until point **t**. Data points beyond this represent predictions or forecasts of the future. Our predictions about future growth are bound by upper and lower limits that define a zone of uncertainty.

One way to begin to anticipate changes in GIS is to break a problem down into its constituent parts. In Section 1.4 (Figure 1.20) we identified the component parts of GIS as hardware, software, network, data, people, and procedures. We devote this next section to crystal-ball gazing about the future of GIS. Of course, there are limits to this form of analysis, since progress in all facets is neither constant nor proportional at all times. "Progress" implies "improvement", and there are various ways in which improvement may

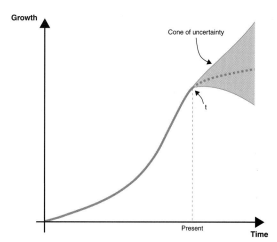

Figure 20.3 Predicting geographic futures

be quantified, as in the measurement of the seemingly inexorable long-term increase in the value of investment in stock market indices like the Dow Jones, FTSE, and Nikkei. Yet this does not mean that it is sensible to anticipate that day-to-day, or even considerably longer period, improvements in such indices are linear – think of the stock market crashes of 1929 and 1987, for example. No-one can predict the scale and rapidity of incremental improvements in hardware and software technology – though speculation about the rate of software development might be expected to be more linear than that of hardware, because software development is cumulative. Yet, even here, who could have predicted the implications of the Internet for software development?

How does the "zone of uncertainty" apply to science? We are conscious of the Kuhnian school of scientific development that views long periods of "normal" science as a series of incremental developments, with much shorter cataclysmic periods of revolution leading to paradigm shifts in thinking. Trends in the development of the component parts of GIS are likely to be similarly prone to volatility induced by paradigm shifts,

Table 20.1 Questions that can be investigated with GIS

Activity Mode	Question	Example
Descriptive	What is . . .?	What features are at this location?
	Where is . . .?	Where are the households with high-income earners?
Prescriptive	Why is . . .?	Why is it that oak forests only occur in sheltered valleys?
	What if . . .?	What would be the impact on consumer spending if a new retail outlet was located at the intersection of Highway 10 and Highway 605?

the vagaries of academic fashions, and the ebb and flow of commercial pressures upon the academy.

Here we offer two contrasting scenarios about the future of GIS. One scenario is pessimistic; the other is optimistic (the lower and upper bounds of the zone of uncertainty in Figure 20.3). In truth the future probably lies somewhere between the two. In predicting the future we are cognizant of Bill Gates' observation that we usually tend to overestimate progress in the next 2–3 years, but underestimate changes over the next 10 years or so.

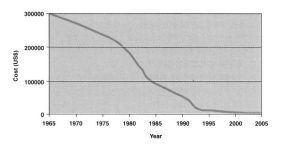

Figure 20.4 The falling cost of a GIS over the last 30 years

20.3.1 The downside: a pessimistic future scenario for GIS

In our pessimistic scenario we focus on the major constraints that could hold back development and place GIS at the lower end of the zone of uncertainty in Figure 20.3.

● Lack of awareness and well-trained staff. Even though GIS is now taught in over 1500 universities, and in many community colleges and schools around the world, there remains a shortage of well-trained staff and users. Many GIS software vendors speak of a lack of suitable IT professionals as a major check on growth. With well-trained staff there will be more, better-run projects. There is a clear need for improvements in technical training to keep pace with the proliferation of geographic data and software. The scope for better modeling is greatly improved with better, more detailed data and better software, but so too, the dangers of misuse and misapplication also multiply. If improvements in the quality of education, training, and management awareness do not keep pace with improvements in the detail and quality of real-world representations, then the future is relatively bleak.

● Continued high costs restrict use. Although the cost of a GIS (defined here as hardware, software, data, and any necessary customization) has fallen by two orders of magnitude in the past four decades, price remains a barrier to penetration of the mass consumer market. Figure 20.4 shows the falling trend in the cost of a GIS. In the early GIS days of the 1960s and 1970s when COTS (commercial-off-the-shelf) solutions did not exist and GIS ran on mainframe computers, the cost of a GIS was $250 000 or more. With the introduction of COTS software and minicomputers in the late 1970s and early 1980s the price fell quickly to around $100 000. The next major period of rapid price fall resulted from the advent of per-

sonal computers and desktop GIS in the early and middle 1990s. Since then prices have continued to fall and in the year 2000 the price for a GIS was around $3000. At the same time as prices have continued to fall the functionality has increased dramatically, increasing by two orders of magnitude over the same time period. There were major jumps in functionality in the early 1980s, with the release of general-purpose workstation COTS GIS software products, and again in the 1990s when desktop GIS came to market (Figure 20.4). For further details of GIS software trends see Chapter 8.

● Limited availability and high cost of data. Geographic information has been described as the "rocket fuel of GIS". Lack of basic framework and thematic coverage information certainly slows down the spread and use of GIS. In some parts of the world such data do not exist; elsewhere they are restricted in their availability for political or security reasons (Section 19.3). Even where available, they are prohibitively expensive for certain applications – what is a feasible cost for a utility company is ruinous for a charity. Pricing is a complex issue (see Sections 17.1.2 and 17.1.5). It has been argued that the $35 million it would cost to purchase 15 cm (6 inch) resolution imagery of the whole coterminous US and the several million dollars it would cost to rent the entire 230 000 constantly up-dated digital maps covering all of Britain in huge detail are obscene. Such charges reflect the size of the market, levels of competition, real costs, and charging for immense collections of map sheets or images rather than browsing and feature-by-feature selection (some people have also argued that profiteering plays a role!). We expect such costs to diminish significantly in the coming years as past costs are treated as "sunk costs", as more alternatives become available, as data collection technology improves, as more sources of update are used, and as more sophisticated pricing models are used. This should lead to a virtuous circle of enhanced use, better feedback on information quality, and larger revenues. We suggest that one of the next grand challenges for GIS is a global address-geocodable street centerline dataset for the whole globe that is available online for less than $10000. It might even be available over the Internet as a GIS service for rent at 10 cents per map view.

● Legal restrictions. One of the key themes of this book has been the need to raise the profile of management in GIS. In Chapter 17 we discussed at length the important legal issues of data ownership, copyright, freedom of information, and litigation. It is possible to envisage a GIS future in which entrepreneurship, innovation, and everyday business activity are stifled by overly restrictive legal issues. The move in the mid-1990s to allow software to be patented is causing significant ripples in the GIS industry with everything from data collection using GPS, to searching geographic databases on the Web, and public access geographic terminals being regulated by law. In such circumstances every use of the patented technology must be licensed from the patent holder (see Section 17.1).

● Networks become clogged. The future of GIS, like many other areas of IT, will be inextricably linked to the development of networks like the Internet. This, again, is a theme we have explored in detail in this book. Those of us who have struggled to use an Internet server during peak Internet usage (typically late afternoon in Europe and morning in the USA, for example) can vouch for the current bandwidth problems. If network bandwidth improvements fail to outstrip demand, users will turn away from using online geographic information services in their investigations.

● Critical theorists come to dominate. From the academic point of view GIS has been a major success for geography and allied disciplines. GIS is without doubt at the heart of the academic discipline of geography and provides a common unifying cause, a leading edge and high profile activity, and a suite of tools to explore the ramifications of many types of geographic theory. Yet as Brian Berry (Box 1.8) and many others have noted, there are those within the academy who do not value the linkage to the scientific mainstream that GIS provides for human geography.

● GIS loses its identity. As GIS technology increasingly becomes embedded and fragmented into specialist areas (e.g. AM/FM, location-based services, resource management, and resource planning) there is a danger that some may lose sight of core GIS values. There is some evidence of this trend in the form of changing attendance at GIS conferences. Gone are the big general conferences of the 1990s such as AutoCarto, GIS/LIS, and EGIS. These have been replaced by vendor-specific, regional, and thematic conferences. Some of these are very successful – for example, in 2000 the largest ever GIS conference was the ESRI worldwide conference in San Diego, California that attracted 10000 delegates (see Figure 16.7). In the case of the vendor-specific conferences they ultimately provide a commercial rather than scientific window on the world.

Even under this pessimistic scenario we still see an important future role for GIS, one in which GIS continues to grow at 5–10% per annum. If some of these impediments can be addressed quickly then rates of growth could be at least double these figures.

20.3.2 The upside: an optimistic future scenario for GIS

In our optimistic scenario we envisage that solutions will be found to many of the potential problems highlighted previously. In this section we focus on some of the main drivers that are likely to fuel GIS growth at annual rates above 20%.

● Hardware advances continue. As Figure 20.5 shows, many of the major advances in GIS have been stimulated by rapid breakthroughs in hardware development. For example, the move from workstations to PCs in the early 1990s changed GIS price:performance ratios, hardware footprints, and the availability of peripheral devices (digitizers, plotters, scanners, etc.). Similarly, the introduction of ink-jet printers in the mid-1990s revolutionized the types of hard copy maps that could be made, making it possible to produce maps featuring continuous tone symbolism and 3D hill shading. Perhaps the best-known law describing changes in hardware performance is Moore's Law named after Gordon Moore, the first Chairman of Intel. He predicted, based on developments in computer processor manufacturing, that hardware price:performance ratio would continue to double approximately every 18 months. Said another way, real prices would fall by 50% every 18 months. Figure 20.6 predicts the trend in computer performance broadly based on Moore's Law. It is likely to be outstripped in the coming years. Based on the evidence of the turn-of-millenium period, the doubling rate seems to have accelerated to every 12 months, making the short-term trend in circuitry more like double exponential than exponential. While processor speed is only one aspect of overall computer performance it does give a useful, if crude approximation of future hardware, and therefore GIS, characteristics.

● Hardware diversity increases. Improvements in hardware performance and developments in screen and storage technology have also stimulated the introduction of a range of new devices capable of handling geographic information. These include Pocket PCs, personal digital assistants (PDAs), pagers, and even wearable computers (Figure 12.4). Aside from the obvious uses of these devices for data collection and delivering up-to-date personal maps, one of the benefits of this improved portability and mobility is that scientists are now able to work in the field and experience the real world first hand, rather than through the lens of secondary data or simulation. The mobile revolution ushers in a new academic and application-led field of mobile GIS and m-commerce. The next generation of GIS will be itinerant and fully distributed.

● Software functionality becomes richer, easier to use, and lower cost. Optimistic extrapolation of the existing situation suggests that GIS software will continue to advance in the areas of capability, usability, and affordability. All the major GIS vendor software development teams are currently hard at work adding new features to the next releases of their products. At the same time they are taking advantage of the general trend toward improved support for geographic data and processing in operating systems (e.g. graphics libraries), databases (e.g. SQL/MM), and Web browsers (e.g. improved XML parsing). As hardware perfor-

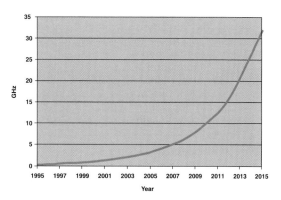

Figure 20.6 Predicted improvements in computer performance based on Moore's Law

mance has improved, software developers have been able to focus on making software easier to use (e.g. the development of graphical user interfaces, context-sensitive help, and wizards, dialogues that automate frequently performed tasks).

● Geographic measurement technologies become all-pervasive. The development of small, low power, relatively accurate, and cost-effective geolocation technologies, such as GPS, GLONASS, and cellular telephone position fixing means that it is currently possible to broadcast in real time the location of almost anything of value. This has already started to change the way we collect and use geographic data, moving us from problems of data scarcity to problems of data overload. This technology opens up many new possibilities and is the basis of the sub-branch of GIS called location-based services as discussed in Section 14.3.1.

● Economies of scale. Today GIS is still a highly technically focused activity. Getting to the next level involves expanding into the mass-market world of consumerism. This is what Moore and McKenna (1999) called "crossing the chasm" in their book of the same name. For GIS to gain critical mass for economies of scale in consumer markets it must become more widespread in markets such as in-car navigation, PDAs, wrist wearable computers that are location-enabled (e.g. using GPS devices), and embedded systems. In our optimistic scenario we envisage this happening within the next five years. Geographic information and maps will be everywhere.

● GIScience becomes the norm. Today GIScience has little following outside academia. We advo-

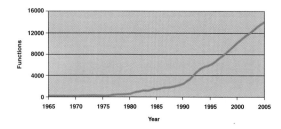

Figure 20.5 The increases in GIS functionality in the last 30 years

cate strongly in this book that widespread adoption of the already substantial advances in GIScience is critical to the future of GIS. We believe also that there is a compelling case for further fundamental research into topics like cognition, uncertainty, modeling, and scientific visualization. If this work is adequately funded we can develop widely accepted theories and models that will help us to understand the world better.

● The combination of all these effects and continued efforts at local, national, and global scales to build spatial data infrastructures (see Chapter 19) results in ever more consistent, widespread, and affordable GI being available. In turn this creates a virtuous circle whereby GIS use increases, decision-making becomes better-informed, and revenues for those who require them (such as commercial enterprises and many governments) are enhanced.

● Role of GIS in science. Through GIS, geography can become the mother of all sciences. As part of the already established move to "new science" GIS can play a pivotal role in goal-oriented, team-based, interdisciplinary, cross-cutting projects that transcend traditional, artificial discipline boundaries. GIS can help us investigate many previously intractable problems (e.g. how to predict the amount of damage caused by hurricanes in central America, and the forces that shaped the surface of other planets in our solar system). We have pointed out earlier that there are many problems which GIS can help to solve or ameliorate: Figure 20.2 illustrates the context in which we all operate – where the human race has transformed half of the land area of the world for its own purposes.

Some of the ideas here may seem a little extravagant for readers who are pragmatists. Remember that this discussion is designed to reflect the optimistic future scenario for GIS. It is most likely that the actual future of GIS lies somewhere between the two extremes of pessimism and optimism. This assumes of course that we can predict the future with any certainty.

20.4 Conclusions

The journey that is this book has taken us on an excursion into the world of Geographic Information Systems and Science. We have tried to define the terms GISystems and GIScience and show how they interact closely. For us both are critically juxtaposed and closely interwoven. One

relies intimately on the other and once separated both are less useful and more prone to misinterpretation. We have also discussed the relationship to GIStudies and GIServices. The former connotes the embedding of GIS in society, the latter a mechanism to deliver GIS more effectively to a wider group of users than has hitherto been technically and commercially feasible.

As we look forward in the early years of the new century, we are aware that the future of GIS lies in the hands of GIS – not systems, science, studies, or services, but geographic information *students* – the next generation of software developers, managers, researchers, and users. We hope that this book will be a contribution to their future and, through their use of GIS, to all our futures. Without GIS we are all too aware that the future of the Third Rock from the Sun will be more uncertain and less well managed.

Questions for Further Study

1. Contrast these two different statements about the relationship between systems and science: "Geographic information systems support and drive science" and "Geographic information science drives the development of systems".
2. What do you think is the most significant impediment to the future success of GIS?
3. To what extent will modern ICT (Information Communication Technologies) render geography irrelevant?
4. Choose an application domain and give specific examples of what is, where is, why is, and what if questions. What datasets would you need to answer these questions successfully?

Online Resources

Articles on Future of GIS:
 www.geospatial-online.com GeoSpatial Online
 www.geoplace.com GeoWorld
 www.gisdevelopment.net
 www.giscafe.com/ GIS Café
 ww.geocomm.com GeoCommunity
 www.tenlinks.com/ TenLinks

GIS research agendas:
 www.ucgis.org/ UCGIS Research Topics
 www.romulus.arc.uniroma1.it/Agile/Agile.html Association of Geographic Information Laboratories for Europe Research Agenda

Reference Links

Longley P A, Goodchild M F, Maguire D J, and Rhind D W (eds) 1999 *Geographical Information Systems: Principles, techniques, management and applications.* New York: John Wiley.
Chapter 2, Space, time, geography (Couclelis H)
Chapter 3, Geography and GIS (Johnston R J)
Chapter 4, Arguments, debates, and dialogues: the GIS-social theory debate and the concern for alternatives (Pickles J)
Chapter 51, GIS for business and service planning (Birkin M, Clarke G P, and Clarke M)
Chapter 54, Enabling progress in GIS and education (Forer P, Unwin D)
Chapter 55, Rethinking privacy in a geocoded world (Curry M R)

References

Curry M 1998 *Digital Places:Living with geographic information technologies.* New York: Routledge.
Moore G A and McKenna R 1999 *Crossing the Chasm: Making and selling high-tech products to mainstream customers.* New York: Harper Collins.
Openshaw S, Cross A, Brunsden M, and Lillie J 1990 Lessons learned from a failed GIS. *Association for Geographic Information Conference Proceedings.* London: AGI, 2.3.1–2.3.5.
Pickles J (Ed) 1995 *Ground Truth: The social implications of geographic information systems.* New York: Guildford Press.
Reeve D E and Petch J R 1999 *GIS, Organizations and People: A socio-technical approach.* London: Taylor and Francis.

Afterword by Joe Lobley

When my father Brian Lobley retired from GIS education, he gave me his library of conference proceedings, card decks, and data files. I burnt the lot. But these guys seem to have learned a little humility at long last and grasped what the real world of commerce has known for years – in short, that it's always risky to stick your neck out and say something about an uncertain world, but it's suicidal not to try. Yep, we're all together in the zone of uncertainty now, for sure, equipped only with our grasp of principles, techniques, and practices. History is bunk. Geography isn't.

INDEX

Note: Page numbers in italics refer to Figures; those in bold refer to Tables